Einstein Studies

Editors: Don Howard John Stachel

Published under the sponsorship
of the Center for Einstein Studies,
Boston University

Volume 1: Einstein and the History of General Relativity
Don Howard and John Stachel, editors

Volume 2: Conceptual Problems of Quantum Gravity
Abhay Ashtekar and John Stachel, editors

Volume 3: Studies in the History of General Relativity
Jean Eisenstaedt and A. J. Kox, editors

Volume 4: Recent Advances in General Relativity
Allen I. Janis and John R. Porter, editors

Volume 5: The Attraction of Gravitation: New Studies in the History of General Relativity
John Earman, Michel Janssen, and John D. Norton, editors

Volume 6: Mach's Principle: From Newton's Bucket to Quantum Gravity
Julian B. Barbour and Herbert Pfister, editors

Julian B. Barbour Herbert Pfister
Editors

Mach's Principle:
From Newton's Bucket to Quantum Gravity

Birkhäuser

Boston • Basel • Berlin

Julian B. Barbour
College Farm
South Newington
Banbury, Oxon OX15 4JG
England

Herbert Pfister
Universität Tübingen
Institut für Theoretische Physik
D-72076 Tübingen
Germany

Library of Congress In-Publication Data

Mach's principle : From Newton's bucket to quantum gravity / edited by
Julian B. Barbour, Herbert Pfister.
p. cm. -- (Einstein studies ; v. 6)
Includes bibliographical references and index.
ISBN 0-8176-3823-7 (alk. paper). -- ISBN 3-7643-3823-7 (pbk. : alk. paper)
1. Mach's principle. 2. General relativity (Physics)
I. Barbour, Julian B. II. Pfister, Herbert, 1936- . III. Series.
QC137.M33 1995 95-22539
530.1--dc20 CIP

Printed on acid-free paper.

The Einstein Studies series is published under the sponsorship of the Center for Einstein Studies, Boston University.

Birkhäuser ®

ISBN 0-8176-3823-7
ISBN 3-7643-3823-7
Typeset by the editors.
Printed and bound by Quinn-Woodbine, Woodbine, NJ.
Printed in the U.S.A.

9 8 7 6 5 4 3 2 1

CONTENTS

4. Other Formulations of Mach's Principle

5. Frame Dragging

6. Experimental Status

7. Critical Reflections

8. Quantum Gravity

Acknowledgments

This volume and the conference at Tübingen (July 1993) on which it is based would not have been possible without the very generous financial support of the Deutsche Forschungsgemeinschaft (DFG). We further thank the Landesregierung Baden–Württemberg in Stuttgart and the Institut Dr. Friedrich Förster in Reutlingen for their contributions. The Max-Planck-House in Tübingen and its manager, Mrs. M. Harders–Opolka, supplied effective organization and the ideal congenial atmosphere for a conference of this size. Special thanks go to Mrs. C. Stiller for perfect secretarial work before, during, and after the conference, and to the local organization team U. Heilig, C. Klein, J. Klenk, and U. Schaudt. We are also grateful to the advisory committee of the conference, B. Bertotti, D. Brill, J. Ehlers, and J. Stachel, for really good advice, and to the editors of the *Einstein Studies Series*, D. Howard and J. Stachel, for their continuous support of the 'Mach project,' for the inclusion of this project in the *Einstein Studies*, and for help with the final editorial work. The camera copy for this volume was typed by Mrs. Kate Draper, to whom we are greatly indebted for her accuracy and unending patience. Our publishers, Birkhäuser, have been supportive, helpful, and understanding. Finally, we should like to express our thanks to Dr Siegfried Wagner for the fascinating talk he gave about Einstein's ancestors and the life of Jews over several centuries in rural Southern Germany. This talk was on the occasion of an excursion from Tübingen to the Jewish Cemetery at Bad Buchau, where many of Einstein's ancestors are buried. Dr Wagner's research is being prepared for publication in a book entitled *Albert Einsteins Ahnen, ihre Herkunft und ihr Schicksal*.

The Editors

1. Introduction and Historical

General Introduction

Hypothesen sind Netze, nur der wird fangen der auswirft. Novalis.

I think it was Hermann Bondi who once said that physics is such a consistent and connected logical structure that if one starts to investigate it at any point and if one pursues correctly every issue that branches away from one's starting point, in the outcome one will be led to understand the whole of physics. With Mach's Principle it seems something like that.

Sir Fred Hoyle, this volume, p. 269.

This volume is based on the conference 'Mach's Principle: From Newton's Bucket to Quantum Gravity,' held July 26–30, 1993, at the Max-Planck-House in Tübingen, Germany. As far as we know, this was the first conference exclusively devoted to Mach's Principle. (Sir Hermann Bondi in his closing remarks: "This conference was a splendid idea, and I am only surprised that nobody thought of having such a conference before.")

The so-called Mach's Principle is surely one of the most elusive concepts in physics: On one hand, Machian aspects have been present either explicitly or implicitly in theoretical astronomy, general physics, and dynamics from their Greek infancy up to the present day (Barbour 1989 and following article). On the other hand, most of practical physics is done, and successfully done, without ever thinking of the 'deep questions' connected with Mach's Principle. (The situation is similar in quantum theory, which functions extremely well using established prescriptions notwithstanding deep and unresolved questions about its interpretation, its measuring process, and its classical limit.)

In this volume, the notion 'Mach's Principle' is understood in as broad a sense as possible. Although it is certainly interesting (see Chap. 1) and may be important (see p. 215) to establish precisely what Mach

Einstein Studies, vol. 6: Mach's Principle: From Newton's Bucket to Quantum Gravity, pp. 1–5

said about absolute and relative elements in physics, and to see how Einstein (who coined the actual expression 'Mach's Principle' in 1918, p. 186) tried to incorporate Mach's ideas in general relativity, it would be ridiculous for a book published in 1995 to narrow these age-old questions about the foundations of physics to the pronouncements of just these two physicists, however eminent, and not to cover the contributions of their contemporaries. It is also very important to consider the development in thinking and the accumulation of experimental facts that have occurred in the intervening period.

The root of Mach's Principle, as understood in this volume, is deeply connected with the question of what constitutes the essence of the method of physics and the concept of a physical system: It is often not sufficiently appreciated how kind nature has been in supplying us with 'subsystems' of the universe which possess characteristic properties (literally in the sense 'proper to the system') that can be described and measured almost without recourse to the rest of the universe. The strategy of dividing the universe into ever smaller and 'simpler' parts has shaped physics, beginning with the investigations of the solar system, which resulted in the concept of a mass point for complicated objects such as planets, going on to atoms and elementary particles, and presumably coming to an end only at the level of the constituents (quarks, subquarks) of elementary particles. On the other hand, it is evident that basic concepts such as 'inertia' and 'centrifugal force' cannot be understood and explained within the context of the subsystems themselves, but at best by taking into account the rest of the universe.

As is well known, Newton 'solved' this conflict by the introduction of the extremely successful concepts of 'absolute space' and 'absolute time.' Newton recognized clearly that only relative quantities can be directly observed but, unlike his relationist contemporaries Huygens, Leibniz, and Berkeley, he was convinced that a scientifically useful notion of motion could not be based on relational quantities. Instead, he sought to demonstrate how absolute quantities could be deduced from relative observations. In this endeavor he was not entirely successful (Barbour 1989).

The most emphatic and most influential physicist to insist on a reformulation or extension of the foundations of physics in purely relational terms was Ernst Mach in the last quarter of the 19th century, though he made only tentative proposals for such a goal. Albert Einstein was very much influenced by Mach's writings, and his general relativity was at least partly conceived in the spirit of realizing Mach's dictum. Indeed, general relativity was the first theory to supply a dynamic spacetime (dependent on the matter distribution) and to indicate at least possibilities of how inertia and centrifugal forces could result from interaction with the distant cosmic objects. On the other hand, general relativity in its present formulation does not, despite its name, fulfill the demand of using solely relational properties between physical objects composed of matter in the strict sense (as opposed to 'generalized matter' in the form of gravitational waves or spacetime curvature). One of the main debates at the conference concerned the question of how far general relativity realizes Mach's Principle: Does this principle make sense for the full theory with its huge manifold of (partly unphysical) solutions, or does it function as a selection principle for special classes of solutions (and if so which?), or has it meaning only in our unique universe? The cosmological context of Mach's Principle goes a long way towards explaining why this principle is so elusive: Cosmology lies somewhere at the edge of the physical method, which usually relies on the possibility of preparing physical systems and confirming results by repeated measurements on ensembles of similar systems. In this respect it is remarkable how much reliable information astronomy and astrophysics have already supplied about our cosmos. In the future we can expect information about still more distant, and therefore earlier, parts of the universe, and in this way information about the cosmos as a whole. This will surely have an impact on 'Machian questions.'

Investigations of the very early cosmos necessarily call for a unification of gravity and quantum theory, which is widely held to be the deepest open problem in contemporary theoretical physics. As it happens, many problems in so-called quantum gravity and quantum cosmology are intrinsically of a Machian character, for instance the goal to treat 'time' no longer as an absolute, external parameter, but to understand it as an intrinsic property of the considered system, i.e., the whole universe. It is clear that this volume cannot do full justice to these rather new and actively developing fields of quantum gravity and quantum cosmology. On the other hand, these may well be the fields in which most activity and progress in Machian questions can be expected from future research.

Although this volume is based on a conference, it is not a usual conference proceedings volume, to which all participants contribute only their latest, very specialized results without much interrelation between them. From the beginning, it was the intention that the conference and this volume should - very much in agreement with the general policy of the *Einstein Studies Series* - cover all aspects of Mach's Principle - historical, philosophical, astronomical, theoretical and experimental - and confront, where necessary, the different views on these aspects. Experts were invited to prepare general overviews, which were distributed to all participants already two months ahead of the conference. In some cases, these overviews were then supplemented by prepared reply talks. It was guaranteed that there was enough time for lively discussions after all talks. In addition, there were scheduled discussion sessions on selected, especially controversial topics. All these discussions were recorded on tape, and were edited by us and the contributors after the conference. Some talks and discussion contributions have been considerably improved and in part even rewritten after the conference. As organizers of the conference, we were very happy that it was possible to gather together in Tübingen nearly all experts worldwide on the different views of Mach's Principle. Only a few prominent names are obviously missing, for example, Boris Al'tshuler, Bruno Bertotti, Jeffrey Cohen, Robert Dicke, Dennis Sciama, and John Wheeler. They had to decline their participation for different reasons, some of them at the last minute. Their influence can nevertheless be easily traced through this entire volume. For example, it turned out that one entire morning session, devoted to the initial-value problem in general relativity and based on Isenberg's paper (p. 188), was intimately related to the Machian ideas of John Wheeler and was exclusively presented by former collaborators of John. The session Chairman, Jayant Narlikar, introduced it as 'Wheeler without Wheeler.'

It should be mentioned that many important historic papers connected with Mach's Principle are scattered in hardly accessible journals or other sources; most of them are originally in German and have never been translated, and some of them have moreover been forgotten for decades. Indeed, one of the more important consequences of the conference was that it brought to light significant papers on Mach's Principle by Hofmann (1904), Reissner (1914, 1915), and Schrödinger (1925) that were virtually unknown, even to experts in the field. Therefore we found it appropriate to collect such papers (partly in extracts) in English translation in this volume.

In summary, we hope that this volume represents a fairly complete status report and reference source on most aspects of Mach's Principle. In order to give greater unity to this collection of contributions, we have not hesitated to give cross references (indicated in square parentheses) to other places in the volume in which the same or related topics are discussed. In various places, especially following the translations and in the chapter introductions, we give commentaries. We have also prepared an index, in which we also attempt to draw the reader's attention to common themes that run through the volume.

Given the topic of the book, it is hardly to be expected that its two editors will be in *complete* agreement on all aspects of Mach's Principle. Indeed, as will be evident from our own contributions, one of us (J.B.B.) believes Mach's Principle is in essence fully contained within general relativity whereas the other (H.P.) has reservations on this score. This divergence of opinion has not been any hindrance to productive and harmonious collaboration; indeed, we feel that the book gains from a certain friendly rivalry, each of us being keen to see the respective viewpoints properly represented. Somehow this seems very appropriate for Mach's Principle - see p. 630. Let the reader decide!

We should also mention a project to publish within the next year or two a book with the provisional title *Relativity and Its Alternatives* (J. Renn *et al.*, eds.). This will cover much ground in common with the present volume; in particular it will include a long paper "The Third Way to General Relativity. Einstein and Mach in Context" by Jürgen Renn. This paper is based on the talk he gave at Tübingen but because of its length unfortunately could not be included in the present volume. The new book may also include translations of some papers with Machian context that also could not be included in this volume for lack of space. In particular, there may be a complete translation of *Absolute oder relative Bewegung?* by Benedict and Immanuel Friedlaender (1896) and also of Reissner's paper of 1915, partial extracts of which are included in this volume (p. 114, p. 309, p. 145ff).

The motto from Novalis - "Hypotheses are nets; only he that casts will catch" - has already been used: by Karl Popper at the head of his book *The Logic of Scientific Discovery*. We are grateful to Domenico Giulini for suggesting its appropriateness in connection with Mach's Principle. (It was also Giulini who drew our attention to the long-forgotten papers of Schrödinger and Reissner.) The idea to use the motto by Fred Hoyle came during the work of compiling the index! A glance at the index confirms the truth of Bondi's remark.

Julian B. Barbour, Herbert Pfister

REFERENCE

Barbour, Julian B. (1989). *Absolute or Relative Motion?*, vol. 1: *The Discovery of Dynamics*. Cambridge: Cambridge University Press.

Mach before Mach

Julian B. Barbour

The debate about motion – Is it absolute or relative? – extends back to antiquity, and 'Machian' attitudes can be readily identified in the writings of Aristotle, but I begin this brief survey at the dawn of the scientific age: with Copernicus and Kepler.

Not surprisingly – since astronomers cannot fail to be aware that observations are relational – both were 'Machians.' Copernicus defined his frame of reference thus: "The first and highest of all is the sphere of the fixed stars, which contains itself and everything, and is therefore immovable. It is unquestionably the place of the universe, to which the motion and position of all the other heavenly bodies are compared."

Kepler's standpoint is particularly interesting, since he was deeply impressed by Tycho Brahe's 'demolition' of the crystal spheres. Kepler posed the problem of astronomy in the famous words: "From henceforth the planets follow their paths through the aether like the birds in the air. We must therefore philosophize about these things differently." His response to the problem was very 'Machian' (Barbour 1989): The planets could not possibly follow such precise orbits by a mere inspection of empty space – they must be both guided and driven in their motion by the real masses of the universe, namely, the sun and the sphere of the fixed stars. This deeply held conviction was a decisive factor in Kepler's discovery of the laws of planetary motion – truly, a pre-Machian triumph of Mach's Principle.

Although Galileo retained many Aristotelian – and hence 'Machian' – concepts, he instinctively believed in motion relative to space. This comes out clearly in his theory of the tides, in the discussion of which he actually uses the expression *absolute motion* (Barbour 1989, p. 400).

The modern debate about motion had a most ironic origin. In 1632, Descartes was about to publish his *Le Monde* when he heard about Galileo's condemnation by the Inquisition. Since Copernicanism was central to his new mechanical philosophy, this put Descartes in a

Einstein Studies, vol. 6: Mach's Principle: From Newton's Bucket to Quantum Gravity, pp. 6–8

quandary. He suppressed *Le Monde* and only ventured to present his new physics in his *Principia Philosophiae* of 1645. To avoid censure, Descartes began by asserting, in a very Aristotelian manner, that both position and motion are relative. A convoluted argument enabled him to wriggle out of potential difficulties with the Inquisition. However, when he came to his laws of motion, he reverted, without explanation or warning, to the instinctively 'absolutist' position he had adopted in *Le Monde*, in which he had advanced something almost identical to Newton's first law of motion as the foundation of his physics.

About 25 years later, Newton spotted the crass discrepancy between Descartes's espousal of relationalism and the use of the law of inertia as the foundation of mechanics. In *De Gravitatione*, which only came to light this century, Newton inveighed against Descartes. He saw that to set up a science of motion one must be able to define velocity as something definite. But if motion is relative and *everything* in the world is in motion - as it is in Cartesian philosophy - Descartes's own relationalism makes a mockery of the Cartesian law of inertia: "That the absurdity of this position may be disclosed in full measure, I say that thence it follows that a moving body has no determinate velocity and no definite line in which it moves." This is the nub - the *fundamental problem of motion* (Barbour 1989, Introduction): If all motion is relative and everything in the universe is in motion, how can one ever set up a determinate theory of motion?

Unlike his contemporaries Huygens and Leibniz, who both cheerfully used the law of inertia as the foundation of dynamics while stoutly maintaining the relativity of motion, Newton felt this problem so acutely that he could not conceive of any dynamics formulated without a rigid framework - absolute space. The Scholium in his *Principia* was simply a coded reworking of *De Gravitatione* in which Newton disdained to mention Descartes by name. Especially revealing is Newton's use of centrifugal force - in Cartesian philosophy the explicatory basis of both light and gravity - to exhibit the reality of absolute motion. Descartes is to be hoist with his own petard. The choice of a bucket was also at least in part mischievous in intent: By Descartes's philosophical concept of motion, the only 'true' motion of the water must be that relative to its *immediate ambience* (the bucket wall). This is why Newton said pointedly: "Therefore this endeavor does not depend upon any translation in respect of the ambient bodies, nor can true circular motion be defined by such translation."

Two centuries later, Mach (unaware of Newton's fixation with Cartesian absurdities) thought Newton naive to suppose the mere bucket wall to have any relevance to centrifugal force and produced one of the great suggestive sayings in the history of physics: "Newton's experiment with the rotating vessel of water simply informs us, that the relative rotation of the water with respect to the sides of the vessel produces *no* noticeable centrifugal forces .... No one is competent to say how the experiment would turn out if the sides of the vessel increased in thickness and mass till they were ultimately several leagues thick." Given the effect of this remark – and the whole absolute–relative debate that Descartes initiated – on Einstein, it may not be too fanciful to suppose that if the Inquisition had condemned Galileo a few months later, and Descartes had published *Le Monde*, Newton might never have thought of the bucket nor Einstein of general relativity!

Let me conclude with a remark about Bishop Berkeley, who in *De Motu* (1721) comments that in empty space motion of two globes around a common center cannot be conceived by the imagination, but that if we "suppose that the sky of the fixed stars is created; suddenly from the conception of the approach of the globes to different parts of that sky the motion will be conceived." For this remark, Berkeley is often credited with having been a true precursor of Mach. Note, however, Berkeley's phrase 'fixed stars.' The stars were still *very fixed* in his mind, as we see from his earlier *Principles of Human Knowledge* (1710, §114):

> Philosophers who have a greater extent of thought, and juster notions of the system of things, discover even the earth itself to be moved. In order therefore to fix their notions, they seem to conceive the corporeal world as finite, and the utmost unmoved walls or shell thereof to be the place, whereby they estimate true motions. If we sound our own conceptions, I believe we may find all the absolute motion we can frame an idea of, to be at bottom no other than relative motion thus defined.

Thus, Berkeley looked *backward* to Kepler and Copernicus just as much as he looked *forward* to Mach. He never confronted the real problem of both Newton and Mach – the definition of determinate velocities if "the heavens began to move and the stars swarmed in confusion" (cf. p. 222).

But the exhortation to "sound our own conceptions" cannot be bettered at the start of our journey to the distant goal of quantum gravity – and perhaps even more remote consensus on Mach's Principle. The references are to my *The Discovery of Dynamics*, cited on p. 5.

Mach's Principle before Einstein

John D. Norton[1]

1. Introduction

The doctrine of the relativity of motion is attractive for its simplicity. According to it, the assertion that a body moves can mean nothing more than that it moves with respect to other bodies. Acceleration has long proved to be the stumbling block for the doctrine, for, in the case of acceleration, the simplest of observations seem to contradict the doctrine. When a test body rotates, for example, it is acted upon by centrifugal forces. The presence of these centrifugal forces seems to be completely independent of whether the test body rotates with respect to bodies immediately surrounding. Thus Newton observed in his famous bucket experiment that these centrifugal forces induced a concavity in the surface of a rotating body of water and did so independently of whether the water rotated with respect to the bucket containing the water. Therefore, using these inertial forces as a marker to indicate whether the body is accelerating, it seems possible to know that a body is accelerating without any concern for whether it accelerates with respect to the other bodies around it. This outcome contradicts the doctrine of the relativity of motion as applied to acceleration.

For about a century now, the most popular escape from this unwelcome refutation has been the following simple idea. Relativists point out that experiments such as Newton's reveal only that inertial forces are not noticeably related to motion with respect to *nearby* bodies. That, however, does not rule out the possibility that inertial forces are caused by acceleration with respect to more distant bodies. If this were the case, then inertial forces would not reveal an absolute acceleration but merely an acceleration relative to these distant masses. The core idea is that the inertial forces acting on an accelerating body arise from an interaction between that body and other bodies. The idea is not so much a proposal of a definite, new physical law; rather it is the prescription

Einstein Studies, vol. 6: Mach's Principle: From Newton's Bucket to Quantum Gravity, pp. 9–57 © 1995 *Birkhäuser Boston, Inc. Printed in the United States.*

that such a law should be found. The law recommended is only loosely circumscribed. It must be such that more distant masses play the decisive role in fixing the inertial forces on a given body, for example.

The proposal's most prominent sponsor was Albert Einstein. In the early years of his work on general relativity, he believed that his theory implemented the proposal, although he completely lost this belief in his later years. Nonetheless, the future of the proposal was guaranteed by the vigorous support of an Einstein who rapidly rose to celebrity status both inside and outside the scientific community. Einstein did not claim the proposal as his own invention. From the earliest moments, he attributed it to Ernst Mach and in 1918 gave a field theoretic formulation of the proposal its now standard name of 'Mach's Principle.' (Einstein 1918).[2]

The story of the role of the principle in Einstein's work, his enchantment with it, and his subsequent disenchantment, has been frequently told because of its enormous importance in the historical development of relativity theory and relativistic cosmology. My purpose in this paper is to explore another side of Mach's Principle, its earliest years prior to its adoption by Einstein, which so profoundly redirected and ruled its future. I will ask: What role did the principle play in Mach's own system? How was it received by Mach's contemporaries? In answering these questions, we shall find a story that is a little different from the one we might expect. With Mach now universally acclaimed as the patron of a growing literature on Mach's Principle and Machian theories, one expects to find in Mach's writings a penetrating voice of prescient clarity that easily transcends the generations that separate us from him. Instead we shall find:

- Mach's own writings that pertain to the principle were vague and ambiguous, bordering on the contradictory. The principle is never clearly stated, but at best obliquely suggested, and it remains unclear whether Mach endorsed the suggestion or condemned it as unscientific.
- It was Mach's disciples and his contemporary and later readers who extracted an unequivocal proposal from his writings. Several even claimed the idea independently of Mach.
- Mach's Principle proved to be an idea that fascinated Einstein so much that he sought to build his general theory of relativity around it. However he was in a minority in his fascination.
- Prior to the advent of general relativity, the principle was a fringe idea, often opposed by those who would become Einstein's most ardent supporters. The philosophical community was largely uninterested in the proposal. As an empirical proposal, it had no foundations because of the failure of every experimental test actually tried. As a product of

philosophical analysis, it smacked of *a priori* physics.

In Sec. 2 of this paper, I will pose the question of precisely what it is that Mach proposed concerning the origin of inertia. In Sec. 3, I will argue that cases can be mounted for each of two plausible answers. In Sec. 4, I will offer a reconciliation. In Secs. 5 and 6, I will assess the broader reaction to the proposal, considering both the favorable and unfavorable responses.

Although use of the term 'Mach's Principle' is anachronistic in much of the time period under consideration, I will use the term here for lack of anything better. Over the years it has come to label a proliferation of different ideas. Here I will understand it to refer to the proposal that the inertia of a body is caused entirely by an interaction with other bodies.

2. What Mach Actually Said

In his first published reference to the principle he attributed to Mach, Einstein (1912, p. 39) formulated it as "...the entire inertia of a point mass is the effect of the presence of all other masses, deriving from a kind of interaction with the latter." A footnote appended to this sentence announced its origin:

> This is exactly the point of view which E. Mach urged in his acute investigations on the subject. (E. Mach, The Development of the Principles of Dynamics. Second Chapter. Newton's Views of Time, Space and Motion.)

The attribution is deliberate and unequivocal. Einstein, who is notorious for the infrequency of citation in his writings, is carefully naming a section of the second chapter of Mach's celebrated *The Science of Mechanics: A Critical and Historical Account of Its Development* (Mach, 1960).

Readers who turn to the relevant section of *The Science of Mechanics*, a critique of Newton's notions of absolute time, space, and motion, will find many assertions reminiscent of the principle Einstein enunciated. But nowhere will they find it stated without distracting qualification or ambiguous hesitation. Indeed if the relevant section of Mach's text was intended to state clearly and advocate forcefully the principle Einstein enunciated, then it has failed. Rather, readers of the relevant section find Mach clearly devoting his expository energies to an attack on Newton's conceptions. The assault is based on two of Mach's favorite themes, which are enunciated clearly and repeatedly. These two themes, rather than some forerunner of Mach's Principle, are what readers find as the principal content of this section of *The Science of Mechanics*. The following remarks from this section are typical:

> No one is competent to predicate things about absolute space and absolute motion; they are pure things of thought, pure mental constructs, that cannot be produced in experience. All our principles of mechanics are, as we have shown in detail, experimental knowledge concerning the relative positions and motions of bodies. Even in the provinces in which they are now recognized as valid, they could not, and were not, admitted without previously being subject to experimental tests. No one is warranted in extending these principles beyond the boundaries of experience. In fact, such an extension is meaningless, as no one possesses the requisite knowledge to make use of it. (Mach 1960, pp. 280) ...
>
> When we say that a body *K* alters its direction and velocity solely through the influence of another body *K′*, we have asserted a conception that it is impossible to come at unless other bodies *A*, *B*, *C* ... are present with reference to which the motion of the body *K* has been estimated. In reality, therefore, we are simply cognizant of a relation of the body *K* to *A*, *B*, *C* ... If now we suddenly neglect *A*, *B*, *C* ... and attempt to speak of the deportment of the body *K* in absolute space, we implicate ourselves in a twofold error. In the first place, we cannot know how *K* would act in the absence of *A*, *B*, *C* ...; and in the second place, every means would be wanting of forming a judgment of the behavior of *K* and of putting to the test what we had predicated – which latter therefore would be bereft of all scientific significance. (Mach 1960, p. 281)

These passages recapitulate the two themes. First is the notion that physical science is or ought to aspire simply to provide economical descriptions of experience. Thus elsewhere Mach (1882) had pronounced "Physics is experience, arranged in economical order" (p. 197), and "The goal which it [physical science] has set itself is the *simplest* and *most economical* abstract expression of facts" (p. 207). The second theme is that Newton's absolute space, time, and motion are idle metaphysical excesses that are superfluous to this goal of economical description. Again elsewhere Mach (1872, 1911) had made the point very clearly. All our statements containing the terms 'space' and 'time' are really only statements of the relation of phenomena to phenomena, and the terms could be struck out without affecting the content of the statements. Mach (1872, 1911, pp. 60–61) even gave a prescription for how this striking out might be effected:[3]

> We can eliminate time from every law of nature by putting in its place a phenomenon dependent on the earth's angle of rotation. The same holds of space. We know positions in space by the affections of our retina, of our optical or other measuring apparatus. And our x, y, z in the equations of physics are, indeed, nothing else than convenient names for these affections. Spatial determinations are, therefore, again determinations of phenomena by means of phenomena.

These two themes comprise Mach's attack on Newton's conception. In his *The Science of Mechanics*, Mach now goes to some pains to emphasize the error that one may fall into if one forgets Mach's lesson and takes Newton's absolute space and time too seriously. Talk of motion of a body K in space is really only an abbreviated description of the change of relations between K and other bodies A, B, C .... If we forget that these abbreviated descriptions do depend essentially on these other bodies and try to anticipate the motion of K 'in absolute space,' that is, if these other bodies were not present, then we will illegitimately extend our science beyond its proper domain. The domain of science is experience. We have no experience of the motion of a body in a space devoid of other bodies. Our extension would cease to be science.[4]

These two themes would be the ones that every modern reader would find pursued by Mach with vigor and clarity in his critique of Newton, were it not for the modern obsession of recovering Mach's Principle from Mach's critique. As a result of this obsession, the modern reading of Mach focuses on passages that are certainly highly suggestive, but, in the last analysis, vague and ambiguous. Typical of them is the most quoted of all passages of Mach's critique (1960, p. 284), which I have broken up into three sentences, labeled s_1, s_2, and s_3, for discussion:

> [s_1] Newton's experiment with the rotating vessel of water simply informs us, that the relative rotation of the water with respect to the sides of the vessel produces *no* noticeable centrifugal forces, but that such forces *are* produced by its relative rotation with respect to the mass of the earth and the other celestial bodies.
> [s_2] No one is competent to say how the experiment would turn out if the sides of the vessel increased in thickness and mass till they were ultimately several leagues thick.
> [s_3] The one experiment only lies before us, and our business is, to bring it into accord with the other facts known to us, and not with the arbitrary fictions of our imagination.

The ambiguity of this famous passage lies in the admissibility of two readings that contradict one another:

First is the reading that returns what we now call Mach's Principle. Sentence s_1 reminds us that, in our search for causes for the centrifugal forces within the bucket, we have overlooked one possibility, the rotation of the water with respect to other bodies. We cannot rule out such a cause, as long as it is a cause that only acts when very large masses are involved. Thus s_2 agrees with Newton that rotation with respect to the walls of the bucket induce no noticeable centrifugal forces. But according to the new physical mechanism conjectured, this would not be so if the walls were substantially increased in mass and size. Sentence s_3 closes by observing that we would never have been tempted with an explanation in terms of absolute space – the "arbitrary fictions of our imagination" – had we recalled that the real business of science is economical description of experience. In this case, the experience is of Newton's experiment and of the other bodies that surround it.

The second reading recalls the two themes of Mach's critique. Since the goal of physical science is economical description of experience, s_1 reminds us of what we should really infer from Newton's experiment. We should conclude merely that there is a correlation between two experiences, the presence of centrifugal forces and rotation with respect to the stars. There is no place for a metaphysical absolute space in such descriptions. Sentence s_2 is a tease to shake the dogmatic belief of a Newtonian. It points out that the Newtonian has inferred far more than what is actually warranted by Newton's experiment. The experiment does not give us enough information to rule out the possibility of an alternative physical theory in which the centrifugal forces are caused by rotation with respect to other bodies. Sentence s_3, however, reaffirms resoundingly that such speculation lies well beyond the compass of science as economical description of experience. This speculation requires us to think of cases in which we do not and cannot have experience: for example, the walls of the bucket enlarged to a thickness of several leagues – "an arbitrary fiction[s] of the imagination" if ever there was one. Therefore Mach will not entertain such speculation.

Thus we have two readings of Mach's famous analysis of Newton's bucket experiment:

- The first escapes Newton's conclusion by proposing a new physical mechanism for the generation of inertial forces that will later be associated with the label 'Mach's Principle.'
- The second effects the escape essentially by insisting that Newton be restricted to describing the experiment only in terms of what is experienced

and pointedly condemns as unscientific the proposal of Mach's Principle.

Our task now is to decide which if either is the correct reading.[5] Our resources are Mach's other writings as well as the interpretations of his contemporaries. Unfortunately we shall see that quite strong cases can be mounted for *both* readings. My accusation of the broader ambiguity of Mach's analysis rests on this unhappy fact. I now proceed to develop the case for each reading of Mach's analysis.

3. Mach Escapes Absolute Space by Urging ...

3.1. ... Mere Redescription. It is clear that a major component of Mach's analysis involved the simple recommendation to redescribe motion in space as experiences that do not invoke the term 'space.' Thus he wrote (1960, pp. 285–86; Mach's emphasis): "When...we say that a body preserves unchanged its direction and velocity *in space*, our assertion is nothing more or less than an abbreviated reference to *the entire universe*."

How are we to decide if in addition to this project of simple redescription Mach is also proposing a new physical mechanism? I shall assume that a proposal for a new physical mechanism must make claims about counterfactual or hypothetical systems, that is, claims about systems which are known not to exist or are not known to exist. Certainly such a proposal cannot approach the proposal of Mach's Principle unless it is prepared to license inferences about such cases as the rotation of a hypothetical bucket with walls several leagues thick or perhaps about the inertial forces induced between two bodies in an otherwise (counterfactually) empty universe.[6]

Under this criterion there would seem to be no possibility that Mach could be proposing a new physical mechanism. For the claim he repeats most in the entire analysis is that we have no business in science speculating about such systems that are beyond our experience. Merely in the passages already quoted above, Mach has made the point *three* times. And it appears elsewhere in his analysis. For example (1960, p. 285):

> The comportment of terrestrial bodies with respect to the earth is reducible to the comportment of the earth with respect to the remote heavenly bodies. If we were to assert that we knew more of moving objects than this their last-mentioned, experimentally given comportment with respect to celestial bodies, we should render ourselves culpable of a falsity.

Or again Mach considers a proposal by C. Neumann, who imagines that a rotating celestial body will still be deformed into oblateness by centrifugal forces even if the other heavenly bodies were absent. Mach (1960, pp. 340–41) insists that this latter assumption is meaningless and objects that one is simply not allowed to assume away these other masses as unimportant when experimenting in thought.[7] But if Mach refuses to allow any consideration of such hypothetical or counterfactual systems, then it is hard to see how he could be proposing a principle that even vaguely resembles the later Mach's Principle. On the contrary he must condemn any such principle as unscientific.

In places Mach does seem to urge a reformulation of the principles of mechanics. He allows for example[8]: "The principles of mechanics can, presumably, be so conceived, that even for relative rotations centrifugal forces arise."

Is Mach suggesting a reconception of mechanics in which the principles are materially changed and a new physical mechanism introduced? Or is the reconception merely a restatement of the same laws in such a way that the idle metaphysical conceptions of space and time are no longer mentioned? We may answer by looking at what Mach proceeds to do. On the pages following, what Mach actually does corresponds to the latter alternative of simple redescription. He seeks ways of restating the law of inertia so that it does not use the term 'space.' This project of redescription proves quite simple for one case (p. 286)

> Bodies very remote from each other, moving with constant direction and velocity with respect to other distant fixed bodies, change their mutual distances proportionately to the time. We may also say, all very remote bodies – all mutual or other forces neglected – alter their mutual distances proportionately to those distances.

Mach then shows (pp. 286–287) how this type of formulation of the law of inertia can be couched in the language of mathematical formulae. The usual form of the principle requires that the acceleration of a body remote from other masses be constant. That is, if the body has absolute spatial coordinates (x, y, z) and the time is t, then

$$\frac{dx^2}{dt^2}=\frac{dy^2}{dt^2}=\frac{dz^2}{dt^2}=0.$$

Mach's goal is to rewrite the law without the absolute spatial coordinates (x, y, z). To achieve this, he considers the distances r, r', r'', ... to the other distant masses m, m', m'', ... from the test mass. In place of the spatial coordinates, Mach uses the mass weighted sum of these distances ($\Sigma mr/\Sigma m$) so that the principle becomes

$$\frac{d^2}{dt^2}\left[\frac{\Sigma mr}{\Sigma m}\right] = 0. \tag{1}$$

The project is clearly just one of redescription of existing laws and not the proposal of a new mechanism.[9] Indeed Mach soon makes it very clear that his new expression for the principle of inertia is not intended to be applied to cases remote from experience (p. 289):

> It is impossible to say whether the new expression would still represent the true condition of things if the stars were to perform rapid movements among one another. The general experience cannot be constructed from the particular case given us. We must, on the contrary, *wait* until such an experience presents itself.

Thus it is possible to present a collection of Mach quotations that drives towards the conclusion that Mach is not advancing what we now know as Mach's Principle, but condemning it. Is this an example of selective and biased quotation? Apparently not in the view of several of Mach's contemporary readers. C. D. Broad (1916) reviewed the supplement that contained a compendium of Mach's additions to the third English language edition of *The Science of Mechanics*. He reported that he now understood more clearly Mach's ambiguous discussion surrounding Newton's bucket experiment. What he understood in that discussion was not a proposal for a new physical mechanism but merely Mach's strictures about redescription:

> There is also a far clearer statement than before of Mach's much quoted remark (in connection with Newton's bucket) that "the universe is not given to us twice, but only once." It is now clear that Mach's meaning is that the Ptolemaic and the Copernican view are simply different ways of describing precisely the same set of facts, and that therefore there is no real difference between the bucket standing still with the fixed stars rotating and the bucket rotating with the fixed stars standing still. This is clearly a necessary result of the relative view, and it is one that is often overlooked.

Broad's remarks were those of a sympathetic reviewer. Far more significant was the evaluation of Paul Carus. Carus was born in Germany in 1852, received a doctorate from the University of Tübingen in 1876, and emigrated to America. There he began working for the Open Court

Publishing Company, editing its journals *Open Court* and the *Monist*. In particular, Carus became the medium, welcomed heartily by Mach, through which Mach's writings were made available in English to the American audience. Carus found a natural empathy with Mach's views[10] and was able to engage Mach in a huge correspondence spanning almost three decades, one of Mach's largest correspondences.[11] What induced Carus to publish on precisely the question that concerns us was a talk given by Philipp Frank in 1909, "Is There Absolute Motion?," published the following year as Frank (1909). Frank clearly attributed to Mach the proposal of a new physical mechanism to explain inertial forces of the type of Mach's Principle. Carus's discussion (1913, pp. 23–40) contains extensive quotation from Frank's lecture and provides the foundation for his denunciation of the suggestion that Mach was proposing a new physical mechanism:

> Another point where we feel justified in doubting Dr. Frank's exposition is the statement that Mach hypothetically assumes a new law of nature as to the efficacy of masses, besides the law of gravitation. The passage in Mach's writings to which Dr. Frank refers[12] does not (in my opinion) suggest the idea of an additional law of nature according to which the distant fixed stars should exercise a mysterious influence on the Foucault pendulum. We will later on let Mach speak for himself. In our opinion it seems that it would be sufficient to ascribe the rotation of the pendulum to its inertia while the earth revolves round itself, and this takes place in the space in which the earth has its motion, viz., the space of the Milky Way system. The pendulum remains in the plane of oscillation in which it started while the earth turns around underneath. ... There seems to me no need of inventing a new force besides gravitation. The law of inertia seems to explain the Foucault pendulum experiment satisfactorily.

Carus supports his reading of Mach with his own selection of Mach quotes, similar to those discussed here, pointing out that Mach's endeavors are devoted to elimination of the terms 'space' and 'time.'

Carus's published argument is based on widely available published writing of Mach. However, because Carus also enjoyed the privileged view of an extensive correspondence with Mach, it is tempting to conjecture that Carus is also silently drawing on this correspondence or even on discussions with Mach during one of Carus's visits to Mach in 1893 or 1907. Whether such correspondence is still extant will have to be decided by a search of the relevant archives. However, the prospect that any such correspondence existed in 1913 seems slight. If it did exist, Carus would almost certainly have published it to buttress his case. As editor of the *Monist*, Carus had clearly been eager to publish a letter by Mach (Carus 1906a) on an earlier article by Carus (1906) on Mach's philosophy. He published it with obvious delight, embedding the letter in the pomp of an introduction and afterword by Carus, and retaining its original German "lest it lose many of the fine points in an English translation." (Carus (1913), however, showed no restraint in presenting extensive passages of Frank (1909) in English translation!)

Finally, whatever their differences over whether Mach did propose a new hypothetical law, both agreed that such a proposal is an anomaly in Mach's broader systematic proclamations in which such hypothesis is abhorred. Thus Frank notes (1909, p. 17; trans. Carus 1913, p. 32): "But Mach in this case stands in the opposite camp as in most other cases where his repugnance to all hypothesis has made him a pioneer in the phenomenological direction." And Carus (1913, p. 32) himself, speaking of Frank's broader reading of Mach, writes provocatively "Strange that Mach, with his reluctance to introduce anything hypothetical except what is absolutely indispensable, should range on the side of the theorists...."

3.2. ... a New Physical Mechanism. Or did Mach intend to recommend more than mere description? Did he intend to propose a new physical mechanism for the origin of inertial forces? Once again a case can be made for this possibility and it too rests on quotations from Mach's writings and on his interactions with colleagues and others. However, if the case for this possibility were to rest only on the first part, Mach's writings for publication, then the case would be considerably weaker than the corresponding case for his advocacy of mere redescription. For none of the writings unambiguously *endorses* a proposal for a new mechanism. Worse, it is not clear which of his writings even talks about such a proposal.

In Mach's critique of Newton's conceptions in *The Science of Mechanics* are several much cited remarks that could be taken as suggesting a new physical mechanism. However, precisely because they are rhetorical flourishes, they admit of many interpretations and do not provide a firm foundation for the case. He exclaims (p. 279): "Try to fix Newton's bucket and rotate the heaven of fixed stars and then prove the absence of centrifugal forces."

But could this not simply mean that Mach takes the case of bucket rotating/stars resting to be exactly the same as the case of bucket resting/stars rotating? Then to try to prove the absence of centrifugal forces, as Mach challenges, is obviously futile since the two cases are really just the one case described differently. Indeed the sentences immediately preceding the exclamation are devoted to arguing that the two cases are really one. Again, there is Mach's famous observation on Newton's bucket experiment (p. 284): "No one is competent to say how the experiment would turn out if the sides of the vessel increased in thickness and mass till they were ultimately several leagues thick."

As we saw above, a consistent continuation in Mach's voice would be "And I [Mach] certainly would not dare to speculate on such an unscientific thing!" – this being a plausible reading of what Mach actually says in the next sentence: "The one experiment only lies before us, and our business is, to bring it into accord with the other facts known to us, and not with the arbitrary fictions of our imagination."

Again, Mach concludes the paragraph preceding with an apparently unequivocal recommendation for a new physical mechanism for inertial forces: "The principles of mechanics can, presumably, be so conceived, that even for relative rotations centrifugal forces arise."[8]

However, the appearance is deceptive, for, as we saw above, this reconception might just be referring to a simple redescription such as leads up to Mach's equation (1) above.

More promising are his later remarks that Barbour (1989, p. 692) identifies as "Mach's clearest statement of the ideal of a *seamless dynamics*" such as would arise were he proposing a new mechanism for inertia. Mach writes (p. 296, Mach's emphasis)

> The natural investigator must feel the need of further insight – of knowledge of the *immediate* connections, say, of the masses of the universe. There will hover before him as an ideal an insight into the principles of the whole matter, from which accelerated and inertial motions result in the *same* way. The progress from Kepler's discovery to Newton's law of gravitation, and the impetus given by this to the finding of a physical understanding of the attraction in the manner in which electrical actions at a distance have been treated, may here serve as a model. We must even give rein to the thought that the masses which we see, and by which we by chance orientate ourselves, are perhaps not those which are really decisive. On this account we must not underestimate even experimental ideas like those of Friedländer [(1896)] and Föppl [(1904, 1904a)], even if we do not yet see any immediate result from them.

Once again I do not see that we can rule out the possibility that these remarks refer to Mach's project of redescription. The understanding of (1) is of immediate connection of the masses since the superfluous mediation of 'space' has been eliminated. And was not the progress from Kepler to Newton (in Machian terms) the discovery of a system of laws that yielded a far more economical summary of not just Kepler's astronomical discoveries but much else besides? It is also very possible, however, that Mach is referring to a new physical mechanism for the origin of inertia. For, as we shall see below, Friedlaender (1896) and Föppl (1904a) both clearly consider such a novel mechanism and conduct experiments to detect it.

If this passage does refer to such a novel mechanism, it still provides no evidence that the proposal of such a mechanism *originated* with Mach or that Mach *endorsed* it. The passage in question was added to the seventh German edition of 1912,[13] presumably in response to Friedlaender (1896) and Föppl (1904, 1904a). Since these works already propose a new physical mechanism for inertia, one can hardly say that the proposal originated with Mach's remarks of 1912. Even Mach's vague suggestion of the use of the theory of electricity as a model had been anticipated and in more precise form. Friedlaender (1896, p. 17) had raised the possibility of applying Weber's law of electrodynamics to gravitation in this context, as does Höfler (1900, p. 126), as we shall see below. Worse, Mach's language suggests that whatever he is introducing is novel and goes beyond what was already said in earlier editions. That is, *after* the Machian ideal of purification from meaninglessness has been achieved, there is a new goal, some "further insight," a speculative "ideal." In one sentence, we are invited "even [to] give rein to the thought [*sogar dem Gedanken Raum geben*] that the masses which we see, and by which we by chance orientate ourselves, are perhaps not those which are really decisive." This invitation would hardly be necessary if we had already made space for that thought in the earlier text of the earlier editions. The thought for which we are to make space might even be a distinctly non-Machian one. If the thought is that the decisive bodies are ones we cannot see, then it contradicts Mach's repeated and forceful pronouncements on the primacy of the observable. If the theoretical and experimental work of the Friedlaenders and Föppl is a part of such non-Machian speculation, then Mach can hardly be giving them unreserved endorsement. Indeed the passage quoted above closes with what seems to be a gentle rebuke: "Although the investigator gropes with joy after what he can immediately reach, a glance from time to time into the depths of what is uninvestigated cannot hurt him."

One could read this as a very kind way for Mach to point out to the Friedlaenders and Föppl that he finds their work to have strayed well beyond science, the domain of economical descriptions of experience, into the murky depths of unscientific speculation.[14]

Remarks published by Mach in 1872 [quoted here from Mach

(1911)] support most strongly his advocacy of a new physical mechanism for the origin of inertia - although they are still subject to the same ambiguities. In the appendix, Mach stresses that in referring motions in the law of inertia to space we should never lose sight of the fact that this is really only an abbreviated reference to other bodies. He then begins to discuss how the motion of these other reference bodies might affect the law of inertia, arriving at the following puzzle (p. 78):

> But what would become of the law of inertia if the whole of the heavens began to move and the stars swarmed in confusion? How would we apply it then? How would it have to be expressed then?

It seems clear enough that Mach's puzzle refers to the problem of stating - redescribing - the law of inertia in a form similar to (1), in the awkward case in which the heavenly bodies adopted chaotic motion. How can Mach be sure that an expression in terms of a simple mass weighted sum of distances ($\Sigma mr/\Sigma m$) will be adequate? This seems to be the same problem that Mach discusses in *The Science of Mechanics* (1960, p. 289) (see above). Mach then proceeded to another example, a free body acted upon by an instantaneous couple so that it rotates. He continues (p. 79)

> Here the body makes very strange motions with respect to the celestial bodies. Now do we think that these bodies, without which one cannot describe the motion imagined, are without influence on this motion? Does not that to which one must appeal explicitly or implicitly when one wishes to describe a phenomenon belong to the most essential conditions, to the causal nexus of the phenomenon? The distant heavenly bodies have, in our example, no influence on the acceleration, but they have on the velocity.

The ambiguity of these remarks resides in the unexplained terms 'influence' and 'causal nexus.' What do they mean? What sort of influence is suggested?[15]

Mach then makes the remarks that most strongly suggest that he is seeking a new physical mechanism. He seems to be conjecturing the form of the law that governs it:

> Now, what share has every mass in the determination of direction and velocity in the law of inertia? No definite answer can be given to this by our experiences. We only know that the share of the nearest masses vanishes in comparison with that of the farthest. We could, then, be able completely to make out the facts known to us if, for example, we were to make the simple supposition that all bodies act in the way of determination

> proportionately to their masses and independently of the distance, or proportionately to the distance, and so on.

This talk of 'share' and 'masses' acting in proportion to their mass and distance might well be a conjecture of some new physical mechanism. However, it can also be read as a part of Mach's project of redescription. As we have seen, Mach gives such a redescription of the law of inertia in terms of the mass weighted sum of distances ($\Sigma mr/\Sigma m$) or its second time derivative $d^2/dt^2(\Sigma mr/\Sigma m)$. The 'share' of each mass m, m', m'' ... in the reformulated law would simply be the magnitude of the term each mass contributes to these sums. The functional dependence of these contributions are then exactly of the type Mach mentions. In the first sum, for example, each mass contributes a term proportional to its mass and to its distance from the test body. And the nearest masses certainly contribute vanishingly small terms in comparison with the remaining masses.

In my reading, one thing makes it clear that Mach intends in this passage to propose only a redescription of the law of inertia and not a new physical mechanism. That is the sentence immediately following the passage quoted above, which closes the paragraph and Mach's discussion: "Another expression would be: In so far as bodies are so distant from one another that they contribute no noticeable acceleration to one another, all distances vary proportionately to one another."

This expression is clearly offered as a variant or, possibly, a special form of the general laws discussed. Yet it is just a redescription of the inertial motion of a collection of noninteracting bodies that avoids mention of space.[16] There is no hint of some new physical mechanism that would enforce the proportional variation of distances.

This discussion is the best evidence in Mach's published writing for his advocacy of a new physical mechanism for the origin of inertial forces. But it does not make a good case. Even in the collective judgments of Mach's sympathetic contemporaries, its intent is unclear. As we have seen, Frank (1909) found it to advocate a new mechanism; Carus (1913) did not. My judgment is also that it is ambiguous, but I think its most natural reading is as a proposal for simple redescription. In my view, this same judgment must also hold of Mach's published corpus on Newton's bucket experiment and the law of inertia. The only unequivocal proposal Mach makes is for a simple redescription of the experiment and the law in a formulation that does not use the term 'space.' It remains unclear whether Mach intended to propose and endorse a new physical mechanism for the origin of inertial forces.

However it is dubious that this verdict represents Mach's real intentions. What speaks loudly against this verdict is that the majority of Mach's contemporaries and confidants understood him to be proposing a new physical mechanism. On this point Carus is in a clear minority. Indeed the view that Mach proposed a new mechanism is a commonplace of the literature from around 1900 and on to the year of Mach's death in 1916. It is mentioned[17] by Friedlaender (1896, p. 9), Höfler (1900, pp. 122–26), Föppl (1904a, p. 383), Frank (1909), Cassirer (1910, p. 177),[18] Petzoldt (1912, p. 1057), Schlick (1915, p. 166), and, of course, Einstein, whose repeated attributions, commencing with Einstein (1912), brought the viewpoint to the broadest audience. If this view were an outright misreading of Mach, then Mach had ample opportunity to correct it. But this correction never came.[19] He even mentioned the work of the Friedlaenders (1896) and Föppl (1904a) in later editions of his *The Science of Mechanics* (1960, p. 296). Surely that is the point at which Mach would issue a correction if both works were misrepresenting his position. Or are the somewhat indirect remarks quoted above ("...a glance ... into the depths of what is uninvestigated...") intended as a gentle rebuke?

It would seem that any corrections that Mach may have issued would have been so gentle as to escape later reporting, or, at least, any reporting of which I am aware. In particular, in a letter of June 25, 1913, Einstein reported to Mach that Einstein's new theory had yielded a new physical mechanism for the origin of inertia and Einstein attributed that idea directly to Mach[20]: "...*inertia* has its origins in a kind of *interaction* of bodies, quite in the sense of your reflections on Newton's bucket experiment."

Yet Einstein's later writings contain no trace of hesitation in continuing this attribution to Mach. Similarly, Frank (1957, p. 153) continues the attribution. Again, in a letter of January 11, 1910 (Blackmore and Hentschel 1985, pp. 66–67) to Mach, Föppl mentions his "treatment of the question of relative motion" – presumably Föppl (1904a). He commented with relief that Mach "at least had no fundamental misgiving [*Bedenken*] to raise against [it]." We might well wonder what Mach did say to evoke such a response!

Fortunately, within Mach's surviving correspondence there is a record of how Mach responded to such attributions. In a letter of September 3, 1904, which contained an enthusiastic response to the fifth German edition of Mach's *The Science of Mechanics*, Petzoldt put to Mach a series of questions and proposals about Mach's ideas on the law of inertia. In particular, he attributed to Mach the idea that the thickened walls of Newton's bucket could induce centrifugal forces and expressed his own doubts on this notion (Blackmore and Hentschel 1985, p. 36, Petzoldt's emphasis):

> I still cannot reconcile myself to your observation (p. 247) on the possible variation of the experiment through the thickening of the bucket walls. With this you still make the appearance of centrifugal forces dependent on the magnitude of the surrounding bodies instead of the (relative) rotations of the bodies. The centrifugal forces are still aroused only through relative rotation against the *locations* of the masses of the earth and the other heavenly bodies. I am inclined, however, very much to the belief, which you also admit, that the heavenly bodies here play only a chance role like the axial rotation of the earth for the determination of temporal processes, and hope for future experiences on the deeper relations of things, without shutting my eyes to your doubt over whether such experiences will ever be accessible to us as people.

Mach's response in a letter of September 18, 1904, is lengthy and unfortunately never actually mentions the walls of Newton's bucket. He does make clear that he dislikes Petzoldt's idea that the *locations* of the masses may be the decisive thing. He objects (Blackmore and Hentschel 1985, p. 39, Mach's emphasis): "A bare, efficacious *location* has been observed by no one."

However, he does clearly leave the impression that Mach's own view of inertia is that it is a matter to be decided by experiment. After explaining that his original thoughts on inertia were formulated before the ascendancy of Faraday's conception of local action and of a medium or material intervening between bodies ("aether, space or whatever it is called"), he continued: (p. 38, Mach's emphasis)

> As long as one attends to bodies alone, one conceives naturally of gravitational processes and inertial motions as determined by them alone or, correspondingly, through other masses. If one now also is not to expect a positive result from the Friedländer fly-wheel experiment,[21] since the mass and velocity of the wheel is too small, then a greatly refined Foucault experiment could still show that a pendulum or gyroscope orients itself not *only* according to the fixed heavenly stars, but also in part is influenced by the earth, which is, after all, a powerful flywheel. However should such an experiment definitely come out negative, that would also be a great gain in insight. ... If I conceive of gravitation as carried through a medium, then I can conceive of the state of this medium still as only determined by the masses of bodies, for the reaction accelerations depend on the masses of the bodies. But if *one* body that is very distant and unaccelerated with respect

> to the others is in motion, then its motion can only be described with reference to the latter. The idea that this motion is *determined* by the latter bodies cannot be dismissed without further ado. In any case, the orientation of the motion through the distant bodies can be a merely *apparent* one. Perhaps the motion is a concern only of the moving body and the medium *alone*. Perhaps each body conducts itself in space like Dirichlet's bodies in a frictionless fluid.[22]

The letter closes with a very brief sketch of an experiment designed to detect the Friedlaenders' effect arising from the earth's motion.

With a response such as this, it is no surprise that Petzoldt (1912, p. 1057) should proceed to attribute to Mach the conjecture that the relative rotation of masses induces centrifugal forces, the same effect sought experimentally by the Friedlaenders and Föppl. However the only definite point that Mach has made is to rule out Petzoldt's proposal with his disparagement of a "bare, efficacious location." His answer strongly suggests that he expects or would welcome a positive outcome of a Friedlaender style experiment. But he has still not positively asserted that he believes that a thickened bucket in Newton's experiment would induce inertial forces – his original passage in *The Science of Mechanics* insists that no one is competent to assert this! And for all our pursuit of Mach's writings, we still do not have a clear statement from Mach that *he* conjectures that the origin of inertia lies solely in an interaction of bodies through some new physical mechanism.

4. A Reconciliation?

This is the puzzle that Mach's writings on inertia pose for us. We must reconcile two facts. Mach's publications contain only a clear advocacy of the view that one ought to *redescribe* Newton's bucket experiment and the law of inertia in such a way that the term 'space' does not arise. If there is a suggestion of a new physical mechanism to explain the origin of inertial forces, then its discussion is vague, and the proposal of a new mechanism might even be condemned as unscientific. On the other hand, Mach must have been aware that the proposal of exactly such a new causal mechanism was routinely attributed to him, but, in spite of ample opportunity, there is no evidence that he ever moved to correct this misattribution – if it did in fact need correcting. In brute form, we are left wondering whether Mach did intend to propose a new mechanism but was simply incompetent in expressing his intention. Or, if he did not intend a new mechanism, we must ask why Mach allowed such widespread misinterpretation of his work.

I can offer two reconciliations, although neither is attractive. The first is that Mach was unwilling to see the need for a new physical mechanism in his system. That is, he was an adherent of the relativist doctrine with respect to motion, which leads to the need for a new mechanism to account for inertial forces. Mach, however, was simply unwilling to embrace this consequence, so willingly embraced by other relativists, and simply tried to avoid committing himself. There is some evidence for this view. It stems from Hugo Dingler, who had been sanctified by an extremely favorable mention from Mach in the penultimate paragraph of his preface to the last edition of *The Science of Mechanics*. He reported in Dingler (1921, p. 157) that Mach's[23]

> ...only salvation [from the problem of centrifugal forces] was to bring the centrifugal appearances into relation with the fixed stars, and, in fact, Mach also accepts this in the last (7th) edition of his *Mechanics* (I cannot really decide how much this was already the case in earlier editions); he was forced to it, even though this also obviously contradicted his sensibilities.

To the last sentence, Dingler appended the crucial footnote

> I thank Herr Dr. Ludwig Mach for the fr[ien]dl[y] communication that this consequence was always "especially tormenting" [*besonders quälend*] for his father, that he knew for a long time of the monstrous conclusions deducible from it, yet did not draw them, but rejected them.

Thus Mach's behavior could be explained by a horror and unwillingness to accept what his system had produced. In this account, his aversion would be so profound that he would be unable to address the horrific consequence squarely in both his writings and in his private correspondence and discussions.

There are two difficulties with this view. First, contrary to Dingler's suggestion, Mach's system offered a perfectly good reason for rejecting a new physical mechanism: It transcended the economical description of experience that was the proper domain of science. With perfect consistency and in clear conscience, Mach could denounce this new mechanism as unscientific, if he disliked it so much, and there would be no need to be tormented. Second, by 1921, Dingler had become an outspoken critic of relativity theory and, as a disciple of Mach, may well have been overeager to seek reasons to remove Mach's support from relativity theory.[24]

A second more plausible reconciliation is the one I favor. It depends on Mach's somewhat idiosyncratic notion of the true nature of causation. In Sec. 3, when seeking to judge whether Mach's proposals advanced beyond mere redescription to a new physical mechanism, I used the criterion that such a mechanism must make claims about counterfactual or hypothetical systems, for that was clearly required if Mach's proposals were to approach what later became Mach's Principle. However Mach's view of physical science as merely economical description of experience rules out exactly such considerations. A causal connection for Mach is merely a functional dependence extracted from experience. He makes this very clear in (Mach 1911, p. 61; Mach's emphasis) when he writes

> The present tendency of physics is to represent every phenomenon as a function of other phenomena and of certain spatial and temporal positions. If, now, we imagine the spatial and temporal positions replaced in the above manner [by phenomena], in the equations in question, we obtain simply *every phenomenon as function of other phenomena.*
>
> *Thus the law of causality is sufficiently characterized by saying that it is the presupposition of the mutual dependence of phenomena.* Certain idle questions, for example, whether the cause precedes or is simultaneous with the effect, then vanish by themselves.
>
> The law of causality is identical with the supposition that between the natural phenomena α, β, γ, δ, ..., ω certain equations subsist.

One cannot overemphasize how different this view is from the common view of causation. Newton's inverse square law of gravity is commonly understood to legislate that the sun causes an acceleration of the earth that varies directly with the inverse square of the distance that separates them. And this is assumed to hold whether the two masses are the sun and earth of our actual universe or a sun and earth in some hypothetical universe devoid of all other matter. As Mach's frequent protestations above show, he does not allow, in general, this assuming away of the other masses of the universe. Now it is not clear whether Mach would want this prohibition to apply in this case. If it does apply, however, then the relevant causal law simply becomes the assertion of the functional relationship between the sun–earth distance and the acceleration of the earth towards the sun that happens to obtain *in our universe alone*.

If we now apply precisely this same thinking to Newton's bucket experiment, we arrive almost verbatim at many of Mach's pronouncements on the experiment and the law of inertia. And we do so without Mach ever proposing the type of new physical mechanism soon to be suggested under the banner of Mach's Principle. If we seek the fundamental causal relation revealed by Newton's bucket experiment, we must recover the functional relation of the actual phenomena – and that is merely

> ... that the relative rotation of the water with respect to the sides of the vessel produces *no* noticeable centrifugal forces, but that such forces *are* produced by its relative rotation with respect to the mass of the earth and the other celestial bodies.

It now follows immediately that, using Mach's definition, the centrifugal forces in the bucket and the mass of the earth and other celestial bodies stand in a causal relation. Speaking loosely, in a way that risks 'idle questions,' we might identify these masses as the cause of the forces. Also, to identify the role that each of the masses play in the functional relation is just to identify their causal role. Mach might well describe this as their 'influence,' a term with obvious causal connotations. Or he might well ask: "What share has every mass in the determination of direction and velocity in the law of inertia?" And if the relevant functional relation is linear in mass, he might well describe the body as 'acting' in proportion to its mass. Further, a result such as (1) appears to non-Machian readers merely to describe a functional relation and nothing more. But to Mach, the very fact that it describes a functional relation between phenomena of our world makes it the statement of a causal relation. Finally, Mach can offer a functional relation such as (1) as the fundamental causal relation pertaining to inertia, that is, the law of inertia, without needing to suggest that this same relation would obtain were the motions of the masses of the universe to be very different. For the functional relation need only obtain for our actual experiences to qualify as a causal relation.[25]

There is an unappealing aspect of this resolution. The resolution rests on the assumption that what Mach meant by causation is very different from what the same term meant for the many proponents of what came to be known as Mach's Principle. Thus, when Einstein wrote to Mach that "*inertia* has its origins in a kind of *interaction* of bodies, quite in the sense of your reflections on Newton's bucket experiment," Einstein's notion of causal interaction extended well beyond the simple functional relations of phenomena. It included relations on hypothetical and counterfactual systems of precisely the type denounced by Mach. What remains unexplained is how Mach could repeatedly allow such misattributions to pass without objection or correction by him.

5. Early Sponsors of Mach's Principle

Whatever may have been Mach's attitude to the principle that came to bear his name, his writing proved to be a continuing inspiration to the advocates of the principle and it prospered under their guidance. Prior to Einstein, the sponsors of the principle formed a scattered group, largely on the fringe of the physics community. Typically, the members of this group thought that the existence of the new physical mechanism was an issue to be settled by experiment.[26] They devoted their energies to devising and executing such experiments - and to the writing of labored but generally inconsequential treatises.

Mach ensured remembrance of two such experiments, those of the Friedlaenders and Föppl, by citing them in his *The Science of Mechanics* (1960, p. 296). The work of the Friedlaenders is described in the short, two-part monograph, Friedlaender (1896). The first part, written by Immanuel Friedlaender, describes how Immanuel's pursuit of the relativity of motion and the problem of centrifugal forces lead him to what we would now call Mach's Principle (p. 14):

> Without knowing that this had already been done by Mach, I have doubted the completeness of these foundations of mechanics for many years now. In particular I have come to the conviction that the appearance of centrifugal forces ought to be explicable also through regular mechanical knowledge [*Erkenntnis*] from the relative motions alone of the systems concerned, without resorting to absolute motion.

In just a few words, Immanuel is able to state clearly the call for a new physical mechanism which would supplement the existing laws of mechanics and explain centrifugal forces in terms of relative motions alone. Yet, ironically, he gives priority for this idea to Mach, even though I have been unable to find a similarly clear formulation of the idea in Mach's writings. Immanuel then proceeded to describe his efforts to detect this mechanism experimentally. He expected that the spinning of a fly wheel would produce forces directed away from its axis through this mechanism, just as the rotation of the heavens about the earth produces centrifugal forces. He proposed to detect these forces with a torsion balance, "the most sensitive of all physical instruments" (p. 15). However, when he sought to carry out these experiments in a rolling mill in Peine in November 1894, this necessary but extreme sensitivity of the balance proved to be his undoing. His results were inconclusive since he was unable to control disturbing influences. He lamented (p. 16): "A sensitive torsion balance is, however, a tricky instrument and a rolling mill certainly not the most comfortable or most favorable location for precision measurement."

Upon the failure of these experiments, he turned to his brother, Dr. Benedict Friedlaender, who only then informed him of Mach's work. Jointly they developed their ideas, upon which Benedict reported in the second part of the monograph. Immanuel concluded by stating his expectation that the correct formulation of the law of inertia ought to lead to "a unified law" which combined both gravitation and inertia as an action of masses. The idea that this new mechanism be integrated with the law of gravitation is not usually attributed to Mach, but is considered Einstein's innovation. Of course, in the Friedlaenders' hands it was merely speculation, but at least we see that the unification Einstein effected was not so completely unanticipated.

Föppl (1904) described his attempt to perform an improved version of the Foucault pendulum experiment. The purpose of the experiment was to reveal the precise disposition of an inertial system, correcting for the acceleration of the laboratory on the surface of the earth. He explained that "Foucault's pendulum experiment is afflicted with such sources of error that its accuracy leaves much to be desired even with careful execution" (p. 5). Föppl described how his experiment employed a carefully suspended gyroscope. Its precessional motion would reveal the disposition of an inertial frame of reference. Föppl hoped his experiment might decide whether (p. 5): "... the terrestrial phenomena of motion is itself influenced by the rotation of the earth in such as way that, for [these motions], the rotation of the earth does not coincide with that [rotation] with respect to the fixed star heaven."

In other words, Föppl is interested in comparing two reference systems. The first is the reference system of the fixed stars. The second is the inertial reference system in the neighborhood of the earth's surface revealed by the motions of bodies, such as the pendulum of Foucault's experiment. These systems are routinely assumed to coincide. Föppl conjectures that they may not because of "a possible, special influence of the rotation of the earth" (p. 5). In the event, Föppl reported that he could detect no deviation from coincidence within the accuracy of his experiment.

The report of this experiment was communicated to the Munich Academy on February 6. It was not until a further communication of November 5 (Föppl 1904a) that we find what led Föppl to conjecture such a special influence. His inspiration was the work of Mach on the relativity of motion. According to Mach, Föppl reported, an inertial system "obtains its orientation from the masses of the system of the universe in some kind of law governed manner." (p. 383). Föppl later (p. 386) considered the bodies of the universe divided into a large and a small group. An inertial system is determined by the combined group. Therefore, if the larger group is used to define a rest system of reference, the inertial frame will execute some motion in it, such as a rotation. This rotation would appear as Coriolis forces in the rest system of the larger group; they would not be regarded as merely artifacts of calculation but as "physically existing forces exerted by the smaller group on each test point." Föppl then explained that these were the considerations that led to the experiment described in his earlier communication. If one takes the fixed stars as the larger group of bodies and the earth as the smaller, then these forces would be the "special influence of the earth" sought.

If Mach ensured remembrance of the work of the Friedlaenders and Föppl, then Einstein similarly ensured remembrance of the work of Hofmann. In (Einstein 1913, §9), he discussed what he called the "hypothesis of the relativity of inertia," the hypothesis that inertial resistance is merely resistance to acceleration with respect to other bodies. As to the origin of the idea, Einstein wrote

> It is well known that E. Mach, in his history of mechanics, first advanced this point of view with all sharpness and clarity, so that here I can simply refer to his exposition. I refer also to the ingenious pamphlet of the Viennese mathematician W. Hofmann, in which the same point of view is advanced independently.

The work referred to is (Hofmann 1904).[27] The forty three page pamphlet is a wordy and labored defense of the relativity of motion. It seeks to escape the inference from centrifugal forces to absolute acceleration by urging that these forces arise from an interaction with the remaining masses of the universe. Unlike Föppl and the Friedlaenders, Hofmann (pp. 28–30) conjectured a new mechanical law that would lead to this interaction and perhaps this is what attracted the description of 'ingenious' from Einstein. Hofmann considered the standard result of traditional mechanics that the kinetic energy (*die lebendige Kraft*) of a body of mass m moving at velocity v is $mv^2/2$. He found this result unsatisfactory since, in the case of two masses m and M in relative motion, the kinetic energy of m with respect to M is not the same as the kinetic energy of M with respect to m. Therefore Hofmann proposed a new, symmetric law for the kinetic energy L of two bodies of mass m and M in relative motion with speed v and at a distance r

$$L = kMmf(r)v^2, \tag{2}$$

where k is a constant and f some function to be determined. For consistency with known results in mechanics, Hofmann indicated that the kinetic energy of a mass of actual experiment derives contributions from all the masses of the universe according to (2), so that (2), upon integration over all these masses, must yield the familiar $mv^2/2$.

Hofmann's law contains a mechanism in which inertial resistance is resistance to acceleration with respect to other bodies; for, in the case of two masses m and M, an attempt to change the relative velocity v will change the kinetic energy and thus require a force. In the case of a body in relative rotation with respect to the bodies of the rest of the universe, one would expect this same mechanism to yield centrifugal forces.

Hofmann did not develop the technical details and formal consequences of his supposition (2) in any systematic or extensive manner. This task was carried out by Reissner (1914, 1915). Reissner gave the usual attribution to Mach. Curiously, however, he made no mention of Hofmann, even though Hofmann's law (2) is the fundamental supposition upon which Reissner's theory is built. Perhaps we should allow for the possibility that Reissner independently arrived at the same supposition. In any case, the years 1914 and 1915 were not the time to construct a theory embodying the relativity of inertia, for such a theory would have no chance of competing with Einstein's general theory of relativity, whose brilliance came to outshine all competitors. By 1916, Reissner (1916) had turned his attentions to work on the latter theory, developing his celebrated solution of Einstein's field equations.

There is a small puzzle associated with the pamphlet. Einstein attributes its positing of the relativity of inertia as independent of Mach. Certainly the pamphlet itself makes no claim either way; no works by other authors are mentioned, and Mach is never named. However there is sufficient similarity between parts of Mach's and Hofmann's analysis to raise suspicion of an unacknowledged debt by Hofmann to Mach. Hofmann, for example, couches part of his discussion in terms of Newton's bucket experiment. He even proposes consideration of what would happen if the water-filled bucket were surrounded by a very heavy ring which is set into as rapid a motion as possible (p. 32) – close indeed to Mach's suggestion of the thickening of the walls of the bucket. Perhaps Einstein's attribution of independence from Mach derives from the failure of the text of Hofmann (1904) to cite Mach. However, Einstein may also have the claim directly from a meeting with Hofmann, which might have happened during Einstein's visit to Vienna for the 85th *Naturforscherversammlung* in September 1913 - (Einstein 1913) is the text of a lecture he delivered at that meeting. Again, Einstein describes Hofmann as a Viennese mathematician. That information could not be gleaned from the pamphlet alone, which simply described Hofmann as a professor and gave no affiliation.

The work of the Friedlaenders, Föppl, and Hofmann enables us to start to assemble an image of the group working around 1900 on what is to become Mach's Principle. First, the group members are on the fringes of the physics community. Only Föppl has any status in this community.[28] And they are an isolated and fragmented group. None of these authors cites any of the others. Indeed, the work of the Friedlaenders and of Hofmann were published in such obscure vehicles that we are now probably only aware of them because they happened to be cited by Mach and Einstein. In any case they are difficult works to procure. Thus we might well conjecture that the works discussed so far are but a random sample of other similar works which may be unknown because of their obscure vehicles of publication or a failure to publish at all.

This conjecture is confirmed by Höfler's (1900, pp. 122–26) report. He described experiments of which he was aware and which were designed to test the relativity of motion. Höfler knew of the Friedlaenders' experiment and described Mach's remark about the thickening of the walls of Newton's bucket as a thought experiment. In addition, he described an experiment due to Johannesson (1896). The experiment, only incompletely described by Höfler, involved rotation in connection with an oil droplet or sphere. Johannesson's results did not correspond at all with Johannesson's expectations. The design of the experiment seems flawed and Höfler devoted a page-long footnote to conjectures on where the deficiencies of the experiment may have been. He made clear that no positive result came from the experiment. Höfler also described another proposal for an experiment by Herr Dr. Karl Neisser.[29] The proposal involved examining the behavior of a gyroscope in air and in atmospheres of reduced pressure. Neisser, a relativist about motion, somehow managed to infer from this doctrine that the behavior of a gyroscope must at least in part be dependent on the relative rotation of the wheel against the air. Therefore he expected that a spinning gyroscope would lose its stability if enclosed in a chamber from which the air is pumped and that it would fall down like a gyroscope that is not spinning. Höfler added a remark to the proofs of his volume that Neisser had informed him that he had been able to perform the experiment, but the expected effect had not occurred. Höfler's report confirms that there was more interest around 1900 in what became Mach's Principle. But it would also seem that these further investigations were not as competently executed.

6. Early Critics of Mach's Principle

When Einstein incorporated Mach's Principle into the foundations of his general theory of relativity, he drew it in from these fringes into a new mainstream. In fact, Einstein's work defined what the new mainstream was to be in the physics of space, time, and gravitation and also, as it happened, a new mainstream in philosophy of space and time. Thus the principle enjoyed an enviable prominence. Einstein incorporated the principle or its precursors into most of his accounts of general relativity in the 1910s and 1920s. And, in his hands, the principle acted as midwife at the birth of modern relativistic cosmology. Einstein's efforts to ensure the place of the principle in his theory in 1917 led to his modification of his gravitational field equations and the introduction of the Einstein universe - not to mention the Einstein–de Sitter controversy.[30] The principle also rapidly entered into a popular and semi-popular literature on relativity, written for a wider, popular audience eager to come to grips with Einstein's great revelations. [See, for example, Born (1924, Ch. VII).] Finally the principle came to enjoy the sponsorship of leading philosophers and became a paradigm of the fruitfulness of the interplay of physics and philosophy. Prominent among these sponsors was Hans Reichenbach, leading figure in the logical empiricist movement, whose works in philosophy of space and time would dominate the discipline for several generations. [See (Reichenbach 1928, Sec. 34).]

6.1. Among the Physicists. The rapidity of the principle's rise and its lasting prominence tend to obscure the fact that it ascended only over a considerable if somewhat quiet opposition that persisted throughout this period as a tenacious skepticism towards the principle. That opposition can be located clearly in two areas: among physicists both before and after the advent of general relativity, and among philosophers, both of the neo-Kantian old guard and of the new generation that spawned logical positivism.

Prior to Einstein's championing of the principle, it is difficult to find broad measures of the overall feeling of the physics community concerning it. Little was said in opposition to it. But it was not a strong position which could expect or demand response from its critics, since, as we have seen, support for the principle lay scattered and disorganized in the fringes of the community. Of course, this fact itself indicates a broader lack of support. However, we have two fairly clear expressions of opposition. Toward the end of the first decade of this century, Ernst Mach and Max Planck engaged in a fairly bitter, polemical exchange (Planck 1909, 1910; Mach 1910). At issue was the reality of atoms, defended resolutely by Planck against Mach's skepticism, and the viability of Mach's notions of economy of thought in science and the elimination of metaphysics. As Planck's assault become more bitter, he decided to mention another area of disagreement with Mach, the relativity of motion, even though this was not the focus of their dispute. He wrote (Planck 1910; taken from Blackmore's translation 1992, p. 145 with minor corrections)

> Where Mach attempts to move forward by relying on his theory of knowledge quite often he runs into error.
>
> Here belongs Mach's strenuously fought for but physically entirely useless thought that the relativity of all translational movements also corresponds to a relativity of all rotary movement, that therefore, one cannot decide at all in principle whether the fixed stars rotate around the earth at rest or the Earth rotates around the fixed stars. The equally general and simple principle that in Nature the angular velocity of an infinitely distant body circling a finite, rotating axis cannot possibly possess finite value is therefore for Mach either false or not applicable. According to Mach's mechanics, one is just as bad as the other.
>
> The conceptual errors about physical matters which this unallowable transfer of the principle of the relativity of rotary movements from kinematics into mechanics has already caused, if they were depicted more closely at this point, would lead us too far astray. It therefore naturally follows that Mach's theory cannot possibly account for the immense progress which is intimately associated with the introduction of the Copernican theory – a circumstance which should suffice by itself to put Mach's theory of knowledge into considerable doubt.

The target of Planck's skeptical ridicule is the relativity of all motion. Since this relativity is the motivation for what soon becomes known as Mach's Principle, Planck's scorn would presumably extend to that principle. It might well be the "conceptual errors about physical matters" to which Planck alludes. Frank (1957, p. 153) did report Planck's remark as aimed directly at this principle.

It is tempting to dismiss Planck's intemperate remarks as a petulant outburst. Even if it was, there is no reason to dismiss its basic sentiments as insincere. Whatever its origin, opposition from Max Planck was very serious. Perhaps it reflected a broader consensus. If not, Planck was sufficiently influential that his views could foster such a consensus. Worse, while we now principally remember Planck for his contribution to quantum theory, he was also one of the earliest well-placed supporters of Einstein's special theory of relativity. He energetically threw in his lot and his prestige with Einstein's theory at a time when the theory's author was still a little-known patent clerk with a proclivity for incorporating bizarre philosophy into his physics.[31] Clearly Planck's opposition to a full relativity of motion did not derive from any ill-considered antipathy to the general idea of the relativity of motion.

Philipp Frank was both physicist and philosopher. As physicist, like Planck, he was one of the early group that took up active research in special relativity. With Hermann Rothe, he first discovered one of the most frequently rediscovered results in special relativity – that the group property and requirement of linearity already powerfully constrain the possible transformation laws between inertial coordinate systems: The only viable options remaining are the Galilei transformation or the Lorentz transformations, with c^2 an undetermined factor (Frank and Rothe 1911). This publication, which was not Frank's first on special relativity, introduced the term 'Galilei transformation.' Frank also had very favorable relations with Einstein: Einstein recommended Frank as Einstein's own successor at the German University in Prague and Frank later published a biography of Einstein (Frank 1947). Thus we might well expect that Frank would have been sympathetic to the view that played such a prominent role in Einstein's thinking. Yet the final outcome of Frank's 1909 lecture, discussed above, is a decision *against* the Machian view, which, in Frank's hands, contains Mach's Principle. Frank (1909) attributed to Mach the view that inertia arises through "a formal, new law of nature about the action of masses" (p. 17). This view allows Mach to retain his relativist position and to answer affirmatively to the question of whether the future behavior of a system of bodies is determined solely by their relative motions and not any absolute motion of the entire system. Frank prefers a view intermediate between the relativism of Mach and antirelativism or absolutism. He considers the absolute motion of mechanics merely a special case of relative motion, that is, it is motion relative to 'fundamental bodies' or 'inertial bodies,' such as the fixed heavenly stars. This somewhat tortured, hybrid position enables him to claim establishment of his conclusion, stated in emphasized text (p. 18): "Physical phenomena do not depend only on the relative motion of bodies; this however still does not admit the possibility of the concept of an absolute motion in the philosophical sense."

Whatever may have been the broader feeling about Mach's Principle in the physics community in this early period, one would expect that, after its endorsement by Einstein, the principle would enjoy the broader support of the physics community, at least through the late 1910s and 1920s, the period of the euphoria over Einstein's discovery of general relativity. Of course, it is widely known that at least one member of the astrophysical community dissented. Willem de Sitter was clearly an enthusiastic supporter of Einstein's general theory of relativity. For example, in 1916 and 1917, when relations between the English and German physics communities were stretched by the bitterness of the Great War, de Sitter took upon himself the task of informing his English colleagues of Einstein's new theory by means of a series of communications to the Royal Astronomical Society. At the same time, however, he found himself disputing sharply Einstein's view that his general theory of relativity satisfied the relativity of inertia or what came to be called Mach's Principle. (See Kerszberg 1989 for a recent account of the controversy.)

There is some evidence that a majority in the physics community at this time did not agree with Einstein's view that Mach's Principle, in some suitable form, was one of the fundamental postulates of general relativity. (Einstein (1918) had listed Mach's Principle along with the principle of [general] relativity and of equivalence as the fundamental postulates of general relativity, when he published a carefully worded defense of his view of the foundations of the theory.) This is an outcome of a survey of expositions of general relativity which I recently completed (see Norton 1993, especially Sec. 4.2). Emphasis on Mach's Principle as a fundamental postulate of general relativity tended to be concentrated in popular and semi-popular expositions. Otherwise, most typically for serious textbook expositions, the principle found no place in the accounts of the foundations of the theory, with Einstein's own expositions comprising the major exception. Or the principle may appear later in discussion devoted to the cosmological problem, as in (Pauli 1921). It is difficult to know what to read into this treatment - or lack of treatment - of the principle. It certainly does not rule out the possibility that many of these authors regarded the principle as an uncontroversial consequence of the theory that they simply did not choose to discuss.

Laue (1921), at least, makes clear that his omission of Mach's Principle was based on reservations concerning the place of the principle in the theory. The goal of Einstein's (1917) famous cosmological paper was to eliminate the need to posit Minkowskian boundary conditions for the metric tensor in general relativity, for Einstein held that such boundary conditions violated the Machian requirement that the inertia of a body be fully determined by other masses alone. His ingenious solution was to abolish spatial infinity by means of the Einstein universe, which became an admissible solution of this field equations after the introduction of the cosmological term. Laue (1921, p. 180) discussed Einstein's proposal in the context of Laue's treatment of Minkowskian boundary conditions:

> According to the fundamental idea of the general theory of relativity, the inertia of a single body should vanish if it is at a sufficient distance from all other masses. For inertia can only be a relational concept, which can be applied only to two or more bodies. ... With the boundary conditions mentioned, however, the inertia [of a single body] continues to exist. Such considerations have led Einstein to the hypothesis of a space which runs back on itself like the surface of a sphere.[32] To us the whole question seems clarified too little physically for us to want to go into the matter. In the following we understand 'infinity' to be regions inside our fixed star system for which the mentioned boundary conditions hold, but which are sufficiently far distant from the bodies of the gravitational field under consideration.

6.2. Among the Philosophers. When it comes to the philosophical community in the period prior to the mid 1910s, it is more difficult to assess the broader view towards what will become Mach's Principle. The principle seems not to have been a major focus of philosophical debate and, for this reason, not to have many supporters or detractors. In 1912, Joseph Petzoldt wrote an article on special relativity and its epistemological connection to relativistic positivism. Because of Petzoldt's close connection and sympathy with Mach and his positivist views and because they had corresponded on precisely this question, we might expect the principle to figure in his article. It is mentioned only briefly in a footnote (p. 1057), and Petzoldt takes no position on it, beyond merely suggesting that further experiments like those of the Friedlaenders and Föppl may settle the question. Perhaps his correspondence with Mach in 1904 had not assuaged the doubts he initially expressed to Mach as quoted above in Sec. 3.2. Frank (1909), in mapping out 'relativist' and 'antirelativist' positions, wrote of the work of Höfler (1900) and more recently Hamel (1909a) as opposing Mach, characterizing their disagreement as a controversy (p. 12) and Höfler as writing a "polemic against Mach's thesis."

However, a reading of the sources Frank cites does not give one the impression of a polemical dispute over the specific question of whether inertia arises from some interaction of accelerated bodies mediated by a new physical mechanism. Hamel [(1909a), and the closely related (1909)] was devoted to developing Hamel's own axiomatic development of mechanics, with the discussion of Mach's views in preliminary surveys of the alternatives. Hamel does not directly address the question of a new physical mechanism for inertia. The closest is a critique of Mach's strictures against absolute space (for example Hamel 1909a, pp. 363–64). Höfler does rehearse lengthy debates over the relativist and absolutist positions. Yet his specific attitude to the possibility of a such a new physical mechanism is very sober and undogmatic. He seems fully prepared to let actual experiment decide. For this reason, presumably, he gave the careful review (discussed above) of experiments designed to detect the mechanism. He then stated his view (or rather buried it in grammar of bewildering complexity!) (pp. 125–26):

> From my point of view I must admit in any case that, in so far as it is allowed at all, or even is ones duty, *before* an experiment, to form ideas over what can reasonably emerge from it, I expect nothing from all such experiments that could become somehow in the future a direct experimental proof for the relativity of rotational motion. I hold this negative expectation not without expressed experiential, even in part experimental foundations. Rather I believe that, f[or] i[nstance], according to the total experiences of mechanics so far, in an axially symmetric* system, such as would be a bucket with miles thick walls in rotation about its geometric axis, no force couples would arise on the water mass inside and therefore according to these mechanical experience so far, it cannot be set into rotation. More precisely: It cannot be set into rotation any more than [a water mass] in a bucket with walls of ordinary thickness, of which we know of course (or at least for the present believe to know), that only the innermost layer is acted upon by friction.

He continued to quote Hertz and Mach to stress the dependence of the question on experiment and the possibility of new experiments overturning the outcomes of old experiments, concluding: "But do I have to give up our current law of inertia, the foundations of our whole present mechanics, for such a 'possibility?'"

It is difficult to fault the good sense of this unadventurous assessment. Let experiment decide, Höfler says. But he notes his skepticism about a positive outcome, since the mechanism sought would have to be quite unlike anything encountered so far in mechanics. The footnote to the word 'symmetric' sought to drive this last point home. Yet, ironically, in the attempt to dismiss them, the footnote ended up anticipating a Machian class of mechanical theories modeled after electrodynamical laws!

> One must at least say that a *geometrically* axially symmetric system is not also *phoronomically* [kinematically] *and dynamically* axially symmetric, even only because it rotates about its axis. But in this case force effects would be ascribed to mass particles propagating in different directions, f[or] i[nstance], antiparallel, and [those effects] should be functions of the direction (and speed?); and also this assumption (an analogy to Weber's electrodynamic hypothesis) is certainly at least suggested by nothing in the current experiences of mechanics and would hardly allow explanation of the current experiences, upon which, after all, the thesis of the relativity of motion depends.

Höfler's work lies in the neo-Kantian mainstream. It is actually an afterword to an edition of Kant's *Metaphysische Anfangsgründe der Naturwissenschaft*, and the two are bound as one volume. Thus it would seem that the neo-Kantians, a dominant force in German language philosophy at this time, had no objection of Kantian principle to the possibility of inertia arising from some new physical mechanism. But that did not guarantee assent from the neo-Kantians. The leading neo-Kantian, Ernst Cassirer (1910, pp. 176–77) attributed to Mach the notion that the fixed stars are "one of the *causative factors* on which the law of inertia is dependent." He felt the view untenable since it amounted to robbing the law of inertia of its status as a law:

> If the truth of the law of inertia depended on the fixed stars as these definite individuals, then it would be logically unintelligible that we could ever think of dropping this connection and going over to another system of reference. The principle of inertia would in this case not be so much a universal principle of the phenomena of motion in general, as rather an assertion concerning definite properties and 'reactions' of a given empirical system of objects; – and how could we expect that the physical properties found in a concrete individual thing could be separated from their real 'subject' and transferred to another? ... [the meaning of principle in this view] corresponds in no way to the meaning and function it has actually fulfilled

in scientific mechanics from the beginning.

While we would not expect unqualified support from neo-Kantians for ideas attributed to Ernst Mach, we would expect such ideas to receive a more sympathetic hearing from members of the Vienna Circle, a discussion group which met in Vienna in the 1920s and out of which the logical positivist movement sprang. Ernst Mach was the spiritual inspiration for the group – Frank (1949, p. 79) called Mach "the real master of the Vienna Circle." Frank himself was one of the longest standing members of the Circle; his early discussion meetings with the mathematician Hans Hahn and economist Otto Neurath starting in 1907 had laid the foundations for the group of the 1920s. Yet as we have seen, Frank (1909) did not endorse the proposal for a new physical mechanism for inertia that he read in Mach's works. This opposition was no longer voiced in Frank's later writings, however. (See, for example, Frank 1947, Ch. 2, §8; 1957, p. 153.)[33]

In 1922, Moritz Schlick was appointed to Ernst Mach's old chair at the University of Vienna, and it was around Schlick that the Vienna Circle organized itself. Thus it is somewhat surprising to discover that the principal burden of Schlick (1915) was to drive a wedge between Mach's analysis of inertia and the treatment given by Einstein in the context of his general theory of relativity.[34] Einstein's approach is praised and Mach's is condemned. Schlick states Mach's escape from Newton's argument in his bucket experiment as follows (p. 166): "Experience does not show us that centrifugal forces do not also arise if the entire fixed heavenly stars were to rotate around it."

Against Mach's view, Schlick levels two objections. The first is aimed at Mach's often stated view that there is no distinction between the cases of the bucket rotating and stars at rest and the case of the bucket at rest and stars rotating, so that (Schlick quoted Mach as saying)

> The experiment [of testing whether rotating stars induce centrifugal forces] cannot be carried out, the idea is completely meaningless, since the two cases do not sensibly differ from one another. I hold the two cases to be the same case and the Newtonian distinction an illusion.

Schlick responds that Mach's proposal is not at all beyond test. He refers to Einstein's work on the relativity of rotation that has led to experimentally testable conclusions. Presumably Schlick means that if rotation relative to the distant stars induces inertial forces, then one would also expect rotation relative to other bodies to induce forces, such as Einstein (1912, 1913) found in his developing general theory of relativity. For example, a rotating shell of matter induces Coriolis forces within it. Schlick's second objection is (p. 166): "...the assertion that the two cases do not differ sensibly, a *petitio principii*, is evoked by ignoring the difference between kinematic and dynamic ways of consideration."

That is, he objects that one can define motion purely kinematically if one wishes; but that does not ensure that all the physical facts associated with motion are reducible to kinematics. Newton's theory supposes otherwise. It posits the possibility of kinematically identical systems which differ dynamically – for example a rotating and non-rotating body. And the difference between the two is a fact of sense experience (p. 168): "We can also ascertain the absolute rotation of a body, according to the Newtonian view, through muscular sensation, for we will find with its help that centripetal forces are needed for the body to keep its shape and to hold together its parts."

Mach's analysis ironically had turned into an exercise in *a priori* physics (p. 167): "It is curious to observe how sometimes exactly the attempt always to stick with just sensible experience leads to clever, *a priori* postulates, since one forgets that experiences can only be isolated from one another in abstraction."

Schlick proceeded to compare Mach's view with that of Einstein in his general theory of relativity. He asked if Einstein's theory amounted to "a great triumph of Mach's philosophy, since it had asserted the relativity of all motion as necessary." Schlick felt it did not represent such a triumph for three reasons:

> The first reason, which is already completely decisive, is one we have already presented, in that we have showed the arguments that led Mach to his conclusion are completely untenable. If, nevertheless, it turns out to be correct, it would result more from an accidental coincidence than a proper verification. With Mach the conclusion arises as a necessity of thought; with Einstein it is posited as a fundamental assumption of a theory and the decision of how far it may be considered valid is finally still left to experience.

The second objection referred to the lack of general covariance of the then current version of general relativity and to Einstein's belief that a generally covariant theory would be physically uninteresting. Thus Einstein's theory contradicted Mach's view, which required the equivalence of all reference systems. This objection could not stand for long, since, in November 1915, Einstein advanced the final generally covariant version of his theory and retracted his objections to general covariance. (See Norton 1984 for an account of this episode.) The third objection repeated parts of the first: Mach had just had a clever idea; but Einstein had built a theory on it. Schlick however was anxious – if not over anxious – to deprive Mach even of much credit for having a clever idea. He called the idea "very obvious" and explained in a footnote (p. 171):

> In order to show just how obvious the idea is, I might perhaps mention that I had already thought of it as a 6th form boy [Primaner] and in conversations stubbornly defended the assertion following from it that the cause of inertia must be assumed to be an interaction of masses. I was delighted, but not at all surprised, to come across the idea again shortly, when I got to know Mach's *Mechanics*.

It is difficult to overlook the unpleasant, dismissive tone of Schlick's remarks. It is somewhat reminiscent of Planck's tone, as is Schlick's general argument. For Planck was clearly happy to endorse a relativity of inertial motion, which formed the foundation of Einstein's special theory of relativity. He was unable to find kind words for Mach's proposal that this relativity be extended to all motion. Thus we may wonder if it is mere coincidence that Schlick studied physics under Max Planck at the University of Berlin, taking his doctorate in 1904. Is there some kind of unhealthy conspiracy against Mach plotted by the students of Mach's opponents? If one wants to, one can always find fragments of evidence for conspiracies. Laue, too, was an assistant of Planck in Berlin, and Frank was a student of Mach's arch rival, Boltzmann! However, I think there is no weight of evidence for such a conspiracy theory. The opposition of Frank and Laue is mild and mildly stated. It is more compatible with seriously considered disagreement. Schlick, however, was more intemperate. He was not prepared to concede anything to Mach. He closed his paper by noting that particular relativistic assertions made by positivists such as Mach were more likely to be refuted than confirmed by advances in physical science. Moreover, the investigations of Mach or other positivists on the concept of time did not pave the way for Einstein's special theory of relativity. "No one anticipated, f[or] i[nstance], the relativization of simultaneity."[35]

7. Conclusion

Mach presents us with a perplexing puzzle in his analysis of Newton's bucket experiment and the law of inertia. On the one hand, in his publications, the only unequivocal proposal is that we eliminate the odious notion of space by redescribing the relevant experiment and law in a way that does not use the term 'space.' If there is a suggestion of a new physical mechanism that would reach from the distant stars to cause the inertial forces in Newton's bucket, then the proposal is made vaguely and we are left to wonder whether Mach endorses it or condemns it as unscientific. On the other hand, if Mach did not wish to propose a new physical mechanism for the origin of inertia, then, in the course of the final two decades of his life, he passed over numerous opportunities to correct many who publicly attributed such a proposal to him.

I favor the view that Mach's published pronouncements cease to be ambiguous when we recognize that Mach held an extremely restrictive view of causation. Specifically, Mach held a causal relation to be nothing more than a functional relation between actual phenomena and prohibited speculation on hypothetical or counterfactual systems as unscientific. All we are allowed to infer from Newton's bucket experiment is that centrifugal forces arise when there is relative rotation between the water in the bucket and the other bodies of the universe. That alone is the causal relation. We have no license to infer to an absolute motion or even what would happen if (counterfactually) the walls of the bucket were made several leagues thick. This reading exonerates Mach of equivocation, ambiguity and inconsistency in his publications. However, it requires that the proposal of a new physical mechanism, as commonly attributed to Mach, is incorrect, and it leaves unexplained why he failed to correct this frequent misattribution in the final decades of his life.

If Mach did not propose such a mechanism, then at least the proposal was widely attributed to him in the 1890s and 1900s. It was then the focus of work of a scattered and disconnected group of investigators, largely on the fringes of the physics community. August Föppl was perhaps the only member of this group with any standing in the physics community. There is some indication of the proposal arising independently of Mach. Immanuel Friedlaender claimed his own version of the proposal came prior to knowing of Mach's work. Einstein attributed independent introduction of the proposal to Hofmann. Because of their lack of cohesion and because they tended to publish in obscure vehicles, it is likely that the full extent of their work is not now appreciated. The known work tends towards actual experimental test of the mechanism (unlike Mach) and labored but rather inconsequential treatises.

It is difficult to gauge the broader view of the proposal for a physical mechanism to explain inertia, prior to its sponsorship by Einstein. The difficulty is that the proponents of the view were largely on the fringes of the physics community and could not expect or demand a response from the mainstream. At least Max Planck, in 1910, spoke out strongly against Mach's insistence on the relativity of motion, while at the same time energetically supporting the relativity of inertial motion in Einstein's special theory of relativity. In 1909, Philipp Frank also weighed the possibility of a new physical mechanism to explain inertia and decided against it. After Einstein's sponsorship of what became Mach's Principle, the notion was widely celebrated by both physicist and philosopher. It seemed to provide a paradigm of fruitful interaction between the two disciplines. However, in the physics community its celebration tended to be focused in the popular and semi-popular expositions of general relativity. In general, as a review of expositions of general relativity from the 1910s and 1920s shows, the broader physics community did not wish to present Mach's Principle as one of the fundamental postulates of general relativity.

Prior to Einstein's sponsorship, the philosophy community devoted little attention to the proposal. If it was noticed at all, it accrued a mention in passing in the more traditional debates over absolute and relative motion. Criticism offered was sober and, in my view, largely justified. For example, if Mach's proposal was to be construed narrowly as urging the replacing of the Newtonian law of inertia by the observation that free bodies move uniformly with respect to the fixed stars, then Cassirer objected that this was a retrograde step for science, for it replaced a general law by an extremely restrictive description of one case. If Mach's proposal was for a new law, then Höfler felt that its merit was to be settled by experiment. But all experiments so far had yielded no positive results. This was hardly an encouraging foundation for overturning the Newtonian principle of inertia, then one of the most successful of scientific laws. In a similar vein, Schlick complained that Mach was engaged in *a priori* physics, an ironical twist given Mach's emphasis on the supremacy of experience. One could, Schlick noted, define motion purely kinematically. But this by no means guaranteed that the complete physics of moving bodies ought to be determined solely by kinematic relations. Newton had certainly supposed otherwise - and the dynamic effects to which he appealed, inertial forces, were matters of direct experience. The centrifugal forces that distinguish rotation are directly sensed by the muscles. Thus Schlick was at pains to distinguish Mach's view, which legislated *a priori* and seemed uninterested in experimental test, from Einstein's view, in which one constructed a definite theory with definite predictions that could be subject to experimental test.

If there is a moral in the early history of Mach's Principle, it lies exactly in Schlick's last point. As long as the relativity of all motion is posited dogmatically and Mach's Principle derived from it as *a priori* physics, then it is moribund. Its promise lies in the realm of empirical science, in the attempt to draw the doctrine of relativity and Mach's Principle into a physical theory that can be subject to experimental test, where one allows that experience may speak against it.[36] It was Einstein's recognition of this point that enabled him to breathe life into Mach's Principle.

NOTES

[1]I am grateful to John Earman, Peter Galison, and Ulrich Maier for assistance in procuring sources for this paper and to Julian Barbour for helpful discussion.

[2]While Einstein is usually credited with christening the principle, Schlick (1915) had already used the term ["das Machsche Prinzip" (p. 170) and "...des Machsches Postulats" (p. 171)]. It is a little unclear precisely to what Schlick's terms referred. They most likely referred to Mach's general proposal for a relativity of all motion, from which, Schlick noted (p. 171), it follows that "the cause of inertia must be assumed to be an interaction of masses."

[3]In his synopsis of his critique of Newton's ideas, Mach (1960, pp. 303–304) gives another example of how the terms 'space' and 'time' can be eliminated in this case from the fundamental propositions of Newton's mechanics.

[4]Mach (1960, pp. 272–73) gives a similar analysis of time. When we say that some process takes time, this is simply an abbreviated way of saying that the process has a dependence on another thing such as the changing position of the earth as it rotates. If we forget that this is all we may mean, we can fall into the error of thinking of time as an independent entity. In fact time has "neither a practical nor a scientific value" and "It is an idle metaphysical conception."

[5]In recent literature, it has been urged that Mach did not intend to propose a new physical law and merely intended a redescription of Newton's theory that preserved its true empirical content. See, in particular, Strauss (1968). Wahsner and von Borzeszkowski (1988, pp. 602–603) also found Mach's goal merely to be redescription of existing Newtonian theory.

[6]This criterion may contradict Mach's own highly restrictive

pronouncements on causality which do not seem to admit such hypothetical or counterfactual claims. However if we rule out the possibility that Mach did allow such claims in his analysis – possibly in contradiction with his own general view of causality – then it seems to me that we have settled the question in advance of whether Mach actually proposed what we now know as Mach's Principle. (Added in proof) Julian Barbour (this volume, p. 218) has identified a remark found only in early editions of Mach's, *Mechanics*, as employing counterfactuals in the way I require. The remark is suggestive but still contains no positive proposal for a new mechanism. Rather it casts doubt on whether a particle in the set up described would move according to Newtonian prescriptions if the fixed stars were absent or not unvarying. The remark makes no positive claim about how the particle would move in this counterfactual circumstance. It does not even deny outright that the motion will not be Newtonian. It is merely "very questionable." These sentiments fit well with Mach's repeated exhortation that we have no business proclaiming what would happen in situations beyond our experience, such as if there were no fixed stars. All such attempts are dubious.

[7]As almost everywhere, Mach's precise point remains clouded by ambiguity. We cannot assume away these bodies, he says, since we cannot assume that "the universe is without influence on the phenomenon here in question." Is Mach assuming there is some influence? If so, he does not say so. What does Mach mean by 'influence' in any case?

[8]The text is a slightly corrected version of the standard English translation (1960, p. 284) "The principles of mechanics can, indeed, be so conceived, that even for relative rotations centrifugal forces arise." I am grateful to Herbert Pfister for pointing out an error in this standard translation of the *wohl* of Mach's original "Die mechanischen Grundsätze können also wohl so gefasst werden, dass auch für Relativdrehungen Zentrifugalkräfte sich ergeben."

[9]Mach continues in a similar vein, using mass weighted sums of distances, to treat the motion of two bodies which do exert a force upon each other.

[10]Carus (1906, p. 332) calls Mach a "kindred spirit" and is "proud to count him among my dearest personal friends," although "there are no doubt differences between Mach's views and mine."

[11]For more details see (Holton 1992, pp. 30–33) and (Thiele 1971).

[12]He refers to a passage from Mach (1911) which will be discussed below.

[13]See Mach (1915, p. 44).

[14]These passages do not exhaust the relevant passages from *The Science of Mechanics*, although I have found none that provide brighter illumination. I leave to readers the task of deciding what Mach intended when he asked of the bodies A, B, C, ... "... whether the part they play is fundamental or collateral" and that "it will be found expedient provisionally to regard all motions as determined by these bodies" (p. 283).

[15]I cannot resist observing that if this consideration is intended to show a Newtonian that the distant masses engage in some causal interaction with the

body in question, then it is an extremely odd argument. Under the Newtonian viewpoint, the reason that distant celestial bodies are so valuable for describing the motion of the rotating body is precisely that there are *no* causal interactions between them and the body.

[16](Added in proof) Julian Barbour (this volume, p. 216) doubts this claim. He points out that distances between inertially moving bodies do not in general vary in direct proportion to one another. In support he cites Mach's own equation for the distance between two inertially moving bodies [Barbour's equation (1)]. I do not find the situation so unequivocal. Since a is constant and $|dr/dt| \leq a$, the equation does give the stated linear dependence in the limit in which r becomes sufficiently large. Mach's words are all too few, but he is considering bodies separated by great distances. Are these distances great enough to bring us towards this limit? (If not, so that the bodies are close but just not interacting, how can Mach escape this equation, whose derivation requires little more than simple geometry?)

[17]Many of these authors were sufficiently close to Mach to meet or enter into correspondence with him, including Petzoldt, Frank, Föppl, and Einstein.

[18]Cassirer is sufficiently unsure of the attribution to indicate that he infers it from Mach's writing by introducing it as "Mach himself must, according to his whole assumption, regard the fixed stars ... as one of the *causal factors*" and felt the need to support the attribution with a lengthy quotation from Mach (1911).

[19]Mach's celebrated July 1913 renunciation of the role of 'forerunner of relativity' in the preface to his *Optics* (1921, pp. vii–viii) is far too vague to be such a correction, since it clearly refers to Einstein's relativity theory in general. Wolters (1987) also urges that this famous renunciation was forged by Ernst Mach's son Ludwig.

[20]Blackmore and Hentschel 1985, p. 121. Otto Neurath also wrote in about 1915 along similar but vaguer lines to Mach (Blackmore and Hentschel 1985, pp. 150–152), although one might no longer reasonably expect a response from an ill Mach who would die in 1916.

[21]The experiment sought inertial dragging effects in the vicinity of a spinning fly-wheel.

[22]Mach here cites a passage in his *The Science of Mechanics* (1960, p. 283, Mach's emphasis) where he reports the result that "...a rigid body experiences resistance in a frictionless fluid only when its velocity *changes*." He conjectures about the possibility of this result as a "primitive fact," introduced prior to the notion of inertia, were our world filled with some hypothetical, frictionless medium, which would be an alternative to "the forlorn idea of absolute space."

[23]Dingler here raises the possibility that Mach's position on this matter may have altered considerably through the years 1883–1912 of the various editions of *The Science of Mechanics*. I have been unable to check this possibility thoroughly. However the task of comparison has been eased considerably by a remarkable and unusual volume (Mach 1915). This volume contains, in

English translation, a compendium of the extensive additions and alterations made in preparation of the 7th German edition of the work. It is interesting to speculate why such a compendium, useless without the earlier volume, should be published at all, rather than simply publishing a complete, updated text. In any case, in examining the volume, I could see no evidence of a significant shift in Mach's viewpoint with respect to the matters at issue here.

[24]Recall also that, on Wolters's (1987) account, Ludwig Mach was hardly a reliable source for his father's views pertaining to relativity theory, since Wolters accuses Ludwig of forging his father's famous renunciation of his role as 'forerunner of relativity theory.'

[25]It is helpful to compare the analysis of Newton's bucket experiment under Mach's view of causation and under a view that leads to some version of Mach's Principle. In both, we arrive at the result that the centrifugal forces in the bucket arise from the rotation of the water relative to the distant masses *A*, *B*, *C*, ... Mach requires that we halt analysis at this point. The other view makes the assumption, decried by Mach, that this one relation can be decomposed into parts. It regards the interaction between the water and the masses *A*, *B*, *C*, ... as the compounding of many smaller interactions between the water and mass *A*, between the water and mass *B*, ... These smaller interactions are understood to obtain in the circumstance in which we have a universe devoid of all matter excepting the water and mass *A*, etc.

[26]This emphasis is quite different from Mach's. He seems less interested in experimental tests. His *The Science of Mechanics* only mentions the possibility of real experimental test in later editions in response to the experiments of the Friedlaenders and Föppl and does so in an equivocal way. He did however propose an experiment to Petzoldt in their correspondence of 1904, as we have seen.

[27]I am grateful to the editors of *The Collected Papers of Albert Einstein* (Draft of 1992) for determining that this was the work referred to by Einstein.

[28]For example, he was the author of (Föppl 1894), one of the most important German language introductions to Maxwell's electrodynamics, and first of the famous series. Many German physicists learned vector analysis from its self-contained exposition of vector analysis.

[29]Neisser is identified as one of the 'Teilnehmer des Kant-Maxwell-Collegs' and the only reference given is to a conference in 1893 on the question "Is absolute motion, if not discernible, at least conceivable?" at the philosophical society of the University of Vienna.

[30]For discussion of the role of the relativity of inertia and Mach's Principle in Einstein's accounts of the foundations of general relativity and of Einstein's later disenchantment with the principle, see (Norton 1993, Sec. 3).

[31]Planck had entered into an encouraging correspondence with Einstein by 1906. He had given a colloquium on the theory in Berlin in the fall of 1905 and encouraged work on the theory, supervising von Mosengeil's doctoral thesis on the theory. Planck also is believed to be the one that approved Einstein's 1905

special relativity paper for publication in *Annalen der Physik* (Miller 1981, p. 2). His immensely important paper (Planck 1908) on relativistic dynamics is credited by Pais (1982, p. 150) as the first paper on relativity authored by someone other than Einstein. See (Stachel 1989, pp. 266–67). Planck's support for Einstein did not wane. He was instrumental in engineering Einstein's move to Planck's own Berlin in 1914.

[32]Laue's footnote merely cites Einstein (1917).

[33]The latter discussion does, however, recapitulate Planck's (1910) objections, but proceeds to allow that Einstein's work eventually vindicated Mach's view.

[34]The same viewpoint is advanced far more briefly and in far more muted voice in (Schlick 1920, pp. 37–40).

[35]Perhaps Schlick might have agreed with Abraham's (1914, p. 520) gibe that Einstein's new theory scarcely fitted Mach's requirement of economy, for it replaces the then standard single gravitational potential with the complication of ten potentials, the components of the metric tensor.

[36]I. Friedlaender's and Föppl's experiments fell short of this goal. While the experiments could in principle reveal a positive effect, a null outcome could not provide a decisive refutation. Since they had no definite theory that fixed the magnitude of the effect, they could not rule out the possibility that a small positive effect lay hidden behind the random error that shrouds all null results.

REFERENCES

Abraham, Max (1914). "Neue Gravitationstheorien." *Jahrbuch der Radioaktivität und Elektronik* 11: 470–520.

Barbour, Julian B. (1989). *Absolute or Relative Motion? A Study from a Machian Point of View of the Discovery and the Structure of Dynamical Theories.* Vol. 1. Cambridge: Cambridge University Press.

Blackmore, John (ed.) (1992). *Ernst Mach – A Deeper Look.* Dordrecht: Kluwer.

Blackmore, John and Hentschel, Klaus, eds. (1985). *Ernst Mach als Aussenseiter: Machs Briefwechsel über Philosophie und Relativitätstheorie mit Persönlichkeiten seiner Zeit; Auszug aus dem letzten Notizbuch (Faksimile) von Ernst Mach.* Wien: Wilhelm Braumüller, Universitäts-Verlagsbuchhandlung.

Born, Max (1924). *Einstein's Theory of Relativity.* Revised edition with G. Leibfried and W. Biem. New York: Dover (1962).

Broad, Charlie Dunbar (1916). "Review of Mach (1915)." *Mind* 25: 118–119; reprinted in Blackmore (1992, pp. 62–63).

Carus, Paul (1906). "Professor Mach's Philosophy." *Monist* 16: 331–356.

Carus, Paul, ed. (1906a). "A Letter from Professor Mach." *Monist* 16: 627–629.

Carus, Paul (1913). *The Principle of Relativity in the Light of the Philosophy of*

Science. Chicago: Open Court.

Cassirer, Ernst (1910). *Substanz, Begriff und Funktionsbegriff*. Berlin; published as first book of William Curtis Swabey and Marie Collins Swabey trans., *Substance and Function and Einstein's Theory of Relativity*. Chicago: Open Court (1923); reprinted New York: Dover (1953).

Dingler, Hugo (1921). *Physik und Hypothese*. Berlin & Leipzig: Vereinigung wissenschaftlicher Verleger, Walter de Gruyter.

Einstein, Albert (1912). "Gibt es eine Gravitationswirkung, die der elektrodynamischen Induktionswirkung analog ist?" *Vierteljahrsschrift für gerichtliche Medizin und öffentliches Sanitätswesen*. 44: 37–40.

Einstein, Albert (1913). "Zum gegenwärtigen Stande des Gravitationsproblems." *Physikalische Zeitschrift* 14: 1249–1262.

Einstein, Albert (1917). "Kosmologische Betrachtungen zur allgemeinen Relativitätstheorie." *Preussische Akademie der Wissenschaften, Sitzungsberichte* (1917), 142–152; trans. as "Cosmological Considerations on the General Theory of Relativity." In H. A. Lorentz *et al.*, *The Principle of Relativity*. Dover (1952), pp. 175–188.

Einstein, Albert (1918). "Prinzipielles zur allgemeinen Relativitätstheorie." *Annalen der Physik* 55: 240–44.

Föppl, August (1894). *Einführung in die Maxwell'sche Theorie der Elektricität*. Leipzig: B. G. Teubner.

Föppl, August (1904). "Über einen Kreiselversuch zur Messung der Umdrehungsgeschwindigkeit der Erde." *Akademie der Wissenschaften, München, Mathematisch-Wissenschaftliche Klasse, Sitzungsberichte*, Sitzung vom 6. Februar, 1904, pp. 5–28. [Partial translation in this volume, p. 312.]

Föppl, August (1904a). "Über absolute und relative Bewegung." *Akademie der Wissenschaften, München, Mathematisch-Wissenschaftliche Klasse, Sitzungsberichte*, Sitzung vom 5. November, 1904, pp. 383–395. [Partial translation in this volume, p. 120.]

Frank, Philipp (1909). "Gibt es eine absolute Bewegung?" Vortrag, gehalten am 4. Dezember, 1909. *Wissenschaftliche Beilage*, 1910, 1–10.

Frank, Philipp (1947). *Einstein: His Life and Times*. trans. George Rosen. rev. 1953. New York: A. A. Knopf; reprinted New York: Da Capo, n.d. An abridged translation of *Einstein: Sein Leben und seine Zeit*. Munich: Paul List (1949); reprinted Braunschweig: Vieweg (1979).

Frank, Philipp (1949). *Modern Science and its Philosophy*. Cambridge: Harvard University Press.

Frank, Philipp (1957). *Philosophy of Science: The Link Between Science and Philosophy*. Englewood Cliffs, N.J.: Prentice-Hall.

Frank, Philipp, and Rothe, Hermann (1911). "Über die Transformation der Raumzeitkoordinaten von ruhenden auf bewegte Systeme." *Annalen der Physik* 34: 825–855.

Friedlaender, Benedict and Friedlaender, Immanuel (1896). *Absolute oder*

Relative Bewegung? Berlin: Leonhard Simion. [Partial translations in this volume, p. 114 and p. 309.]

Hamel, Georg (1909). "Über die Grundlagen der Mechanik." *Mathematische Annalen* 66: 350–97.

Hamel, Georg (1909a). "Über Raum, Zeit und Kraft als apriorische Formen der Mechanik." *Jahresbericht der Deutschen Mathematiker-Vereinigung* 80: 357–85.

Höfler, Alois (1900). *Studien zur gegenwärtigen Philosophie der Mechanik.* Als Nachwort zu *Kant's* [sic] *Metaphysische Anfangsgründe der Naturwissenschaft. Veroffentichung der Philosophischen Gesellschaft an der Universität zu Wien, Band IIIb.* Leipzig: Verlag von C. E. M. Pfeffer.

Hofmann, W. (1904). *Kritische Beleuchtung der beiden Grundbegriffe der Mechanik: Bewegung und Trägheit und daraus gezogene Folgerungen betreffs der Achsendrehung der Erde und des Foucault'schen Pendelversuchs.* Wien und Leipzig: M. Kuppitsch. [Partial translation in this volume, p. 128.]

Holton, Gerald (1992). "Ernst Mach and the Fortunes of Positivism in America." *Isis* 83: 27–60.

Johannesson, Paul (1896). "Das Beharrungsgesetz." *Jahresbericht des Sophien-Realgymnasiums in Berlin*, Ostern.

Kerszberg, Pierre (1989). "The Einstein–de Sitter Controversy of 1916–1917 and the Rise of Relativistic Cosmology." in *Einstein and the History of General Relativity, Einstein Studies*, Vol. 1. Don Howard and John Stachel, eds. Boston: Birkhäuser, pp. 325–366.

Laue, Max (1921). *Die Relativitätstheorie. Vol. 2: Die allgemeine Relativitätstheorie und Einsteins Lehre von der Schwerkraft.* Braunschweig: Friedrich Vieweg und Sohn.

Mach, Ernst (1872). *Die Geschichte und die Wurzel des Satzes von der Erhaltung der Arbeit.* Prag: Calve'sche Buchhandlung.

Mach, Ernst (1882). "The Economical Nature of Physical Inquiry." in *Popular Scientific Lectures*, 5th ed. T. J. McCormach, trans. LaSalle, Illinois: Open Court, 1943, pp.186–213.

Mach, Ernst (1910). "Die Leitgedanken meiner naturwissenschaftlichen Erkenntnislehre und ihre Aufnahme durch die Zeitgenossen." *Physikalische Zeitschrift* 11: 599–606; abbr. trans. as Blackmore (1992), pp. 133–37.

Mach, Ernst (1911). *History and Root of the Principle of the Conservation of Energy,* P. Jourdain, trans. Chicago: Open Court. [Translation of Mach (1872)].

Mach, Ernst (1915). *The Science of Mechanics: A Critical and Historical Account of its Development. Supplement to the Third Edition, Containing the Author's Additions to the Seventh German Edition.* P. E. B. Jourdain, trans. and annot. Chicago and London: Open Court.

Mach, Ernst (1921). *Die Prinzipien der physikalischen Optick.* Leipzig, 1921; trans. as *The Principles of Physical Optics.* 1926; reprinted New York:

Dover.

Mach, Ernst (1960). *The Science of Mechanics: A Critical and Historical Account of Its Development*, 6th ed. T. J. McCormach, trans. LaSalle, Illinois: Open Court.

Miller, Arthur (1981). *Albert Einstein's Special Theory of Relativity: Emergence (1905) and Early Interpretation (1905–1911)*. Reading, MA: Addison–Wesley.

Norton, John D. (1984). "How Einstein found his Field Equations: 1912–1915." *Historical Studies in the Physical Sciences*, 14: 253-316; reprinted in *Einstein and the History of General Relativity, Einstein Studies*, Vol. 1. Don Howard and John Stachel, eds. Boston: Birkhäuser (1989) pp.101–159.

Norton, John D. (1993). "General Covariance and the Foundations of General Relativity." *Reports on Progress in Physics* 56: 791–858.

Pais, Abraham (1982). *'Subtle is the Lord...': The Science and the Life of Albert Einstein*. Oxford: Clarendon.

Pauli, Wolfgang (1921). "Relativitätstheorie." In *Encyklopädie der mathematischen Wissenschaften, mit Einschluss ihrer Anwendung*. Vol.5, *Physik*, Part 2. Arnold Sommerfeld, ed., Leipzig: B.G. Teubner, 1904–1922, pp.539–775. [Issued November 15, 1921] English translation *Theory of Relativity*. With supplementary notes by the author. G. Field, trans. London: Pergamon, 1958; reprint New York: Dover.

Petzoldt, Joseph (1912). "Die Relativitätstheorie im erkenntnistheoretischen Zusammenhange des relativistischen Positivismus." *Deutsche Physikalische Gesellschaft. Verhandlungen* 14: 1055–1064.

Planck, Max (1908). "Zur Dynamik bewegter Systeme." *Annalen der Physik* 26: pp. 1–34.

Planck, Max (1909). "Die Einheit des physikalischen Weltbildes." *Physikalische Zeitschrift* 10: 62–75; abbr. trans. as Blackmore (1992), pp. 127–32.

Planck, Max (1910). "Zur Machschen Theorie der physikalischen Erkenntnis." *Physikalische Zeitschrift* 11: 1186–90; abr. trans. as Blackmore (1992), pp. 141–46.

Reichenbach, Hans (1928). *Philosophie der Raum-Zeit-Lehre*. Berlin: W. de Gruyter; Maria Reichenbach and John Freund, trans., *Philosophy of Space and Time*. New York: Dover, 1957.

Reissner, Hans (1914). "Über die Relativität der Beschleunigungen in der Mechanik." *Physikalische Zeitschrift* 15: 371–75.

Reissner, Hans (1915). "Über eine Möglichkeit die Gravitation als unmittelbare Folge der Relativität der Trägheit abzuleiten." *Physikalische Zeitschrift* 16: 179–85.

Reissner, Hans (1916). "Über die Eigengravitation des elektrischen Feldes nach der Einsteinschen Theorie." *Annalen der Physik* 50: 106–120.

Schlick, Moritz (1915). "Die philosophische Bedeutung des Relativitätsprinzips." *Zeitschrift für Philosophie und Philosophische Kritik* 159: 164–171.

Schlick, Moritz (1920). *Space and Time in Contemporary Physics*. New York:

Oxford University Press.
Stachel, John, ed. (1989). *Collected Papers of Albert Einstein. Volume 2. The Swiss Years: Writings, 1900–1909.* Princeton: Princeton University Press.
Strauss, M. (1968) "Einstein's Theories and the Critics of Newton." *Synthese* 18: 251–284.
Thiele, Joachim (1971). "Paul Carus und Ernst Mach: Wechselbeziehungen zwischen deutscher und amerikanischer Philosophie um 1900." *Isis* 62: 208–219.
Wahsner, Renate and von Borzeszkowski, Horst–Heino (1988). "Nachwort" to Mach, Ernst, *Die Mechanik in ihrer Entwicklung dargestellt: Historisch-kritisch dargestellt.* Berlin: Akademie-Verlag, pp. 563–647.
Wolters, Gereon (1987). *Mach I, Mach II, Einstein und die Relativitätstheorie: Eine Fälschung und ihre Folge.* Berlin & New York: Walter de Gruyter.

Discussion

Nordtvedt: Did you consider all the experiments a failure in the sense that they saw nothing, or did they have problems? Null experiments are good experiments even when they see no effects, and perhaps you were being a bit hard on the experimentalists, particularly Föppl.
Norton: I did not mean to say that the experiments were bungled. Rather what I meant was that the results had to be inconclusive since the experimenters had no idea of the magnitude of the effect sought. Therefore a null result could not eliminate the possibility of a positive effect smaller than their experimental error. Of course the experiments were not uninformative, since they did place an upper bound on the size the effect could have.
Ehlers: I'd like to have your reaction to the following: If one takes the redescription interpretation, then it seems to me that although Mach hinted at possibilities of redescription, he would not have been able to reconstruct the whole body of Newton's theory. Newton was, I think, much more of a mathematical physicist than Mach. Mach was perhaps more an intuitive and empirically oriented physicist. The Newtonian system needs a basis for concepts such as velocities, accelerations, and so on, some framework, relatively to which these concepts are well defined. Even people like Euler struggled with the question: How can you give a meaning to the concept of velocity if you don't have some space to which you refer it? It requires a considerable amount of abstraction to consider velocity as meaningful without absolute space, namely, only have a certain class of inertial frames. So my question is: Could a redescription be given which does not lose an essential part of

the Newtonian system as a quantitative mathematical theory?

The second remark is a comment only. I think if one, as a physicist, compares these two interpretations, namely, the redescription interpretation and the interpretation that one would like to have a new mechanics, then if one cannot decide, as a historian, which of the two interpretations is truer to the text, then I think it matters that physicists are interested in history, not so much because they want to know what has been said by such and such a person but which useful suggestions are contained in earlier works. The second view, namely, looking for a new mechanics, is fruitful and interesting for bringing physics further, whereas the redescription point of view, in that sense, is not of interest. Therefore I feel, even if one cannot decide, that for a physicist the other point of view is more fruitful and interesting.

Norton: Briefly, on the second point, as a historian, I'm fairly constrained by what happens [laughter], at least, I try to be. As a physicist you try to be constrained by the world. If the two can coincide, and I can find useful things happening, all the better, but I have to stick with what was there.

On the first point, I think you can redescribe everything that Newton had in his science without talk of absolute space and time. It's simply a matter of doing what Mach prescribed. You work through Newton's texts replacing every metaphysical claim by a statement of the observational content of the claim. Whether the resulting description will be economical is the real question. And this, I think, is what has always troubled Mach's system. There was a tension between the need for the descriptions to be restricted to observation and for them to be economical. We see this clearly in the case of Mach's skepticism over atoms. We like them since they do provide a very economical systematization of many physical phenomena. But the price of the economy is talk of entities that transcend observation. So it is with spacetime structures; they are unobserved, but, as you point out, they do enable just the systematization we want. In the end, I think this problem was a major part of the transition from the simple positivism of Mach to the logical positivism of those who followed Mach. It was the realization that one cannot be so narrow and restrict all talk to experience. You also need theoretical terms. Then follows the long debate over what to do with these theoretical terms. Are spacetime structures real entities or merely convenient aids to prediction?

Rindler: Did you say Föppl and Friedlaender had no idea of the magnitude of the effect they were looking for? Why didn't they have an idea? In those days there were a number of people who had played with

Maxwellian theories of gravitation. Dennis Sciama later pointed out that the Maxwellian type of gravitational theory has various Machian features. My question is, surely somebody before Dennis Sciama must have thought of that. Why isn't it that people used some kind of a Maxwellian estimate for the magnitude of the Machian effects they were looking for before they did those experiments? Of course, this would have totally discouraged them from even trying.

Norton: You're referring, I take it, to the literature in gravitation theory towards the end of the nineteenth century. They were trying to start modeling extra terms for Newton's theory on the basis of electrodynamics. I believe that a Weber-like law was one of them; there are many different variants. However, I did not find any cases of experimentalists using such laws to estimate the magnitude of the effect sought. As you point out, that is odd.

Renn: I think the answer to the question as to why the scientists who were looking around the turn of the century for Machian effects did not come up with precise ideas on the magnitude of these effects can be found in the split of two conceptual traditions, that of mechanics and that of electrodynamics, which I discuss at some length in my contribution [see p. 5]. Without much exaggeration one can say that those interested in electrodynamic theories of gravitation did not link this interest with a critique of mechanics along the lines of Mach and vice versa. A short footnote in the paper of the Friedlaender brothers, referring to a Weberian theory of gravitation, and the work of Einstein are exceptional in establishing the link between these two traditions.

Editorial Note (J.B.B.): The reader may be puzzled by the limited discussion recorded above of the issue raised by Norton of whether Mach truly intended a physically new theory of inertia or merely a redescription of Newtonian theory in relational terms. In fact, there was a fairly extended discussion at Tübingen around the passage by Mach reproduced in its entirety on p. 110 (beginning line 5) and discussed by Norton on pp. 16–17. However, examination of the discussion transcript showed that quite a large proportion of the comments, which were made without benefit of the complete exact text for examination, were either irrelevant or misleading, though Kuchař did make the important point that, irrespective of the physical significance Mach may have read into Eq. (1) on p. 17, the equation itself is mathematically incomplete, since it is a single scalar equation and therefore insufficient to describe either absolute or relative motion (cf. my comments on p. 217). Since the issue of whether Mach merely intended a redescription is discussed in some length in my own contribution (pp. 215–218) and Notes 1 and 2 on p. 230) and both Norton (in his Notes 6 and 16) and von Borzeszkowski and Wahsner (pp. 65–66) have responded to my comments, there seems little point in reproducing here the Tübingen discussion.

Mach's Criticism of Newton and Einstein's Reading of Mach: The Stimulating Role of Two Misunderstandings

Horst–Heino v. Borzeszkowski and Renate Wahsner

In the present paper we will give some arguments in favor of the thesis that the so-called Mach's Principle owes its existence to two misunderstandings, namely first to Mach's misunderstanding of fundamentals of Newtonian mechanics, mainly of the Newtonian notion of space, and second to a misreading of Mach by Einstein. The latter was admittedly a reading of genius, but nevertheless a misreading.

To start with, it should be mentioned that in order to discuss this matter it is not sufficient to study appropriate passages of Mach's *Mechanik*. Rather, one has also to analyze the other critical-historical treatises and, first of all, the philosophical work of Mach. Since this, however, is not the place for discussing Mach's philosophy in detail, here we shall make only a few remarks summarizing some aspects of Mach's philosophy that are of interest in the context of this topic. (For a detailed consideration, see Wahsner and v. Borzeszkowski 1988.)

Mach's main intention was to free mechanics, optics, and other physical branches from metaphysics, so that the real nature of physics becomes visible. In the search for a way toward his aim, he arrived at the conclusion that one has to study the history of physics. In his 1872 pamphlet *Die Geschichte und die Wurzel des Satzes von der Erhaltung der Arbeit*, which was programmatic for his life's work, Mach said:

> Denn metaphysisch pflegen wir diejenigen Begriffe zu nennen, von welchen wir vergessen haben, wie wir dazu gelangt sind. Man kann jedoch nie den thatsächlichen Boden unter den Füssen verlieren oder gar mit den Thatsachen in Collision gerathen, wenn man stets auf den Weg

Einstein Studies, vol. 6: Mach's Principle: From Newton's Bucket to Quantum Gravity, pp. 58–66

> zurückblickt, den man gegangen. [For the notions that we usually call metaphysical are the ones for which we have forgotten how we arrived at them. However, we can never lose the real ground from under our feet or, worse, come into collision with the facts if we always look back over the way that we have gone.] (Mach 1872, p. 2)

As far as mechanics is concerned, Mach considered it correct but, for historical reasons, represented by Newton in a manner containing a lot of metaphysical elements. Therefore, he intended to remove these elements by reformulating mechanics, and the result was his critical-historical account of mechanics: *Die Mechanik in ihrer Entwicklung. Historisch-kritisch dargestellt* (Mach 1883; later editions published in 1889, 1897, 1901, 1904, 1908, 1912).

As is well known, Mach was dissatisfied in particular with Newton's definition of mass and his representation of the axioms of mechanics. Therefore, he started by reformulating them.

First he replaced Newton's definition of masses, saying that the ratio of the masses m_1 and m_2 of two bodies is equal to the ratio of their weights, $m_1/m_2=G_1/G_2$, by the definition that states: The ratio of the masses m_1 and m_2 of two bodies is equal to the negative and inverse ratio of the accelerations b_1 and b_2 caused by their mutual interaction, $m_1/m_2=-b_2/b_1$.

From this point of view, Mach considered Newton's second law as a convention and the third law as a consequence of his definition of mass. So for him there remained only the task to answer the question as to the first Newtonian law. Mach's answer was: This law is a fact first perceived by Galileo, but it has only a definite meaning when one can answer the question as to the reference system one needs in order to determine the motion.

He argued as follows. When one says "a system or a body on which no forces act is either at rest or in uniform motion," one has to ask "uniform motion, relative to what?" Newton's answer was "to absolute space." But this is a metaphysical element that can be replaced by the totality of the cosmic masses, more precisely, by the fixed stars realizing a rigid reference system.

To demonstrate this, Mach showed that the center of mass of an N-body system on which no external masses act realizes a system to which the motion described by mechanics can be referred. Assuming then that this N-body system is the system of cosmic masses (on which *per definitionem* no external forces act), he has in this way determined a cosmic reference system. For Mach this was a proof that one can free mechanics of the metaphysical element 'absolute space' by replacing it with something that is nearer to experience. In this way, inertia seemed to him caused by cosmic masses.

Mach was encouraged to use this formulation by his analysis of Newton's arguments concerning the behavior of water in a rotating bucket. According to Newton, the curved surface of water arising when the water was rotating with respect to the heavens showed that inertial forces are caused by the motion with respect to the absolute space. In contrast to Newton, Mach believed that this experiment is no proof of the existence of an absolute space, since one can ask what would happen if the whole of the heavens rotated around the bucket. He believed that it should lead to the same result, namely to a curved surface of water as a consequence of the rotating cosmic masses. So one cannot – he argued – distinguish between relative and absolute motions by experience. One can, however, talk about the real (relative) motion with respect to the cosmic masses.

Now it is not intended to discuss here the problem of the extent to which Mach did really provide a formulation of mechanics which could be used in physical work (for a detailed discussion of this, see, for instance, Bunge 1966). The point we want to stress is rather that the starting point of Mach's considerations was a misunderstanding of mechanics. When Mach started he believed that the space of Newtonian mechanics is a rigid background given once and for all like a stage in front of which physical processes unfold. He did not see that the so-called absolute space is the totality of all inertial systems and thus is not a metaphysical ghost but a constructive element like the quantity *mass* and other notions that are determined by the entire system of classical mechanics.

To be fair, it should be mentioned that in Mach's day classical mechanics was taught in a version which indeed was loaded with metaphysical ballast. Furthermore, when the first edition of Mach's *Mechanik* was published, the clarifying papers of Carl Neumann (Neumann 1870) and Ludwig Lange (Lange 1886) were not well known or even not yet published. Finally, the meaning of inertial systems was only understood when the role of Galileo's principle of relativity was cleared up, and this was only done in the context of the discussion around Einstein's special theory of relativity.

Mentioning these objective reasons for Mach's misreading of classical mechanics, however, one has also to state that it was too a consequence of his philosophical standpoint, i.e., of his empiro-pragmatic philosophy. This philosophy replaced the system (the physical theory) by a catalog of experimental data and their mutual relations. Mach wrote:

> Wenn man eine *vollständige Theorie* als das Endziel der Forschung bezeichnen wollte, ... müssten [wir] unter diesem Namen vielmehr eine *vollständige systematische Darstellung der Thatsachen* begreifen ... Das Ideal aber, dem jede wissenschaftliche Darstellung wenn auch sozusagen asymptotisch zustrebt, ... ist ein *vollständiges übersichtliches Inventar der Thatsachen eines Gebietes.* (Mach 1896, p. 461) [If we wish to say that *a complete theory* is the final aim of research ... we [must] understand by this word a *complete and systematic representation of the facts* ... But the ideal to which every scientific representation tends (even though only so to speak asymptotically) ... is *a complete and clear inventory of the facts of a domain* (quoted with slight alteration from Mach 1986, p. 415).]

Therefore, Mach did not and could not realize in what manner a physical theory determines its notions. He overemphasized the role of that what he called the real *(das Tatsächliche)*, so that his expurgation of metaphysics from physics degenerated into a liquidation of basic epistemological prerequisites of physics (Wahsner and v. Borzeszkowski 1988, pp. 595–597).

Because of his missing insight into the inevitability of transcendental assumptions of physics, it was difficult for Mach to incorporate results of authors like Lange clarifying the notion *inertial system* into later editions of his *Mechanik.* In the second edition of 1889, one finds, for instance, an Appendix with remarks on Lange's 1886 paper, but no change of the main text of his book. Subsequent editions then incorporate the supplements and other insertions into the main text. Here one feels that he has a lot of problems accepting Lange's definition of inertial systems without changing his own criticism of Newtonian mechanics. His way out of this dilemma is to say that Lange's answer to the question as to the reference system of mechanics and thus to the notion of space is purely mathematical, while his own is physical.

Let us now turn to Einstein and his attitude to Mach's ideas. As is well known, Einstein did not refer to Mach during the first period of the foundation of the theory of general relativity. Only when he arrived at the conclusion that the principle of relativity should be extended to arbitrarily moving reference systems and that gravitation is to be described by the metric tensor of a curved spacetime did he begin to talk of the relativity of inertia (this was about in 1912). In a 1912 paper he writes:

> Es legt dies die Vermutung nahe, dass die ganze Trägheit eines Massenpunktes eine Wirkung des Vorhandenseins aller übrigen Massen sei, auf einer Art Wechselwirkung mit den anderen beruhend. [This makes it plausible that the entire inertia of a point mass is the effect of the presence of all other masses, deriving from a kind of interaction with the latter.] (Einstein 1912, p. 39).

And in 1918 he even used the expression *Mach's Principle* (Einstein 1918).

Although Einstein's attitude to Mach's ideas changed in his later years (see, for example, Pais 1982; Wahsner and v. Borzeszkowski 1988), this principle played a stimulating and constructive role in physical discussions in the course of years. While initially the question as to validity of the principle in the theory of general relativity was in the center of interest, later this principle became the point of departure for the construction of alternative gravitational theories. In this connection, different authors were working with different formulations of this principle. In an analysis of this situation, it was stated (Goenner 1981) that this is due to the fact that Mach did not propose a definite *ansatz* for an induction of inertia by cosmic masses, so that Mach's principle says more about Einstein's and other authors' reading of Mach than about Mach's intention.

The thesis in favor of which we will give arguments here goes a step further. It says that Mach did not only not create a cosmic principle of the type gathered by Einstein from Mach's *Mechanik* but such a principle is even in conflict with Mach's ideas.

To this end, let us return to Mach's philosophy. As mentioned above, as a consequence of his empiro-pragmatic standpoint, Mach could not understand the status of a physical theory. The analysis of the discussion between Mach and Boltzmann, Planck, Hertz, and Einstein (Wahsner and v. Borzeszkowski 1988, pp. 604–642) makes this especially clear. Thus Mach could not grasp in what a manner a physical theory determines its notions and, in particular, not understand that Newton's axioms determine simultaneously the physical dynamics and the systems of reference to which this dynamics refers or has to be referred. Since he could only conceive of a catalog of single statements and facts but not of a theory, he could only ask whether a statement under consideration is a fact or not. In this scheme there is no room left for the space notion of classical mechanics. Therefore, he did not think of another physical theory as an answer to his criticism of Newtonian mechanics. He did not think at all in terms of theories, and Einstein's theory of general relativity had to seem to him further from experience than Newton's theory. To repeat, the aim he had was to reformulate classical mechanics so that its notions and statements were nearer to experience.

This is one line of argument showing that Mach did not think of a new cosmic principle that would lay the foundation of a new theory. But there are also more explicitly formulated arguments that one can find in Mach and which show the same.

When Mach had shown that the law of inertia can also be referred to the cosmic masses, he added that this reading implies the same difficulties as Newton's. For, in the Newtonian version one has to refer to *absolute space*, on which one cannot get a hold. In the other case, only a limited number of masses is accessible to our knowledge but not the totality of cosmic masses. In Mach's words:

> In dem einen Fall können wir des absoluten Raumes nicht habhaft werden, in dem anderen Fall ist nur eine beschränkte Zahl von Massen unserer Kenntnis zugänglich, und die angedeutete Summation ist also nicht zu vollenden (Mach 1912, p. 230). [In the one case we are unable to come at an absolute space, in the other a limited number of masses only is within the reach of our knowledge, and the summation indicated can consequently not be fully carried out (Mach 1960, p. 289).]

For Mach, these obstacles were of a fundamental nature, so that he did not believe that one could overcome them by modifying the physical theory. According to him, the universe as a whole is not tractable as a physical system. Notions like *energy of the universe* or *entropy of the universe* have no tangible sense because they imply applications of measurement notions to an object that is not accessible to measurement (Mach 1896, p. 338). The only thing he wanted to do was to bring to our attention the fact that the law of inertia (and other physical laws) is based on experience, on experience that is never complete and, even more, that can never be completed.

To conclude, Einstein introduced a cosmic principle into physics, and the irony of the story is that this was initiated by Mach and called by his name, although the possibility of cosmology as a physical discipline was the very thing that Mach himself denied.

REFERENCES

Bunge, Mario (1966). "Mach's Critique of Newtonian Mechanics." *American Journal of Physics* 34: 585-596.

Einstein, Albert (1912). "Gibt es eine Gravitationswirkung, die der elektrodynamischen Induktion analog ist?" *Vierteljahresschrift für gerichtliche Medizin und öffentliches Sanitätswesen* 44: 37-40.

Einstein, Albert (1918). "Prinzipielles zur allgemeinen Relativitätstheorie." *Annalen der Physik* 55: 240-244.

Goenner, Hubert (1981). "Machsches Prinzip und Theorien der Gravitation." In *Grundlagenprobleme der modernen Physik*. Jürgen Nitsch, Joachim Pfarr und Ernst-Walther Stachow, eds. Mannheim/Wien/Zürich: B. I. Wissenschaftsverlag. pp. 85-101.

Lange, Ludwig (1886). *Die geschichtliche Entwicklung des Bewegungsbegriffs und ihr voraussichtliches Endergebnis. Ein Beitrag zur historischen Kritik der mechanischen Prinzipien.* Leipzig: W. Engelmann.

Mach, Ernst (1872). *Die Geschichte und die Wurzel des Satzes von der Erhaltung der Arbeit.* Prag: Calve'sche Buchhandlung.

Mach, Ernst (1883, 1888, 1897, 1901, 1904, 1908, 1912). *Die Mechanik in ihrer Entwickelung. Historisch-kritisch dargestellt.* Leipzig: F. A. Brockhaus.

Mach, Ernst (1896, 1900). *Die Principien der Wärmelehre. Historisch-kritisch entwickelt.* Leipzig: Johann Ambrosius Barth.

Mach, Ernst (1960). *The Science of Mechanics: A Critical and Historical Account of Its Development.* LaSalle, Illinois: Open Court.

Mach, Ernst (1986). *Principles of the Theory of Heat Historically and Critically Elucidated.* (*Vienna Circle Collection*, Vol. 17), Dordrecht: D. Reidel.

Neumann, Carl (1870). *Über die Principien der Galilei-Newtonschen Theorien.* Leipzig: Teubner.

Pais, Abraham (1982). *'Subtle is the Lord.' The Science and the Life of Albert Einstein.* Oxford/New York: Oxford University Press.

Wahsner, Renate and v. Borzeszkowski, Horst-Heino (1988). "Nachwort der Herausgeber." In *Ernst Mach, Die Mechanik in ihrer Entwicklung. Historisch-kritisch dargestellt.* Renate Wahsner und Horst-Heino v.Borzeszkowski, eds. Berlin: Akademie-Verlag. pp. 563-647.

Discussion

Norton: I wanted to see if you have had any more luck with the historical puzzle than I have had. We seem to agree that Einstein is misreading Mach. I have tried to get some idea of where Einstein got the reading from. Is it possible that he just read Mach by himself, we know he read it in his early years, and produced this reading or is it possible that he had some help? Was there some intermediate source? Did he read about Mach elsewhere? For a while I conjectured that Philip Frank had played some intermediate role on the basis of Frank's 1909 paper that I cited and the fact that Frank was, I believe, Einstein's successor at Prague, but I couldn't come up with anything. I don't know if you might have come across something.

Borzeszkowski: I don't know. Maybe there was such an intermediate stage, but I think the main reason is that Einstein read only Mach's *Mechanics*, and then he did what a physicist should do. He tried to win from it a constructive idea.

One finds it also in other connections that when authors discussed epistemological questions and they were talking about philosophy, Einstein read it in a physicalizing manner, and, to repeat, this can be useful for purely physical considerations. One encounters, however, another situation when one wants to discuss such matters as we did this morning, namely the relation to philosophy and historical context. Then one has, of course, to analyze the whole edifice of thoughts of the author one refers to.

Bondi: You have not mentioned the very interesting statement of Mach's "the universe is only given once," which I think influenced Einstein. It certainly influenced me. To me, it means all our physics is learned in the presence of just the universe we've got and of no other.

Borzeszkowski: Yes, I wanted to mention it, but I didn't due to the shortage of time. Because it is a further hint that Mach did not really mean that one should construct a physics that starts from a cosmological principle. In his *Wärmelehre*, for instance, he says that one can't use the notions which we know from physics like energy and so on which one applies to several different finite systems to the universe as a whole, because it is only once given. More precisely, Mach says (*Wärmelehre* 1900, p. 338) that sentences on the energy, entropy, and so on of the universe have no conceivable sense since they contain applications of measuring notions to an object which is not accessible to measurement. I completely agree.

Von Borzeszkowski and Wahsner: Two comments on Julian Barbour's comments (pp. 215–218). (i) With Eq. (1) describing the change of the relative distance between two bodies moving purely inertially, Mach presents a further simple implication of Newton's laws – here, in particular, of the first law. This passage shows once more, first, that he considered Newtonian mechanics to be true and, second, that he believed that the laws and statements of this physics can be reformulated so that, instead of absolute, relative distances occur. That this, to some extent, is possible, C. Neumann (*Über die Principien der Galilei–Newtonschen Theorie,* Leipzig 1870) had already shown by demonstrating that, choosing an arbitrary body *alpha*, mechanics can be written in Jacobian coordinates. Roughly speaking, Mach intended to rewrite Newtonian mechanics by replacing the body *alpha* by something one can call 'the totality of cosmic masses,' maybe, the center of cosmic masses.

(ii) Mach's criticism of Lange that one finds in some editions of his *Mechanik* shows that he did not mention that, accepting - as he did - Lange's construction of an inertial system, for reasons of logical self-consistence, a fourth force-free material point *must* follow with respect to one of Lange's inertial systems a straight line (uniformly). The passage here under consideration shows again Mach's initial misunderstanding, not only of Newton but also of Lange. This led him to a dim formulation. One should not, however, forget that Mach himself dropped this passage later. In later editions, in particular in the last edition supervised by the author, Mach agrees with Lange. There his point then was to state that Lange's point of view need not be the last word. Mach could imagine a physics describing Friedlaender–Föppl effects. Anyone looking for a passage in Mach that can be read as something like Einstein's version of Mach's Principle should take this one (cf. Chap. 2, Sec. 6, Subsec. 11 in the 7th edition).

Einstein's Formulations of Mach's Principle

Carl Hoefer

It is well known that Einstein first used the term 'Mach's Principle' in his 1918 paper on the general theory, "Prinzipielles zur allgemeinen Relativitätstheorie." In that paper Einstein expresses his current understanding of the requirements of Mach's ideas on inertia:

> *Mach's Principle:* The *G*-field is *without remainder* determined by the masses of bodies. Since mass and energy are, according to results of the special theory of relativity, the same, and since energy is formally described by the symmetric energy tensor ($T_{\mu\nu}$), this therefore entails that the *G*-field be conditioned and determined by the energy tensor.[1]

What is less well known is that Einstein struggled with other ways of understanding Mach's ideas on inertia in the context of the general theory, only arriving at his 1918 conception after failing adequately to cash out Mach's ideas in other ways in the years from 1912 to 1917; and that Einstein had to abandon this 1918 formulation of Mach's Principle by the middle of that year.

In this paper I will discuss some of the history of Einstein's work on Mach's Principle, by identifying several distinct ways that Einstein adopted, at various times, of formulating the general idea that we now call 'Mach's Principle.'[2] Many important details of Einstein's work in 1916 on Mach's Principle are not widely known and deserve greater attention from those interested in Machian ideas on inertia. In addition to highlighting some very puzzling aspects of Einstein's work on Mach's Principle, I will also contend that, compared with the 1918 formulation, Einstein's 1917 formulation in his cosmological paper (Einstein 1917) was correct in some crucial respects, even though it conflicts with much current usage of the term 'Mach's Principle' in the physics community.

Einstein Studies, vol. 6: Mach's Principle: From Newton's Bucket to Quantum Gravity, pp. 67–90

1. Pre-1916 Formulations and Puzzles

I want to begin with a quote from 1912, which I believe is the earliest expression of Einstein's understanding of Mach's Principle. This is from the paper "Gibt es eine Gravitationswirkung, die der elektrodynamischen Induktionswirkung analog ist?":

> This suggests the hypothesis that the whole inertia of any material point is an effect of the presence of all other masses, depending on a kind of interaction with them.[3]

A footnote citing Mach's *The Science of Mechanics* follows this sentence in the text. This is only the first of many such passages to appear in Einstein's writings between 1912 and 1918. In 1912, Einstein had not yet made the move to working on gravitation through field equations linking a metric tensor to material tensors, using what was then called the absolute differential calculus. When he did make this move, it affected his expressions of Mach's Principle in two main ways. First, it immediately suggested to Einstein that the metric tensor should be 'determined' by the tensor describing matter and energy (the core of the 1918 formulation). Second, it led Einstein to equate the achievement of generally covariant field equations for the metric with a complete relativization of motion – and therefore, presumptive satisfaction of Mach's Principle. These connections will be discussed further below.

The most important differences between Einstein's understanding of Mach's Principle in the 1913–1915 period, and the 1918 formulation, are two: First, in the early period there is (*apparently*) no recognition of the fact that an empty spacetime with Minkowski structure is incompatible with Mach's Principle; and second, Einstein substantially equated general covariance, Mach's Principle, and the equivalence principle. This equation was responsible for several conceptual problems that plagued Einstein prior to November, 1915, and also helped shape the next stage of Einstein's thinking on Mach's Principle, in 1916. An examination of Einstein's understanding of covariance, the equivalence principle, and their relations to Mach's Principle, is therefore necessary for understanding Einstein's later formulations.

The next passage of interest is from Einstein's 1913 exposition of the *Entwurf* theory that appeared in the *Vierteljahrsschrift der Naturforschenden Gesellschaft Zürich*. In the paper Einstein describes the highest goal of a theory of gravitation as being the task of determining the (components of the) metric, when the 'field-creating' material contents of the world are given. Here we see the core of the 1918 formulation expressed. Later, Einstein takes up Mach's ideas explicitly and claims that the *Entwurf* theory overcomes the 'epistemological defect' of absolute acceleration that Mach criticized:

> The theory sketched eliminates an epistemological defect, emphasized particularly by E. Mach, that affects not only the original relativity theory but also Galilean mechanics. It is plausible to suppose that the concept of the acceleration of a material particle can no more have an absolute meaning ascribed to it than the concept of velocity.... One must demand that the occurrence of an inertial resistance be tied to the acceleration of the body under consideration relative to other bodies....[4]

Einstein appears to claim that Mach's Principle is fully implemented in the *Entwurf* theory, since the epistemological defect of earlier mechanics has been overcome. In other passages from this period, Einstein's claims are more modest, describing the Machianization of inertia as still not yet complete in the *Entwurf* theory. Still, this claim of having overcome the epistemological defect, which is repeated in other writings of 1913–1914 (and in the 1916 review paper on the final GTR), is remarkable to modern readers. The reason is that it is clear that Minkowski spacetime is a solution of the *Entwurf* equations and that Einstein realized this. But Minkowski spacetime is the most clearly anti-Machian spacetime possible: It has a well-defined inertial and metrical structure, without any matter being present that could be said to 'determine' or explain that structure. We will come back to this point below; first, some further intriguing passages from this period.

In a long exposition of the whole of his current general relativity theory of October 1914, Einstein invokes the Machian conception of inertia in a way that suggests strongly that he viewed it as the same as the principle of equivalence. After mentioning the Newtonian argument from centrifugal effects to the existence of absolute motion, Einstein expresses the Machian response particularly clearly:

> We need not necessarily trace the existence of these centrifugal forces back to a[n absolute] movement of K'; we can instead just as well trace them back to the rotational movement of the distant ponderable masses in relation to K', whereby we treat K' as 'at rest.'... On the other hand, the following important argument speaks for the relativistic perspective. The centrifugal force that works on a body under given conditions is determined by precisely the same natural constants as the action of a gravitational field on the same body [i.e., its mass], in such a way, that we have no means to

> **differentiate a 'centrifugal field' from a gravitational field.... This quite substantiates the view that we may regard the rotating system K' as at rest and the centrifugal field as a gravitational field.[5]**

The reader will recognize the spirit (if not the letter) of the equivalence principle in Einstein's reference to the fact that one natural constant is involved in the definition of both gravitational and centrifugal forces, and in the idea of our being able to regard the system K' (here rotating, rather than uniformly accelerated) as being at rest in a gravitational field. To regard a centrifugal force field as being a gravitational field is, on the one hand, a natural Machian move: Inertial forces of all kinds are produced by interaction with other masses, and hence are just gravitational forces. On the other hand, it is an extension of the equivalence principle, extending what Einstein had claimed about an inertial system K and a uniformly accelerated system K', to systems K' accelerated in different ways.

The connection between general covariance and the equivalence principle can be seen as follows. The equivalence principle shows that we can extend the validity of the equations of a theory of motion to reference frames that are uniformly accelerated, so long as we regard them as in the presence of a uniform gravitational field. But the extension of the validity of the equations of a theory to *all* reference frames, including uniformly accelerated ones, is just what is achieved by general covariance. Therefore, as Einstein wrote in 1916, "*The requirement of general covariance of equations embraces the principle of equivalence as a quite special case.*"[6] To modern readers, it seems clear that this reasoning confuses reference frames with coordinate systems, and that the purely formal requirement of general covariance is in fact unrelated to the equivalence principle. But this is somewhat unfair. Einstein's understanding of general covariance was more robust than the modern view, and this is clear from his 1918 *Prinzipielles* paper. In particular, he viewed the absence of prior absolute spatiotemporal structure in GTR (a feature not shared by generally covariant formulations of other theories) as crucially part of what he understood by 'general covariance.' Therefore, GTR did implement general covariance in a way that does not *necessarily* make it misleading to say that the equations of the theory apply in accelerated reference frames just as they do in unaccelerated frames (in so far as such frames can be meaningfully defined and related to coordinate systems) – and so such frames may be considered 'at rest' in a kind of gravitational field.

Having now seen the links between Mach's Principle and the

equivalence principle, and between the equivalence principle and general covariance, there remains the question of the link between general covariance and Mach's Principle. This latter link proceeds through the idea of an extended principle of relativity: General covariance is sufficient to ensure that no reference systems are privileged, this ensures the extension of the principle of relativity to arbitrary motions, and this is what is demanded by the Machian criticism of absolute motion. In the *Entwurf* period, this line is never pursued to completion by Einstein because of his uneasy conviction that, despite lacking general covariance, the *Entwurf* equations do implement Mach's Principle and a general relativity of motion. A discussion from the 1914 paper "Die formale Grundlage der allgemeinen Relativitätstheorie" (Einstein 1914a) illustrates Einstein's dilemma. After arguing for the special principle of relativity from the fact that from a kinematical standpoint all coordinate systems should be considered equal, Einstein continues:

> This argument, however, immediately provokes a counter-argument. The kinematic equivalence of two coordinate systems, namely, is not restricted to the case in which the two systems, K and K', are in uniform relative translational motion. The equivalence exists just as well, from the kinematical standpoint, when for example the two systems rotate relative to one another. One feels therefore forced to the assumption that the previous relativity theory is to be generalized in a far-reaching way, so that the apparently incorrect privileging of uniform translations as opposed to other sorts of relative motions disappears.[7]

Einstein follows this passage immediately with a discussion of the Newtonian argument against such a widened relativity: the argument from inertial effects of rotation for the absoluteness of such motion. Einstein goes on in the usual way to give the Machian response, thus showing the identification of Machianization of inertia and a general relativity of motion in his thinking. But what about general covariance? Because of the non-general covariance of the *Entwurf* equations, Einstein postpones the question until after he has had a chance to explain the reasons for this apparent failure.

> The question now naturally arises, what kinds of reference systems and transformations we should regard as 'justifiable.' This question will however first be answered much later (section D). In the meantime we shall take up the standpoint that all coordinate systems and transformations are to be allowed, so long as they are compatible with the always-presupposed conditions of continuity.[8]

At least three puzzles arise in the above passages on Mach, relativity, and covariance:

(1) Einstein throughout this period viewed the extension of the *covariance* of a theory to be crucial to an extension of the relativity of motion in the theory. But the *Entwurf* theory had covariance properties that Einstein recognized to be far short of what would be required for an extension of relativity to acceleration and, in particular, rotation. For a brief period in 1914, Einstein thought that covariance over a wide class of transformations, including uniform rotations, pertained to the theory's equations; but this proved to be an error.

(2) Einstein equated the extension of covariance to cover acceleration transformations with the principle of equivalence, on the one hand; and on the other hand, he equated an extended equivalence principle with the implementation of Mach's ideas on the origin of inertial forces (as seen in the quote just above). But Einstein maintained that Mach's ideas were implemented in the *Entwurf* theory, without clearly explaining how this could be the case without a corresponding general (or at least wide-ranging) covariance of all the theory's equations.

(3) Einstein articulated Mach's ideas as the demand that the metric field $g_{\mu\nu}$ should be fully determined by the material distribution $T_{\mu\nu}$. And Einstein was aware that in the Minkowski spacetime of the Special Theory (with $T_{\mu\nu}=0$), Mach's ideas are violated and spacetime has a structure of its own. But Einstein repeatedly referred to the Minkowski metric as the proper case of *no* gravitational field being present, and never discussed the obvious problem that this metric is a valid solution of the *Entwurf* equations.

I believe that some of these puzzles can be resolved through the following story on Einstein's thinking in this period.

The *Entwurf* theory is not generally covariant, and even at the time Einstein was prone to view this as a serious defect of the theory given its goal of generalizing the relativity of motion. A letter to Lorentz from August 1913 makes this clear, as Einstein writes:

> *But the gravitation equations themselves unfortunately do not have this property of general covariance.* Only their covariance under linear transformations is established. But now, the entire trust in the theory rests on the conviction that acceleration of a reference system is equivalent to a gravitational field.[9]

And in a letter just two days after this one, Einstein refers to the lack of general covariance as an "ugly dark spot" on the theory.

But Einstein had two arguments that he believed justified and explained the lack of general covariance.[10] One was the now-notorious 'hole argument.' The other was Einstein's temporary belief that this failure was justified by the necessity of restricting coordinate systems to those in which the conservation law below holds[11]:

$$\sum_{\nu} \frac{\partial (T_{\sigma\nu} + t_{\sigma\nu})}{\partial x_{\nu}} = 0. \tag{1}$$

As an ordinary partial differential equation, this equation is not generally covariant if the quantities $T_{\sigma\nu}$ and $t_{\sigma\nu}$ are tensors; given that it holds in a coordinate system, it will then hold also only in coordinate systems related to the first by a linear transformation. By laying the blame for failure of general covariance at the feet of the conservation law, Einstein was able to persuade himself that this failure of the equations to hold in arbitrary systems did not bring with it a failure of the Machian idea that the metrical structure of spacetime should be determined by the material distribution. For, as Einstein (1914b) pointed out in the *Scientia* article of 1914, there is no prior selection of certain coordinate systems or frames as privileged; instead, the specification of the material distribution appears subsequently to pick out certain coordinate systems, namely those in which (1) hold. Einstein held - and this was his belief in the Machian character of the *Entwurf* theory - that the material distribution determines the metric field. So, if one imagines that the material distribution had instead been laid out differently on the spacetime manifold, the metric (and, hence, the class of privileged coordinate systems, since the metric restricts the systems in which the conservation law can hold) would 'follow' the material distribution; and this shows that they are only frames privileged by the actual material distribution, not frames privileged in an absolute way as in Newtonian mechanics.[11]

There is still some tension left, since Einstein never describes how the covariance limitations imposed by the hole argument affect the question of relativity of motion and the existence of privileged coordinate systems. Einstein never fully dropped the linkage of covariance to the relativity of motion in the back of his mind (even after 1918); rather, he was greatly relieved in November 1915 when he finally achieved the generally covariant equations of GTR. Nevertheless, this interpretation clears up some puzzles about Einstein's thinking on Mach's Principle in the *Entwurf* period.

But the problem remains that the material distribution does not fully determine the class of inertial systems in the *Entwurf* theory, despite

Einstein's claims. This is shown by the compatibility of Minkowski spacetime with the field equations. The *Entwurf* theory faces the same problem of models with absolute or quasi-absolute inertial structure that the final GTR faces.

How did Einstein reconcile the Minkowski spacetime solution with the allegedly Machian character of the theory? This is a problem that carries over into the early period after the discovery of the final GTR field equations, since at that time too Einstein claimed that his theory overcame the problem of the Machian epistemological argument against privileged frames.[12] I believe that insofar as there is a solution to this puzzle, it is the same for both periods: Einstein was aware of the difficulty, even though he did not mention it in his published writings of the time; and he intended to overcome it by finding suitable *boundary conditions* to impose on physically realistic solutions, conditions that would rule out the empty Minkowski spacetime. At the latest, Einstein was already working on such conditions by May 1916, as they are mentioned in a letter to Besso of that month. Einstein may have been working on them much further back, in late 1915 or early 1916, and he may well have had the idea in the *Entwurf* period.

2. 1916: Mach's Principle as Boundary Conditions

In the next stage of Einstein's work on Mach's Principle, then, there are two kinds of formulations of Mach's Principle to be found. First, continuing expressions of the type "The metric $g_{\mu\nu}$ should be completely determined by the material distribution $T_{\mu\nu}$." And second, mathematical expressions of Mach's Principle through the idea of Machian boundary conditions that would supplement the field equations and eliminate non-Machian solutions. In the face of Einstein's recognition in 1916 that the field equations alone do not satisfy Machian demands, the Machian boundary conditions sought would have amounted to the implementation of Mach's Principle in GTR.[13]

Einstein thought that Mach's Principle should be implemented in GTR through boundary conditions rather than by some more general mathematical constraint, because of the way in which he saw the problematic models as violating Mach's Principle. Minkowski spacetime and Schwarzschild's solution both violate Mach's Principle because they display metrical/inertial structure that cannot be attributed to a material distribution. In the case of the Schwarzschild solution, this structure is evident at large r, where spacetime is essentially Minkowskian and the central mass evidently is not responsible for that structure. Instead, on Einstein's way of thinking, the Minkowski boundary conditions imposed in deriving the solution are to blame for that absoluteness. The same perspective can be used in thinking of empty Minkowski spacetime. The metrical structure at any point is a result of the 'absolutist' boundary conditions, plus the local (lack of) material distribution.

Given this perspective, the way to avoid violations of Mach's Principle is to come up with boundary conditions that *do not* impart any absolute structure to spacetime, so that the structure of spacetime at finite distances from the center is attributable only to the global matter distribution, not to the boundary conditions as well. Einstein expresses this idea in a letter to Willem de Sitter, from June 1916:

> I am sorry to have plagued you with too much emphasis on the question of boundary conditions.... But I must add that I have *never* thought about a *temporally* finite extension of the world; and even spatially, the *finite extension* is not what matters. Rather, my need to generalize drove me to the following view: It is possible to give a spatial envelope (massless geometrical surface) (in four dimensions, a tube) outside of which a gram weight has as little inertia as I choose to specify. Then I can say that inside the envelope, inertia is determined by the masses present there; and to be sure, *only* by these masses.[14]

Very little survives about Einstein's work on such conditions. All we have to go on are the clues from the above verbal expression, the 1917 *Betrachtungen* paper discussion (Einstein 1917), and brief reports by de Sitter in two 1916 articles. The boundary conditions that de Sitter provides, as well as others that he himself proposes as cashing out Mach's Principle, turn out to be in themselves meaningless, for reasons I will discuss below. But a mathematical reconstruction of what Einstein's calculations may have involved in this period might be able to shed more light on Einstein's temporary belief in a boundary conditions approach to Mach's Principle.

De Sitter reported in September 1916 that Einstein 'found' that the following set of boundary conditions at infinity satisfies the demand for the complete relativization of inertia[15]:

$$\left.\begin{matrix} 0 & 0 & 0 & \infty \\ 0 & 0 & 0 & \infty \\ 0 & 0 & 0 & \infty \\ \infty & \infty & \infty & \infty^2 \end{matrix}\right\} \qquad (2)$$

De Sitter gives no explanation of the exponent on g_{44}; he does remark that the invariance of these values is restricted to transformations in which $x^{4'}$ is a function of x^4 alone.

This set of boundary values, as well as two others proposed by de Sitter in 1917, were intended to cash out an interesting idea that Einstein seems to have held for a while in 1916: The General Theory needed supplementation by *generally covariant boundary conditions* in order to secure the complete relativization of inertia.[16]

The idea is that the boundary values of $g_{\mu\nu}$ at infinity should be such that they are left unchanged by a wide group of coordinate transformations (at least those corresponding to rigid motions – a requirement well short of *general* covariance). Choosing boundary values of either 0 or ∞ for the relevant metric components seems superficially to be a good way to approach this idea, but this alone falls well short of guaranteeing anything about what will happen under a coordinate transformation. With a given metric expressed in some coordinate system, for example, a component that approaches zero or infinity in some limit may well fail to do so after a transformation such as a linear acceleration or rotation. Whether this is so or not depends on the given metric. In fact, it is not too strong to say that boundary conditions such as (2) are completely meaningless, until they are linked to one or more concrete metrics.

Unfortunately, this set of boundary conditions is not accompanied by any discussion by de Sitter of how they arise, i.e., what sort of actual functions might be compatible with the field equations, and also take these limiting values. Without a concrete example, of course, there is no way to verify that they do in fact represent a boundary region in which inertia 'disappears,' in some intrinsic sense. Further, much general work delimiting the class of metrics that can take such boundary values (with a given definition of 'at infinity') would be necessary in order to establish that the boundary conditions correctly capture Machian demands.

Einstein may have subscribed to these boundary conditions for *perhaps* as long as four or five months, from before September 1916, to December 1916, at which point Einstein had already turned to the cosmological constant and his closed universe, abandoning the idea of Machian boundary conditions.

A letter to Besso from December 1916 shows Einstein giving, in brief, the argument against boundary conditions and in favor of a closed world, that he would repeat in more detail in the *Kosmologische*

Betrachtungen paper. In this letter Einstein formulates what he sees as the dilemma facing him:

> It's certain that infinitely large differences of potential would have to give rise to stellar velocities of very significant magnitude, and these would surely have already have manifested themselves long ago. Small potential differences in combination with an infinite [spatial] extent of the world demand the emptiness of the world at infinity (constancy of the $g_{\mu\nu}$ at infinity given appropriate choice of coordinates [Minkowski conditions]), in contradiction with a meaningfully understood relativity. Only the closure of the world frees us from this dilemma. [17]

The technical part of the argument against the boundary conditions (equation) based on stellar velocities is made only somewhat more clear in Einstein's *Kosmologische Betrachtungen* paper (Einstein 1917). I present the entire relevant excerpt below (the English translation only, due to its length):

> The opinion which I entertained until recently, as to the limiting conditions to be laid down in spatial infinity, took its stand on the following considerations. In a consistent theory of relativity there can be no inertia *relatively to "space,"* but only an inertia of masses *relatively to one another*. If, therefore, I have a mass at a sufficient distance from all other masses in the universe, its inertia must fall to zero. We will try to formulate this condition mathematically.
>
> According to the general theory of relativity the negative momentum is given by the first three components, the energy by the last component of the covariant tensor multiplied by $(-g)^{1/2}$
>
> $$m\sqrt{-g}\, g_{\mu\alpha}\frac{dx_\alpha}{ds}, \tag{3}$$
>
> where, as always, we set
>
> $$ds^2 = g_{\mu\nu}dx_\mu dx_\nu. \tag{4}$$
>
> In the particularly perspicuous case of the possibility of choosing the system of coordinates so that the gravitational field at every point is spatially isotropic, we have more simply
>
> $$ds^2 = -\mathrm{A}(dx_1^2 + dx_2^2 + dx_3^2) + \mathrm{B}dx_4^2. \tag{5}$$
>
> If, moreover, at the same time
>
> $$\sqrt{-g} = 1 = \sqrt{\mathrm{A}^3\mathrm{B}}$$
>
> we obtain from (4), to a first approximation for small velocities,
>
> $$m\frac{\mathrm{A}}{\sqrt{\mathrm{B}}}\frac{dx_1}{dx_4},\quad m\frac{\mathrm{A}}{\sqrt{\mathrm{B}}}\frac{dx_2}{dx_4},\quad m\frac{\mathrm{A}}{\sqrt{\mathrm{B}}}\frac{dx_3}{dx_4}$$

for the components of momentum, and for the energy (in the static case)

$$m\sqrt{B}.$$

From the expressions for the momentum, it follows that $m(A/\sqrt{B})$ plays the part of the rest mass. As m is a constant peculiar to the point of mass, independently of its position, this expression, if we retain the condition $(-g)^{1/2}=1$ at spatial infinity, can vanish only when A diminishes to zero, while B increases to infinity. It seems, therefore, that such a degeneration of the coefficients $g_{\mu\nu}$ is required by the postulate of relativity of all inertia. This requirement implies that the potential energy $m\sqrt{B}$ becomes infinitely great at infinity. Thus a point of mass can never leave the system; and a more detailed investigation shows that the same thing applies to light-rays. A system of the universe with such behavior of the gravitational potentials at infinity would not therefore run the risk of wasting away which was mooted just now in connexion with the Newtonian theory....

At this stage, with the kind assistance of the mathematician J. Grommer, I investigated centrally symmetrical, static gravitational fields, degenerating at infinity in the way mentioned. The gravitational potentials $g_{\mu\nu}$ were applied [*angesetzt*], and from them the energy-tensor $T_{\mu\nu}$ of matter was calculated on the basis of the field equations of gravitation.[18] But here it proved that for the system of the fixed stars no boundary conditions of the kind can come into question at all, as was also rightly emphasized by the astronomer de Sitter recently.

For the contravariant energy-tensor $T^{\mu\nu}$ of ponderable matter is given by

$$T^{\mu\nu}=\rho\frac{dx_\mu}{ds}\frac{dx_\nu}{ds},$$

where ρ is the density of matter in natural measure. With an appropriate choice of the system of coordinates the stellar velocities are very small in comparison with that of light. We may, therefore, substitute $(g_{44})^{1/2}dx_4$ for ds. This shows us that all components of $T^{\mu\nu}$ must be very small in comparison with the last component T^{44}. But it was quite impossible to reconcile this condition with the chosen boundary conditions. In the retrospect this result does not appear astonishing. The fact of the small velocities of the stars allows the conclusion that wherever there are fixed stars, the gravitational potential (in our case $\sqrt{B}$) can never be much greater than here on earth. This follows from statistical reasoning, exactly as in the case of the Newtonian theory. At any rate, our calculations have convinced me that such conditions of degeneration for the $g_{\mu\nu}$ in spatial infinity may not be postulated.[19]

This discussion contains all the evidence there really is, about why

Einstein abandoned his boundary-conditions formulation of Mach's Principle. When one considers that this transition was Einstein's motivation for introducing the λ-term, and thus beginning modern cosmology with his 1917 closed model of the universe, the absence of discussion of this passage in the literature is quite remarkable.[20] A reconstruction of Einstein's calculations on the question of boundary conditions would significantly enhance our understanding of the early history of GTR.

Without yet having such a reconstruction in hand, it is still possible to note some doubtful aspects of Einstein's reasoning in the 1917 paper. Because Einstein at this time (with everyone else) did not *fully* grasp the difference between coordinate effects (for example, singularities) and intrinsic effects, his results concerning large potentials and large stellar velocities have to be clearly reconstructed before they can be accepted as sound. Even if large stellar velocities are derived in some intrinsically meaningful sense, it has further to be shown whether these velocities would be observable from the earth, and whether they would be velocities in a static space, or 'velocities' like the velocities of recession of distant galaxies, which are a function of the expansion of spacetime. There are ample reasons to doubt that Einstein's arguments against the boundary-conditions approach are sound - though there are also ample reasons to doubt that the approach itself makes sense to begin with.

3. 1917: The Closed-Universe Formulation

The boundary-conditions expression of Mach's Principle was replaced by the demand of a closed universe, and not by any explicitly mathematical reformulation of the principle. Instead, Einstein's 1917 - early 1918 understanding of Mach's Principle can be cashed out only verbally, as comprised of two demands:

(I) The universe should be finite and closed, i.e., have *no* boundary region; then, the local metric cannot be thought of as determined in part by boundary conditions of space (but rather only by the global matter distribution). This, Einstein thought, would assure that the metric is fully determined by the matter distribution in spacetime.

(II) The modified field equations should allow *no* matter-free, singularity-free solutions. This is necessary to ensure that the theory as a whole (not just some subset of models) is Machian in character.

Demand (I) emerges clearly from the *Kosmologische Betrachtungen* paper and the December 1916 letter to Besso. Both (I) and (II) are present in the 1918 discussion in the *Prinzipielles* paper but are stated even more explicitly by Einstein in a letter to de Sitter of March 24, 1917:

> In my opinion it would be dissatisfying, if there were a conceivable world without matter. The $g^{\mu\nu}$-field should rather *be determined by the matter, and not be able to exist without it.* This is the heart of what I understand by the demand for the relativity of inertia. One could just as well speak of the "material conditionedness of geometry." As long as this demand was not fulfilled, for me the goal of general relativity was not yet completely achieved. This was first achieved through the introduction of the λ term.[21]

In demand (I) we see the residue of Einstein's conviction that the non-Machian character of certain models is a product of their absolutist boundary conditions: if there is no boundary region, Einstein assumes, there is no room for a non-Machian determination of the metric. This reasoning can only hold, of course, if spacetime is nonempty; this explains the importance of demand (II) for Einstein.

As is now well known, the introduction of the λ-term failed to achieve condition (II). In early 1917, de Sitter found a $T_{\mu\nu}=0$ solution to the new field equations. Einstein struggled for over a year to show either that the solution was physically unacceptable due to a singularity, or not really matter-free after all; he gave up the struggle in June 1918, and in an important sense this marks the end of Einstein's advocacy of Mach's Principle.[22]

4. Post-1918 Formulations

After accepting the failure of the modified field equations to meet demand (II), Einstein's attempts to implement Mach's Principle in GTR ended, and his enthusiasm for Mach's Principle began a steady decline that culminated, near the end of his life, in complete repudiation of the principle.[23] The decline can be explained in part as due to the evident failure to make GTR perfectly Machian, and in part as due to Einstein's growing interest in unified field theories, in which a realistic (as opposed to reductionistic) attitude towards the metric field is presupposed. But Einstein did not cease to discuss Mach's ideas on inertia, in a positive manner, for many years after 1918. Instead, he tended to emphasize the respects in which it seems that Machian ideas are fulfilled in the general theory, and to advocate his closed-universe cosmology as fulfilling the

Machian demands due to its lack of a boundary region. The discussion in the textbook *The Meaning of Relativity* is representative:

> The theory of relativity makes it appear probable that Mach was on the right road in his thought that inertia depends upon a mutual action of matter.... What is to be expected along the line of Mach's thought?
> 1. The inertia of a body must increase when ponderable masses are piled up in its neighborhood.[24]
> 2. A body must experience an accelerating force when neighboring masses are accelerated, and, in fact, the force must be in the same direction as the acceleration.
> 3. A rotating hollow body must generate inside of itself a 'Coriolis field,' which deflects moving bodies in the sense of the rotation, and a radial centrifugal field as well.

Two pages later Einstein continues, discussing his cosmological model

> Although all of these effects are inaccessible to experiment, because κ is so small, nevertheless they certainly exist according to the general theory of relativity. We must see in them a strong support for Mach's ideas as to the relativity of all inertial actions. If we think these ideas consistently through to the end we must expect the *whole* inertia, that is, the *whole* $g_{\mu\nu}$-field, to be determined by the matter of the universe, and not mainly by the boundary conditions at infinity.[25]

After this passage, Einstein discusses his closed universe model, citing with particular favor its lack of boundary conditions.

The differences between Einstein's expression of Mach's Principle here and his 1917 expression described above are subtle but important. Here the emphasis is on Machian-seeming effects that are present in GTR, and on *one* model that seems both to satisfy the core demand that the metric be determined by the energy tensor and to be physically reasonable. There is no demand that the theory exclude empty (and hence anti-Machian) models in general. And with the absence of this demand, there is no trace of *any* precise mathematical expression of Mach's Principle.

Einstein seems content with the possibility that the actual world is well described by a model that appears to be Machian by the light of intuition. This attitude toward Mach's Principle is quite common among current relativists who, like Einstein, are sympathetic to Mach's ideas on inertia. Aside from a few, such as Wheeler and Raine, who do try to formulate an explicit mathematical version of Mach's Principle, most physicists are content to rely on their intuitions about Machian effects in the absence of mathematical criteria and to follow Einstein in regarding a closed matter-filled universe as automatically Machian.[26]

5. Conclusion: The Correct Formulation

I will end with some critical remarks about Einstein's formulations of Mach's Principle and current assumptions among working physicists.

The widespread assumption that a closed, matter-filled cosmology such as Einstein's spherical cosmology must satisfy Mach's Principle is questionable. It is based on the reasoning discussed above, that since in anti-Machian models the trouble seems to come from the boundary conditions, if one eliminates the boundary region one eliminates the problem. But this reasoning is clearly fallacious. There is a missing premise: The only way a model can fail to be Machian is to have an empty boundary region in which an absolute spatiotemporal structure is posited. This premise is by no means intuitively obvious, and it could only be established if we had a general, *mathematical* explication of Mach's Principle and could show that this premise follows from it. Such a mathematical version of Mach's Principle would itself have to be supported by arguments showing that it correctly captures the core of Mach's ideas on the origin of inertia. It would entail a restriction of the class of models of GTR and delimit exactly those models in which inertia is fully determined by matter–energy. I do not know if such a mathematical expression is possible for GTR; but its attainment is necessary before we can claim that any given model does satisfy Mach's Principle.

In the meantime, it seems more clear that we can (as did Einstein) rule out some models of GTR as definitely anti-Machian and use these judgments as constraints on any explication of Mach's Principle. The clearest case is that of empty spacetimes. Since they do not contain any matter–energy and have a definite spatial (and hence inertial) structure, they clearly run contrary to the core of Mach's ideas on the origin of inertia. Therefore, I believe that Einstein was absolutely right to demand (II) as a necessary condition for the relativization of inertia, or satisfaction of Mach's principle by a gravitation theory. Demand (II) should remain our most secure touchstone in theorizing about how to create a Machian gravitation theory.

I stress this point because apparently it has become a minority view among physicists working on Mach's Principle.[27] The reasons have to do, I believe, with work done by Wheeler and others on initial-value formulations of Mach's Principle (which do not rule out empty solutions), and also with the widespread view that gravitational field energy [$t_{\sigma\nu}$ of Eq. (1)] should count as part of the energy that helps to determine the metrical structure of spacetime. If gravitational stress–energy were a tensor quantity (and hence well-defined and localizable), this attitude would clearly be appropriate; the Machian demand would then be, roughly, that the combination of material and gravitational stress–energy uniquely determines the whole metric field $g_{\mu\nu}$. But since this is not the case, the status of gravitational stress–energy as a second kind of matter-energy in the universe is dubious. Furthermore, it is one thing to suppose that gravitational stress–energy present on a hypersurface or thin-sandwich in a matter-containing world should be included as part of the material distribution that determines future inertial structure, but it is quite another to suppose that a matter-free universe could count as Machian in virtue of some conditions satisfied by the gravitational waves present. At any rate, it seems to me that much further work clarifying the status of gravitational stress–energy is needed if we are to abandon the initially compelling view that a matter-free ($T_{\mu\nu}=0$) universe is automatically anti-Machian.

Acknowledgments. Citations from Einstein's personal correspondence are reproduced with the kind permission of the Albert Einstein Archives at the Hebrew University of Jerusalem.

NOTES

[1]Einstein (1918), pp. 241–242. "*Machsches Prinzip*: Das *G*-Feld ist *restlos* durch die Massen der Körper bestimmt. Da Mass und Energie nach den Ergebnissen der speziellen Relativitätstheorie das Gleiche sind und die Energie formal durch den symmetrischen Energietensor ($T_{\mu\nu}$) beschrieben wird, so besagt dies, daß das *G*-Feld durch den Energietensor der Materie bedingt und bestimmt sei." Throughout, all translations are my own unless otherwise noted.

[2]A discussion of some of the history of Einstein's work on Mach's ideas can be found in Kerszberg (1989a, b). Some of the material in this paper is also covered in (Hoefer 1993).

[3]Einstein (1912), p. 39. "Es legt dies die Vermutung nahe, daß die ganze Trägheit eines Massenpunktes eine Wirkung des Vorhandenseins aller übrigen Massen sei, auf einer Art Wechselwirkung mit den letzteren beruhend."

[4]Einstein (1913a), p. 290. "Durch die skizzierte Theorie wird ein erkenntnistheoretischer Mangel beseitigt, der nicht nur der ursprünglichen Relativitätstheorie, sondern auch der Galilei'schen Mechanik anhaftet und insbesondere von E. Mach betont worden ist. Es ist einleuchtend, daß dem

Begriff der Beschleunigung eines materiellen Punktes ebensowenig ein absolute Bedeutung zugeschrieben werden kann wie demjenigen der Geschwindigkeit .... [Es wird] gefordert werden müssen, daß das Auftreten eines Trägheitswiderstandes an die Relativbeschleunigung des betrachteten Körpers gegenüber andern Körpern geknüpft sei...."

[5]Einstein (1914), pp. 1031–2. "Die Existenz jener Zentrifugalkräfte brauchen wir nämlich nicht notwendig auf eine Bewegung von K' zurückzuführen; wir können sie vielmehr ebensogut zurückführen auf die durchschnittliche Rotationsbewegung der ponderabeln fernen Massen der Umgebung in bezug auf K', wobei wir K' als 'ruhend' behandeln.... Für die relativistische Auffassung spricht anderseits folgendes wichtige Argument. Die Zentrifugalkraft, welche unter gegebenen Verhältnissen auf einen Körper wirkt, wird genau durch die gleiche Naturkonstante desselben bestimmt wie die Wirkung eines Schwerefeldes auf denselben, derart, daß wir gar kein Mittel haben, ein 'Zentrifugalfeld' von einem Schwerefeld zu unterscheiden.... Dadurch gewinnt die Auffassung durchaus an Berechtigung, daß wir das rotierende System K' als ruhend und das Zentrifugalfeld als ein Gravitationsfeld auffassen dürfen."

[6]Einstein (1916), p. 641. Here I use John Norton's translation (Norton 1989b, p. 26). I am greatly indebted to this paper of Norton's for the above points on the equivalence principle. Needless to say, Norton might not agree with my interpretation on all points.

[7]Einstein (1914), p. 1031. "Dies Argument fordert aber ein Gegenargument heraus. Die kinematische Gleichberechtigung zweier Koordinatensysteme ist nämlich durchaus nicht auf den Fall beschränkt, daß die beiden ins Auge gefassten Koordinatensysteme K und K' sich in gleichförmiger Translationsbewegung gegeneinander befinden. Diese Gleichberechtigung vom kinematischen Standpunkt aus besteht z.B. ebensogut, wenn die Systeme relativ zueinander gleichförmig rotieren. Man fühlt sich daher zu der Annahme gedrängt, daß die bisherige Relativitätstheorie in weitgehendem Mass zu verallgemeinern sei, derart, daß die ungerecht scheinende Bevorzugung der gleichförmigen Translation gegenüber Relativbewegungen anderer Art aus der Theorie verschwindet."

[8]Einstein (1914), p. 1032. "Es erhebt sich nun naturgemäß die Frage, was für Bezugssysteme und Transformationen wir in einer verallgemeinerten Relativitätstheorie als "berechtigte" anzusehen haben. Diese Frage wird sich jedoch erst viel später beantworten lassen (Abschnitt D). Einstweilen stellen wir uns auf den Standpunkt, daß alle Koordinatensysteme und Transformationen zuzulassen seien, die mit den bei physikalischen Theorien stets vorausgesetzten Bedingungen der Stetigkeit vereinbar sind."

[9]EA 16-434. "*Aber die Gravitationsgleichungen selbst haben die Eigenschaft der allgemeinen Kovarianz leider nicht.* Nur deren Kovarianz *linearen* Transformationen gegenüber ist gesichert. Nun beruht aber das ganze Vertrauen auf die Theorie auf der Überzeugung, daß Beschleunigung des

Bezugsystems einem Schwerefeld äquivalent sei."

[10]See Norton's (1989a) pp. 126–132, for an enlightening discussion of Einstein's arguments against general covariance in 1913 and 1914.

[11]Einstein (1913b), p. 1258.

[12]Einstein (1923b), p. 113.

[13]Early 1916 is the latest date that can be placed on Einstein's full recognition of the non-Machian character of the field equations: The Schwarzschild solution showed that even solutions containing matter might be radically non-Machian. But as I indicated earlier, Einstein's recognition of the problem was probably complete at an earlier stage.

[14]EA 20-539. "Es thut mir leid, Ihnen gegenüber zu viel Nachdruck auf die Frage der Grenzbedingungen gelegt zu haben. ... Aber ich muss doch sogleich hinzufügen, dass ich an eine *zeitlich* endliche Ausdehnung der Welt *niemals* gedacht habe; auch bei dem Räumlichen kommt es auf eine *endliche Ausdehnung* nicht an. Sondern es trieb mich mein Verallgemeinerungsbedürfnis mir zu folgendes Auffassung:

Es sei möglich eine räumliche Hülle (masselose geometrische Fläche) (in vierdimensionalen einen Schlauch) anzugeben, ausserhalb welcher ein Grammgewicht eine so geringe Trägheit hat, als ich nur immer will. Dann kann ich sagen, dass innerhalb der Hülle die Trägheit durch die dort vorhandenen Massen bedingt sei; und zwar *nur* durch diese."

[15]De Sitter (1916), p. 531.

[16]It is impossible to be sure that Einstein's search for boundary conditions is accurately described in this way; only de Sitter uses the term 'generally covariant boundary conditions,' in texts that survive. But since de Sitter and Einstein were in intensive correspondence in the period from June–December 1916 (only some of which correspondence survives), it is likely that de Sitter's reports on Einstein's thinking are accurate.

[17]Speziali, p. 97. Speziali dates this letter as probably mid-December, 1916, but the dating is not certain. "Sicher ist, dass unendlich grosse Potentialdifferenzen zu Sterngeschwindigkeiten von sehr bedeutender Grösse Anlass geben müssten, die sich wohl schon lange eingestellt hätten. Kleine Potentialdifferenzen im Verein mit unendlicher Ausdehnung der Welt verlangen Leersein der Welt im Unendlichen (Konstanz der $g_{\mu\nu}$ im Unendlichen bei passender Koordinatenwahl), im Widerspruch mit einer sinnvoll aufgefassten Relativität. Nur Geschlossenheit der Welt befreit aus dem Dilemma.

[18]This sentence tempts one to suppose that Einstein had constructed a complete solution to the field equations that embodied the boundary conditions (5), thus constructing a (possibly) Machian cosmological model before the famous Einstein universe of the *Betrachtungen* paper. Of course, Einstein's calculations may well have fallen far short of that.

[19]Einstein (1923a), pp. 180–182.

[20]I have searched many of the most important current and older textbooks on GTR for any discussion of this passage – in vain. One reason for the lack

of attention to this episode may be the fact that, like de Sitter, most relativists were and are hostile to the goal of Machianizing GTR. This was especially true in the late teens and '20s (long before the work of Wheeler, Brans, Dicke, and others revived interest in the '50s and '60s). By the time later physicists returned to Mach's Principle in GTR, this episode had been completely written out of the textbook history of general relativity.

[21]EA 20-548. "Es wäre nach meiner Meinung unbefriedigend, wenn es eine denkbare Welt ohne Materie gäbe. Das $g^{\mu\nu}$-Feld soll vielmehr *durch die Materie bedingt sein, ohne dieselbe nicht bestehen können.* Das ist der Kern dessen, was ich under der Forderung von der Relativität der Trägheit verstehe. Man kann auch ebensogut von der 'materiellen Bedingtheit der Geometrie' sprechen. Solange diese Forderung nicht erfüllt war, war für mich das Ziel der Allgemeinen Relativität noch nicht ganz erreicht. Dies wurde durch das λ-Glied erst herbeigeführt."

[22]Einstein conceded the singularity-free nature of de Sitter's metric in a postcard to Felix Klein, EA 14-449.

[23]Einstein to Felix Pirani, EA 17-448.

[24]Julian Barbour (1992) has recently argued, persuasively I believe, that this effect is not really a consequence of Mach's ideas.

[25]Einstein (1922), pp. 100, 103.

[26]At the 1993 Tübingen conference, Julian Barbour organized several straw-polls of the attendees concerning their views on GTR and Mach's Principle. My claims here are based partly on these straw polls [p. 106], as well as the evidence of current literature.

[27]Again, this fact emerged clearly from straw-polls and discussions at the 1993 Tübingen conference.

REFERENCES

Barbour, Julian B. (1992). "Einstein and Mach's Principle." In *Studies in the History of General Relativity*, Vol. 3 of *Einstein Studies*. Jean Eisenstaedt and A.J. Kox, eds. Boston: Birkhäuser, pp. 125–153.

de Sitter, Willem (1916). "On the Relativity of Inertia in Einstein's Theory." *Proc. of the Section of Sciences, Koninklijke Akademie van Wetenschappen* 19: 527–532, September 30.

Einstein, Albert (1912). "Gibt es eine Gravitationswirkung die der elektrodynamischen Induktionswirkung analog is?" *Vierteljahrsschrift für gerichtliche Medizin und öffentliches Sanitätswesen* 44: 37–40.

Einstein, Albert (1913a). "Physikalische Grundlagen einer Gravitationstheorie." *Vierteljahrsschrift der Naturforschenden Gesellschaft Zürich* 58: 284–290.

Einstein, Albert (1913b). "Zum gegenwärtigen Stande des Gravitationsproblems." *Physikalische Zeitschrift* 14: 1249–1266.

Einstein, Albert (1914a). "Die formale Grundlage der allgemeinen Relativitätstheorie." *Sitzungsberichte der Preussischen Akademie der Wissenschaften*,

pp. 1030–1085, Part 2.
Einstein, Albert (1914b). "Zum Relativitäts-Problem." *Scientia (Bologna)* 15: 337–348.
Einstein, Albert (1916). "Über Friedrich Kottlers Abhandlung 'Über Einsteins Äquivalezhypothese und die Gravitation'." *Annalen der Physik* 51: 639–642.
Einstein, Albert (1917). "Kosmologische Betrachtungen zur allgemeinen Relativitätstheorie." *Sitzungsberichte der Preussischen Akademie der Wissenschaften*, pp. 142–152. Translation: Einstein (1923a).
Einstein, Albert (1918). "Prinzipielles zur allgemeinen Relativitätstheorie." *Annalen der Physik* 55: 241–244.
Einstein, Albert (1922). *The Meaning of Relativity*. Princeton: Princeton University Press.
Einstein, Albert (1923a). "Cosmological Considerations on the General Theory of Relativity." In *The Principle of Relativity*. Papers by Albert Einstein *et al.* New York: Dover, pp. 177–188.
Einstein, Albert (1923b). "Foundations of the General Theory of Relativity." *ibid.*
Hoefer, Carl (1993). "Einstein's Struggle for a Machian Gravitation Theory." *Studies in History and Philosophy of Modern Physics* 25: 287–335.
Kerszberg, Pierre (1989a). "The Einstein-de Sitter Controversy of 1916–1917 and the Rise of Relativistic Cosmology." In *Einstein and the History of General Relativity*, Vol. 1 of *Einstein Studies*. Don Howard and John Stachel, eds. Boston: Birkhäuser, pp. 325–366.
Kerszberg, Pierre (1989b). *The Invented Universe: The Einstein-De Sitter Controversy (1916–1917) and the Rise of Relativistic Cosmology*. Oxford: Oxford University Press.
Norton, John (1989a). "How Einstein Found his Field Equations." In *Einstein and the History of General Relativity*, Vol. 1 of *Einstein Studies*. Don Howard and John Stachel, eds. Boston: Birkhäuser, pp. 101–159.
Norton, John (1989b). "What was Einstein's Principle of Equivalence?" In *Einstein and the History of General Relativity*, Vol. 1 of *Einstein Studies*. Don Howard and John Stachel, eds. Boston: Birkhäuser, pp. 5–47.
Speziali, Pierre, ed. (1972). *Albert Einstein, Michele Besso: Correspondence 1903–1955*. Paris: Hermann.

Discussion

Earman: People sometimes talk as if there is a dichotomy between universes that are spatially open and universes that are spatially closed. Now, of course, there are universes you can slice up with open sections and there's another way of slicing with closed sections, so is it enough for Mach's Principle that there exists *a* way of slicing it with spatially closed sections?

Hoefer: My way of thinking about Mach's Principle, and I can only speak for my own understanding, is that closed *vs* open and infinite *vs* noninfinite has nothing to do with Mach's Principle. I have never seen any reason to connect Mach's Principle with any kind of demand on the topology.

Barbour [comment after conference]: I believe the argument for closure is rather obvious. Mach said that motion is with respect to the universe *as a whole*. Now motion of one body relative to a *finite* universe is easy to define but to define motion, as a definite quantity, relative to an infinite universe is not at all easy. Virtually all actual implementations of Machian ideas have assumed the universe is finite. Also in Wheeler's geometrodynamic approach, in which the three-geometry is the basic concept, two slightly different *closed* three-geometries in principle determine a complete spacetime by themselves (thin-sandwich principle). However, if space is infinite, boundary conditions have to be imposed arbitrarily. The dynamics of the universe is no longer self-contained. It was this sort of arbitrariness Einstein sought to avoid.

Hoefer [response to above comment]: Dr. Barbour's comment illustrates exactly the widespread conceptions of the relation of finitude/closure to Mach's ideas that I believe to be misconceptions. Motion is not more difficult to define relative to an infinite universe than to a finite universe, if by 'defining' we simply mean a nonmetrized description via a coordinate system. If we mean something stronger – specifying relative velocities or accelerations for pairs of bodies, for example – then problems do arise, in relativity in general, but especially difficult problems arise for the Machian. Spatiotemporal structure is needed to characterize these motions, yet the structure of spacetime is, for the Machian, supposed to arise out of those very relative motions. This problem is not resolved, conceptually speaking, just by assuming closure of space.

Wheeler's approach to Mach's ideas illustrates what I mean. As Barbour points out, the idea is that two different closed three-geometries determine the whole structure of spacetime. Is it an acceptable Machian strategy for the relativist to help herself to the whole geometry of these slices, which is clearly *more* than just a summary of relative motions of bodies at a time (for example, they may contain gravity waves)? I don't know whether it is or not – but I do claim that their being *closed* three-geometries does not automatically validate them for Machian use. Nor, if one used open, infinite spatial slices, do I see that this would automatically violate any Machian ideas. Boundary conditions would have to be imposed, but would they have to be *arbitrary*? Einstein thought that there might be naturally Machian boundary conditions, and while his attempt to work out this idea was a failure, I have argued that this doesn't show that the idea itself is necessarily mistaken.

Bondi [response to same comment]: I disagree with Barbour. As I see it, any radius of curvature significantly greater than the Hubble distance is of little relevance, whether it is positive, negative or infinite.

Barbour: In response to both Bondi and Hoefer, I still maintain it is far easier to define a definite relative motion of either mass particles or fields that can be used in an action function in the case of a finite or closed universe. As regards the role of gravitational degrees of freedom, when Mach first criticized Newton, the 'ontology' of the world was mass points in Euclidean space. Einstein changed the ontology and worked with fields and dynamic geometry, but he never seems to have asked himself seriously this question: What precisely is the Machian problem in the new context of fields and dynamic geometry? The Poincaré-type criteria of Machianity that I develop in my paper (p. 214, Sec. 3) translate immediately into the new context, but frankly it seems to me anachronistic in a world of fields and dynamic geometry to say only matter, and not other degrees of freedom, can determine the inertial frames.

Gravitational waves are just as observable as matter fields. The fact that there is no proper energy–momentum tensor of the gravitational field presents no problem in the formulation of the thin-sandwich problem, which operates exclusively with 3-metrics (and fields if they are also present). See *gravitational degrees of freedom, role in Mach's Principle* in the Index and also the immediately following comments of Ciufolini made at Tübingen.

Ciufolini: This discussion is related to the view of John Wheeler. He thinks that a model universe is in agreement with Mach's Principle if it has a Cauchy surface that is a closed manifold, that is compact and without boundary; that is a model that admits a closed Cauchy surface. I think Jim [Isenberg] has done some work on that.

Isenberg: Yes that is essentially Wheeler's view: That if a spacetime admits a closed Cauchy surface, and if it also satisfies Einstein's equations (with the constraints thus imposed on the initial data on any Cauchy surface) then the spacetime should be Machian. Oh, he also includes that topological restriction on the Cauchy surface.

Ciufolini: So according to Wheeler the corresponding initial-value formulation clarifies the origin of inertia...

Isenberg: There is no ambiguity about whether a Cauchy surface is open or closed.

Ciufolini: Yes, but in the spatially nonclosed case you have to admit some kind of prior geometry such as asymptotic Minkowskian geometry.

Hoefer: Well, I have always been puzzled about this, exactly why that demand expresses anything Machian.

Narlikar: From the last transparency, it was not clear to me that Einstein wanted singularities or no singularity.

Hoefer: Singularity free. The confusion arises because Einstein desperately wanted the de Sitter solution to have some kind of singularity, because it was a matter-free solution and his demand for a

physically reasonable solution was that it be singularity free.
Renn: You mentioned the problem of the *Entwurf* theory not being fully covariant. That was actually only one of its 'Machian problems,' so to say. Another problem is related to Einstein's claim that his theory would also do justice to the requirement that inertial mass is created by the presence of other masses in the universe. Max Abraham, who wrote a critical review of the *Entwurf* theory in 1914, actually calculated the effect that other masses have on the mass of a given body according to this theory. He found this effect to be so small that he concluded that Einstein's claim can only be maintained if the existence of invisible matter is assumed, an assumption he considered absurd.
Hoefer: I was curious when you mentioned that earlier. I am not clear about that, why the smallness of the effect should be a stumbling block. Any effect at all would seem to fulfill your Machian expectations.
Lynden-Bell: No, no. All of it has to be.
Hoefer: You mean removing all the rest of the mass from the universe only subtracts a negligible amount of inertia?
Ehlers: It seems to me that the term inertia was used in a somewhat unclear fashion even in the quotations which you showed. One could either think that by saying that the inertia of a particle should be determined by the cosmic masses it is to be interpreted as saying a local piece of the inertial trajectory of the particle, or one could interpret the term as meaning the value of the inertial mass, and these are rather different requirements. I am not sure which requirement was considered as a Machian requirement at that period by Einstein.
Hoefer: Well Einstein thought both that the inertial mass should be a product of the presence of other masses and also that the local piece of the inertial trajectory should be determined by the distribution of masses. I believe Professor Barbour has argued that the first requirement shouldn't be thought as a true Machian requirement.
Barbour: That is certainly my view. I am delighted with Jürgen's question. That's one of the things I'm hoping we will discuss in the session this afternoon (p. 91).
Norton: I have a brief remark on why Einstein thought the theory was Machian. As early as 1912 and 1913, he could derive the weak field effects associated with the dragging of inertial frames by accelerating masses. Even though his theory was not generally covariant at this early stage, he did believe (erroneously) that it was covariant under transformations to rotating frames of reference. That problem was fixed in November 1915, when he found the generally covariant version of his theory.

General Discussion: What is the Machian Program?

Because Mach's Principle is surrounded by so much controversy, the final session of the first day of the Tübingen conference was devoted to a general discussion, led by Barbour, on the theme *What is the Machian Program?* The edited transcript of the discussion, to which a few comments made at other times during the conference have been added, follows. The editors feel that the discussion session did achieve its purpose – to identify all the main issues associated with Mach's Principle. At the end of the discussion, a straw poll on certain issues was held. The questions and results of the poll are given at the end of the discussion transcript [p. 106]. At the end of the final day of the conference, the straw poll was repeated to see if any significant changes of opinion had occurred. The results of that poll too are given.

Barbour: There are at least four questions that I feel we should discuss, the first of which has already been raised by John Norton [p. 9].

> *Question* 1: What was Mach actually advocating? Was he advocating a mere redescription of Newtonian theory without any change in its physical content, or was he advocating a genuinely new theory?

This next question has already been precisely formulated by Jürgen Ehlers [p. 90]:

> *Question* 2: Should the Machian principle be something to do with a cosmic derivation of the inertial mass, some sort of formula where m, the inertial mass, is equal to some integral stretched over the entire universe, something like that, or is it just to do with a cosmic derivation of the local inertial frames of reference?

Einstein Studies, vol. 6: Mach's Principle: From Newton's Bucket to Quantum Gravity, pp. 91–106

My own view is that it's the latter. I think Einstein brought in a red herring by requiring a cosmic derivation of the inertial mass. This is an important issue because it determines the sort of theory we want to find. Indeed, for me the main purpose of this meeting, and especially this session, is: Can we establish, if it's possible, what are the *true criteria of 'Machianity'*? When can we say that a theory is truly Machian? In fact, as I argue in my contribution [p. 214, Sec. 3], I believe that Poincaré has given us a very useful and precise criterion.

The third issue has not yet been mentioned today and has seldom been raised in the literature. When you read Mach's *Mechanics*, the first five or six pages of his critique of Newtonian mechanics are not about motion; they are about *time*. He starts by making a big issue about time.

In fact, Mittelstaedt (1976) even wondered whether one should not formulate a *Second Mach's Principle*, which is to do with relativity of time. This would then match the First Mach's Principle, or the first Machian requirement if you like, to do with the relativity of motion. When we get into quantum gravity, I think we shall see it's extremely important, and that it is a Machian issue. Therefore:

> *Question* 3: Is there a *Second Mach's Principle* to do with the relativity of time?

Finally, if we do accept that the Machian requirement is to show how the local inertial frames of reference are determined by the universe at large, then: What agents do that determining?

> *Question* 4: In the context of general relativity, must the local inertial frames of reference be determined completely by the energy—momentum tensor of matter in the narrow sense, or can the gravitational degrees of freedom themselves contribute to the determination of the local frames of reference?

Hoyle: Specified under what mathematical conditions? On a Cauchy surface?

Barbour: That is fair enough; a Cauchy surface. I think this is an important issue, because quite a lot of interpretations of Mach's Principle, which stem from Einstein himself [p. 180], suggest that $g_{\mu\nu}$, which determines the local frames of reference, should be determined by the matter alone. Therefore, if you are going to give a Machian interpretation of general relativity, is Einstein right about the agents that can determine $g_{\mu\nu}$? There's quite a large body of opinion that thinks it must be pure matter only and not the gravitational degrees of freedom.

Ehlers: Not to anticipate the debate, but it seems to me that Question 4 as it stands does not make sense, because if you have a tensor $T_{\mu\nu}$ and not a metric, then this does not meaningfully describe matter. There's no theory of physics so far which can describe matter without already the metric as an ingredient of the description of matter, and therefore within existing theories the statement that the matter by itself determines the metric is neither wrong nor false, but meaningless.

Hoyle: I agree entirely with that. If you're to specify a field, it's got to be given on a free surface or a Cauchy surface.

Renn: The remark of Jürgen Ehlers concerning the metric as an ingredient of the description of matter corresponds almost literally to what Einstein said in his letter to Felix Pirani, from 2 February 1954. Einstein drew the conclusion that one should no longer speak of Mach's Principle at all.

Barbour: I myself agree very largely with you've said, Jürgen [Ehlers], but I think that nevertheless there are people who are attempting to make sense of these sort of things, and that is actually how Einstein himself formulated it when he coined the expression 'Mach's Principle' [p. 186]. However, before we get into this discussion, can anyone add any major issues to the four I've listed.

Kuchař: I don't know what it is that we are doing here. Are we trying to interpret what Mach has said historically, or are we trying to say what he should have said?

Barbour: I would say that one quarter is trying to establish what he actually said – that is the historical part – and then the rest is trying to establish what Einstein should have done and whether he succeeded.

Kuchař: Well, I would say it's pouring new wine into old bottles.

Jones: Mach pointed out that the inertial frames we observe do not rotate relative to the stars as we see them, and I would say any theory has to explain why that seems to be the case.

Lynden-Bell: None of us believes it's true though ...

Barbour: You think it's only approximate, Donald?

Lynden-Bell: Yes, I think it's only approximate, and I think most people think it's only approximate.

Jones: Yes, but it's approximate to a very high degree of accuracy.

Lynden-Bell: No more accurate than you would expect.

Jones: No, I think there are actually observations to show that it's quite accurate.

Lynden-Bell: Quite accurate, but no more accurate than one would expect.

Hoyle: But do the stars count? Couldn't you talk about the microwave background? That would be more accurate.

Bondi: I want to make two points: 1) If we want this conference to lay down criteria for Machianity, history is interesting but shouldn't be the final judge; 2) I don't regard Question 4 as controversial. I think the answer is obvious.

Barbour: And the answer is?

Bondi: I mean that if I have a Cauchy surface and the same $T_{\mu\nu}$ but give different gravitational wave situations in between differences will develop, so I don't think it's a controversial question.

Raine: I'm not quite sure that what you've actually written down really encapsulates what you are trying to say because once you actually try and spell out what you really mean you start having to talk about particular theories and getting down to specifics. What you really are trying to say is in some sense this: Are there, at the beginning of the universe, some free gravitational modes that have to be put in as well as all the matter modes, or should we somehow eliminate the free gravitational degrees of freedom in the Big Bang.

Barbour: That is certainly one aspect of the question. I was also thinking of Wheeler's interpretation, which is done on instantaneous surfaces by means of the constraints. Wheeler argues strongly that the effective energy density of the gravitational field should also determine the local inertial frames of reference. So I think that there are at least two theories where these things are discussed a bit more precisely. Can we note you've registered that point and the issue may need more precise formulation, Derek?

Ciufolini: Can I rephrase Question 4 and instead of 'matter' use mass 'energy' to make Professor Bondi happy?

Barbour: What do you mean by 'energy' though? You don't mean $T_{\mu\nu}$, I think.

Ciufolini: I mean that when one talks of energy one should somehow include the energy of the gravitational field and, in particular, the energy of gravitational waves.

Barbour: Well, this was my alternative view, that both of them should count. This is John Wheeler's viewpoint. You think it should be that?

Ciufolini: Yes.

Barbour: Well, by saying gravitational degrees of freedom, that was what I meant. I am avoiding talking about energy of gravitational waves because that is so difficult.

Hoefer: It seems to me that the sentiment here is that you can't do anything just with $T_{\mu\nu}$ neglecting things like gravitational waves, but on the other hand I am inclined philosophically to think that for the theory to be perfectly Machian it would have to be that, just with $T_{\mu\nu}$ [see p. 83].

Barbour: That's certainly how Einstein was interpreting it around 1917–18, and that is what I was trying to capture with this question.

Bondi: The way you formulated Question 4, it appears as a question of what is correct mathematically, physically correct within general relativity, which is the technical question to which I, and I think many other people here, think the answer is clear. The hidden agenda which has been brought out is: Is a theory in which the gravitational degrees of freedom play a role thereby non-Machian?

Barbour: You are right. There is the technical question, on which there is little disagreement. However, it still may be possible as Derek Raine has intimated, to formulate some sort of condition where at the start of the evolution there are, in some sense, no gravitational waves. So there is a group of people who are trying to eliminate gravitational waves and think that it is technically possible to make that meaningful, notwithstanding what you've said, and that that is necessary. Then the second question is: Does the fact that the gravitational degrees of freedom play a role make general relativity non-Machian? I share John Wheeler's view that their role in the determination of the local frames of reference by no means disqualifies general relativity as a Machian theory.

Isenberg: I think that it is important to think about other theories of gravity as well as Einstein's theory and consider whether you can get away without having gravitational degrees of freedom and yet have the theory agree with experiment and observation.

Kuchař: I don't know what Derek [Raine] meant, but isn't Penrose's proposal that the initial singularity be such that the Weyl tensor vanishes at it just a formulation that answers sort of your question?

Isenberg: Well, that's just a restricted class of universes. He's saying that there are a number of solutions of Einstein's equations where we can look at the initial data, and the gravitational degrees of freedom are turned off.

Kuchař: No, I would say Mach's Principle is used as a selection principle, which is traditionally how it was used many times.

Barbour: Paul Tod at Dennis Sciama's birthday celebration a year or two ago gave a very interesting paper on just that question, that perhaps the Green's function formulation that Derek [Raine] will be talking about is realized in Roger Penrose's idea.

Raine: I will also be talking about that tomorrow [p. 286].

Assis: On Question 2, I would say that in my mind both things would be Machian and should be looked for in a theory in which one tries to implement Mach's Principle. There should be a cosmic derivation of inertial mass and a cosmic derivation of the local inertial frame of reference.
Barbour: Would you be prepared to see one of them go and say that it would be a nice optional extra rather than absolutely essential? Has it got to be both?
Assis: In my mind, both together, because, for instance, when Mach makes the definition of inertial ratios, of inertial masses, the opposite ratio of the accelerations, it's not only about the mass. The accelerations are relative to the distant stars, and so the two things are intimately connected.
Barbour: I have difficulty in accepting the idea about the inertial mass, because it seems to me that Mach was totally happy with the idea that the inertial mass was something intrinsic to the body. It is true you only see inertial mass when a body interacts with another body, but the inertial mass is just determined by the mutual accelerations that the two particles impart to each other.

Suppose we have a situation in which two bodies interact, one is taken as unit of mass, and we find that the other has two units of mass when they interact with a certain mass background. If we take away half the mass in the universe and let those two bodies interact again, then surely the mass ratio is not going to change, so I see no way in which you can change the inertial mass unless you say you have got an independent definition of what you mean by a force. Then you could perhaps define what you mean by inertial mass, but for me it's a nonissue, as regards both what Mach himself wanted and what is called for physically.
Jones: What you say would be true if you consider only gravitational forces, but if you bring in any other forces like electromagnetism, then you have the e/m ratio, which could be different.
Barbour: Something like that may turn up, but the masses alone are not going to give you anything meaningful. You've got to be talking about more than one force.
Nordtvedt: You do. You have other forces.
Ehlers: I would even go so far as to say that the requirement that the inertial mass should come about by the interaction with distant masses is contrary to the use which so far one has made of the concept of mass in physics, namely, mass is a characteristic, a number, which one assigns to a body insofar as the body can be considered as isolated from the rest,

and in that approximation and only then it is useful. However, nowadays when one considers quarks, quark masses are not well defined in particle physics precisely because of confinement, and all the useful uses of mass, at least in particle physics, arise whenever you consider particles as essentially free. That is the role of mass, and if you look at general relativity and you consider a system of two stars, forming a double star, then you can meaningfully assign masses separately to the bodies precisely to the extent that you can consider one body as separated from the field. As soon as you can no longer, you would perhaps have a Machian situation, but at the same time mass would no longer be unambiguously defined.

Ciufolini: In Question 4, do you mean by the gravitational degrees of freedom the energy of gravitational waves and also the possibility of asymptotic flatness?

Barbour: I would allow that, but I think a lot of people wouldn't. They would want closure, so they wouldn't consider the second possibility.

Ciufolini: The two possibilities are different, so they should be distinguished. If you assume conditions in which you have a compact manifold and you also have gravitational waves, that's one possibility, and another possibility is that you have asymptotic flatness, so, according to some views, one is Machian and the other is not.

King: Question 2 on cosmic derivation of the inertial mass: Regardless of whether Mach wanted that or not, it's not terribly interesting anymore, because it's been ruled out by experiment, and there's two flavors of that: You could have just *G* in Newtonian gravitation changing, or you could have an anisotropic mass if it's cosmically determined, but both of those possibilities in a sense have been ruled out.

Assis: By which experiments?

King: The Brans-Dicke theory is the best example of *G* varying, and that's been essentially ruled out.

Barbour: Do you have a figure on that?

Will: The most recent experiments looking for anisotropoy in energy levels of atoms, using basically trapped atoms, are in the region of about a part in 10^{26}.

Barbour: So for anisotropy there is a very low bound. What about variation of the gravitational constant?

Will: Parts in 10^{11} per year, still from Viking radar.

Barbour: As I understand it, that is not much different from what one would expect anyway, since the gravitational constant could certainly decrease as the reciprocal of the Hubble 'radius.'

Will: It's on the edge.

Barbour: So I think it's not totally ruled out that the gravitational constant is changing. [See also Nordtvedt's contribution, p. 422].

Lynden-Bell: I was just worried about one thing. We've heard a lot about the relationships of Einstein and Mach: How long did they meet? How long did they talk? Can someone who knows much more about the history than I do tell us?

Renn: As Gereon Wolters (1987) has reconstructed from an entry by Einstein in a contemporary scratch notebook, he visited Mach on a day in the last week of September 1910. Einstein had traveled to Vienna to discuss his appointment at the German University in Prague with the authorities, and he took the occasion to see Ernst Mach. There are three reports of this encounter, all due directly or indirectly to Einstein. There is no indication in these reports that Einstein and Mach discussed problems of relativity, although they might have, of course, as Einstein was rereading Mach's *Mechanics* in this period. In any case, according to Einstein's notebook entry, the meeting was short: He planned to meet Mach at 4:30 p.m. and had the next appointment already at 5:45 p.m.

Lynden-Bell: Well, you must know when it is, therefore, and whether what Einstein says is Mach's Principle is likely to be something that he's heard, or could have heard, directly from Mach.

Renn: No, he certainly had it from his reading of Mach, as is confirmed by his later recollections.

Lynden-Bell: Long before he met Mach?

Renn: Long before, according to one recollection even before 1907, so he had quite some freedom to give it his own interpretation.

Kuchař: In fact, your account of the discussion between Mach and Einstein reminds me of the account of the only encounter of Newton and Huygens, who in fact met in Cambridge and then went to London by a coach, and no one knows what they discussed [laughter].

Norton: We only have fleeting glimpses of this; there's an entry in a notebook where you see Einstein's going to visit Mach, and there's a very brief correspondence where Einstein attributes the Mach Principle to Mach. A major focus of discussion this morning was whether that attribution was correct [see the papers of Norton (p. 9), v. Borzeszkowski and Wahsner (p. 58), and Barbour (p. 214)]. Clearly Mach never straightened it out if it wasn't, though I believe that it wasn't. The only extensive discussion that's close to contemporary that I would know of is Philipp Frank, who in his biography of Einstein gives a lengthy story of how they met and what they talked about, and atoms was the topic of discussions, and apparently Mach was prepared to move at last on atoms if some decisive experiment could come up.

Assis: To go back, giving a modern argument why it seems to me that a cosmic derivation of inertial mass would be interesting in any theory trying to implement Mach's Principle, it is because of the remarkable fact that the inertial mass is proportional to the gravitational mass and not to the electric charge or to a nuclear property of the body and so on.

So if we only assume that is a coincidence, that is all right, but if we try to understand this proportionality, which is to me remarkable, then it is natural to try and find some derivation of the inertial mass as some kind of gravitational interaction with the distant bodies.

Barbour: Thank you for bringing that in. I should have included this issue of the identity of the two masses.

Perhaps I could mention here the very interesting result that Reissner obtained in his theory using a Weber-type potential, from which he can actually also derive gravitational type forces that come out of inertial forces. He argues that gravitation is really a manifestation of the inertial effects of rotating bodies with rapid internal motion [p. 142, Note].

This seems to be a hint you can get some way in that direction.

Kuchař: I have one comment about the program that can be regarded as implementing Mach's ideas, namely, to eliminate spacetime and turn everything into a relational theory of the matter. There is a complementary program in relativity, which is to eliminate matter and leave only the spacetime. This idea goes back to Rainich, and it was later developed by John Wheeler, Charlie Misner, and many other people. I would say that the Rainich program is much easier to implement than the Mach program. It's much easier to eliminate matter, leave the spacetime as the only dynamical entity, and still have viable physics, than to do it the other way around.

Earman: Can you give some indication of why that's the case?

Kuchař: Yes, because of the universality of the gravitational interaction. All matter leaves an imprint on the gravitational field, but the gravitational field leaves only a partial imprint on the motion of the matter. Furthermore, general relativity is a field theory, rather than a Machian action-at-a-distance, which enables one to reconstruct matter dynamics more or less locally from its universal imprint on the metric structure.

Isenberg: Just a comment on what Karel's [Kuchař] remarks about the geometrodynamics program. My understanding is that if you include the Yang–Mills fields, then the geometrodynamics program doesn't work in the traditional sense. Maxwell fields and Abelian fields are okay, but non-Abelian Yang–Mills fields cause trouble.

Renn: I would like to make a philosopher's comment on Question 1.

I have the impression that the distinction between a redescription and a new theory is not really so obvious as it may appear, and I want to give you some examples illustrating this point: Is Hamilton's formulation of classical mechanics a redescription of Newton's mechanics or is it a new theory? Is Hertz's reformulation of Maxwell's electrodynamics a redescription or a new theory? Now, regarding these two examples one might in fact argue that the emphasis is on formal improvement rather than on conceptual development. But what about the example of Lorentz's electron theory - can one not perhaps consider special relativity merely as a reformulation of Lorentz's theory? At least many of Einstein's and Lorentz's contemporaries have taken this point of view. My philosophical argument is that if one introduces a new formalism to describe an old theory, one may achieve at first just a more or less equivalent reformulation - but eventually the new formalism will allow one to draw consequences that transcend the horizon of the original formulation.

Ehlers: I wonder how you would answer to the following proposal. If I have two formulations of a theory, I would be inclined to call them physically equivalent if the empirically or observationally testable predictions of both are the same. Isn't that a reasonable proposal?

Renn: No. I think the proposal is unsatisfactory because it does not provide a criterion for distinguishing between conceptually different theories and differences in the stage of elaboration of one and the same theory. For instance, initially special relativity and Lorentz's electron theory could be considered to be physically equivalent in the sense of your definition, that is, they agreed on all empirical accounts known at the time; but then special relativity would lead to consequences which are inconceivable in the conceptual framework of Lorentz's theory, such as the idea to generalize the relativity principle. In other words, what may have initially appeared as a mere reformulation eventually turned out to have fundamentally new implications. But even when these new consequences became visible, special relativity and Lorentz's theory could still be considered equivalent on the basis of your definition.

Ehlers: That I don't understand, and you would have to show me.

Renn: Is what you doubt my claim that there is a version of Lorentz's electrodynamics which was empirically equivalent to special relativity? I have to refer to the historical literature for evidence to this effect. In order to proceed further let me just assume that I can in fact substantiate my claim that such different formulations with the same empirical content may indeed exist. Their difference would then primarily consist in the distinct conceptual implications to which the two formulations give rise.

Ehlers: Yes, I will say as soon as you can derive from the two different formulations empirical consequences which follow in one formulation and not in the other, then of course I would consider them different.

Renn: The crucial point is, however, the time index by which you must label the different formulations. What begins as the physically equivalent (in your sense of the term) reformulation of a theory at one point in time may end up as a conceptual revolution later on. In fact, I would go so far as to claim that conceptual revolutions in physics begin, as a rule, with reformulations of preexisting theories. The principle of inertia was found by reformulating a consequence of Aristotelian dynamics; classical electrodynamics emerged from a theory that was originally formulated in mechanical terms; Einstein reformulated Planck's radiation law, introducing the revolutionary concept of light quanta, and so on.

You may always claim, of course, that all later consequences were already implicitly present in the original formulation. But this view ignores the fact that the elaboration of a theory involves its application to new problems, which may have consequences that cannot be predicted by purely logical means at the outset of the development. In other words, the development of a theory is always also the development of its concepts and can hence not be sufficiently described by formal logic. The notion of 'implicit consequences' only buries this problem.

Barbour: Surely Feynman was always making the point that for any existing theory you should have as many different conceptual formulations as is possible so that you can see different ways to generalize the theory and find some new theory. However, I was thinking of something more definite than that. You certainly can rewrite Newtonian theory in purely relative terms. There's no question of it. The question is what time derivatives go into the equations of motion. If you allow some *third* time derivatives in the rewriting of Newtonian theory, you can recast Newtonian theory in purely relative terms. It's completely equivalent in its observational predictions to standard Newtonian theory (see Poincaré's comments, pp. 111–112), and that is a mere redescription, and in fact I think that is what Lange did do, and Mach praised him for it but said that's still not what I want, I want something different from that [pp. 217–218]. Now, in fact, those theories of Reissner [p. 134] and Schrödinger [p. 147], the ones that Bruno Bertotti and I and several others, including Liebscher and Assis [p. 159], rediscovered, definitely make new predictions. They have a perihelion advance. That is a new theory, and it's Machian, so it is possible to go over to a theory which has new content. No question of it.

Ciufolini: As regards Question 1, I think that when Mach was writing about increasing the the mass of the walls of the bucket he was thinking of something like the possibility of dragging of inertial frames or at least dragging of a test particle due to the mass current, the angular momentum, and therefore this was a new effect and therefore he was looking for a new theory.

Bondi: May I come in here with a quotation from Eddington, who in discussing variational principles said in any such principle we divide possible states into three: what actually happens, those that are near enough to what happens so we compare it with them - that's what the variational principle is - and those so different that we don't allow them to enter. Now I think this is precisely the question about your inertial mass. If you allow an almost empty universe, I rebel against the idea that you put one particle there and that fixes all inertial frames and everything has the same inertia. If on the other hand you only permit as a comparison universes reasonably similar to our own, you may well only have the cosmic derivation of the frames without a complete derivation of the mass, and I think it is the universal comparison, if I may use the term, that defines the answer to Question 2. If I come back to a final point, it goes back to what we discussed earlier today. If there are two bodies of unequal masses revolving about each other, then I believe any advocate of cosmic derivation of inertial mass will think that in a different universe the ratio of the radii of the orbits would be the same. But the sizes of the orbits would be different.

Assis: Let's go to Question 3. It seems to me that this is a very important issue, the nonexistence of time, but very few models or theories have tried to implement that, and I would like you to make a few more comments on that.

Barbour: Thank you. I do think that there is a Second Mach's Principle, which has to do with the fact that there is no external time and that any 'time' we use has to be got from some motion that is in some way observed. We have to get a measure of time from the motions that are happening within the universe, and I think this is a very important point. It's *half* of Mach's criticism of Newtonian mechanics.

Lynden-Bell: If there are comments on time like that, that it does not exist, do we try and say the nonexistence of space in that spirit or not?

Barbour: That's a very deep question. I would like to establish the nonexistence of space, but I don't yet see any way to do it. It's very hard to formulate a dynamical theory unless you've got some structure. If you are going to formulate dynamical theory, your variables must correspond to *something*, be it a Riemannian three-geometry or particles

in Euclidean space or something. It is much easier to do without time (see my contribution in this volume, p. 214). After all, for two millenia the astronomers used the rotation of the earth to tell the time, but still made models of the bodies of the solar system in space.
Lynden-Bell: Yes, but there's something very nonrelativistic about your separating out time, and that's what I was trying to get at.
Kuchař: I would rather say it's a technically ill-defined question, because for any system with an external time, you can adjoin that time to the rest of the dynamical variables and formulate everything internally on an extended configuration space, which is truly a configuration spacetime. The Second Mach's Principle is then implemented on that space, and the change of everything, including the formal time variable, is driven by a super-Hamiltonian constraint. I would say that your question has two ingredients. There is the philosophical ingredient - What variables qualify for time? - and there is a technical ingredient - Does the theory satisfy the Second Mach's Principle? The answer to the technical question, I think, is pretty clear: Whatever system you have, you can always cast it into the mold in which it satisfies the Second Mach's Principle.
Barbour: Now as Karel well knows, we have been arguing about this for fifteen years [laughter]. An external time is a totally heterogeneous element. The time you put in parametrized particle dynamics, the model Karel has in mind, is simply not there. You can't see it. As Mittelstaedt says, *die Zeit ist nicht wahrnehmbar*. When astronomers look through telescopes, they actually *see* the separations of bodies. They don't see time, so I think it's quite wrong just to adjoin formally something which you call time and claim you have a Machian theory of time. In the real world, there are just relative positions of bodies, and that is something quite different from a heterogeneous and invisible time, so I take that as a challenge to construct a theory that uses only things we truly see.
Kuchař: It's the same question as that of when a theory is generally covariant. As Kretschmann (1917) pointed out, if you take more variables and toss them into the theory, you can always make it generally covariant. You can argue from simplicity that those elements which were tossed in are in some sense heterogeneous, but simplicity is a tacky subject.
Barbour: I would say you put your finger precisely on the criterion: You are not allowed to throw in these heterogeneous elements. The kinematic framework must contain only the relative distances, if we are talking about a Newtonian-type theory. I quite agree that one must distinguish the philosophical question - perhaps one might call it the

ontology – from the purely technical requirement of reparametrization invariance. For me the ontology is decisive, and I am not sure it is such a tacky issue.

Nojarov: I think it is very complicated about time, but time is ultimately related with motion, velocity, and you are right when you say that the astronomers see only positions. You don't have any time, but if you record change in positions, which we see, and they are different, you have to introduce time in order to describe a process.

Barbour: In my paper [p. 214], I try to explain how everything can be done with different configurations.

Norton: Before things get out of hand, I just want to mention one thing – what Mach said about inertia is I think ambiguous, but having recently reread what Mach said about time I don't think it's ambiguous. For that reason, lest we perpetuate another myth, I'd urge that in Question 3 we strike out the words Second Mach's Principle and write in Barbour's Principle [laughter].

Ehlers: Maybe it would help if the word 'time' there would be specified. I understand you are saying that the time metric should not be imposed *a priori* as a fundamental structure but should come out of the theory. The time metric and metric statements about time should be derived within the theory and not put in *a priori*, but wouldn't you also agree that you need an ordering of configurations in order to start talking at all about changes?

Barbour: Certainly at the classical level. I am not trying to do away with the idea that there is a sequence of configurations.

Ehlers: The time order has to be used as a primitive concept in order to start talking meaningfully.

Barbour: At the level of the classical theory. But in the quantum domain that goes, and there is nothing at all, so you must do something drastic, I believe, to recover notions of time in the quantum theory [p. 501].

Isenberg: I think it's interesting in this context to think about certain spacetimes that are flat and yet are spatially compact and even have an intrinsic notion of time based on the mean curvature. These are the ones that are compactified using some funny topology, and yet at each point of spacetime there's a unique constant mean-curvature surface which goes through it. There's nothing moving in it, in a sense, and yet you do have this nice notion of time. These are just certain models, but they are certainly solutions of Einstein's equations.

Barbour: Is that all that different from the astronomers using the rotation of the earth as a clock? If you've got configurations, you can always take one coordinate to label the time if not put a metric on it. The question then is whether that's a sensible time. For example, if you took the mean extrinsic curvature as time seriously, you would have a nonconservative dynamics.

Zeh: Julian, I was surprised to hear you agree with Ehlers, because you once introduced the analogy with a deck of cards that can be shuffled, so even their ordering is merely a consequence of their intrinsic structure. Why did you now agree that this ordering has to be there from the beginning?

Barbour: As long as the universe has got a lot of particles in it and is in a generic area of its configuration space, then just the successive configurations of the particles, if you took snapshots of them, would be sufficient for you to order them in a curve. I agreed with Jürgen [Ehlers] in that I conceive of classical physics as an extremal history in the configuration space, that it's a one-dimensional sequence of configurations. That *is* classical physics. To that extent one thing does follow another (though, in fact, one direction of change is as good as the opposite one). Certainly, if you give all the snapshots in a jumbled heap, I could, just by examination, establish that they did form a one-dimensional sequence and put them in the correct order as long as there are no nongeneric configurations.

Zeh: There is no *absolute* order!

Barbour: The order is in the configurations. Everything is in the configurations. It's not absolute in that sense.

Kuchař: I just want to make the point that when we discussed Question 3 we did it within the model theories which were Newtonian. There was only one foliation on which evolution took place. It's pretty clear to everyone who ever studied the Hamiltonian formulation of general relativity that there is no such privileged foliation, that one should work with the many-fingered time concept, that there is no single ordering which one can put in, or, as Wheeler put it, spacetime is a sheaf of geodesics in superspace. Moreover, if we ask Question 3 about time, we should also ask it about space. In general relativity, space and time come in a single package.

REFERENCES

Kretschmann, Erich (1917). "Über den physikalischen Sinn der Relativitätspostulate, A. Einsteins neue und seine ursprüngliche Relativitätstheorie." *Annalen der Physik* 53: 575–614.

Mittelstaedt, Peter (1976). *Der Zeitbegriff in der Physik*. Mannheim:

Bibliographisches Institut.
Wolters, Gereon (1987). *Mach I. Mach II. Einstein und die Relativitätstheorie: Eine Fälschung und ihre Folgen*. Berlin and New York: de Gruyter.

Results of Straw Polls

The four following questions were put to the participants at the end of the general discussion session on the first day of the conference (Entrance Poll) and repeated at the end of the conference (Exit Poll). It should be noted that a few participants of the Exit Poll did not participate in the Entrance Poll and also that several participants had to leave before the Exit Poll.

Question	*Answer*	*Entrance Poll*	*Exit Poll*
1) Was Mach advocating a mere redescription (MRD) of Newtonian Mechanics without any change of physical content or was he advocating a genuinely new theory (GNT)?	MRD	12	10
	GNT	26	16
2) Is general relativity perfectly Machian?	Yes	2	3
	No	30	21
3) Is general relativity with appropriate boundary conditions of closure of some kind perfectly Machian?	Yes	14	9
	No	18	14
4) Is general relativity with appropriate boundary conditions of closure of some kind very Machian?	Yes	19	14
	No	18	7

2. Nonrelativistic Machian Theories

Introduction

The passages from Mach with which this chapter opens have been chosen so that the reader can judge the extent to which the later papers in the chapter truly implement Mach's ideas in the context of the *nonrelativistic* physics in which they were formulated. For further quotations from Mach look under *Mach* in the separate Quotations Index (p. 636). Discussions of precisely what Mach had in mind are given by Norton (p. 9), von Borzeszkowski and Wahsner (p. 58), and Barbour (p. 214). The passages from Poincaré (p. 111) are important because of their precision and the attention they draw to the initial-value problem in dynamics. Mach's writings imply in very qualitative terms that only relative distances should be used in the formulation of mechanics, but as Poincaré makes clear Newtonian mechanics can be written in purely relative terms if one allows time derivatives of the relative separations *of higher than the second order* in the equations of motion. As Poincaré notes, "for the mind to be fully satisfied," this should not be the case, and he says one should require that the future be uniquely determined by what may be called [though Poincaré does not even mention Mach] *Machian initial data*: the masses of the particles, their separations, and the rates of change of those separations. This requirement then provides a precise criterion of *Machianity* (p. 92 and p. 218ff). All the nonrelativistic theories presented in this chapter meet this requirement, which is implemented by the construction of a Lagrange function which depends on the quantities listed by Poincaré and nothing else.

Before the Tübingen conference, virtually no one knew that Machian models had been proposed several times in the early part of this century. The simple and decisive step, first taken by Hofmann (p. 128) and

Einstein Studies, vol. 6: Mach's Principle: From Newton's Bucket to Quantum Gravity, pp. 107–108 © 1995 *Birkhäuser Boston, Inc. Printed in the United States.*

Reissner (p. 134) and independently rediscovered many times since then (as recounted by Assis, p. 159), is to replace the Newtonian kinetic energy, which is a sum over individual masses, by a sum over products of pairs of masses multiplied by the square of the relative velocity of the corresponding pair and some function of the separation. However, all the Machian kinetic energies of this type proposed in the early models lead to an anisotropic inertial mass. In 1925, Schrödinger came within a hair's breadth of being able to rule out such models on observational grounds (see p. 157).

When such models were rediscovered 50 years later, the problem of mass anisotropy could not be ducked, and this led to the formulation of an alternative scheme based on the notion of the *intrinsic derivative* (p. 223ff). In the original form of this model, the kinetic energy is a sum over individual masses, as in Newtonian theory, but with a 'Machian correction' that depends on all the masses in the universe. For this reason, mass anisotropy is avoided. Very interestingly, as Lynden-Bell shows (p. 172), it turns out that this model can, when the Machian correction is calculated and substituted back in the Lagrange function, be cast in the same basic form as the Hofmann–Reissner–Schrödinger models with kinetic energy in the form of a sum over products of masses. Thus, in all the models inertia arises as a kind of interaction (cf. Einstein, p. 180) and "accelerated [under gravitational forces] and inertial motions result in the *same* way," as Mach anticipated (p. 110).

Because the Lagrange function of general relativity can be cast in a form using a generalized intrinsic derivative (p. 223ff), the present writer believes general relativity is perfectly Machian. That, however, is the subject matter of the next chapter and, no doubt, considerable controversy. But as regards nonrelativistic theories, the situation seems to be clear beyond dispute: Mach's qualitative idea was cast into a precise form by Poincaré in 1902 and has been implemented in the framework of a certain class of theories many times since then; at least one theory of this class does not lead to mass anisotropy and is locally indistinguishable from Newtonian theory. Only historical accident and the overwhelming influence of Einstein obscured these facts for so long.

J.B.B.

Selected Passages: Mach, Poincaré, Boltzmann

Ernst Mach

But if we think of the earth at rest and the other celestial bodies revolving around it, there is no flattening of the earth, no Foucault's experiment, and so on - at least according to our usual conception of the law of inertia. Now, one can solve the difficulty in two ways; either all motion is absolute, or our law of inertia is wrongly expressed. Neumann preferred the first supposition, I, the second. The law of inertia must be so conceived that exactly the same thing results from the second supposition as from the first. By this it will be evident that, in its expression, regard must be paid to the masses of the universe.... Now what share has every mass in the determination of direction and velocity in the law of inertia? No definite answer can be given to this by our experiences. We only know that the share of the nearest masses vanishes in comparison with that of the farthest. We would, then, be able completely to make out the facts known to us if, for example, we were to make the simple supposition that all bodies act in the way of determination proportionately to their masses and independently of the distance, or proportionately to the distance and so on (Mach 1872).

The universe is not *twice* given, with an earth at rest and an earth in motion; but only *once*, with its *relative* motions, alone determinable.... The principles of mechanics can, presumably [see p. 48, Note 8], be so conceived, that even for relative rotations centrifugal forces arise.

Newton's experiment with the rotating vessel of water simply informs us, that the relative rotation of the water with respect to the sides of the vessel produces *no* noticeable centrifugal forces, but that such forces *are* produced by its relative rotation with respect to the mass of the earth and other celestial bodies. No one is competent to say how the experiment would turn out if the sides of the vessel increased in thickness

Einstein Studies, vol. 6: Mach's Principle: From Newton's Bucket to Quantum Gravity, pp. 109–113

and mass till they were ultimately several leagues thick....

When, accordingly, we say, that a body preserves unchanged its direction and velocity *in space*, our assertion is nothing more or less than an abbreviated reference to *the entire universe*....

Instead, now, of referring a moving body to space, that is to say to a system of coordinates, let us view directly its relation to the bodies of the universe, by which alone such a system of coordinates can be determined. Bodies very remote from each other, moving with constant direction and velocity with respect to other distant fixed bodies, change their mutual distances proportionately to the time. We may also say, all very remote bodies - all mutual or other forces neglected - alter their mutual distances proportionately to those distances. Two bodies, which, situated at a short distance from one another, move with constant direction and velocity with respect to other fixed bodies, exhibit more complicated relations. If we should regard the two bodies as dependent on one another, and call r the distance, t the time, and a a constant dependent on the directions and velocities, the formula would be obtained: $d^2r/dt^2=(1/r)\,[a^2-(dr/dt)^2]$. It is manifestly much *simpler* and *clearer* to regard the two bodies as independent of each other and to consider the constancy of their direction and velocity with respect to other bodies.

Instead of saying, the direction and velocity of a mass μ in space remain constant, we may also employ the expression, the mean acceleration of the mass μ with respect to the masses m, m', m''.... at the distances r, r', r''.... is $=0$, or $d^2(\Sigma mr/\Sigma m)/dt^2=0$. The latter expression is equivalent to the former, as soon as we take into consideration a sufficient number of sufficiently distant and sufficiently large masses. The mutual influence of more proximate small masses, which are apparently not concerned about each other, is eliminated of itself (Mach (1883).

The natural investigator must feel the need of further insight - of knowledge of the *immediate* connections, say, of the masses of the universe. There will hover before him as an ideal an insight into the principles of the whole matter, from which accelerated and inertial motions result in the *same* way. The progress from Kepler's discovery to Newton's law of gravitation, and the impetus given by this to the finding of a physical understanding of the attraction in the manner in which electrical actions at a distance have been treated, may here serve as a model. We must even give rein to the thought that the masses which we see, and by which we by chance orientate ourselves, are perhaps not those which are really decisive. On this account we must not underestimate even experimental ideas like those of Friedländer and Föppl, even if we do not yet see any immediate result from them (Mach 1912).

Henri Poincaré

Consider any material system whatever. We have to consider on the one hand the 'state' of the various bodies of this system – for example, their temperature, their electrical potential, etc.; and on the other hand their position in space. And among the data which enable us to define this position we distinguish the mutual distances of these bodies that define their relative positions, and the conditions which define the absolute position of the system and its absolute orientation in space. The law of the phenomena which will be produced in this system will depend on the state of these bodies, and on their mutual distances; but because of the relativity and the inertia of space, they will not depend on the absolute position and orientation of the system. In other words, the state of the bodies and their mutual distances at any moment will solely depend on the state of the same bodies and on their mutual distances at the initial moment, but will in no way depend on the absolute initial orientation. This is what we shall call, for the sake of abbreviation, *the law of relativity*....

To apply the law of relativity in all its rigour, it must be applied to the entire universe; for if we were to consider only a part of the universe, and if the absolute position of this part were to vary, the distances of the other bodies of the universe would equally vary; their influence on the part of the universe considered might therefore increase or diminish, and this might modify the laws of the phenomena which take place in it. But if our system is the entire universe, experiment is powerless to give us any opinion on its position and its absolute orientation in space....

I have spoken above of the data which define the position of the different bodies of the system. I might also have spoken of those which define their velocities. I should then have to distinguish the velocity with which the mutual distances of the different bodies are changing, and on the other hand the velocities of translation and rotation of the system; that is to say, the velocities with which its absolute position and orientation are changing. For the mind to be fully satisfied, the law of relativity would have to be enunciated as follows: The state of bodies and their mutual distances at any given moment, as well as the velocities with which those distances are changing at that moment, will depend only on the state of those bodies, on their mutual distances at the initial moment, and on the velocities with which those distances were changing at the initial moment. But they will not depend on the absolute initial position of the system nor on its absolute orientation, nor on the velocities with which that absolute position and orientation were changing at the initial moment. Unfortunately, the law thus enunciated does not agree with experiments....

We have seen that the co-ordinates of bodies are determined by differential equations of the second order, and that so are the differences of these co-ordinates. This is what we have called the generalised principle of inertia, and the principle of relative motion. If the distances of these bodies were determined in the same way by equations of the second order, it seems that the mind would be entirely satisfied. How far does the mind receive this satisfaction, and why is it not content with it? To explain this we had better take a simpler example. I assume a system analogous to our solar system, but in which fixed stars foreign to this system cannot be perceived, so that astronomers can only observe the mutual distances of the planets and the sun, and not the absolute longitudes of the planets. If we deduce directly from Newton's law the differential equations which define the variation of these distances, these equations will not be of the second order. I mean that if, outside Newton's law, we know the initial values of these distances and of their derivatives with respect to time – that would not be sufficient to determine the values of these same distances at an ulterior moment. A datum would still be lacking, and this datum might be, for example, what astronomers call the area-constant....

Our universe is more extended than theirs, since we have fixed stars; but it, too, is very limited, so we might reason on the whole of our universe just as these astronomers do on their solar system. We thus see that we should be definitively led to conclude that the equations which define distances are of an order higher than the second.... The values of the distances at any given moment depend upon their initial values, on that of their first derivatives, and something else. What is that *something else*? If we do not want it to be merely one of the second derivatives, we have only the choice of hypotheses. Suppose as is usually done, that this something else is the absolute orientation of the universe, or the rapidity with which this orientation varies; this may be, it certainly is, the most convenient solution for the geometer. But it is not the most satisfactory for the philosopher, because this orientation does not exist (Poincaré 1905).

Ludwig Boltzmann

Quite independently of this there is the question whether the mechanical equations here developed and therefore also the law of inertia might perhaps be only approximately correct and whether, by formulating them more correctly, the improbability or rather inhomogeneity of having to adopt into the picture a co-ordinate system as well as material points would disappear of itself.

Here Mach pointed to the possibility of a more correct picture, obtained by assuming that only the acceleration of the change of distance between any two material particles is determined mainly by the neighbouring masses, its velocity being determined by a formula in which very distant masses are decisive. This naturally avoids the adopting of any co-ordinate system into the picture, since now it is only a question of distances. Of course, Mach does not avoid introducing other difficulties, for example that the world is finite, a kind of action at a distance for the greatest distances and so on....

In all these considerations we started from the presupposition that the world is finite. If one conceives the world as infinite, concepts such as the world's centre of gravity, invariable axis, principal inertial axes and so on become quite empty. One would then have to assume that the law of inertia is determined by a formula according to which masses that are nearby have vanishing influence on the formulation of the law of inertia, that those at distances like Sirius have the greatest such influence and those at much greater distance still again next to none (Boltzmann 1904).

REFERENCES

Ludwig Boltzmann (1904). *Vorlesungen über die Principe der Mechanik*, II. Leipzig: J. A. Barth. Quoted from: *Theoretical Physics and Philosophical Problems, Selected Writings*. B. McGuiness, ed. Dordrecht: Reidel (1974).

Mach, Ernst (1883). *Die Mechanik in ihrer Entwickelung. Historisch-kritisch dargestellt*. Leipzig: F. A. Brockhaus. Quoted from Mach (1960).

Mach, Ernst (1911). *History and Root of the Principle of the Conservation of Energy*. Chicago: Open Court, pp. 77–80. German original 1872.

Mach, Ernst (1912). *Die Mechanik in ihrer Entwicklung. Historisch-kritisch dargestellt*. Sixth ed. Leipzig: F. A. Brockhaus.

Mach, Ernst (1960). *The Science of Mechanics: A Critical and Historical Account of Its Development*. LaSalle, Illinois: Open Court [translation of Mach (1907)].

Poincaré, Henri (1905). *Science and Hypothesis*. London: Walter Scott Publ. Co., pp. 75–78 and 118. French original 1902.

Absolute or Relative Motion?[1]

Benedict Friedlaender

No one has yet carried out and tested that recasting [of the Newtonian law of inertia]. The law of inertia in the usual manner of representation can be transparently described by saying that every body opposes any change of its velocity (conceived as absolute) with a resistance proportional to its mass in the corresponding direction. Here, the remaining bodies of the universe are completely ignored; in fact, a point that must be especially emphasized is that the concept of mass is, except for its derivation from gravity (mg), derived precisely from the facts of inertia. Every change in velocity, i.e., every acceleration, for example, in the simplest case the imparting of motion to a body previously at rest until it reaches a certain velocity, is held to be opposed by a *resistance*, the overcoming of which requires the quantity of energy that is afterwards present in the corresponding body, namely that contained in the considered motion as 'kinetic energy.' It is here to be noted once more that translational motion of a single body in space otherwise regarded as empty is a nonsense, namely, it does not differ from its opposite, rest. Thus, the creation of such a chimera should not require any energy; therefore if in contrast the actual world does agree with our prerequisites of thought, the relevant question should be that of the other bodies with respect to which motion is to be created, in a word, it is that of what relative motion of previously nonmoving bodies is to be created. Accordingly, inertia is to be grasped relatively; one could formulate the law of relative inertia as follows: All masses strive to maintain their *mutual* state of motion with respect to speed and direction; every change requires positive or negative energy expenditure, that is, work is either required – in the case of an increase in velocity – or is given up – in the case of a decrease in velocity. The resistance to changes in velocity would then, as soon as we regard all motions as relative, be expressed not only in the one body that, as we are accustomed to say, we 'set in motion' (that is, set in motion relative to the earth) but also in all the

Einstein Studies, vol. 6: Mach's Principle: From Newton's Bucket to Quantum Gravity, pp. 114–119

others that we regard as being at rest in accordance with the usual conception....

The application of the thought indicated here is very simple but unusual to a high degree. For if we consider the resistance to acceleration that some body exhibits, we do not have the slightest thought of other masses that are nearby! But if we do so and hold firmly to the guiding thought that the masses strive to maintain their *relative* velocity, it turns out that (for motion on a straight line of body *A* relative to body *B* as the simplest case)

accelerated approach and decelerated withdrawal

must have a repulsive effect, and

accelerated withdrawal and decelerated approach

must have an attractive effect....

Let us now apply these considerations to our flywheel and the torsion balance placed before it [see Immanual Friedlaender's account, p. 309].

Let the circle *AFCBDF'A* [Fig. 1] represent the rim of the flywheel and *P* a readily movable body or mass point within the rim of the flywheel, as close as possible to its plane, namely a part of the mass of

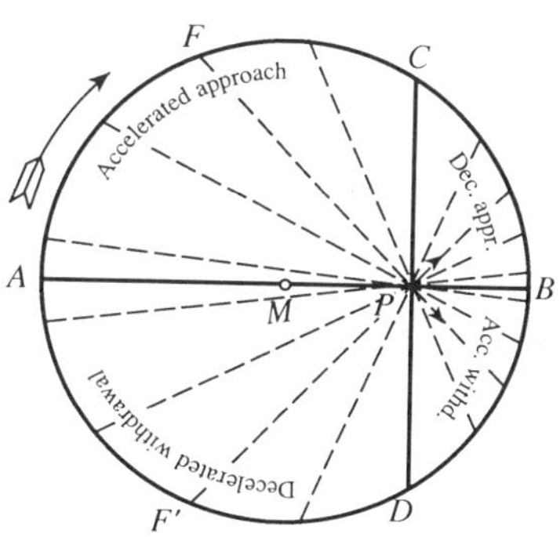

Fig. 1

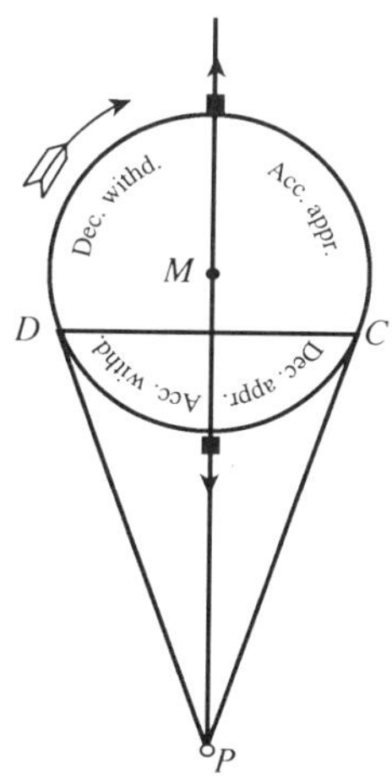

Fig. 2

the arm of the torsion balance. For simplicity, let us assume that the point *P* actually lies within the plane of the flywheel.... Let us now join the point *P* to the middle point *M* of the flywheel, extend this line until it meets the rim at *A* on the left, at *B* on the right and also erect the perpendicular on *AB* at *P*, cutting the rim above at *C*, below at *D*; then it is clear that every mass point of the rim on its way from *A* over *C* to *B* approaches the point *P* and then on the way from *B* over *D* to *A* recedes from it. However, the approach on the semicircle *AB* is accelerated up to *C* and then decelerated to *B*; similarly, the withdrawal on the semicircle *BA* is accelerated to *D* and decelerated from *D* to *A*. In view of the simplicity of the situation we can dispense with the analytic proof. But since in accordance with what we have said accelerated approach and decelerated withdrawal act in the same sense, namely both repulsively, while decelerated approach and accelerated withdrawal both act attractively, we see that we can divide the rim into two parts that differ in their effect, namely, the part left of *CD*, which repels, and the part right of *CD*, which attracts the point *P*.... Therefore, on the basis of the conception of the relativity of inertia an acceleration *away from the axis* is imparted to the point *P*, as our conception of the invertibility of centrifugal force requires. The relative rotation between the wall of Newton's bucket and the water contained in it would indeed generate appreciable centrifugal forces if the wall contained mass in sufficient amount to be no longer practically non-existent compared with the mass of the earth.[1)]...

If the ideas sketched here are correct, many consequences will follow, some of which will admittedly seem very strange. The same amount of gunpowder, acting on the same cannon ball in the same cannon, would impart to the projectile a greater velocity on, for example, the moon than on the earth; naturally, however, the greater velocity would not represent a greater but the same amount of energy as the smaller velocity that the projectile receives on our more massive planet. This would reveal itself in the fact that, despite the greater velocity, the penetration capacity would not be greater than on the earth. For the $(1/2)mv^2$ as measure of the so-called kinetic energy would not be the complete expression, lacking allowance for the environment in accordance with mass and distance, namely, the specification of the masses for which the velocity 'v' holds....

[1)]We originally said 'universe' but now the 'earth.' It is to be assumed that the earth will probably play a much larger role than the more distance masses of the universe.

The phenomena of the tides would also have to have a treatment different from the usual one; namely, in our figure we merely need to take our point *P* outside the circle [Fig. 2] and draw tangents from it to *C* and *D*; the circle then represents the earth, the point *P* the moon or the sun, and *PC* and *PD* the axial section of a cone tangent to the earth from the moon, treated as a point. One will then see that the earth will be divided by the approximately circular plane *CD*, which appears in the figure as a line, into two parts that, on account of the distance, are very nearly equal; of these, the half below *CD*, i.e., the part turned toward the moon, will be attracted, while the part above *CD*, away from the moon, must be repelled; the mobile water follows these attractive and repulsive forces and excites both the tidal waves that in the time between two culminations of the moon circle the earth....

To indicate the extent to which the problem of motion that we have raised and hypothetically solved is related to that of the nature of gravitation but also at the same time shows some similarity with the known manner in which electrical forces act, let us mention the following parallels: A body that approaches or recedes from a second body would have no influence on it as long as the velocity of approach, which is to be taken as positive or negative, remains unchanged; in contrast, any change of the velocity would have the effect previously shown.

It is well known that the presence of a current in a conductor is not sufficient to generate an induction effect - there must be a change of either the current strength or the distance; in our case the change of distance by itself would not be sufficient to generate the attractive or repulsive effects, i.e., motion itself is not sufficient, the velocity must change....

On the basis of our conception it is naturally also necessary to modify the interpretation of the astronomical facts.... In accordance with the conception of the relativity of all motions, including therefore central motions, a revolution of the earth can be completely replaced by an axial rotation of the sun *insofar as only these two bodies come into consideration*. The circumstance that the earth, despite the 'attraction,' does not plunge into the sun, or the moon into the earth, is of course explained on the basis of the usual conception by the motion of revolution of the smaller celestial body, while, for example, the axial rotation of the sun with respect to the universe plays no role at all. If our conception is correct, the so-called axial rotations are not irrelevant for the equilibrium of the world systems but must be equally taken into account like all other factors. Incidentally, the assumption of an attraction of the earth by the sun is not a felicitous interpretation of the factual situation insofar as the so-called *attractive* forces can only be adduced from reduction of distance; naturally, this is not to say that the sun would not attract the earth if the relative motions of the two bodies were other than they actually are. However, as the facts stand, actual reduction of distance does not occur; in accordance with everything we know, it would indeed occur in the case of relative rest of the bodies and bring about the fall of the earth into the sun. The attraction is compensated by the existing relative motions, and this would correspond to the usual conception if it would take into account the relative motions instead of operating with the phantom of absolute rotation and inertia treated correspondingly as absolute.

It is also readily seen that in accordance with our conception the motions of the bodies of the solar system can be regarded as pure *inertial motions*, whereas in accordance with the usual conception the inertial motion, or rather its gravitationally continually modified tendency, strives to produce a rectilinear tangential motion....

Berlin, January 1896.

NOTES

[1]Translated by Julian B. Barbour from: Friedlaender, Benedict and Friedlaender, Immanuel (1896). *Absolute oder Relative Bewegung?* Teil II: *Ueber das Problem der Bewegung und die Umkehrbarkeit der Centrifugalerscheinungen auf Grund der relativen Trägheit*. Berlin: Leonhard Simion, pp. 24–33. Mach refers briefly to the Friedlaenders' booklet in the editions of his *Mechanik* from 1897. See the Notes on p. 311.

COMMENTARY

As my coeditor comments (p. 315), the Friedlaenders' booklet is the first really interesting contribution to the problems of inertia and frame dragging after Mach's initial comments. The above extracts from the part by Benedict should be read in conjunction with the description of the conceptually very beautiful experiment described by Immanuel in his attempt to measure a putative Machian centrifugal force generated near the axis of a rapidly rotating flywheel (p. 309).

Besides the actual experiment, the booklet is noteworthy for two further reasons:

1) The brothers get rather closer to Mach in the actual formulation of a law of relative inertia. In fact, on the basis of simple heuristic arguments very similar to ones used repeatedly by Einstein himself, Benedict is able to show how centrifugal forces will arise in the context of a theory of relative inertia near the axis of a rapidly rotating flywheel. This seems to me to be work of high quality and a genuine technical advance, even if Benedict somewhat spoils

the achievement by his clearly unrealistic belief (footnote, p. 116) that the earth is much more important for the determination of terrestrial inertia than the distant masses of the universe. In fact, his simple intuitive arguments bear a remarkable similarity to those used by Kepler [see p. 6] in the *Astronomia Nova* (1609) to justify the use of physical forces (rather than invisible space) to explain why the planets follow such precise orbits. Even the illustrations are similar. Both investigators were groping for observable determinants of motion to replace invisible mathematical frameworks (crystal spheres and absolute space, respectively). It is also worth noting the similarity between Benedict's *qualitative* formulation of the law of relative inertia (p. 114) and the one given eight years later by Hofmann (p. 128). Like Hofmann, and unlike Mach, who despised the notions of analytical mechanics (p. 217), Benedict is looking for a 'Machian kinetic energy.' All he lacked was the final decisive step in which that energy is represented explicitly as a two-body interaction dependent solely on relative quantities (p. 108). Lastly, it is worth noting that in his somewhat bizarre attempt to explain the tides Benedict gets very close to discovering the internal-motion mechanism of generation of a gravity-type force that Reissner found in 1915 (p. 142, Note). Had Benedict included a $1/r$ distance dependence of his force of relative inertia and considered the effect of the rotating earth *on the moon*, he would have been able to predict the existence of a weak attractive force with strength proportional to the product of the *masses* of the two bodies, just as in the case of gravity. See also my comment below.

2) Unlike Mach, the Friedlaenders quite clearly anticipate Einstein in asserting that there is an intimate relationship between inertia and gravity (in 1904, Föppl actually explicitly denied any such connection, see p. 124); very significantly, they also draw attention to an analogy between their concept of relative inertia and inductive effects in electromagnetism, the possibility of which Einstein noted in 1912 (p. 180). In this connection, it is a pity that they were not just a little bit more explicit about gravitation; both brothers make some tantalizing suggestions. In the final passage translated here Benedict seems almost to anticipate Einstein's geodesic unification of gravity and inertia (cf. pp. 317–317), though he has not made any explicit claim that Newtonian gravity receives a new explanation in terms of a theory of relative inertia. One would also like to understand the significance of Immanuel's use of the expression "relative rotations" in his comments on gravity (p. 309). Did the brothers somehow have some intuition for the Reissner mechanism mentioned above, or were they, again like Einstein, interpreting gravitation in a generalized sense in which inductive forces are added to Newtonian gravity? The second possibility seems more likely.

J.B.B.

On Absolute and Relative Motion[1]

August Föppl

The most acute observations on the physical significance of the law of inertia and the related concept of absolute motion are due to Mach. According to him, in mechanics, just as in geometry, the assumption of an absolute space and, with it, an absolute motion in the strict sense is not permitted. Every motion is only comprehensible as a relative motion, and what one normally calls absolute motion is only motion relative to a reference system, a so-called inertial system, which is required by the law of inertia and has its orientation determined in accordance with some law by the masses of the universe (Weltsystem).

Most authors are today in essential agreement with this point of view, as expressed most recently by Voss[1)] and Poincaré[2)] in particular. A different standpoint is adopted by Boltzmann,[3)] who does not believe he can simply completely deny an absolute space and, with it, an absolute motion. Here, however, I shall proceed from Mach's view and attempt to add some further considerations to it.

Mach summarizes his considerations in the following sentence[4)]: "The natural standpoint for the natural scientist is still that of regarding the law of inertia provisionally as an adequate approximation, relating it in the spatial part to the heaven of fixed stars and in the time part to the rotation of the earth, and to await a correction or refinement of our knowledge from extended experience." Now it seems to me not entirely impossible that just such an extended experience could now be at hand. In a recent publication of K. R. Koch[5)] on the variation in time of the

---

1)A. Voss, Die Prinzipien der rationellen Mechanik. Enzyklop. d. math. Wissensch., Band IV, 1, p. 39 (1901).

2)H. Poincaré, Wissenschaft und Hypothese. Deutsch von F. und L. Lindemann, Leipzig (1904).

3)L. Boltzmann, Prinzipe der Mechanik, II, p. 330, Leipzig (1904).

4)E. Mach, Mechanik, 4. Aufl. p. 252, Leipzig (1901).

5)K. R. Koch, Drudes Annalen der Physik, Band 15, p. 146 (1904).

Einstein Studies, vol. 6: Mach's Principle: From Newton's Bucket to Quantum Gravity, pp. 120–127

strength of gravity we read: "Accordingly, the assumption of a genuine variation of gravity, or, more precisely, its difference between Stuttgart and Karlsruhe, seems to me appropriate." We shall naturally have to wait and see if this assertion stands up to further testing; at the least, we must now reckon with the real possibility that it is correct.

An explanation of such a phenomenon, if it is correct, would be very difficult on the basis of known causes. This circumstance encourages me to come forward now with a consideration that I have already developed earlier and long ago led me to the assumption that small periodic variations of gravity of measurable magnitude should be considered as a possibility.

Experience teaches us first that the inertial system required by the law of inertia can be taken to coincide with the heaven of the fixed stars to an accuracy adequate for practical purposes. It is also possible to choose a reference system differently, for example, fixed relative to the earth, in order to describe the phenomena of motion. However, it is then necessary to apply to every material point the Coriolis additional forces of relative motion if one is to predict the motions correctly. One can therefore say that the inertial system is distinguished from any other reference system by the fact that in it one can dispense with the adoption of the additional forces. Rectilinear uniform translation of the chosen reference system can be left out of consideration here as unimportant.

However, it is obvious that the fixing of the inertial system relative to the heaven of the fixed stars cannot be regarded as fortuitous. Rather, one must ascribe it to the influence, expressed in some manner, of the masses out of which it is composed. We can therefore pose the question of the law in accordance with which the orientation of the inertial system is determined when the instantaneous form and relative motion of the complete system of masses, i.e., the values of the individual masses, their separations, and the differential quotients of these separations with respect to the time, are regarded as given.

The logical need for such a formulation of the problem if one wishes to avoid the assumption of an absolute space was also felt by Boltzmann when he referred in passing to the possibility[6)] that the three principal axes of inertia of the complete universe could provide the required orientation. If this rather natural idea could be maintained, the conceptual difficulties would indeed be overcome. However, I believe that the proposal is not admissible. Let us imagine, for example, a universe that is otherwise arranged like ours but with the only difference

[6)]loc.cit., p. 333.

that there are no forces at all between the individual bodies in the universe. Then for the inertial system valid for this universe, all the bodies in it would move along straight lines. However, a calculation that is readily made shows us that under this assumption the principal axes of inertia of the complete system would in general execute rotations relative to the inertial system. It is therefore necessary to look for a different condition that can enable us to understand the fixing of the inertial system.

If first we assume that all the bodies of the universe are at rest relative to each other except for a single mass point that I suppose is used to test the law of inertia, and which I will call the 'test point,' then in accordance with the experiences we already have one could not doubt that the test point would, when no forces act on it, describe a straight path relative to a reference system rigidly fixed to the masses. In this case, the inertial system would be immediately fixed in space.

We can now imagine the case in which the bodies of the universe consist of two groups, one of which is 'overwhelmingly' large compared with a smaller group and in which the masses within each group do not change their relative separations, whereas the smaller group, regarded in its totality, does carry out at the considered time a motion, say a rotation, relative to the larger group. If only one of the two groups were present, the inertial system would be fixed relative to it. Since the two work together, and one of the groups has been assumed to be much more 'powerful' than the other, the inertial system will now be indeed very nearly at rest relative to the first group, but it will still execute a small motion relative to that group, which, of course, will be the consequence of the influence of the second, smaller group.

Given such a situation, what would be the most expedient way to proceed? I believe that one cannot be in doubt. One would fix the reference system exclusively using the first, overwhelming group and calculate as if this were the inertial system but take into account the influence of the second group by applying in this case to every test point the very weak additional forces of the relative motion that the chosen reference system executes relative to the true inertial system. If one makes such a decision, then these Coriolis forces no longer appear as mere computational quantities that arise from a coordinate transformation but as physically existing forces that are exerted by the masses of the smaller group on every test point and arise because these masses have a motion relative to the chosen reference system.

To develop this idea further, one could start by investigating the case in which the second, smaller group that I just mentioned is represented by a single body. One then has the task of determining the magnitude and direction of the force, which will depend on the velocities of the single body and the test point relative to the reference system determined by the remaining bodies of the universe and on the separation between the single body and the test point. If we suppose that this problem has been solved for a single body, then, using the superposition law, we can also obtain the influence of a whole group of moving bodies.

The securely established observational results that are currently available are certainly not adequate to solve this fundamental problem; however, one does not therefore need to doubt that on the basis of further observations we could arrive at a solution.

After these preliminary considerations, I now turn to the case that corresponds to reality. Using the circumstance that the constellation of the fixed stars changes little in the course of several years or centuries, we can suppose that a reference system that more or less coincides with the inertial system is fixed relative to three suitably chosen stars. However, in order to take into account the small deviations that still remain, one must suppose that to each test point there are applied Coriolis forces, which, as we have just described, are to be interpreted as forces that depend on the velocities of the individual bodies in the universe and the velocity of the test point.

We are now in the position - and on this I put considerable value - to specify a condition meeting our requirement for causality that must be satisfied by the true inertial system required by the law of inertia. Namely, the true inertial system is the reference system for which all the velocity-dependent forces that arise from the individual bodies of the universe are in balance at the test point. Even if in practice it is clear that we have not gained very much through this statement, it does appear to me that we have thereby obtained a very suitable basis for forming a clear concept of what is known as absolute motion in the framework of mechanics. There is at the least a prospect opened up of a way of determining the inertial system once the law that establishes the velocity-dependent forces has been found. In other words, it will be possible to construct the absolute space that appears in the law of inertia without having to sacrifice the notion that ultimately all motions are merely relative.

In fact, in all these considerations my main aim is to make it at least plausible that if one is to find a satisfactory solution to the questions that relate to the law of inertia it will be necessary to assume the existence of forces between the bodies in the universe that depend on their velocities relative to the inertial system. If this is accepted, then there follows the task of looking for possible phenomena whose relationship to the expected general law of nature could be such that the law governing the velocity-dependent forces could be deduced. These forces, which for brevity I shall in what follows simply call 'velocity forces,' have nothing to do with gravitational forces, which arise concurrently with them, and specifically they can – and probably will – follow a quite different law as regards the distance dependence compared with the gravitational forces.

At this point I should like to make a remark in order to divide this communication into two quite separate sections. I believe that I can defend with complete definiteness and confidence what I have said up to now. However, I regard what follows as merely an attempt that could very well fail; nevertheless, it is an attempt that at the least has a prospect of success and therefore must be brought forward at some time.

It seems to me that the most promising way of proving the existence of the postulated velocity forces and finding the law in accordance with which they act is to observe with the greatest possible accuracy phenomena associated with motions near the earth that occur with great velocity. Just as the discovery of gravitation had as its starting point the observation of free fall, here too the first step to the solution of the puzzle could be obtained through observations of terrestrial motions and their correct interpretation. The immediate vicinity of the earth's mass opens up some prospect of proving the existence of velocity forces more accurately than would be possible with the finest astronomical observations, which, as experience teaches, are certainly only very weak under normal circumstances.

This thought led me some time ago to make the gyroscope experiments that I reported to the Academy very nearly a year ago.[7)] I expected then, as I explicitly said, to establish a behavior of the gyroscope that did not agree with the usual theory in the hope that the observed deviation could be attributed to the velocity forces I seek and that these would therefore be made accessible to experimental research. Now certain indications of a deviation were indeed discernible, but as a careful and conscientious experimentalist I could not put any weight on them and I was forced, as I did, to declare a negative result of the experiment as regards the direction that it was intended to follow in the first place. In the meanwhile, I have made some further experiments with the same apparatus, though admittedly few, since they are very laborious and time consuming. However, the result could do nothing but

[7)]Sitzungsberichte 1904, p. 5.

strengthen me in the view that the accuracy that can be achieved with this experimental arrangement is not sufficient to prove the existence of the velocity forces if they exist at all.

[Most of pp. 390–392 of Föppl's paper, which considers the possible implications of what were evidently ephemeral rogue experimental results, including the observation of Koch mentioned in the beginning of the paper, is here omitted. The translation ends with Föppl's final speculations about velocity forces.]

Consider a planet that circles its central star in agreement with the first two laws of Kepler. Let the law of the velocity forces be of the form that the planet is subject to an attraction by its sun that is proportional to the velocity component orthogonal to the radius vector and inversely proportional to the first power of the distance. One immediately recognizes that under these circumstances one would not need any gravitational force in addition to the velocity force in order to explain the motion of the planet that is given by the observations. The astronomers of a solar system with only a single planet would have indeed no means to decide whether Newton's gravitational force or the velocity force adopted in the indicated manner were correct if they wished to restrict themselves to observation of the orbit alone. However, the difference would immediately be apparent when they took into account observations on their planet.

There is in accordance with Newton's gravitational law too a daily period of variation of the gravity force that gives rise to the contribution of the sun to the motion of the tides but is too weak to be established by pendulum observations. However, if the astronomers of that solar system were to make the attempt to replace Newton's law of gravitation by the law of the velocity forces that we have mentioned, they would have to expect a much greater daily period, which, for the same relationships between our earth and the sun, would be about 180 times greater than would be expected in the other case.

It should also be remarked here that the velocity law, which was chosen at random, is in fact only one of infinitely many that would all achieve the same, namely, the explanation of the motion of a single planet around its sun in agreement with Kepler's first two laws without having to invoke in addition Newton's gravitational force. All one needs to do is to allow the velocity component in the direction of the radius vector, which was hitherto assumed to be without influence, to participate as well in accordance with some arbitrary law and then arrange the law according to which the orthogonal velocity component acts on the force of attraction in such a way that the required motion results. There is also no need to make a restriction to the first power of the velocity; one could also consider the second or other powers.

When a solar system has more than one planet, it is naturally much more difficult to explain all the planetary orbits merely with the help of velocity forces, since it is now necessary to satisfy Kepler's third law as well. So far as I can see, one would then be forced to make decidedly artificial assumptions. Even if one could achieve success in a simpler manner than it now appears to me, it would still be questionable if one could also explain the disturbances of the planetary orbits, the motion of the moon, etc.

However, one should not forget the aim of this discussion. It is in no way my intention to replace Newton's law by a law of velocity forces. I only want to make it plausible that under certain circumstances the velocity forces by themselves could have effects very similar to those of the gravitational forces. If this is then granted, it immediately follows that in such an event it would be very difficult to separate out from the astronomical observations the part due, on the one hand, to gravitational forces and, on the other, to the velocity forces.

On the basis of this consideration, I believe it is best not to be deflected by the admittedly very weighty objections of the astronomers from seeking phenomena that could be related to velocity forces. If it does prove possible, following this entirely independent research approach, to derive a law of the velocity forces, it will still be possible to make, as the best test of the admissibility of the result, an accurate comparison with the astronomical observations, taking into account the error limits that are relevant.

Naturally, I would not recommend such a procedure if I did not have great confidence in the very existence of the velocity forces, even though I must leave it as an open question whether they have a magnitude such that they are measurable in motions accessible to our perception. If one will admit an absolute space, then, of course, every ground for the adoption of velocity forces disappears. However, in this point at least – that I do not recognize an absolute space – I am in agreement with the majority of natural scientists, and I therefore hope that I shall receive recognition among them, at least for the conclusions drawn in the first part of this communication.

NOTES

[1]First published in *Sitzungsberichte der Bayerischen Akademie der Wissenschaften, mathematisch-physikalische Klasse* (1904) 34: 383–395 (submitted November 5, 1904). Translated by Julian B. Barbour. There is an extended and very positive critique of both of Föppl's papers (this and the one from which we give a partial translation on p. 312) in a supplement (*Anhang*) to Mach's 1908 edition of the *Mechanik*. In the 1912 edition, only the brief mention on p. 111 is made.

COMMENTARY

The main interest of this paper is that it represents a clear formulation of the Mach problem by a scientist of high standing (cf. Norton's comments, p. 34 and p. 50, Note 8) made at a time (1904) before the explosion of Einstein and special relativity onto the scene – and all the complications both introduced. Several details should be noted: 1) There is no suggestion that there is a need for some cosmic derivation of inertial mass (Einstein's red herring, p. 91–92). 2) The problem is seen entirely as that of showing how inertial frames of reference, whose effective existence is demonstrated by dynamics, are determined at a given point "in accordance with some law by the masses of the universe" (p. 120, first paragraph). 3) Föppl states explicitly the physical quantities that must enter the law "in accordance with which the orientation of the inertial system is determined"; they are: "the individual masses [of the complete system of masses], their separations, and the differential quotients of these separations with respect to the time" (p. 121, penultimate paragraph). The essential identity of this listing and Poincaré's formulation in *Science and Hypothesis* of the problem of predictability of classical dynamics and of stating the relativity principle in a form with which the mind can be truly satisfied is noteworthy (it will be seen that Föppl cites Poincaré at the start of his paper). It is obvious that the ability to predict the future uniquely from purely relative instantaneous data, as required by Poincaré, will bring with it the ability to determine the inertial frames of reference, as required by Föppl. A further point of interest in this connection is the quotation from Mach given by Föppl at the bottom of p. 120. I have not been able to find this in the English translation of the *Mechanics* published by Open Court in 1960. The quotation makes clear that Mach was solely concerned with the *law of inertia*, not *inertial mass*, and that he definitely saw the problem as consisting of two parts, a *spatial* part and a *time* part (cf. the two Machian requirements, p. 92 and p. 102ff).

Otherwise the paper is somewhat disappointing; unlike Hofmann, Föppl is hesitant to attack the problem head on (cf. the commentary on Hofmann's paper, p. 133), and, to me at least, several of his suggestions seem rather unphysical, especially in the final part.

J.B.B.

Motion and Inertia[1]

Wenzel Hofmann

The hitherto existing concept of inertia is regarded as *absolute* because it is defined without any reference to any other body apart from the one that is actually being considered, but I cannot accept this absolute character; much rather, I am of the opinion that the concept of inertia, like the concept of motion, is to be regarded as exclusively *relative*.

Namely, a body can be in a state of rest or motion only with respect to some other body. If we now say that the considered body has the tendency to maintain its state, this cannot mean anything other than that it strives to remain in rest or motion relative to that second body. Thus, the inertia of the considered body consists of a relationship between it and the body with respect to which the state of rest or motion has been established.

In order to explain my ideas on this matter through an example, I suppose that in infinite space there is nothing else apart from two material points A and B that steadily move away from each other. It is then obvious that for these two points the law of inertia cannot be formulated in any other way than that the two bodies have a tendency to maintain their relative motion.

I must now be able to assert that the inertia of A is expressed in the fact that it continually increases its distance from B and, conversely, B has the tendency to increase its distance from A.

Considered in this way, we find that the inertia of each of the two points consists of a relation to the other.

If I now imagine several material points A, B, C, etc., then, for example, A will want to follow the inertial tendency with respect to each of the other points; the behavior of point A must then be established as the resultant of the individual inertial tendencies.

This consideration can then be extended to whole groups of material points, and to bodies. I must therefore be allowed to make the following statement:

Einstein Studies, vol. 6: Mach's Principle: From Newton's Bucket to Quantum Gravity, pp. 128–133 © 1995 *Birkhäuser Boston, Inc. Printed in the United States.*

Every body is subject to the law of conservation of its relative state of motion or rest with respect to all the other bodies in space; its actual behavior is then the resultant of all the individual influences.

However, the extent to which this resultant depends on the mutual separations, the sizes of the various bodies, and their mutual disposition can hardly be found by speculation but only on the basis of experience (perhaps through experiments made specifically for this purpose). However, if one is committed to the view that the inertia of a mass is to be regarded as relative, then one must certainly abandon rectilinear inertia, since then the orbit of the considered body will depend on so many varying conditions.

I should only like to emphasize once more that, in my opinion, inertial effects are relationships of the masses to each other, and that therefore these relationships occur mutually and, like the mutual motions of masses, exhibit the character of reciprocity, that is, A expresses its inertia relative to B in the same way as B with respect to A.

However, a characteristic difference in establishing the relative behavior of bodies in respect of motion or inertia is that when one establishes a relative motion of a point this can be referred to 'one' arbitrary, freely chosen reference system, whereas when one is observing the inertia of a body one must always take into account the simultaneous influence of all the other masses in space.

It is now an exceptionally broad, but also difficult undertaking to search out the laws in accordance with which the masses execute their mutual inertial tendencies. The greatest difficulty is certainly the fact that in any experiments one might set up the observed mass can never be considered in its dependence on a single mass but always with respect to the totality of all existing masses.

It cannot enter my head to want to develop here a complete theory of inertia; that cannot be the work of a single person done in a few weeks.

What should be done here is draw attention to the inadequacy of the law of inertia currently regarded as valid and simultaneously give a stimulus and indication of the sense in which possible studies aimed at a better foundation of the law of inertia should be made.

However, some basic principles can already be established.

To this end, I return once more to the example in which I assumed the existence in infinite space of just two masses in relative motion that they strive to maintain as a result of inertia. If I choose mass A as reference system, then B must exhibit inertia relative to mass A, and conversely the law of inertia must be valid for mass A with B as reference system. However, in each instant the relative position of the two bodies must be the same in both cases whatever reference system we take as basis of the observation, since the behavior of the two masses cannot depend on the point of view that an observer associates with a phenomenon.

The reciprocity of the inertial effects of two masses on each other can also be illuminated from another standpoint. By virtue of its inertia, every body that is in motion has the capacity to do work; we call the corresponding quantity its *vis viva* (lebendige Kraft).

The *vis viva* of a moving mass is therefore an inertial phenomenon, and it is therefore natural to examine the law of inertia from this point of view too.

To this end, I will suppose that there are in space nothing else but two unequal masses M and m which are in a state in which they approach each other, doing so, in fact, in such a way that the distance between them is reduced by the amount v in the unit of time.

If I take M as the reference system, then m is the mass in motion and in the system M it can exhibit a certain *vis viva*, i.e., it can do a certain amount of work.

However, if I change my point of view and choose m as the reference system, then the mass M has a *vis viva* with respect to m.

It is now very interesting to pose this question: What is the relationship of these two amounts of *vis viva* that are acquired by the different masses that have the same velocities (the velocity in each of the two cases is equal v).

In accordance with the familiar expression $L=mv^2/2$, we should have to say that the greater mass generates the greater *vis viva*, since the changing of the reference system does not change the velocity.

But that is not the case, as we shall see from the following consideration: Suppose that the two masses M and m finally collide as a result of the mutual approach to each other; then the work capacity of these two masses can be actually realized. Let us suppose that at the point of impact an instrument is set up that consumes the existing kinetic energies and simultaneously records them, for example, an elastic spring that is compressed by the two masses and frozen in this state; then the instrument gives directly (for example, in the tension achieved in the spring) the measure of the work that has been done.

If we first allow the mass M as the reference system, then the energy stored in the spring is the work done by m in the system M.

However, in the other case, namely, when we regard m as the reference system and M as the mass that is in motion, the same energy of the stressed spring is to be regarded as the measure of the work done by M in the system m. Therefore, the amounts of *vis viva* achieved by the unequal masses M and m are the same.

From this we draw an important conclusion: "If two mass systems M and m are in relative motion, the vis viva of M relative to m is equal to the vis viva of m relative to M.

Since the validity of this law is independent of the magnitude of both masses M and m, it must also be true when, for example, $M=\infty$; that is, the *vis viva* that an arbitrary body can exert through its motion relative to all masses situated outside it is equal to the *vis viva* with which these bodies can act on the first one if it is chosen as the reference system.

From all the foregoing, we can conclude that all inertial phenomena are to be traced to the mutual relationships of the masses to each other, so that the effects of inertia that are achieved are independent of the mass that is chosen as the reference system.

I should like to give this law the name *reciprocity of inertia*.

The equation $L=mv^2/2$ appears to be in contradiction to these discussions; however, this is resolved by the following consideration:

The inertial effect of a mass M relative to another mass m is a function of both masses; the expression of the amounts of *vis viva* that they exhibit relative to each other must therefore contain both masses.

Let us call the *vis viva* that two mass units possessing a relative velocity equal to the length unit can exert on each other k; then the *vis viva* associated with the two masses M and m that have relative velocity v can be expressed by the equation $L=k \cdot M \cdot m \cdot v^2$, where I take it as proven that the velocity exerts its influence in the quadratic relationship.

In addition, in setting up this equation we have made no allowance for a possible, indeed probable influence of the separation r of the two masses. If however such an influence could be established by experiments, the equation would then read: $L=k \cdot M \cdot m \cdot f(r) \cdot v^2$.

[half a page omitted]

The conclusions that can be drawn from this principle concerning the phenomenon of centrifugal force in rotating masses are particularly interesting.

If we suppose some rotating body K, then we must regard the experimentally established phenomenon of centrifugal force present in it as an inertial relationship of the rotating mass relative to all the masses outside the considered body that do not take part in the rotation.

Then in accordance with the principles that we have developed earlier, the same inertial tendencies must occur if I regard the first body

K as the reference system and, as a result, suppose that all the other masses are rotating around it [cf. Mach, p. 109].

[11 pages omitted]

If we apply this view in the case of a pendulum oscillating at the pole, we must draw the following conclusion: The pendulum must exhibit inertia simultaneously with respect to the earth, the sun, and all the remaining heavenly bodies; the overwhelming influence is surely to be ascribed to the infinitely great mass of the heaven of the fixed stars, while the inertial influence of the sun and the earth must be regarded as subsidiary. Nevertheless, it is to be assumed that the mutual separation of the masses does have an influence, and therefore even the sun and earth will not be entirely without effect, since they are at a shorter distance from the swinging pendulum. The various inertial influences to which the pendulum is subject must be expressed in such a way that the pendulum, when subject to the influence of the mass of the earth alone, must exhibit an unchanged position relative to the earth; the influence of the mass of the sun, however, ought to cause the plane of the oscillations to rotate once in a solar day, while the inertial effects of the remaining heaven of the fixed stars ought to cause a complete rotation of the plane of the pendulum already within a sidereal day. Under these circumstances, it is natural to assume that both the sun and the earth ought to have a retarding effect on the rotation of the plane of the pendulum. Therefore, we should expect the complete rotation of the plane of the pendulum to require a time that is somewhat greater than one sidereal day.

It would therefore be very interesting to consider the Foucault pendulum experiment from this point of view in order to establish experimentally the influence exerted by the earth, sun, and the other masses on the pendulum.

Indeed, we can go further; it is a small step from this to wish to learn the effect of smaller terrestrial masses on the freely swinging pendulum.

To this end, what one should do is set as large a mass as possible in the most rapid possible rotation underneath a Foucault pendulum; the rapid rotation could then to a certain degree paralyze the overwhelming influence of the mass of the earth and the remaining celestial bodies.

If it proved possible in this way to change the rate of rotation of the plane of oscillation of the Foucault pendulum, this would not only be a proof of 'relative' inertia but also be a means to establish experimentally the extent to which these inertial influences depend on the magnitude of the masses, their mutual separations, and their velocities.

NOTES

[1]Partial translation from p. 26 to the end (with breaks) of: Hofmann, Wenzel (1904). *Kritische Beleuchtung der beiden Grundbegriffe der Mechanik: Bewegung und Trägheit und daraus gezogene Folgerungen betreffs der Achsendrehung der Erde und des Foucault'schen Pendelversuchs*. Vienna and Leipzig: M. Kuppitsch Wwe. Translated by Julian B. Barbour. The German original makes very extensive use of wide-spaced type for emphasis which has not been reproduced. Hofmann is described as a K. K. Professor (K. K. ≡ Kaiserlich und Königlich). Einstein refers to him as a mathematician who developed his ideas independently of Mach (see Norton's comments, pp. 32–34). Because of space limitations, we have not been able to include Hofmann's discussion (on pp. 32–33 of his booklet) of Newton's bucket experiment, in which, like Mach (p. 109), he comments that a very much larger mass than Newton's bucket might well have an effect on the motion of the water: There is also a proposal for a Friedlaender-type experiment (p. 309).

COMMENTARY

Although, as Norton comments (p. 32), Hofmann's booklet is *very* wordy (the five pages translated here are about an eighth of the total), the above passage is noteworthy for the clarity and simplicity of its argument and for the fact that Hofmann perfectly anticipated the later work and motivation of no less a person than Schrödinger (p. 147) (and Reissner, p. 134). His intuition for the heart of the problem seems to be surer than Föppl's (p. 120), who was writing at the same time (1904). Unlike Föppl, with his provisional assumption of velocity forces manifested with respect to a not quite perfectly determined inertial frame of reference (p. 122), Hofmann goes straight for effects that depend directly on purely relative quantities.

There is a striking similarity between Mach's "... the mean acceleration of the mass μ with respect to the masses m, m', m'', ... at the distances r, r', r'' .... is $=0$, or $d^2(\Sigma mr/\Sigma m)/dt^2=0$" (p. 110) and Hofmann's central conclusion in italics on p. 129. However, whereas Mach bungled the mathematics, writing down a scalar equation where a vector equation was needed, Hofmann went on correctly to write down a scalar Machian kinetic energy, from which vector equations of motion will follow. So far as I know, Hofmann was also the first person to state clearly that in a relational theory of inertia the kinetic energy cannot be a sum of contributions of individual masses but must be a sum over *products* of all possible pairs of masses (pp. 107–108). In fact, it seems clear that Hofmann's is the earliest known implementation of the Machian idea in a physically and mathematically transparent form. Even Poincaré failed to achieve that, despite having correctly formulated the problem (pp. 111–112) two years earlier in 1902. *Note added in proof.* See Note 2 on p. 230.

J.B.B.

On the Relativity of Accelerations in Mechanics[1]

Hans Reissner

The relativity postulate can, so far as I can see, be extended for accelerated states of motion in two directions, namely, through the two following essentially different requirements:

1. The complete equivalence that holds in Newtonian mechanics between external forces (in particular gravitational forces) and inertial forces is also to be implemented for electromagnetic–optical and thermodynamic phenomena (Einstein's equivalence hypothesis).

2. It is to be required that not only absolute velocity but also any absolute motion whatever, in particular acceleration, must be undetectable. This requirement was already formulated by Mach, but it has not yet been carried out; much rather, modern researchers have repeatedly denied the justification of such a postulate.

In contrast to this view, I wish to show now that the implementation of this last postulate can indeed be arranged very easily; it is true that the foundations of Newtonian mechanics are changed, but its consequences are not greatly affected.

For this first demonstration, only mass points will be considered, although it will later be necessary to support the results by a limiting process from a continuum.

It will also be assumed that all velocities are small compared with the velocity of light, so that questions such as the speed of propagation of the interaction of masses through empty space will not arise.

I proceed now to the establishment of a mechanics of relative accelerations as follows:

It has no meaning to speak of the acceleration or kinetic energy of an independent mass point, but the following statement does have meaning.

The kinetic energy in relative-acceleration form. Two points with gravitating masses m_1 and m_2 separated by the distance r possess a kinetic

Einstein Studies, vol. 6: Mach's Principle: From Newton's Bucket to Quantum Gravity, pp. 134–146

energy that is proportional to the product of the two masses and, further, is a quadratic function of their relative velocity $\dot{r}$, and may also contain the separation r. Thus

$$T = m_1 m_2 \dot{r}^2 f(r). \tag{1}$$

It is then to be required that the energy theorem holds for this isolated system of two mass points, namely:

$$T - U = \text{const}, \tag{2}$$

where $-U$ is the potential energy of the gravitation of these two masses, which are conceived as being alone in the world, and

$$U = \gamma \frac{m_1 m_2}{r}. \tag{3}$$

This assumption already contains the statement that a single mass point possesses neither potential nor kinetic energy, namely, when m_1 or $m_2 = 0$ holds.

The energy theorem (2) now yields

$$\dot{r}^2 f(r) - \gamma r^{-1} = \text{const}, \quad \dot{r}^2 = \frac{\gamma}{r f(r)} + \frac{\text{const}}{f(r)}.$$

Now since the velocity $\dot{r}$ must be able to take any arbitrary finite prescribed value for an infinitely large separation of the two masses, $f(r)$ must either tend asymptotically to a constant at infinity or simply be a constant. For the moment, there is no reason not to make this last simplest assumption, and we therefore set

$$T = \delta m_1 m_2 \dot{r}^2$$

and in accordance with the energy theorem

$$\dot{r}^2 = \frac{\gamma}{\delta r} + \frac{\text{const}}{\delta}.$$

Thus $\dot{r}_\infty = (\text{const}/\delta)^{1/2}$ is the relative velocity that the two masses would have when separated by a great distance for the realized initial condition (the value of the constant).

Thus, the total energy of the two masses is here divided into a potential energy and a kinetic energy. Whether and how these energy forms can, as in relativity theory, be unified at velocities that are of the same order as the velocity of light can hardly be established purely mechanically.

The force in relative-acceleration form. The equation of motion of two mass points that exist alone is now to be derived in the Lagrangian form from the expression (2) for the energy:

$$\frac{d}{dt}\left[\frac{\partial}{\partial \dot{r}}(T-U)\right]-\frac{\partial}{\partial r}(T-U)=0.$$

This expression can be decomposed into the contribution of the gravitational force,

$$K_g=\frac{\partial U}{\partial r},$$

and the contribution of the inertial force:

$$K_i=\frac{d}{dt}\left(\frac{dT}{d\dot{r}}\right)-\frac{\partial T}{\partial r}=2\delta m_1 m_2 \ddot{r}. \quad (4)$$

This law of motion, which is possible only if at least two masses are present, thus states the following:

Two mass points (when they exist alone in the world) oppose a relative acceleration $\ddot{r}$ with a resistance of magnitude $2\delta m_1 m_2 \ddot{r}$.

In contrast, there is no sense in speaking of an acceleration of these points in any other direction.

If one wishes, one may also say that a force cannot arise in any other direction.

Finally, one can say that the space of two mass points has one dimension.

Space first becomes multidimensional through the addition of other mass points; in other words, accelerations and velocities in a direction different from the line joining the first two mass points first become detectable when further masses are present.

In order to proceed further, let us now make the simplest assumption that when further mass points are present not only the gravitational forces but also the inertial forces are added geometrically.

From this assumption we must then be able to deduce universal inertia of masses and, in particular, Newton's law of motion and inertia in a form which shows that Newton's law is an extraordinarily good approximation of the relative-acceleration law of motion and inertia.

To set up such a law, a coordinate system must now be chosen that in some way is fixed relative to the mass points that are present, which incidentally are assumed to be very numerous. The following question must then be answered:

Does there exist in the relative-acceleration mechanics assumed above a coordinate system for which the Newtonian law of motion, which has hitherto been regarded as absolute, holds with a sufficiently good approximation?

The resultant inertial force in the direction of the X axis that acts on a mass point of mass m_1 with coordinates x_1, y_1, z_1 as a consequence of the action of the remaining masses $m_2, \ldots, m_n, \ldots, m_r$ has in accordance with our assumption the value

$$X = -2m_1\delta\sum m_n \ddot{r}_{1n}\frac{x_n - x_1}{r_{1n}}. \tag{4a}$$

If we express r_{1n} and $\ddot{r}_{1n}$ in terms of the coordinates, we obtain

$$\ddot{r}_{n1} = \frac{1}{r_{n1}}\left[(\dot{x}_n - \dot{x}_1)^2 + \cdots + (x_n - x_1)(\ddot{x}_n - \ddot{x}_1) + \cdots\right]$$
$$-\frac{1}{r_{n1}^3}\left[(x_n - x_1)(\dot{x}_n - \dot{x}_1) + \cdots\right]^2.$$

If we place the origin of the coordinate system for the considered instant coincident with m_1 with respect to position and velocity and separate this expression into the part that contains the factor $\ddot{x}_1$ and the part that contains all the remaining terms, we obtain the inertial force in the form

$$X = 2m_1\ddot{x}_1\delta\sum m_n\frac{x_n^2}{r_{n1}^2} + 2m_1\ddot{y}_1\delta\sum m_n\frac{x_n y_n}{r_{n1}^2}$$
$$+ 2m_1\ddot{z}_1\delta\sum m_n\frac{x_n z_n}{r_{n1}^2} - 2m_1\delta\sum m_n\frac{\ddot{r}_{n0}x_n}{r_{n1}}. \tag{4b}$$

We now have a coordinate system whose origin coincides as regards position and velocity with point 1 but with respect to it has accelerations $\ddot{x}_1$, $\ddot{y}_1$, $\ddot{z}_1$; $\ddot{r}_{n0}$ is now to be the acceleration of separation of the point n relative to this coordinate origin.

The usual Newtonian equations of motion then hold with respect to the three axes if the following equations hold exactly or with sufficiently good accuracy:

$$\sum m_n\frac{x_n y_n}{r_{n1}^2} = \sum m_n\frac{x_n z_n}{r_{n1}^2} = \sum m_n\frac{y_n z_n}{r_{n1}^2} = 0, \tag{5}$$

$$\delta\sum m_n\frac{x_n^2}{r_{1n}^2} = \delta\sum m_n\frac{y_n^2}{r_{n1}^2} = \delta\sum m_n\frac{z_n^2}{r_n^2} = -\tfrac{1}{2}, \tag{6}$$

$$\sum m_n\frac{\ddot{r}_{n0}x_n}{r_{n1}} = \sum m_n\frac{\ddot{r}_{n0}y_n}{r_{n1}} = \sum m_n\frac{\ddot{r}_{n0}z_n}{r_{n1}} = 0. \tag{7}$$

Equations (5) will hold when the mass distribution of the radiating and nonradiating celestial bodies possesses a certain polar symmetry.

Equations (6) give the meaning of the constant δ if one sets

$$\sum \frac{m_n x_n^2}{r_{n1}^2} = \frac{1}{3}\sum m_n \frac{x_n^2 + y_n^2 + z_n^2}{r_{n1}^2} = \frac{1}{3}\sum m_n$$

in the relation

$$-2\delta = \frac{1}{\frac{1}{3}\sum m_n},$$

and let us see that in accordance with these assumptions proportionality of inertial and gravitational mass is guaranteed.

The smallness of the constant δ of the law of interaction of inertia shows that although two masses do have a resistance to mutual acceleration this effect is immeasurably small compared with the inertial force of all the remaining masses.

On the other hand, the value $-2\delta = 3/\Sigma m$ shows that Eq. (4) does not represent an elementary law since in the interaction of any two masses all the remaining masses have an influence on δ. In contrast, the Newtonian gravitational law, with a fixed value of the gravitational constant γ that is independent of the presence of the remaining masses, appears as a true elementary law.

Admittedly, this behavior can also be inverted. For example, if the masses are introduced in gravitational units, the kinetic energy must be written in the form

$$T = \frac{\delta}{\gamma} m_1 m_2 \dot{r}^2,$$

and one would then obtain

$$-2\delta = \frac{3}{\sum m}.$$

One could then regard δ as a constant that is independent of the number of masses and γ as dependent on Σm.

The treatment of the inertia of masses presented above need not differ greatly in its consequences from the Newtonian behavior, but it does meet the Machian requirement of the undetectability of acceleration relative to empty space; in particular, it gives the following answer to the objection of the supporters of absolute mechanics that the centrifugal forces associated with the rotation of bodies prove absolute rotation:

For a start, these centrifugal forces correspond to equally great and opposite centrifugal forces that are distributed over all the remaining masses of the world, this happening in such a way that exactly the same forces and counterforces would arise if we were to regard the rotating body as in rest and the heaven of the fixed stars as rotating. In accordance with the assumptions made above, there is no difference between the two representations of the motion [cf. p. 109].

Significant centrifugal forces arise whenever one body rotates relative to a certain average position of all the other bodies. It then exerts on all the other bodies centripetal forces, but because these are distributed over all the masses of the world they cannot be noted.

The three equations (7) determine for every point of space and every instant the behavior of three inertial planes relative to the masses of the world. In fact, strictly speaking the behavior is different for each point and for each instant, since the masses in the world are accelerated among each other.

However, it will certainly follow from astronomical statistical considerations that the spatial and temporal variations will be undetectable for spaces and times that are very large on the terrestrial scale.

In order to say more about the determination of the relative-acceleration inertial system, it will be necessary to investigate the inertial torques exerted on a small body in accordance with the above assumptions and compare them with Euler's gyroscope equations of absolute-acceleration mechanics.

The kinetic energy of a mass point and the line element of space. The law of inertia for a mass point can be obtained by setting the inertial force K_i equal to zero. However, it is more convenient to use here the conservation law for the energy.

In accordance with Eq. (2), the energy theorem gives

$$T-U=\text{const},\quad \sum\sum m_r m_s(\delta\dot{r}_{rs}^2-\gamma r_{rs}^{-1})=\text{const}$$

and is now to be used to give a relation for the mass point m_1. To this end, we write the last equation as follows:

$$m_1\sum m_s\left[\delta\dot{r}_{1s}^2-\frac{\gamma}{r_{1s}}\right]+\sum_2\sum_2 m_r m_s\left[\delta\dot{r}_{rs}^2-\frac{\gamma}{r_{rs}}\right]=\text{const},$$

where the subscript 1 no longer occurs in the second sum.

Since we wish to obtain the effect of inertia alone, without gravitation, on the mass point m_1, we may omit the summand γr_{1s}^{-1} in the first term and obtain

$$m_1\delta\sum m_s\dot{r}_{1s}^2=-\sum\sum m_r m_s\left[\delta\dot{r}_{rs}^2-\frac{\gamma}{r_{rs}}\right]+\text{const}.$$

The first term on the right-hand side is the total energy (gravitational and kinetic) of all the other masses and may be set equal to a constant because of the slight significance of the particle m_1 for the remaining world.

If, further, we express r_{1s} in Cartesian coordinates, the above equation becomes

$$m_1\delta\sum m_s\frac{[(x_s-x_1)(\dot{x}_s-\dot{x}_1)+(y_s-y_1)(\dot{y}_s-\dot{y}_1)+(z_s-z_1)(\dot{z}_s-\dot{z}_1)]^2}{r_{s1}}=\text{const},$$

or, after $\dot{x}_1^2$, $\dot{x}_2$, $\dot{y}_1^2$, etc., have been taken in front of the corresponding sums,

$$m_1\delta\left[\dot{x}_1^2\sum\frac{m_s(x_s-x_1)^2}{r_{s1}^2}+\dot{y}_1^2\sum\frac{m_s(y_s-y_1)^2}{r_{s1}^2}+\dot{z}_1^2\sum\frac{m_s(z_s-z_1)^2}{r_{s1}^2}+\right.$$

$$+2\dot{x}_1\dot{y}_1\sum m_s\frac{(x_s-x_1)(y_s-y_1)}{r_{s1}^2}+2\dot{x}_1\dot{z}_1\sum m_s\frac{(x_s-x_1)(z_s-z_1)}{r_{s1}^2}+$$

$$+2\dot{y}_1\dot{z}_1\sum m_s\frac{(y_s-y_1)(z_s-z_1)}{r_{s1}^2}$$

$$-2\dot{x}_1\sum m_s\frac{\dot{x}_s(x_s-x_1)^2+\dot{y}_s(y_s-y_1)(x_s-x_1)+\dot{z}_s(z_s-z_1)(x_s-x_1)}{r_{s1}^2}$$

$$-2\dot{y}_1\sum m_s\frac{\dot{y}_s(y_s-y_1)^2+\dot{z}_s(z_s-z_1)(y_s-y_1)+\dot{x}_s(x_s-x_1)(y_s-y_1)}{r_{s1}^2}$$

$$\left.-2\dot{z}_1\sum m_s\frac{\dot{z}_s(z_s-z_1)^2+\dot{x}_s(x_s-x_1)(z_s-z_1)+\dot{y}_s(y_s-y_1)(z_s-z_1)}{r_{s1}^2}\right]. \tag{8}$$

If this expression is multiplied by the square of the time element $(dt)^2$, a homogeneous quadratic function in the four quantities dx, dy, dz, dt arises with coefficients g_{xy}, g_{xt}, etc., that are functions of position, and one sees that this function has the same nature as the line element of the Einstein–Grossmann gravitational space. This expression only goes over into the expression of Newtonian mechanics, $m_1(ds/dt)^2=\text{const}$, when, as in Eq. (6),

$$\delta\sum m_s\frac{(x_s-x_0)^2}{r_{s0}^2}=\delta\sum m_s\frac{(y_s-y_0)^2}{r_{s0}^2}=\delta\sum m_s\frac{(z_s-z_0)^2}{r_{s0}^2}=\text{const}$$

and the remaining factors are set equal to zero.

This last will, for statistical reasons, be true to a very good approximation relative to the first factors, but nevertheless it will not hold exactly.

One can now either, as has up to now been done here, attribute these deviations in the motion of the mass point from the motion of absolute-acceleration mechanics to the inertial influence of the remaining masses, or one can blame this deviating behavior on the geometrical structure of empty space.

Indeed, H. A. Lorentz similarly attributed the contraction of bodies and the change in the rate of clocks in the case of a change in velocity to ponderomotive electromagnetic forces, whereas Einstein and Minkowski made the properties of the space–time world responsible for them.

It seems to me that Einstein in the assumption[1)] for the line element of the gravitational space had the intention also in the future for acceleration-relative physics to express the influence of the gravitational field through purely geometrical properties of the space–time world and that my assumption (8) above for the kinetic energy of a mass point, which agrees in its form[1)] with Einstein's Hamilton function H and the line element ds of space of the generalized relativity theory and gives the coefficients $g_{\mu\nu}$ as functions of the coordinates, provides a first physical example for such a non-Euclidean line element.

If the expression (8) is regarded as a line element, the isotropy of space must then indeed be given up, but the law of inertia of uniform motion in the shortest path in the absence of forces that has hitherto been used remains valid.

It would then appear that space and its geometrical properties are first created by the presence of masses and that the coefficients of the resulting velocity (8), $g_{\mu\nu}$ in the case of Einstein, prescribe the nature of the measurement of length and time at each point.

The expressions given above contain in their coefficients the distances of distant mass points and therefore for the time being are based on the notion of action at a distance.

It now clearly appears to be desirable that the inertia of masses should also be represented from the point of view of a field effect without explicit knowledge of the position and relative motion of the distant masses. For this it would be necessary that the coefficients of the *vis viva* (8) (the $g_{\mu\nu}$ in Einstein's case) should be represented as integrals of differential equations (generalized Laplace equations). However that may be, it is certainly not the gravitational potential alone that must be considered here, since the coefficients of the form $\Sigma m(xy/r^2)$ cannot be derived from the potential $\Sigma(m/r)$, as follows from the fact that for one and the same gravitational potential $\Sigma(m/r)$ very different values of

[1)] A. Einstein und M. Grossmann, Entwurf einer verallgemeinerten Relativitätstheorie und einer Theorie der Gravitation, 1913 (Teubner), p.7.

$\Sigma m(xy/r^2)$ can be constructed by suitable mass distributions.

According to Einstein, they should not be derivable from a scalar function at all.

That which we have attempted to formulate here for the inertial forces of ponderable masses will certainly also have to be required for all forces, for example, for the ponderomotive forces that act on electrical charges. Thus, it will also be necessary to express Maxwell's electromagnetic equations in such a form that a single electron experiences no effects as a consequence of its own field alone but only in the presence of either other electrical charges or ponderable masses. Was this perhaps also Einstein's desired collateral result when he put the speed of light equal to the gravitational potential up to a constant? In the case of an isolated electric particle, this constant would then have to be able to become zero, so that in the absence of gravitating masses the Maxwell equations would have no content anymore.

The outline presented here represents, of course, only the part of the relativity of acceleration that deals with mechanics alone. The further postulates that absolute acceleration should also not be detectable by optical, electrical, thermodynamic, or elastic means and also that inertial forces should not be distinguishable from gravitational forces through these last means are what will make this extension of the theory of relativity truly fruitful – with a success that in my opinion must be awaited with the most eager expectation.

Charlottenburg

NOTES

[1] Translated (with corrections of some obvious errors and misprints) from *Physikalische Zeitschrift* (1914) 15: 371–375 (submitted March 21, 1914). Translated by Julian B. Barbour. A further paper by Reissner, "On a Possibility of Deriving Gravitation as a Direct Consequence of the Relativity of Inertia," appeared in *Physikalische Zeitschrift* (1915) 16: 179–185 (submitted April 1, 1914). In it Reissner specialized the general Hofmann–Reissner expression (1) for the Machian kinetic energy to the form $\frac{1}{2}m_1m_2\dot{r}^2/r$ and showed that in such a theory microscopic internal motion of bodies gives rise to a gravitational-type attraction. Because of shortage of space, we publish below a translation of only a few selected passages from the second paper that deal with the general Machian program. A complete translation may appear in the book mentioned on p. 5. Reissner's idea for the generation of gravity from Machian inertia was rediscovered in: Barbour, Julian B. (1975). "Forceless Machian Dynamics." *Nuovo Cimento* 26B: 16–22. Essentially the same idea was rediscovered independently in: Cook, R. J. (1976). *Nuovo Cimento* 35B: 25–34.

Selected Passages from Reissner's 1915 Paper

From p. 179: Acceleration could be referred to an absolute space as long as one could take a nonmoving luminiferous ether as reference system. Nevertheless, Mach's *Mechanics* already declared in 1883 that the idea that a distinguished frame of reference independent of material processes could exist was absurd and gave some hints of a conception of acceleration relative to space as being an average relative to all the remaining masses.[1)]

In particular, Mach already considered the argument of the supporters of absolute mechanics that it is permitted to speak of absolute centripetal accelerations because they are detectable through centrifugal forces, and he pointed out that centrifugal forces are only observed in rotating systems that have a very small extension compared with the heaven of the fixed stars.

However, quite recently Abraham and Mie advanced against Einstein's requirement of covariance of the physical laws with respect to arbitrary transformations of the frame of reference the fact that such covariance contradicts the observed inertial forces.[2)]

Only recently, after I had in the meanwhile made clear the possibility of a relative-acceleration mechanics in a specific case, has Abraham withdrawn his fundamental objection.

In the paper to which I am referring, I have, apparently for the first time, stated and quantitatively formulated the fact that relativity of acceleration can be implemented only when the centrifugal forces of a rotating body correspond to centripetal forces of all the remaining masses, so that there is no dynamical difference between rotation of the body with respect to all the remaining masses or rotation of all the remaining masses relative to the body.

---

1)E. Mach, Die Mechanik in ihrer Entwicklung, 6, Aufl., 1908, pp.250-253.

2)Discussion contribution of G. Mie to Einstein's lecture published in this journal (14: 1264 (1913)); Abraham, Die neue Mechanik, Scientia Jan. 1914, Sur le problème de la relativité, Juli 1914.

From p. 181: It is also to be noted that the mass [in the theory with two-body Machian kinetic energy $\frac{1}{2}\mu_s\mu_t\dot{r}_{st}^2/r_{st}$]

$$m_t = \frac{\mu_t}{3}\sum \mu_s r_{st}^{-1}$$

of a point also cannot be a universal constant in a scalar theory but is rather a function of position. However, for those forces that also turn out to be proportional to the mass this variability will not be manifested. Also this variability is common to all relative-acceleration theories.[1)]

On the other hand, the fact that classical mechanics with the mass as a constant scalar quantity yields such good service is surely to be regarded as an indication that we find ourselves in a part of space with a sufficiently symmetric mass distribution, unless it should turn out that variable properties of our measuring instruments should make the tensorial and variable inertial mass appear to be scalar and unchanging. Since, however, the generalized relativity theory permits our measuring instruments to establish bending of light rays and a displacement of spectral lines in a gravitational field, the second possibility appears to me less probable. In accordance with Eq. (4), we must also consider whether there are not indications of the inertial forces being greater in the plane of the Milky Way than in the direction perpendicular to it.

[1)]In Nordström's theory, for example, the mass is set equal to $m=\mu(\text{const}-\Sigma\mu/r)$.

From end p. 183: It is to be desired that the results so far achieved should be incorporated in a field theory that also encompasses variations in time and the relativity postulate.

Now it is certainly the case ... that such incorporation cannot be achieved in Nordström's scalar theory of gravitation, since in that theory the inertial mass decreases on approach to other masses, whereas in accordance with our assumption, as in Einstein's theory, it increases. The character of our above assumption also appears to point more toward a tensor theory.

However, I have not yet succeeded in achieving a complete accommodation to the generalized relativity scheme of Einstein and Grossmann. It seems to me that this is difficult for the following reason.

The complete differential equations of the gravitational field and the complete covariant stress–energy tensor of the mass current in Einstein's latest publications, which together form the generalization of the Laplace–Poisson potential equation, present a mathematically very difficult problem. Admittedly, Einstein himself manages to gain valuable results from them by taking the line element of the previous relativity theory as a first approximation and obtains from the energy tensor of this first approximation a correction, assumed to be small, by means of the differential equations of the field, which now become linear.

By adopting this procedure he gives up, knowingly, an insight into the mechanical building up of the initial value of the line element, which he takes as given, although it should follow from the differential equations. But precisely the assumptions made here provide this physical insight, if only for equilibrium of the field, and therefore it is perhaps only after a different method of integration of the general Einstein field equations that our assumptions can be accommodated. I believe in such an accommodation because my results relating to the dependence between inertia, potential function, and velocity of light have a very similar structure to those of Einstein, and Einstein's scheme must have a very great compass.

COMMENTARY

It will be seen that Reissner's first paper consists basically of two parts. In the first he presents a theory of relative inertia that is essentially identical to the part of Hofmann's paper published 10 years earlier that we have translated on pp. 128–133. Although there is slightly more formal development, Reissner obtains no results that were not intuitively clear to and explicitly anticipated by Hofmann. The question must arise of whether Reissner knew of Hofmann's work. He can hardly have been unaware of Hofmann's existence, since Hofmann's booklet was cited by Einstein in his 1913 lecture in Vienna (cited by Reissner in the second footnote on p. 143), which Reissner had attended. It will be noted that in 1915 (p. 143) Reissner believed he had the priority for a result that Hofmann had already stated in more or less identical terms. The charitable explanation for Reissner's mistake is most likely the correct one: Einstein's brief reference to Hofmann (p. 32) gives no indication that Hofmann had given a precise mathematical formulation that went beyond anything that can be found in Mach, so there is no reason to suppose Reissner felt it was necessary to obtain a copy of Hofmann's obscure and not properly cited booklet.

The second and final part of Reissner's 1914 paper opens up territory entirely foreign to Hofmann's booklet and reflects the dramatic impact of Einstein's (and Minkowski's) contributions to the debate. It is clear that Reissner was extremely impressed (not to say overawed) by Einstein's work and in both the 1914 and 1915 papers he is constantly 'looking over his shoulder' at what Einstein is doing and trying to interpret his own ideas in terms of the conceptual formalism that Einstein was developing for general relativity.

It is interesting that Reissner opens his 1914 paper with the statement that, so far as he can see, his requirement "that not only absolute velocity but also any absolute motion whatever, in particular acceleration, must be undetectable" is "essentially different" from Einstein's approach to extension of the relativity principle even though Einstein himself had strongly implied the identity of the two approaches (p. 180ff). Reissner does not elaborate on his claim and in fact in later passages in both papers tries to show that his ideas might well serve to illustrate what happens in a general relativistic framework. However, this is hardly possible, since, as we now know, a Hofmann–Reissner type theory leads to anisotropy of inertia in clear disagreement with both experiment and general relativity.

Particularly interesting in this connection are Reissner's comments on anisotropy of inertia in his 1915 paper (p. 144), which should be read in conjunction with Rindler's comments on pp. 56–57. Reissner missed an opportunity to make interesting *quantitative* statements about anisotropy of inertia by not calculating the specific effects of the anisotropy of inertia with the particular Machian two-body kinetic energy $\frac{1}{2}m_s m_r \dot{r}_{st}^2/r_{st}$ that he introduced in his 1915 paper. It appears that Schrödinger (p. 153) was the first to realize that in a theory of Hofmann–Reissner type with $1/r$ dependence the mass anisotropy induced by the sun and Galaxy would in principle have observable effects in solar-system dynamics. In this connection, it is interesting that in the early history of general relativity neither Einstein nor anyone else seems to have discussed explicitly the situation in general relativity with regard to this effect; given Einstein's conviction that inertia arises from interaction with other masses, it is perhaps surprising that he did not discuss the possible influence of the Galaxy on solar-system dynamics.

Other points of interest are that, in 1914, Reissner judged "modern researchers" to be hostile to the Machian idea (p. 134), whereas in 1904 Föppl considered that "most authors are today in essential agreement with this point of view" (p. 120). Presumably Planck's deep antagonism had something to do with this (see Norton's paper, pp. 36–37). It is also interesting to note that according to Reissner Abraham withdrew his objection to the Machian idea having been persuaded by Reissner's 1914 paper (p. 143). I am not aware that Einstein mentions Reissner's papers explicitly anywhere (but see p. 186). It would be interesting to know if there is anything in his correspondence.

On Reissner's part it seems clear that he felt the main interest of his papers was that they would provide a simple model of illustration of Einstein's theory, which Reissner believed to be Machian. With hindsight we can see that this expectation was based on an entirely incorrect idea of how, in the framework of general relativity, the metric tensor, indeed space itself, would be generated by matter [see lines 8–13 after Eq. (4) on p. 136, the paragraph beginning: "It would then appear ..." in the middle of p. 141, and the third extract from the 1915 paper (pp. 144–5)]. Einstein himself expressed a very similar idea in 1918 (p. 186). In this connection, see Ehlers's comments (p. 93) on the mathematical impossibility, within the framework of general relativity, of first specifying a matter distribution and then determining a metric tensor from that distribution.

J.B.B.

The Possibility of Fulfillment of the Relativity Requirement in Classical Mechanics[1]

Erwin Schrödinger

It is well known that classical point-particle mechanics with central forces, the foundations of which were developed in the clearest form by L. Boltzmann,[1)] was already criticized by E. Mach[2)] because it does not satisfy the relativity requirement clearly suggested by epistemological considerations - its laws do not hold for *arbitrarily* moving coordinate systems but only for a group of so-called inertial systems, which have a uniform translational motion relative to each other. Empirically, it is found that the inertial systems are coordinate axes that on the average are at rest relative to the heaven of the fixed stars or have a uniform translational motion relative to it, but the foundations of classical mechanics do not in any way indicate a reason for this.

The general theory of relativity too in its original form[3)] could *not* yet satisfy the Machian requirement, as was soon recognized. After the secular precession of the perihelion of Mercury was deduced, in amazing agreement with experiment, from it, every naive person had to ask: With respect to *what*, according to the *theory*, does the orbital ellipse perform this precession, which according to *experience* takes place with respect to the average system of the fixed stars? The answer that one receives is that the theory requires this precession to take place with respect to a coordinate system in which the gravitational potentials satisfy certain boundary conditions at infinity. However, the connection between these

[1)]L. Boltzmann, Vorlesungen über die Prinzipe der Mechanik, Leipzig, J. A. Barth, 1897.

[2)]E. Mach, Die Mechanik in ihrer Entwicklung, Leipzig, F. A. Brockhaus, 3. Aufl. 1897. Vgl. bes. Kap. II. 6.

[3)]A. Einstein, Ann. d. Phys. 49. S. 769. 1916.

Einstein Studies, vol. 6: Mach's Principle: From Newton's Bucket to Quantum Gravity, pp. 147–158 © 1995 *Birkhäuser Boston, Inc. Printed in the United States.*

boundary conditions and the presence of the masses of the fixed stars was in no way clear, since these last were not included in the calculation at all.

The way to overcome this difficulty is today suggested by cosmological theories which require a spatially closed world and thereby avoid boundary conditions altogether. Because of the conceptual difficulties that these cosmological theories still present,[4] and not least because of the mathematical difficulty of their understanding, the solution of an important epistemological problem, which is immediately clear to any scientifically educated person, is thus transferred to a field in which few can follow it and in which it is truly not easy to distinguish between truth and fantasy. I do not doubt that when the solution is finally reached in the sense of those theories it will not only be satisfactory to a high degree but also will permit representation in a form that allows true insight to a wide circle. However, given the present status, it is perhaps not without value to ask whether the Machian relativity requirement could not be satisfied, and the determination of the inertial systems by the heaven of the fixed stars made comprehensible, in a simple manner by a simple modification of classical mechanics.[5]

The expression for the *potential* energy in point-particle mechanics and, in particular, the expression for the Newtonian potential already satisfies the Machian postulate since it only depends on the separation of the two mass points and not on their absolute position in space. Since it has proved itself, it can therefore also be retained from the standpoint of that postulate, if only as a first approximation for a law that in reality is much more complicated. The situation is different with regard to the *kinetic* energy. In accordance with classical mechanics, it is determined by the absolute motion *in space*, whereas in principle only *relative* motions, separations, and variations of separations of mass points are observable. One must therefore see if it is possible in the case of the kinetic energy, just as hitherto for the potential energy, to assign it, not to mass points individually, but instead also represent it as an energy of

---

[4]H. Weyl, Raum, Zeit. Materie, 5. Aufl. § 39. – Berlin. J. Springer. 1923. Cf. also the paper "Massenträgheit und Kosmos" by the same author in the *Naturwissenschaften* (1924, 12. Jahrgang).

[5]The solution of this problem is in fact already contained in the representation of the law of inertia given by Mach. The main reason why it has received so little recognition is presumably mainly because Mach believed he had to adopt a mutual inertial influence that is *independent of the distance* (*loc. cit.*, p. 228 ff.).

interaction of any two mass points and let it depend only on the separation and the rate of change of the separation of the two points. In order to select an expression from the copious possibilities, we use heuristically the following analogy requirements:

1. The kinetic energy as an interaction energy shall depend on the masses and the separations of the two points in the same manner as does the Newtonian potential.

2. It shall be proportional to the square of the rate of change of the separation.

For the total interaction energy of two mass points with the masses μ and μ' with separation r we then obtain the expression

$$W = \gamma \frac{\mu\mu' \dot{r}^2}{r} - \frac{\mu\mu'}{r}. \tag{1}$$

The masses are here measured in a unit such that the gravitational constant has the value 1. The constant γ, which for the moment is undetermined, has the dimensions of a reciprocal velocity. Since it should be universal, one will expect that, apart from a numerical factor, this will be the velocity of light, or that γ will be reduced to a numerical factor when the light second is chosen as the unit of time. We shall have cause later to set this numerical factor equal to 3.

Let us now suppose a mass point μ in the neighborhood of the center of a hollow sphere of radius R that has a uniform mass density σ distributed over it. We refer all expressions to a coordinate system in which the hollow sphere is at rest. Let the mass point move in this coordinate system, its spatial polar coordinates be ρ, ϑ, φ, and those of a surface element of the sphere be R, ϑ', φ'. The distance r of the point from the surface element is given by

$$\begin{aligned} r^2 &= R^2 + \rho^2 - 2R\rho\cos(R\rho) = R^2 + \rho^2 - \\ &\quad 2R\rho[\cos\vartheta\cos\vartheta' + \sin\vartheta\sin\vartheta'\cos(\varphi-\varphi')]. \end{aligned} \tag{2}$$

The total *potential* energy is the same in every position, and we do not consider it. By differentiation we obtain

$$\begin{aligned} r\dot{r} &= \rho\dot{\rho} - R\dot{\rho}[\cos\vartheta\cos\vartheta' + \sin\vartheta\sin\vartheta'\cos(\varphi-\varphi')] - R\rho[-\sin\vartheta \\ &\quad \cos\vartheta'\dot{\vartheta}' + \cos\vartheta\sin\vartheta'\cos(\varphi-\varphi')\dot{\vartheta} - \sin\vartheta\sin\vartheta'\sin(\varphi-\varphi'\dot{\varphi}]. \end{aligned} \tag{3}$$

Since we can choose an arbitrary orientation of the coordinate system, it is sufficient to make the calculation for $\vartheta=0$. Further, we want to calculate only the main terms that remain when $\rho \ll R$. We can then omit the terms with ρ except when they are multiplied by $\dot{\vartheta}$ or $\dot{\varphi}$. In this approximation, we also have $r=R$. That gives

$$\dot{r} = -\dot{\rho}\cos\vartheta' - \rho\dot{\vartheta}\sin\vartheta'\cos(\varphi-\varphi'). \tag{4}$$

Then in accordance with (1)

$$W=\frac{\gamma\mu\sigma R^2}{R}\int_0^{2\pi}d\varphi'\int_0^{\pi}\sin\vartheta'\,d\vartheta'\,[\dot\rho^2\cos^2\vartheta'+$$
$$2\rho\dot\rho\dot\vartheta\sin\vartheta'\cos\vartheta'\cos(\varphi-\varphi')+$$
$$\rho^2\dot\vartheta^2\sin^2\vartheta'\cos^2(\varphi-\varphi')=\frac{4\pi\gamma\mu\sigma R}{3}(\dot\rho^2+\rho^2\dot\vartheta^2). \tag{5}$$

This is exactly the value of the kinetic energy in accordance with classical mechanics on the condition that the usual mass m of our point (in grams) must be given by

$$m=\frac{8\pi\gamma\sigma R}{3}\mu. \tag{6}$$

Since now, on the other hand, in accordance with the assumption made for the potential energy

$$m=\frac{\mu}{\sqrt{k}}, \tag{7}$$

where k is the usual gravitational constant, we must have

$$\frac{1}{\sqrt{k}}=\frac{8\pi\gamma\sigma R}{3}. \tag{8}$$

Alternatively, if we introduce for σ the usual surface density s,

$$s=\frac{\sigma}{\sqrt{k}}, \tag{9}$$

we obtain

$$\frac{4\pi sR^2}{R}=\frac{3}{2k\gamma}, \tag{10}$$

and this is a relation that we shall have to discuss later.

If the masses are expressed in grams, the total interaction energy becomes

$$W=\frac{\gamma kmm'}{r}\dot r^2-\frac{kmm'}{r}. \tag{1'}$$

If a mass point m (planet) moves in the neighborhood of a large mass m' (sun), it will also be necessary to take into account not only the kinetic energy (5) of the mass point with respect to the 'mass horizon' but also its potential and kinetic energy (1′) with respect to m'. For the total energy of the 'single-body problem' we obtain

$$W=\left[\frac{m}{2}+\frac{\gamma kmm'}{r}\right]\dot r^2+\frac{m}{2}r^2\dot\varphi^2-\frac{kmm'}{r}. \tag{11}$$

The presence of the sun has, in addition to the gravitational attraction, also *the* effect that the planet has a somewhat greater inertial mass 'radially' than 'tangentially.' Using the area law, which is unchanged,

$$r^2\dot{\varphi} = f, \tag{12}$$

and making the substitution

$$r^{-1} = \xi, \tag{13}$$

we obtain after elimination of the time from (11) and (12) in the usual manner

$$(1+2\gamma km'\xi)\left[\frac{d\xi}{d\varphi}\right]^2 + \xi^2 - \frac{2km'}{f^2}\xi - \frac{2W}{mf^2} = 0. \tag{14}$$

Setting

$$\xi = \eta + \frac{km'}{f^2}, \quad C = \frac{2W}{mf^2} + \frac{k^2m'^2}{f^4}, \tag{15}$$

we obtain

$$d\varphi = \frac{d\eta\sqrt{1 + 2\gamma k^2m'^2/f^2 + 2\gamma km'\eta}}{\sqrt{C-\eta^2}}, \tag{16}$$

which deviates from the usual form in the square-root factor in the numerator. It is easy to show that when applied to planetary orbits this factor merely represents a small correction provided γ has the order of magnitude of the reciprocal of the square of the velocity of light. We can therefore use the approximation

$$\varphi = \left[1 + \frac{\gamma k^2m'^2}{f^2}\right]\sin^{-1}\eta - \gamma km'\sqrt{C-\eta^2} + \text{const.} \tag{17}$$

Whereas the second term on the right-hand side represents only an extraordinarily small *periodic* perturbation, the first term gives a secular perihelion precession of magnitude

$$\Delta = \frac{2\pi\gamma k^2m'^2}{f^2} \tag{18}$$

per revolution, in the sense of the revolution (φ passes through the angle $2\pi+\Delta$ before η and, therefore, also r returns to the same value and in the same phase of motion). Now in accordance with the well-known expressions

$$km' = \frac{4\pi^2a^3}{\tau^2}, \quad f = \frac{2\pi ab}{\tau}, \tag{19}$$

and therefore

$$\frac{k^2m'^2}{f^2} = \frac{4\pi^2a^4}{b^2\tau^2} = \frac{4\pi^2a^2}{\tau^2(1-\varepsilon^2)}$$

(τ, a, b, ε are the orbital period, the semimajor and semiminor axes, and the numerical eccentricity of the ellipse). This gives

$$\Delta = \frac{8\pi^3\gamma a^2}{\tau^2(1-\varepsilon^2)}. \tag{20}$$

We obtain agreement with the perihelion precession derived in the general theory of relativity,[6)] i.e., in the case of Mercury agreement with experiment also, if we set

$$\gamma = \frac{3}{c^2}. \tag{21}$$

The expression (1) then takes the more definite form

$$W = \frac{3\mu\mu'\dot{r}^2}{r} - \frac{\mu\mu'}{r}, \tag{1''}$$

when the time and mass units are so chosen that the velocity of light and the gravitational constant are both equal to 1. Equation (10) becomes

$$\frac{4\pi sR^2}{R} = \frac{c^2}{2k} = 6.7 \cdot 10^{27} \text{ c.g.s.} \tag{10'}$$

If one assumes that the 'mass horizon' is made up of individual mass points and allows irregularly distributed velocities among them, which nevertheless do not have a greater order of magnitude relative to suitably chosen coordinate systems as the velocities with which the experiments are being made at the center of the sphere, the change in the result (5) in the case of sufficiently large R is nothing more than, first, that this result holds relative to that coordinate system among those considered with respect to which the center of gravity of the horizon masses is *at rest* and, secondly, that there is an additional constant term which arises from the radial velocities of the horizon masses but which has no influence on the motion.

Further, it is clear that the surface-type distribution of the horizon masses can also be replaced by a spatial distribution that is spherically symmetric around the observation point on a large scale provided the conditions are such that the innermost shells of this spatial distribution for which R is not yet sufficiently large in order to justify the approximations made above make only vanishing contributions to the total inertial effect. Let d be spatial density of this distribution in g/cm^3 and R be its

---

6) A. Einstein, *loc. cit.*, final page.

outer radius; then in place of (10′) we obviously obtain

$$\int_0^R \frac{4\pi\rho^2 d}{\rho} d\rho = 2\pi R^2 d = \frac{c^2}{2k} = 6.7 \cdot 10^{27} \text{ c.g.s.}, \tag{10''}$$

where we have performed the integration for a d that is constant inside R. This strange relation states that the (negative) potential of all masses at the point of observation, calculated with the gravitational constant *valid at the point of observation*, must be equal to half the square of the velocity of light.

A rough approximation of the integral in (10″) for the radiating masses of our stellar system gives the value 10^{16} c.g.s. It has here been assumed that a sphere of radius $R=200$ parsec (1 parsec$=3.09 \cdot 10^{18}$ cm) is uniformly filled with stars with the mass of the sun in such a way that 30 such stars are present in a sphere of radius 5 parsec. It follows from this that only an entirely vanishing fraction of the inertial effects observed on the earth and in the planetary system arises from the interaction with the masses of our Milky Way system. With regard to the admissibility of the ideas developed here, this is a very encouraging result. For were the conditions to be only slightly different in order of magnitude, it would only be possible with great difficulty to explain the absence of every *anisotropy* of the terrestrial and planetary inertia. A mass distribution like that established for the radiating stars would have to have the consequence that bodies are subject to a greater inertial resistance *in* the galactic plane as at right angles to it. The circumstance that we are probably not exactly in the middle of this mass distribution would also have to have similar consequences. The orders of magnitude established above appear to me to depress the inertial anisotropy that arises from the asymmetric distribution of the masses of our Milky Way system *just about* under the limit of astronomical observability, as one can roughly estimate by comparison with the anisotropy of the mass of Mercury, which is still readily established.

However, it seems that the question of why our inertial systems are free of rotation precisely with respect to *our* stellar system (or it with respect to them) reappears if the inertial systems are not primarily 'anchored' in that system but rather in much more distant stellar masses. The reason, or, better, the actual state of affairs, is evidently, from our entirely naive and elementary standpoint, that empirically only comparatively small relative stellar velocities occur at all, namely, velocities that are significantly smaller than the velocity of light. Our expression (1″) does not let us recognize any reason for this state of affairs.

However, such a reason appears quite naturally if to the knowledge of the mechanics of our solar system, which is all that we have so far used, we add as a purely empirical basis, the observations of a significant growth of the inertia as the velocity of light is approached (deflection experiments with electrons). These experiments show that the expression (1″) is only to be regarded as an approximation for small velocities, and requires a correction for large $\dot{r}$, i.e., comparable with unity. If we regard the 'relativistic' energy formula as expression of the observations,

$$\text{Kin. En.} = mc^2 \left[\frac{1}{\sqrt{1-\beta^2}} - 1 \right], \tag{22}$$

it is easy to give a modification of (1″) that for arbitrary velocities leads precisely to (22). Let us set

$$W = \frac{\mu\mu'}{r} \left[\frac{2}{(1-\dot{r}^2)^{3/2}} - 3 \right]; \tag{1‴}$$

we here substitute $\dot{r}$ in accordance with (4) and perform the calculation analogous to (5) [omitting the second bracket term in (1‴), which yields only a constant]:

$$W = \frac{2\mu\sigma R^2}{R} \int_0^{2\pi} d\varphi' \int_0^{\pi} \frac{\sin\vartheta' \, d\vartheta'}{(1-[\dot{\rho}\cos\vartheta' + \rho\dot{\vartheta}\sin\vartheta' \cos(\varphi'-\varphi)]^2)^{3/2}}.$$

If we set here in the first place

$$x = \cos\vartheta', \quad y = \sin\vartheta' \cos(\varphi'-\varphi),$$

then x and y pass over the surface of the unit circle *twice* when ϑ' and φ' pass over their complete range. We find

$$W = 4\mu\sigma R \iint^{x^2+y^2\leq 1} \frac{dx\,dy}{(1-[\dot{\rho}x+\rho\dot{\vartheta}y]^2)^{3/2}\sqrt{1-x^2-y^2}}.$$

Let us now introduce for x and y 'planar polar coordinates' r and ψ and recognize that in place of r it is expedient to take immediately

$$\sqrt{1-r^2} = z$$

as variable. That gives

$$W = 4\mu\sigma R \int_0^{2\pi} d\psi \int_0^1 \frac{dz}{(1-a^2+a^2z^2)^{3/2}}$$

$$= 4\mu\sigma R \int_0^{2\pi} \frac{d\psi}{1-a^2} = 4\mu\sigma R \int_0^{2\pi} \frac{d\psi'}{1-v^2\cos^2\psi'}$$

with the abbreviations

$$a = \dot{\rho}\cos\psi + \rho\dot{\vartheta}\sin\psi, \quad v = \sqrt{\dot{\rho}^2 + \rho^2\dot{\vartheta}^2}.$$

One now sees, most readily through a series expansion of the last integral (or by direct calculation or by integration in the complex domain), that finally

$$W = \frac{8\pi\mu\sigma R}{\sqrt{1-v^2}} = \frac{8\pi\mu\sigma R}{\sqrt{1-\dot{\rho}^2-\rho^2\dot{\vartheta}^2}}, \tag{23}$$

which in accordance with (6) and (21) agrees with the variable part of (22), since in the present calculation we have from the very beginning taken the velocity of light as unity.

Let us mention in passing that to the expression (1‴) there corresponds the Lagrange function

$$L = \frac{\mu\mu'}{r}\left[\frac{2}{\sqrt{1-\dot{r}^2}} - 4\sqrt{1-\dot{r}^2} + 3\right] \tag{24}$$

which satisfies the equation

$$\dot{r}\frac{dL}{d\dot{r}} - L = W = \frac{\mu\mu'}{r}\left[\frac{2}{(1-\dot{r}^2)^{3/2}} - 3\right]. \tag{25}$$

If L is integrated in accordance with (24), in the same way as W earlier for the interaction of our mass point with the hollow sphere, we obtain, up to a constant, the well-known relativistic Lagrange function of a mass point:

$$L = -mc^2\sqrt{1-\beta^2}\,, \tag{26}$$

where β is again the ratio of the velocity of the mass point to the velocity of light.

The most serious objection that can be raised against the conceptual possibilities presented in this note is that they appear to rely on the principle of instantaneous action at a distance that nowadays is quite unacceptable. It is obvious that today no one, the present author included, will be persuaded to regard the assumptions (1), (1″), etc., truly in this sense. But in just the same way as we may be convinced that a star that is many light years away exerts on a terrestrial second-pendulum a tiny and apparently instantaneous effect through its gravitational field, even when gravitation in truth only propagates with the velocity of light, in just the same way we are allowed, I believe, to calculate with the $\dot{r}$-dependent terms of our expressions without sinning against the basic principle of a finite propagation velocity of all influences provided the conditions are *such* that on the average it does not matter whether we calculate with the instantaneous or the retarded state of motion of the distant world mass.

In other cases, one would admittedly encounter at first certain difficulties if one wanted to take into account seriously the retardation time. It is then in principle impossible to specify $\dot{r}$. One could define it purely empirically by the observed Doppler effect, but for two observers on two different mass points that send each other light signals the effect is not the same 'in the same instant.' The kinetic energy of the interaction, which at the start we have put together in a single term, then necessarily breaks up again into two terms. One may also note that the reason for the difference of the Doppler effect, if the two world masses have approximately the same mass, can only be recognized in the existence of all the remaining world bodies, which accordingly must define an inertial system for light just as well as for point-particle motion.

I believe it is probable that through further development of these ideas one will finally, after certain modifications, arrive at the general theory of relativity. For this represents a framework that can hardly be completely overthrown by any future theory but today is by no means yet filled out with concrete and lively concepts. I regard the concept used here – that change of the relative, not the absolute, state of motion of bodies requires expenditure of work – to be at least an allowed and useful intermediate stage that makes it possible to understand, in a simple and yet basically sound manner, a simple empirical state of affairs by means of concepts that are familiar to everyone.

Zürich, Physikalisches Institut der Universität

NOTES

[1]Translated from *Annalen der Physik* 77: 325-336 (1925) (submitted June 16, 1925). Translated by Julian B. Barbour. We are grateful to the editorial board of the *Annalen der Physik* for permission to publish this translation. In the recently published collected works of Schrödinger, a typewritten note, signed by Schrödinger, is appended to this paper. In the note Schrödinger expresses profuse apologies to Reissner for unconsciously plagiarizing his idea of deriving inertia in a Machian manner by two-body interactions (p. 134). Schrödinger says that he quite definitely knew of Reissner's first paper but is not certain about the second. He calls both of Reissner's papers "very interesting" and hopes that his own paper will still be of some interest on account of his somewhat different standpoint and treatment.

COMMENTARY

Like Reissner, Schrödinger sees the main value of his paper in its potential to serve as a simple illustration of general relativity, which he believes will eventually be shown to be Machian in a satisfying manner. It is worth noting that the opinion which Schrödinger attributes to Mach in the opening sentence reads much more like some of the early formulations of Einstein (p. 180).

The idea that represents the heart of Schrödinger's approach (pp. 148–149) is a rerun of Hofmann's arguments from 20 years earlier but with a *brevity* of which Hofmann was clearly incapable (what Hofmann says in paragraphs, Schrödinger says in sentences). It is noteworthy that Schrödinger's two requirements 1) and 2) (top of p. 149) identify exactly the same variables (masses, relative separations, rates of change thereof) that were listed by Poincaré and Föppl as being the only ones allowed to appear in a satisfactory relational mechanics. The conceptual advance made by Schrödinger is just the same as that of Hofmann and Reissner – *implementation* of the Poincaré–Föppl requirements by a Machian kinetic energy containing products of the masses of all possible pairs of particles. [Incidentally, models can be constructed using three-body interactions, see (Bertotti and Easthope 1978).]

The real novelty of Schrödinger's paper is his explicit working out of solar-system dynamics in a model cosmology, from which he can draw potentially very interesting conclusions (of the type Rindler mentions, p. 56–57): In such a theory the advance of Mercury's perihelion appears as a Machian effect, and from its magnitude one can make a nontrivial estimate about the distribution and amount of matter in the universe, in particular that there must be vastly more matter in the universe than is visible in our Galaxy (no longer a dramatic prediction in 1925). Schrödinger's cautious conclusion that the inertial anisotropy which arises from the "asymmetric distribution of the masses of our Milky Way system" is "*just about* under the limit of astronomical observability" is quite ironic, since it was based on what we now know is a gross underestimate of the mass of the Galaxy, which Schrödinger took to be $\sim 2\cdot 10^6$ solar masses against the present estimate of $1.4\cdot 10^{11}$ solar masses. The effect of the error in the mass was offset somewhat by Schrödinger's underestimate of the radius of the Galaxy (200 parsec against the present estimate of 12000 parsec), but he still underestimated the mass-anisotropy effect of the Galaxy by about 1000 times. As Nordtvedt (1975) showed 50 years later, such putative mass-anisotropy effects of the Galaxy would be very readily observable in the solar system (comparable with the perihelion advance of Mercury). (See also my comment at the end of the first paragraph of p. 146.)

There are several other points of interest in Schrödinger's paper. For example, like Reissner (p. 144, middle of page) Schrödinger insists that a proper formulation of Machian interaction must be based on modern field-theoretical notions, and to avoid instantaneous action at a distance he insists that retardation effects must be taken into account. Admittedly, he is then forced to recognize the difficulties to which Ehlers draws attention on p. 466. In fact, it seems to me that in his penultimate paragraph Schrödinger casts serious doubt on the whole approach of his paper. So far as I know, the Green's function approach described by Raine (p. 274) is the only one in which allowance for retardation has been directly attempted. Hoyle and Narlikar seek to avoid fields by using symmetric advanced–retarded potentials (p. 250). In contrast, Wheeler (see Isenberg's paper, p. 188), followed by Bertotti and myself (p. 214), seeks to implement the Machian requirement in a field-theoretical framework through the initial-value constraints of general relativity. That this may be the route to the Machian goal was also noted some years ago by Lindblom and Brill (see Pfister's comments on p. 324) and was recently advocated by Lynden-Bell, Katz, and Bičák (1995).

Finally, it may be noted that Schrödinger, like Einstein and Reissner before him, clearly thinks that in a Machian approach the inertial properties of spacetime are to be determined by the matter degrees of freedom alone. On the evidence of the Tübingen workshop, there is a growing awareness among physicists that in the context of general relativity the elimination of a role of gravitational degrees of freedom can be questioned on physical grounds and seems extremely difficult to realize mathematically (see *Gravitational degrees of freedom, role in Mach's Principle* in the Index).

J.B.B.

REFERENCES

Bertotti, Bruno, and Easthope, Paul (1978). "The Equivalence Principle According to Mach." *International Journal of Theoretical Physics* 17: 309–318.

Lynden-Bell, Donald, Katz, Joseph, and Bičák, Jiri (1995). "Mach's Principle from the Relativistic Constraint Equations." *Monthly Notices of the Royal Astronomical Society* 272: 150–160.

Nordtvedt, Kenneth (1975). "Anisotropic Gravity and the Binary Pulsar PSR 1913+16." *Astrophysical Journal* 202: 248–249.

Weber's Law and Mach's Principle

André K. T. Assis

1. Introduction

Recently we applied a Weber's force law for gravitation to implement quantitatively Mach's Principle (Assis 1989, 1992a). In this work we present a brief review of Weber's electrodynamics and analyze in greater detail the compliance of a Weber's force law for gravitation with Mach's Principle.

2. Weber's Electrodynamics

In this section we discuss Weber's original work as applied to electromagnetism. For detailed references of Weber's electrodynamics, see (Assis 1992b, 1994).

In order to unify electrostatics (Coulomb's force, Gauss's law) with electrodynamics (Ampère's force between current elements), W. Weber proposed in 1846 that the force exerted by an electrical charge q_2 on another q_1 should be given by (using vectorial notation and in the International System of Units):

$$\mathbf{F}_{21} = \frac{q_1 q_2}{4\pi\epsilon_0} \frac{\hat{\mathbf{r}}_{12}}{r_{12}^2} \left[1 - \frac{\dot{r}_{12}^2}{2c^2} + \frac{r_{12}\ddot{r}_{12}}{c^2} \right] . \tag{1}$$

In this equation, $\epsilon_0 = 8.85 \cdot 10^{-12}$ F/m is the permittivity of free space; the position vectors of q_1 and q_2 are $\mathbf{r}_1$ and $\mathbf{r}_2$, respectively; the distance between the charges is

$$r_{12} \equiv |\mathbf{r}_1 - \mathbf{r}_2| = [(x_1 - x_2)^2 + (y_1 - y_2)^2 + (z_1 - z_2)^2]^{1/2};$$

$\hat{\mathbf{r}}_{12} = (\mathbf{r}_1 - \mathbf{r}_2)/r_{12}$ is the unit vector pointing from q_2 to q_1; the radial velocity between the charges is given by $\dot{r}_{12} \equiv dr_{12}/dt = \hat{\mathbf{r}}_{12} \cdot \mathbf{v}_{12}$; and the

Einstein Studies, vol. 6: Mach's Principle: From Newton's Bucket to Quantum Gravity, pp. 159–171

radial acceleration between the charges is

$$\ddot{r}_{12}=\frac{d\dot{r}_{12}}{dt}=\frac{d^2r_{12}}{dt^2}=\frac{[\mathbf{v}_{12}\cdot\mathbf{v}_{12}-(\hat{\mathbf{r}}_{12}\cdot\mathbf{v}_{12})^2+\mathbf{r}_{12}\cdot\mathbf{a}_{12}]}{r_{12}},$$

where

$$\mathbf{r}_{12}\equiv\mathbf{r}_1-\mathbf{r}_2,\ \mathbf{v}_{12}\equiv\frac{d\mathbf{r}_{12}}{dt},\ \mathbf{a}_{12}\equiv\frac{d\mathbf{v}_{12}}{dt}=\frac{d^2\mathbf{r}_{12}}{dt^2}.$$

Moreover, $c\equiv(\epsilon_0\,\mu_0)^{-1/2}$ is the ratio of electromagnetic and electrostatic units of charge ($\mu_0=4\pi\cdot10^{-7}$ N/A^2 is the permeability of free space). This quantity c was first measured experimentally by W. Weber and Kohlrausch in 1856, when they found $c=3.1\cdot10^8$ m/s. This was one of the first unambiguous and quantitative indications of an essential interconnection between electromagnetism and optics.

In 1848, Weber presented a potential energy U_{12} from which he could derive his force by $\mathbf{F}_{21}=-\hat{\mathbf{r}}_{12}dU_{12}/dr_{12}$:

$$U_{12}=\frac{q_1q_2}{4\pi\epsilon_0}\frac{1}{r_{12}}\left[1-\frac{\dot{r}_{12}^2}{2c^2}\right]. \tag{2}$$

There is a Lagrangian L and a Hamiltonian H from which we can also derive his electrodynamics. For a system of two charges q_1 and q_2 of masses m_1 and m_2 interacting through Weber's force, we have a kinetic energy T_{12} and a Lagrangian energy S_{12} given by:

$$T_{12}=m_1\frac{\mathbf{v}_1\cdot\mathbf{v}_1}{2}+m_2\frac{\mathbf{v}_2\cdot\mathbf{v}_2}{2}=\frac{m_1v_1^2}{2}+\frac{m_2v_2^2}{2}, \tag{3}$$

$$S_{12}=\frac{q_1q_2}{4\pi\epsilon_0}\frac{1}{r_{12}}\left[1+\frac{\dot{r}_{12}^2}{2c^2}\right]. \tag{4}$$

Note the change of sign in front of $\dot{r}_{12}^2$ in U_{12} and S_{12}.

Weber's Lagrangian and Hamiltonian are then given by

$$L\equiv T_{12}-S_{12}, \tag{5}$$

$$H\equiv\left[\sum_{k=1}^{6}\dot{q}_k\frac{\partial L}{\partial\dot{q}_k}\right]-L=T_{12}+U_{12}, \tag{6}$$

where $\dot{q}_k$, with k ranging from 1 to 6, represents the velocity components, namely, $\dot{x}_1$, $\dot{y}_1$, $\dot{z}_1$, $\dot{x}_2$, $\dot{y}_2$, and $\dot{z}_2$, respectively.

Weber's force can be obtained from S_{12} by the usual procedure. For instance, the x-component of $\mathbf{F}_{12}$ is given by

$$F_{21}^{x}=\frac{d}{dt}\frac{\partial S_{12}}{\partial \dot{x}_1}-\frac{\partial S_{12}}{\partial x_1}=\frac{q_1q_2}{4\pi\epsilon_0}\frac{x_1-x_2}{r_{12}^3}\left[1-\frac{\dot{r}_{12}^2}{2c^2}+\frac{r_{12}\ddot{r}_{12}}{c^2}\right]. \tag{7}$$

The main properties of Weber's electrodynamics are:

A. It complies with Newton's action and reaction law, which means conservation of linear momentum for an isolated system of particles interacting through Weber's force and through other forces which also follow the law of action and reaction.

B. The force is always along the straight line connecting the two charges, which means conservation of angular momentum.

C. The force can be derived from the velocity dependent potential energy U_{12}, which means conservation of the total energy $E\equiv T_{12}+U_{12}$. Although Weber presented U_{12} in 1848, he proved the conservation of energy for his electrodynamics only in 1869 and 1871. In 1847, only one year after Weber had presented his force law (1), Helmholtz published his famous paper on the conservation of energy. In this work he showed that a force which depends on the distance and velocities of the interacting particles does not conserve energy, even if the force is a central one. This was the main objection that, from his first paper on electromagnetism of 1855/56, Maxwell advanced against Weber's electrodynamics and the reason that, in his own words, prevented him from considering Weber's theory as an ultimate one (Maxwell 1965a, 1965b). Maxwell was wrong, but he only changed his mind in 1871, after Weber's proof (Harman 1982). When he wrote the *Treatise* in 1873, he presented the new point of view that Weber's electrodynamics is consistent with the principle of conservation of energy (Maxwell 1954). Helmholtz's proof of 1847 does not apply to Weber's electrodynamics because Weber's force depends not only on the distance and velocity of the charges but also on their accelerations. This general case was not analyzed by Helmholtz at that time.

Other properties of Weber's law are:

D. When there is no relative motion between the interacting charges ($\dot{r}_{12}=0$ and $\ddot{r}_{12}=0$), we recover Coulomb's force and Gauss's law. So all electrostatics is embodied in Weber's electrodynamics.

E. Weber suceeded in deriving Faraday's law of induction (1831) from his force (Maxwell 1954).

F. Weber derived his force from Ampère's force (1823) exerted by the current element $I_2d\mathbf{l}_2$ on $I_1d\mathbf{l}_1$:

$$d^2\mathbf{F}_{21} = -\frac{\mu_0}{4\pi}\frac{I_1I_2}{r_{12}^2}\hat{\mathbf{r}}_{12}\left[2(d\mathbf{l}_1 \cdot d\mathbf{l}_2) - 3(\hat{\mathbf{r}}_{12} \cdot d\mathbf{l}_1)(\hat{\mathbf{r}}_{12} \cdot d\mathbf{l}_2)\right]. \tag{8}$$

Alternatively we can postulate Weber's law and derive Ampère's force between current elements as a special case of Weber's electrodynamics. From Ampère's force (8) Maxwell derived what is known as Ampère's circuital law in 1856, twenty years after Ampère's death. Maxwell was the first to derive the circuital law even without the term with the displacement current.

The force between current elements usually found in the textbooks is due to Grassmann (1845) utilizing the Biot–Savart magnetic field $d\mathbf{B}_2$ of 1820, namely

$$d^2\mathbf{F}_{21} = I_1 d\mathbf{l}_1 \times d\mathbf{B}_2 = I_1 d\mathbf{l}_1 \times \left\{\frac{\mu_1}{4\pi}\frac{I_2 d\mathbf{l}_2 \times \hat{\mathbf{r}}_{12}}{r_{12}^2}\right\}$$

$$= -\frac{\mu_0}{4\pi}\frac{I_1I_1}{r_{12}^2}\left[(d\mathbf{l}_1 \cdot d\mathbf{l}_2)\hat{\mathbf{r}}_{12} - (d\mathbf{l}_1 \cdot \hat{\mathbf{r}}_{12})d\mathbf{l}_2\right]. \tag{9}$$

Ampère's force (8) complies with the action and reaction law in the strong form for any independent orientation of each current element, while this is not valid in general for Grassmann's force (9). Both expressions give the same result for the force of a closed current loop of arbitrary form on a current element of another circuit. In the last ten years many experiments have been performed trying to distinguish (8) and (9) in situations involving a single circuit (for instance, measuring and calculating the force and tension on a mobile part of a closed circuit due to the remainder of the circuit). Although most experiments seem to favor Ampère's force over Grassmann's one, the situation is not yet completely clear, and more experiments and theoretical analysis are desirable before a final conclusion can be drawn. For references on this topic see (Assis 1989, 1992b, 1994).

It should be remembered that Maxwell knew both expressions, (8) and (9). When comparing these assumptions he said that "Ampère's is undoubtedly the best, since it is the only one which makes the forces on the two elements not only equal and opposite but in the straight line which joins them" (Maxwell 1954, vol. 2, § 527, p. 174).

The last property of Weber's law to be discussed here is undoubtedly one of the most important of them. It is also closely related to Mach's Principle:

G. The law depends only on the relative distance between the particles, r_{12}, on the relative velocity between them, $\dot{r}_{12} = dr_{12}/dt$, and on the relative radial acceleration between them,

$$\ddot{r}_{12}=d\dot{r}_{12}/dt=d^2r_{12}/dt^2.$$

This is what we call a relational theory. These terms have the same value in all frames of reference, even for noninertial ones.

This is a distinguishing feature of Weber's electrodynamics. In the other formulations of electromagnetism the terms in the velocity and acceleration of the particles which are relevant depend on the velocities or accelerations of the charges either relative to a material medium like the ether, or relative to an inertial frame of reference. This last situation is typical of Lorentz's force law, $\mathbf{F}=q\mathbf{E}+q\mathbf{v}\times\mathbf{B}$, where $\mathbf{v}$ is the velocity of the charge q relative to an arbitrary inertial frame of reference (and not, for instance, relative to the laboratory or to the magnet which generated the magnetic field $\mathbf{B}$).

After this short review we shall discuss the relation of Weber's electrodynamics to Mach's Principle.

3. The Mach–Weber Model

In order to implement quantitatively Mach's Principle we need to modify Newton's law of gravitation by including terms which depend on the velocity and acceleration between the interacting bodies. This was never done by Mach himself. In our opinion the best model in this direction seems to be some kind of Weber's law for gravitation. In the first place this would comply with Mach's idea that only relative positions and motions are important, as this force depends only on r_{12}, $\dot{r}_{12}$, and $\ddot{r}_{12}$. It also depends on the accelerations of the source and test bodies. So it has embodied in it the possibility of deriving $m\mathbf{a}$, the centrifugal and Coriolis forces as real gravitational forces arising from the relative acceleration of the test body and the remainder of the universe.

Here we list some (but not all) people who have worked with this model. The first to propose a Weber's law for gravitation seems to have been G. Holzmüller in 1870 (North 1965, p. 46). Then Tisserand, in 1872, studied a Weber's law for gravitation and its application to the precession of the perihelion of the planets (Tisserand 1872, 1895). Weber himself and Zölner obtained this law as applied to gravitation around 1876, when implementing the idea of Young and Mossotti of deriving gravitation from electromagnetism (Assis 1992b; Woodruff 1976). Later on Paul Gerber obtained essentially the same potential energy up to second order in $1/c$ (Gerber 1898). He obtained this law independently, following ideas of retarded time, without discussing

Weber's work. He also studied the precession of the perihelion of the planets. Gerber's work was criticized by Seeliger (Seeliger 1917), who was aware of Weber's electrodynamics. The work of Tisserand applying a Weber's law for gravitation in celestial mechanics was also discussed by Poincaré in a course which he delivered at the Faculté des Sciences de Paris during 1906–1907 (Poincaré 1953, see especially p. 125 and Chap. IX, pp. 201–203, "Loi de Weber"). None of the authors tried to implement Mach's Principle with these force laws.

Although Mach dealt with many branches of physics (mechanics and gravitation, optics, thermodynamics), we are not aware that he ever mentioned Weber's electrodynamics. We also do not know any reference of Einstein to Weber's force or potential energy. The first to suggest a Weber's law for gravitation in order to implement Mach's Principle seems to have been I. Friedlaender in 1896 (Friedlaender and Friedlaender 1896, p. 17, footnote, p. 310 in this volume). They seem to have been also the first to suggest that inertia should be related to *gravitation*. Höfler in 1900, although opposing Mach, mentioned Weber's electrodynamics when discussing Mach's Principle (Norton 1995). Hofmann in 1904 suggested a kinetic energy that depended on the product of the masses, on a function of the distance between the interacting masses, and on the square of their relative speed, which is somewhat similar to Weber's potential energy when applied to gravitation (this volume, p. 128). In this century we have Reissner and Schrödinger considering relational quantities in gravitation to implement Mach's Principle (Reissner 1914, 1915, this volume, p. 134; Schrödinger 1925, this volume, p. 147). They arrived independently at a potential energy very similar to that of Weber, apparently without being aware of Weber's electrodynamics. In 1933, we have Przeborski discussing Weber's law and other expressions in connection with Newton's second law of motion, although not analyzing Mach's Principle directly (Przeborski 1933). More recently we have Sciama (1953). Although he made an analogy between gravitation and electromagnetism, he did not work with a relational force law, and his expression did not even comply with Newton's action and reaction principle. He also did not mention Weber's electrodynamics. Brown was closer to this idea, although his force law is different from Weber's one (Brown 1955, 1982). Moon and Spencer published an important work on this topic (Moon and Spencer 1959), although they did not consider Weber's law or relational quantities. Edwards worked explicitly with relational quantities and with analogies between electromagnetism and gravitation (Edwards 1974). Once more Weber's electrodynamics is not mentioned. Barbour and Bertotti opened new lines of research working not only with relational quantities but with intrinsic derivatives and with the relative configuration space (RCS) of the universe (Barbour 1974; Barbour and Bertotti 1977, 1982). Eby worked along this line and studied the precession of the perihelion of the planets (Eby 1977). Although he worked essentially with a Weber's Lagrangian, he did not mention Weber's work. Treder, von Borzeszkowski, van der Merwe, Yourgrau, and collaborators have worked with and discussed explicitly a Weber's force applied to gravitation. References to their original works and to other authors can be found in (Treder 1975; Treder, von Borzeszkowski, van der Merwe, and Yourgrau 1980). Ghosh worked with closely related ideas, although he was not aware of Weber's force (Ghosh 1984, 1986, 1991). More recently we have Wesley and a direct use of Weber's law (Wesley 1990). He also worked with a potential similar to Schrödinger's potential energy (Schrödinger 1925), without being aware of that work.

Although we could quote many other authors and papers, we stop here. This short list gives an idea of the continuing effort and research that has been performed by many important people along this line (trying to implement quantitatively Mach's Principle by some kind of Weber's law). We are following these ideas, although we were not aware of many of these works when we began. Here we present how we deal with this subject (Assis 1989, 1992a).

Our basic idea is to begin with a gravitational potential energy between two particles given by

$$U_{12} = -H_g \frac{m_{g1} m_{g2}}{r_{12}} \left[1 - \frac{\xi}{2} \frac{\dot{r}_{12}^2}{c^2} \right] \exp(-\alpha r_{12}). \tag{10}$$

In this expression, H_g is an arbitrary constant, $m_{g1,2}$ are gravitational masses, ξ is a dimensionless constant, and α gives the characteristic length of the gravitational interaction. Newton's potential energy is (10) with $H_g = G$, $\xi = 0$, and $\alpha = 0$.

The first to propose an exponential decay in the gravitational potential energy were Seeliger and Neumann, in 1895–1896. What they proposed would be equivalent to (10) with $H_g = G$ and $\xi = 0$. An exponential term in Newton's gravitational force (but not in the potential) had been proposed much earlier by Laplace, in 1825. For references and further discussion see (Assis 1992a; North 1965, pp. 16–18; Laplace 1969; Seeliger 1895). In this century there is a remarkable paper by W. Nernst proposing an exponential decay in gravitation (Nernst 1937). These exponential decays have been proposed as an absorption of gravity due to the intervening medium, in analogy with the propagation of light. In this case α would depend on the amount and distribution of the intervening matter in the straight line between m_{g1} and m_{g2}. Alternatively it has also been proposed to solve some gravitational paradoxes arising in an infinite and homogeneous universe (indefinite value of the potential or of the gravitational force). In this last situation α may be considered as a universal constant irrespective of the medium between m_{g1} and m_{g2}.

To our knowledge we were the first to propose the exponential decay in a Weberian potential (Assis 1992a).

To simplify the analysis in this work, we will consider the arbitrary constant H_g as equal to Newton's gravitational constant G. Moreover we will treat α as a constant irrespective of the medium between the particles 1 and 2. Its value will be taken as $\alpha=H_0/c$, where H_0 is Hubble's constant (Assis 1992a). We will also take $\xi=6$, as in our previous work (Assis 1989, 1992a).

The force exerted by m_{g2} on m_{g1} can be obtained utilizing $\mathbf{F}_{21}=-\hat{\mathbf{r}}_{12}dU_{12}/dr_{12}$. This yields:

$$\mathbf{F}_{21}=-G\frac{m_{g1}m_{g2}}{r_{12}^2}\hat{\mathbf{r}}_{12}\left[1-\frac{\xi}{2}\frac{\dot{r}_{12}^2}{c^2}+\xi\frac{r_{12}\ddot{r}_{12}}{c^2}\right.$$

$$\left.+\frac{H_0}{c}r_{12}\left(1-\frac{\xi}{2}\frac{\dot{r}_{12}^2}{c^2}\right)\right]\exp(-H_0r_{12}/c). \tag{11}$$

We now integrate this expression for a particle of gravitational mass m_{g1} interacting with an isotropic, homogeneous and infinite universe. Its average gravitational matter density is represented by ρ_0. In order to integrate we utilize spherical coordinates and replace m_{g2} by $\rho_0 r_2^2 \sin\theta_2\, dr_2 d\theta_2 d\varphi_2$. We integrate from $\varphi_2=0$ to 2π, from $\theta_2=0$ to π, and from $r_2=0$ to infinity. The procedure is the same as in (Assis 1989, 1992a). We perform the integration in a frame of reference relative to which the universe as a whole (the set of distant galaxies) has an overall translational acceleration $\mathbf{a}_u$ and is rotating with an angular velocity $\boldsymbol{\omega}_u(t)$. Relative to this arbitrary frame of reference, the particle m_{g1} is located at the position $\mathbf{r}_1$ and has a velocity $\mathbf{v}_1=d\mathbf{r}_1/dt$ and acceleration $\mathbf{a}_1=d^2\mathbf{r}_1/dt^2$. The final result of the integration is found to be

$$\mathbf{F}_{u1}=-Am_{g1}\left[\mathbf{a}_1+\boldsymbol{\omega}_u\times(\boldsymbol{\omega}_u\times\mathbf{r}_1)-2\boldsymbol{\omega}_u\times\mathbf{v}_1-\frac{d\boldsymbol{\omega}_u}{dt}\times\mathbf{r}_1-\mathbf{a}_u\right]. \tag{12}$$

In this expression

$$A=\frac{4\pi}{3}H_g\frac{\xi}{c^2}\rho_0\int_0^\infty r_2\exp(-\alpha r_2)dr_2=\frac{4\pi}{3}G\xi\frac{\rho_0}{H_0^2}. \tag{13}$$

In Newtonian mechanics, this expression is zero.

To complete the formulation of a Machian dynamics, we need the principle of dynamical equilibrium (Assis 1989). According to this principle, the sum of all forces of any nature (gravitational, electromagnetic, elastic, nuclear, etc.) on any particle is always zero in all coordinate frames, even when the particle is in motion and accelerated. We represent by $\Sigma_{j=1}^{N}\mathbf{F}_{j1}$ the resultant force acting on m_{g1} due to N local bodies j (like the gravitational force of the earth and the sun, contact forces, electromagnetic forces, friction forces, etc.). The principle of dynamical equilibrium can then be expressed as:

$$\sum_{j=1}^{N}\mathbf{F}_{j1}+\mathbf{F}_{u1}=0. \tag{14}$$

Utilizing (12) this can be written as

$$\frac{\sum_{j=1}^{N}\mathbf{F}_{j1}}{A}-m_{g1}\boldsymbol{\omega}_u\times(\boldsymbol{\omega}_u\times\mathbf{r}_1)+2m_{g1}\boldsymbol{\omega}_u\times\mathbf{v}_1$$

$$+m_{g1}\frac{d\boldsymbol{\omega}_u}{dt}\times\mathbf{r}_1+m_{g1}\mathbf{a}_u=m_{g1}\mathbf{a}_1. \tag{15}$$

This is essentially Newton's second law of motion with 'fictitious' forces. In the Mach–Weber model these are real gravitational forces which arise in any frame of reference in which the universe as a whole has a translational acceleration $\mathbf{a}_u$ and is rotating as a whole with an angular velocity $\boldsymbol{\omega}_u$. The proportionality between Newton's inertial and gravitational masses (the principle of equivalence) is derived at once in this model as the right-hand side of (15) arose from the gravitational interaction (12) of m_{g1} with the isotropic matter distribution surrounding it. The constant A must be exactly equal to 1, and this is known to be approximately true since the 1930s with Dirac (Assis 1989, 1992a). Equation (15) takes its simplest form in a frame of reference in which the universe at large is essentially stationary ($\mathbf{a}_u=0$, $\boldsymbol{\omega}_u=0$, $d\boldsymbol{\omega}_u/dt=0$). This explains the coincidence (in Newtonian mechanics) that the frame of the fixed stars is the best inertial frame we have, namely, a frame in which there are no fictitious forces (Schiff 1964).

ACKNOWLEDGMENTS

The author wishes to thank Dr. J. B. Barbour for calling his attention to the papers by Reissner and Schrödinger. He is also grateful to the Organizing Committee of the International Conference on Mach's Principle for financial support.

REFERENCES

Assis, André K. T. (1989). "On Mach's Principle." *Foundations of Physics Letters* 2: 301–318.

Assis, André K. T. (1992a). "On the Absorption of Gravity." *Apeiron* 13: 3–11.

Assis, André K. T. (1992b). "Deriving Gravitation from Electromagnetism." *Canadian Journal of Physics* 70: 330–340.

Assis, André K. T. (1994). *Weber's Electrodynamics*. Dordrecht: Kluwer.

Barbour, Julian B. (1974). "Relative-Distance Machian Theories." *Nature* 249: 328–329. Misprints corrected in *Nature* 250: 606 (1974).

Barbour, Julian B. and Bertotti, Bruno (1977). "Gravity and Inertia in a Machian Framework." *Nuovo Cimento B* 38: 1–27.

Barbour, Julian B. and Bertotti, Bruno (1982). "Mach's Principle and the Structure of Dynamical Theories." *Proceedings of the Royal Society of London A* 382: 295–306.

Brown, G. B. (1955). "A Theory of Action-at-a-Distance." *Proceedings of the Physical Society B* 68: 672–678.

Brown, G. B. (1982). *Retarded Action-at-a-Distance*. Luton: Cortney Publications, Chapter 7, "The Law of Gravitation and the Origin of Inertia."

Eby, Peter B. (1977). "On the Perihelion Precession as a Machian Effect." *Lettere al Nuovo Cimento* 18: 93–96.

Edwards, W. F. (1974). "Inertia and an Alternative Approach to the Theory of Interactions." *Proceedings of the Utah Academy of Sciences, Arts, and Letters* 51 (2): 1–7.

Friedlaender, Benedict and Friedlaender, Immanuel (1896). *Absolute oder relative Bewegung?* Berlin: Leonhard Simion. [Partial transl. in this volume p. 114 and p. 309.]

Gerber, Paul (1898). "Die räumliche und zeitliche Ausbreitung der Gravitation." *Zeitschrift fur Mathematik und Physik II* 43: 93–104.

Ghosh, Amithaba (1984). "Velocity Dependent Inertial Induction: An Extension of Mach's Principle." *Pramana Journal of Physics* 23: L671–L674.

Ghosh, Amithaba (1986). "Velocity-Dependent Inertial Induction and Secular Retardation of the Earth's Rotation." *Pramana Journal of Physics* 26: 1–8.

Ghosh, Amithaba (1991). "Velocity-Dependent Inertial Induction: A Possible Tired-Light Mechanism." *Apeiron* 9–10: 35–44.

Harman, P. M. (1982). *Energy, Force, and Matter – The Conceptual Devel-*

opment of Nineteenth-Century Physics. Cambridge: Cambridge University Press, pp. 96–97.

Laplace, Pierre S. (1969). *Celestial Mechanics*, Vol. 5. New York: Chelsea Publishing Company; reprint of Vol. V of the *Traité de Mécanique Céleste*, first published in 1825 and reprinted in 1882 as Volume 5 of Laplace's collected works, Book XVI, Chapter IV, pp. 448–452.

Mach, Ernst (1960). *The Science of Mechanics: A Critical and Historical Account of Its Development*. LaSalle, Illinois: Open Court.

Maxwell, James C. (1954). *A Treatise on Electricity and Magnetism*. New York: Dover, Volume 2, Chapter 23.

Maxwell, James C. (1965a). "On Faraday's Lines of Force." In *The Scientific Papers of James Clerk Maxwell*. W. D. Niven, ed. New York: Dover, pp. 155–229, see especially p. 208.

Maxwell, James C. (1965b). "A Dynamical Theory of the Electromagnetic field." In *The Scientific Papers of James Clerk Maxwell*, W. D. Niven, ed. New York: Dover, pp. 526–597, see especially p. 527.

Moon, Parry and Spencer, Domina E. (1959). "Mach's Principle." *Philosophy of Science* 26: 125–134.

Nernst, Wilhelm (1937). "Weitere Prüfung der Annahme eines stationären Zustandes im Weltall." *Zeitschrift fur Physik* 106: 633–661.

Newton, Isaac (1952). *Mathematical Principles of Natural Philosophy*. In *Great Books of the Western World*, Vol. 34. Chicago: Encyclopaedia Britannica, translated from Latin by Andrew Motte and revised by Florian Cajori.

North, John D. (1965). *The Measure of the Universe – A History of Modern Cosmology*. Oxford: Clarendon Press.

Norton, John D. (1995). "Mach's Principle before Einstein." This volume, p. 9.

Poincaré, Henri (1953). "Les Limites de la Loi de Newton." *Bulletin Astronomique* 17: 121–269.

Przeborski, M. A. (1933). "Sur les Forces Dépendent des Accélérations." *Comptes Rendues de l'Academie des Sciences de Paris* 197: 300–302.

Reissner, Hans (1914). "Über die Relativität der Beschleunigungen in der Mechanik." *Physikalishe Zeitschrift* 15: 371–375. [Transl. in this volume, p. 134.]

Reissner, Hans (1915). "Über eine Möglichkeit die Gravitation als unmittelbare Folge der Relativität der Trägheit abzuleiten." *Physikalishe Zeitschrift* 16: 179–185. [Extracts translated in this volume, p. 134.]

Schiff, Leonard I. (1964). "Observational Basis of Mach's Principle." *Reviews of Modern Physics* 36: 510–511.

Schrödinger, Erwin (1925). "Die Erfüllbarkeit der Relativitatsforderung in der klassischen Mechanik." *Annalen der Physik* 77: 325–336.

Sciama, Dennis W. (1953). "On the Origin of Inertia." *Monthly Notices of the Royal Astronomical Society* 113: 34–42.

Seeliger, Hugo (1895). "Ueber das Newton'sche Gravitationsgesetz." *Astronomische Nachrichten* 137: 129–136.

Seeliger, Hugo (1917). "Bemerkung zu P. Gerbers Aufsatz: 'Die Fortpfanzungsgeschwindigkeit der Gravtitation'." *Annalen der Physik* 53: 31–32.

Tisserand, François Félix (1872). "Sur le Mouvement des Planètes autour du Soleil, d'après la Loi Électrodynamique de Weber." *Comptes Rendues de l'Academie des Sciences de Paris* 75: 760–763.

Tisserand, François Félix (1895). *Traité de Mécanique Céleste*. Paris: Gauthier-Villars, Vol. 4, Chap. 28 (Vitesse de propagation de l'attraction), pp. 499–503, "Loi d'Attraction Conforme à la Loi Électrodynamique de Weber."

Treder, Hans J. (1975). "The Post-Newtonian Effect of Gravitation and the Retardation of the Gravitational Potential in Classical and Relativistic Theories of Gravitation." *Annalen der Physik* 32: 338–350.

Treder, Hans J., von Borzeszkowski, H. H., van der Merwe, Alwyn and Yourgrau, W. (1980). *Fundamental Principles of General Relativity Theories*. New York: Plenum, Chapter 2.

Wesley, James P. (1990). "Weber Electrodynamics: Part III. Mechanics, Gravitation." *Foundations of Physics Letters* 3: 581–605.

Woodruff, A. E. (1976). "Weber, Wilhelm Eduard." In *Dictionary of Scientific Biography*. C. C. Gillispie, ed. New York: Scribner, Vol. 14, pp. 203–209.

Discussion

Vucetich: When you introduce the expression for the force, you explicitly introduce Hubble's constant, which is not a constant, generally, but varies in time. So do you get a varying gravitational constant?

Assis: The Hubble constant in this model is introduced as the term in the exponential decay. So if you write that expression in terms of Hubble's constant, you have two choices: If the universe is expanding and so on, you get that the Newtonian gravitational constant is related to Hubble's constant. If one is varying, the other is also varying. But if the universe has no expansion, and Hubble's law of red shift has another origin, like tired light or any other thing, then Hubble's constant and Newton's gravitational constant will be constant in time. So that depends on the origin of the red shift.

Lynden-Bell: You take a totally isotropic universe.

Assis: No, I assume you can always divide the universe into two parts - one anisotropic, and one isotropic.

Lynden-Bell: Yes, but I think if you take a small but significant thing like the center of the Galaxy, or the Great Attractor, or something like that, which is far away, and in the system, you'll find that the mass is slightly anisotropic.

Assis: Not necessarily, because this anisotropy may also appear in the other constants, which you apply in the force. So, like Dicke (1961, 1964) said, the effects may cancel out in the end.

Lynden-Bell: Well, that might happen, but I think in a purely gravitational situation, I don't think it does [see, for example, (Nordtvedt 1975)].

Brill: If you try to implement into your scheme the principle of relativity, according to which influences take a finite time to propagate, would you then need to introduce advanced potentials? If I start accelerating now, then I see the distant universe accelerate now; but because what I see now happened earlier, the universe must have started accelerating a long time ago.

Assis: Yes, what I would say is that only recently have people begun to introduce retardation in Weber's law. There is a paper by Wesley (1990), who introduced that since 1987. Not only in electrodynamics, but also in gravitation. And so the situation is still open with regard to what we will get with retardation in Weber's law applied to gravitation and electrodynamics. But this is a new area of research which is being performed nowadays, so I can't answer it now.

REFERENCES

Dicke, Robert H. (1961). "Experimental Tests of Mach's Principle." *Physical Review Letters* 7: 359–360. Reprinted in (Dicke 1964).

Dicke, Robert H. (1964). *The Theoretical Significance of Experimental Relativity*. London: Blackie and Son Ltd.

Nordtvedt, Kenneth (1975). "Anisotropic Gravity and the Binary Pulsar PSR 1913+16." *Astrophysical Journal* 202: 248–249.

Wesley, James P. (1990). "Weber Electrodynamics." *Foundations of Physics Letters* 3: 443–490.

A Relative Newtonian Mechanics

Donald Lynden-Bell

1. Introduction

Newton was well aware of the difficulty of defining absolute space. In the *Principia* (Newton 1687, Cajori 1934) he writes (my italics):

> It is indeed a matter of *great difficulty* to discover and effectually to distinguish the *true* motions of particular bodies from the *apparent*, because the parts of that immovable space in which those motions are performed *do by no means come under the observation of our senses*. Yet the thing is not altogether desperate.

To Newton absolute space was part of reality. To Leibniz,[1] it was a useful invention for simplifying calculations of the relationships between bodies (Alexander 1984).

Can we rebuild Newtonian mechanics without the concept of *absolute* space? In this paper, a purely relative Lagrangian is found that yields the same results as Newtonian mechanics in all cases when the angular momentum of the whole universe is zero.

2. Relativity of Translation

This is most readily done from the Lagrangian

$$L = T - V, \qquad (1)$$

where

$$T = \sum_i \frac{1}{2} m_i \left[\frac{d\mathbf{r}_i}{dt} \right]^2, \qquad (2)$$

and

Einstein Studies, vol. 6: Mach's Principle: From Newton's Bucket to Quantum Gravity, pp. 172–178 © 1995 *Birkhäuser Boston, Inc. Printed in the United States.*

$$V = -G\sum_{i<}\sum_{j} m_i m_j / r_{ij}. \tag{3}$$

Why is one of these terms a sum over all particles and the other a sum over all pairs of particles? Why does the first change when we choose a moving or rotating frame while the other does not?

Relative to axes moving with velocity $\mathbf{v}(t)$, the kinetic energy is

$$T_v = \sum \frac{1}{2} m_i \left[\frac{d\mathbf{r}_i}{dt} - \mathbf{v}\right]^2 = T - M\mathbf{u}\cdot\mathbf{v} + \frac{1}{2}M\mathbf{v}^2, \tag{4}$$

where

$$M = \sum m_i \text{ and } \mathbf{u} = \frac{1}{M}\sum_i m_i \frac{d\mathbf{r}_i}{dt}. \tag{5}$$

Now minimize T_v over all possible choices of $\mathbf{v}(t)$, and we find

$$\mathbf{v} = \mathbf{u}, \tag{6}$$

and the minimum value, T^*, of T_v is, after a little manipulation and writing $\mathbf{r}_{ij} = \mathbf{r}_j - \mathbf{r}_i$,

$$T^* = \sum_{i<}\sum_{j} \frac{m_i m_j}{2M}\left[\frac{d\mathbf{r}_{ij}}{dt}\right]^2. \tag{7}$$

Notice that T^* is a double sum like V and only involves the relative velocities, so, like V, it is invariant under the transformation $\mathbf{r}_i \to \mathbf{r}_i + \mathbf{\Delta}(t)$. Hence the new Lagrangian, $L^* = T^* - V$, gives no equation for the motion of the center of mass. We can take it to move how we like! Lagrange's equations are

$$m_i \frac{d^2\mathbf{r}_i}{dt^2} = \sum_{j\neq i} \frac{Gm_i m_j \mathbf{r}_{ij}}{r_{ij}^3} + \frac{m_i}{M}\sum_j m_j \frac{d^2\mathbf{r}_j}{dt^2}. \tag{8}$$

If we decide to choose axes so that the mass center of the universe moves uniformly in a straight line, then in such axes the final term vanishes, so we recover the Newtonian equations. However, we are at liberty to choose axes that move more jerkily, in which case the final term remains. The new Lagrangian may be written

$$L^* = \frac{1}{M}\sum_{i<}\sum_{j} m_i m_j \left[\frac{1}{2}\left(\frac{d\mathbf{r}_{ij}}{dt}\right)^2 - \frac{GM}{r_{ij}}\right]; \tag{9}$$

in this form it is remarkable that G only occurs in the product GM. The initial M^{-1} is irrelevant as the variation of ML^* gives the same result as the variation of L^*. Then G and the mass of the universe occur only in the product GM. This suggests, but in no way proves, that the value of the gravitational constant is in some way determined by the mass of the universe.

The relative motions of the bodies are the same in all such axes and those are the observables. However T^* is not purely relative; nor is L^*, so although this dynamics is invariant to arbitrary time-dependent displacements of the axes, it is not invariant under transformation to rotating axes. Rotation remains absolute.

3. Relativity under Rotation

To remove absolute rotation from Newtonian mechanics, we make a similar modification of T^*. Relative to axes that rotate at angular velocity $\mathbf{\Omega}(t)$, the kinetic energy is

$$T_\Omega^* = \sum_{i<}\sum_j \frac{m_i m_j}{2M}\left[\frac{d\mathbf{r}_{ij}}{dt} - \mathbf{\Omega}\times\mathbf{r}_{ij}\right]^2. \tag{10}$$

Minimizing this over all choices of $\mathbf{\Omega}(t)$, we find, calling the $\mathbf{\Omega}$ of the minimizing frame $\mathbf{\Omega}^*(t)$,

$$\sum_{i<}\sum_j \frac{m_i m_j}{2M}\left[\frac{d\mathbf{r}_{ij}}{dt} - \mathbf{\Omega}^*\times\mathbf{r}_{ij}\right]\times\mathbf{r}_{ij} = 0 \tag{11}$$

which yields

$$\underline{I}\cdot\mathbf{\Omega}^* = \mathbf{J}, \tag{12}$$

where $\underline{I}$, the moment of inertia about the barycenter, is

$$\underline{I}(t) = \sum_{i<}\sum_j \frac{m_i m_j}{M}(\underline{\delta} r_{ij}^2 - \mathbf{r}_{ij}\mathbf{r}_{ij}), \tag{13}$$

$\underline{\delta}$ is the unit tensor (Kronecker's symbol), and $\mathbf{J}$, the angular momentum of the universe about its barycenter, is

$$\mathbf{J} = \sum_{i<}\sum_j \frac{m_i m_j}{M}\mathbf{r}_{ij}\times\frac{d\mathbf{r}_{ij}}{dt}. \tag{14}$$

The minimum kinetic energy is given by

$$T^{**} = \sum_{i<}\sum_j \frac{m_i m_j}{2M}\left[\frac{d\mathbf{r}_{ij}}{dt}\right]^2 - \frac{1}{2}\mathbf{J}\cdot\underline{I}^{-1}\cdot\mathbf{J}. \tag{15}$$

To demonstrate more explicitly that this kinetic energy is independent of the rotating frame chosen to write it in, we write $\mathbf{r}_{ij} = r_{ij}\hat{\mathbf{r}}_{ij}$, where r_{ij} is the purely relative distance and $\hat{\mathbf{r}}_{ij}$ is the unit vector in the initial axes. Its rate of change may be written in those axes

$$\frac{d\hat{\mathbf{r}}_{ij}}{dt} = \boldsymbol{\omega}_{ij} \times \hat{\mathbf{r}}_{ij}. \tag{16}$$

We may then rewrite the kinetic energy as

$$T^{**} = \sum_{i<}\sum_{j} \frac{m_i m_j}{2M}\left\{\left(\frac{dr_{ij}}{dt}\right)^2 + [(\boldsymbol{\omega}_{ij} - \boldsymbol{\Omega}^*) \times \mathbf{r}_{ij}]^2\right\}, \tag{17}$$

which clearly depends only on the *relative* angular velocity $\boldsymbol{\omega}_{ij} - \boldsymbol{\Omega}^*$, which is independent of the frame. It is the angular velocity of $\hat{\mathbf{r}}_{ij}$ as seen by an observer in the $\boldsymbol{\Omega}^*(t)$ frame. It is related to the angular momentum in the initial frame by (12).

Now consider the mechanics that follows from the Lagrangian $L^{**} = T^{**} - V$. In L^{**} the quantity $\mathbf{J}$ is not fixed but is a shorthand for expression (14), and $\mathbf{r}_{ij}$ and $d\mathbf{r}_{ij}/dt$ are to be varied. Expressions (7) and (15) only differ in the last term $-\mathbf{J}\cdot\underline{I}^{-1}\mathbf{J}/2$. The first-order variation of this term is $-\delta\mathbf{J}\cdot\underline{I}^{-1}\cdot\mathbf{J} - \mathbf{J}\cdot\delta\underline{I}^{-1}\cdot\mathbf{J}/2$. Whatever the δ terms are, this is zero if $\mathbf{J}$, the angular momentum of the universe in the Newtonian frame, is zero. Thus in a Newtonian universe with zero angular momentum the new dynamics given by L^{**} precisely agrees with Newton's.

The new dynamics is non-Newtonian in that it has no absolute space and is purely relative. It does not predict motions of bodies but rather their separations and relative orientations as functions of time. To compare its predictions with Newtonian theory, we must predict those relative quantities according to Newtonian theory. From the discussion of the last paragraph, we see that when Newtonian theory is applied to a universe with $J=0$ in absolute Newtonian axes, then the relative dynamics' Lagrangian L^{**} is equivalent to L^*, which was shown to give exactly the same relative motions as Newtonian mechanics in our discussion in Sec. 1, Relativity of Translation. For such universes the new relative dynamics gives the same predictions as Newton's. However, now suppose that absolute space exists and Newtonian mechanics governs even universes that rotate relative to absolute axes. In such universes $\mathbf{J} \neq 0$. If, perhaps wrongly, we were to apply the new relative mechanics, will its predictions differ from those found from Newtonian mechanics? The answer is yes!

Translating what the new relative mechanics predicts into the more familiar Newtonian language, its answer for the relative motions is that given by the following procedure. Take the given initial positions and velocities in absolute axes permanently zeroed at the barycenter. Take off from the resulting initial velocities the rotational velocity $\boldsymbol{\Omega}^* \times \mathbf{r}_i$, where $\boldsymbol{\Omega}^* = \underline{I}^{-1}\cdot\mathbf{J}$. This, of course, gives the initial velocities relative to axes rotating with $\boldsymbol{\Omega}^*$. Now work out the motions that Newton would have predicted from these doctored initial conditions *had he forgotten that these axes rotate!* These motions clearly do not obey true Newtonian mechanics unless $\boldsymbol{\Omega}^*=0$ (i.e., unless $\mathbf{J}=0$)! However, the relative motions predicted by adopting this procedure and then calculating the resulting *relative* motions are exactly those predicted by the new relative mechanics. In this sense the new mechanics differs from Newton's if the universe rotates in absolute space.

However, the evidence suggests that the universe does not rotate; so applied to it the new mechanics predicts the same relative motions as Newton's. Furthermore, the new mechanics has extra symmetries and does away with absolute space, so it is perhaps preferable to the Newtonian description. If we adopt it as a revised correct 'Classical Mechanics,' then there is no meaning to the motion of the center of mass of the universe and no meaning to the rotation or angular momentum of the universe because all motion is relative!

One part can, of course, rotate relative to another. The Lagrangian L^{**} is invariant to arbitrary time-dependent rotations of axes $\boldsymbol{\Omega}(t)$ and to arbitrary time-dependent translations $\boldsymbol{\Delta}(t)$. It is convenient in practice to choose a frame in which parts of the system that are far distant from any subsystem that concerns us have as little effect as possible. This is conveniently done by choosing axes such that for the universe at all times

$$\sum_{i<}\sum_{j} \frac{m_i m_j}{M} \mathbf{r}_{ij} \times \dot{\mathbf{r}}_{ij} = 0$$

and

$$\sum_i m_i \dot{\mathbf{r}}_i = M\mathbf{u} = \text{const.}$$

In such a frame the equations are Newtonian, and we can consider the isolated subsystems as independent except for the influence of the universe, which is perceived only through these inertial axes.

Although rediscussions of Newtonian mechanics such as this may clarify some seventeenth-century issues, they are no substitute for a full general relativistic discussion of the relativity of inertia, some steps towards which are given in this book. Many past attempts [for example, those of Al'tshuler (1967), Lynden-Bell (1967), Sciama *et al.* (1969), and Raine (1975, 1981)] are unsatisfactory since they neglect energy and momentum in gravitational waves as a source of inertia.

4. Retrospect

The considerations above are in the spirit of the seventeenth century and cannot be converted readily into a relativistic theory (Barbour and Bertotti 1982; Earman 1989). However, in the course of trying the first step I found an interesting quantity in special relativity that needs further physical interpretation. It is placed here so others may be intrigued and interpret physically.

In special relativity the total energy E and momentum $\mathbf{P}$ are the sums of those of the independently moving components

$$E=\sum_i \epsilon_i, \quad \mathbf{P}=\sum \mathbf{p}_i; \tag{18}$$

$E^2-\mathbf{P}^2c^2$ is invariant to changes to moving axes. Hence the minimum of E, E^*, is attained in axes for which $\mathbf{P}=0$. Now put $\Sigma m_i=M$ and consider

$$E^{*2}-(Mc^2)^2. \tag{19}$$

After some manipulation I find, writing $\epsilon_{ij}^{*2}=(\epsilon_i+\epsilon_j)^2-(\mathbf{p}_i+\mathbf{p}_j)^2c^2$,

$$E^{*2}-(Mc^2)^2=\sum_{i<}\sum_{j}\{\epsilon_{ij}^{*2}-[(m_i+m_j)c^2]^2\}. \tag{20}$$

If one calls $E^{*2}-(Mc^2)^2$ the 'kinergy,' then Eq. (20) says that the kinergy of the system is the sum of the kinergies of all pairs of the particles that constitute the system. Note that the kinergy is *not* the square of the kinetic energy in the center-of-mass frame.

This article is closely related to that published in the Symposium in honor of Michael Feast (Lynden-Bell 1992).

Note added 3 October 1994. For the general-relativity version of this work, the reader is referred to the paper by Lynden-Bell, Katz, and Bičák (1995) "Mach's Principle from the Relativistic Constraint Equations." *Monthly Notices of the Royal Astronomical Society* 272: 150–160.

NOTE

[1]Huygens, Berkeley, Mach, and Einstein were all critical of absolute space! Bondi gives a fine discussion of the issues. The development here is in the spirit of Barbour and Bertotti (1977) and turns out to be mathematically equivalent to the quasi-Newtonian example given in §3 of Barbour and Bertotti (1982), although that equivalence is not apparent.

REFERENCES

Alexander, H. G., ed. (1984). *The Leibniz–Clarke Correspondence*. New York: Barnes and Nobel.

Al'tshuler, Boris (1967). "Integral Formulation of Einstein's Equations and a Covariant Formulation of Mach's Principle." *Soviet Physics JETP* 24: 766–769.

Barbour, Julian B. and Bertotti, Bruno (1977). "Gravity and Inertia in a Machian Framework." *Nuovo Cimento* 38B: 1–27.

Barbour, Julian B. and Bertotti, Bruno (1982). "Mach's Principle and the Structure of Dynamical Theories." *Proceedings of the Royal Society London* 382: 295–306.

Berkeley, George (1721). *De Motu*.

Bondi, Hermann (1952). *Cosmology*. Cambridge: Cambridge University Press, pp. 27–33.

Cajoir, Florian, ed. (1934). *Sir Isaac Newton's Mathematical Principles*. Berkeley and Los Angeles: University of California Press.

Earman, John (1989). *World Enough and Space–Time*. Cambridge, MA: MIT Press.

Einstein, Albert (1920). In *Sidelights on Relativity*, reprinted 1983. New York: Dover.

Huygens, Christiaan (1888–1950). *Oevres Complètes de Christiaan Huygens*, 22 vols. The Hague: Martinus Nijhoff. (See especially Vol. 16, pp. 189–251, Vol. 18, p. 657ff., and Vol. 21, p. 503ff.)

Lynden-Bell, Donald (1967). "On the Origins of Space–Time and Inertia." *Monthly Notices of the Royal Astronomical Society* 135: 413–428.

Lynden-Bell, Donald (1992). "The Relativity of Acceleration." In *Variable Stars and Galaxies*, Symp. in honor of Michael W. Feast, Brian Warner ed., *Astronomical Society of the Pacific Conference Series* 30: 1–9.

Mach, Ernst (1960). *The Science of Mechanics: A Critical and Historical Account of Its Development*. LaSalle, Illinois: Open Court.

Newton, Isaac (1687). Scholium. In: *Philosophiae Naturalis Principia Mathematica*. London: Royal Society.

Raine, Derek J. (1975). "Mach's Principle in General Relativity." *Monthly Notices of the Royal Astronomical Society* 171: 507–528.

Raine, Derek J. (1981). "Mach's Principle and Space–Time Structure." *Reports on Progress in Physics* 44: 1151–1173.

Raine, Derek J. and Heller, Michael (1981). *The Science of Space–Time*. Tucson, Arizona: Pachart.

Sciama, Denis W., Waylen, Peter C., and Gilman, Richard C. (1969). "Generally Covariant Integral Formulation of Einstein's Field Equations." *Physical Review* 187: 1762–1770.

3. General Relativity as a More or Less Machian Theory

Introduction

This chapter opens with selected comments by Einstein that relate to the Machian issue. Further comments will be found under *Einstein* in the separate Quotations Index at the bottom of p. 636; see also Hoefer's paper, p. 67. Note especially the difference between Einstein's formulation of the relativity principle in terms of form invariance of the laws of nature in different coordinate systems and Poincaré's coordinate-free formulation in terms of the physical quantities that need to be posed in an initial-value problem for the evolution of the universe as a whole (p. 111–112). Poincaré's formulation is much closer to Mach's instinctive conviction and supplies a precise criterion, which is not easy to garner from Einstein's statements.

It is striking how many *different* formulations Einstein gives: form invariance of the laws of nature in different coordinate systems (1907, 1909, 1911); inertial mass must arise from interaction with other masses (1912, 1913, 1917); the principle of sufficient reason - observable effects must have observable causes (1914, 1916); general covariance needed because all measurements are merely verifications of coincidences (1916, 1918); the metric tensor must be completely determined by the matter in the universe (1918); a dynamical theory should not contain entities that act but which are not acted upon (p. 458).

The papers by Barbour and Isenberg basically attempt to analyze the dynamical structure of general relativity from Poincaré's initial-value standpoint. Both of these papers and King's contribution assume that in the determination of inertial frames of reference gravitational degrees of freedom are on an equal footing to matter degrees of freedom.

J.B.B.

Einstein Studies, vol. 6: Mach's Principle: From Newton's Bucket to Quantum Gravity, p. 179

Selected Passages on Machian Ideas

Albert Einstein

Hitherto we have applied the principle of relativity, i.e., the assumption that the laws of nature are independent of the state of motion of the reference system, only to systems of reference free of acceleration. Is it conceivable that the principle of relativity is also valid for systems that are accelerated relative to each other?.... In what follows, we shall assume the complete physical equivalence of a gravitational field and a corresponding acceleration of the reference system.

This assumption extends the principle of relativity to the case of uniformly accelerated translational motion of the reference system (Einstein 1907).

The treatment of the uniformly rotating rigid body seems to me to be of great importance on account of an extension of the relativity principle to uniformly rotating systems along analogous lines of thought to those that I tried to carry out for uniformly accelerated translation in the last section of my paper published in the *Zeitschrift für Radioaktivität* [sic] (Einstein (1909).

This assumption of exact physical equivalence [of a uniform gravitational field and uniform acceleration] makes it impossible for us to speak of the *absolute acceleration* of the system of reference, just as the usual theory of relativity forbids us to talk of the *absolute velocity* of a system (Einstein 1911).

In itself, this result is of great interest. It shows that the presence of the inertial shell K increases the inertial mass of the material point P within it. This makes it plausible that the *entire* inertia of a mass point is the effect of the presence of all other masses, resulting from a kind of interaction with them. This is exactly the standpoint for which E. Mach has argued persuasively in his penetrating investigations of this question (Einstein 1912).

Einstein Studies, vol. 6: Mach's Principle: From Newton's Bucket to Quantum Gravity, pp. 180–187

The theory [Einstein–Grossmann] that has been outlined eliminates an epistemological shortcoming that is present in not only the original theory of relativity but also in Galilean mechanics and has been emphasized by E. Mach in particular. It is clear that the concept of acceleration of a material point can no more be given an absolute significance than can that of velocity. Acceleration can only be defined as relative acceleration of a point relative to other bodies. This circumstance indicates that it is meaningless to ascribe to a body a resistance relative to acceleration as such (inertial resistance of bodies in the sense of classical mechanics); much rather, it must be required that the appearance of an inertial resistance be tied to the relative acceleration of the considered body relative to other bodies. It must be required that the inertial resistance of a body can be increased by bringing unaccelerated ponderable masses into the neighborhood of the body (Einstein 1913a).

To avoid misunderstanding, it should be said once more that I do not, any more than Mach, assert a logical necessity of the relativity of inertia. But a theory in which the relativity of inertia does hold is more satisfying than the theory currently used, since in the latter an inertial system is introduced whose state of motion is not determined by the state of the observable objects, i.e., is not determined by anything accessible to observation, while on the other hand it is supposed to play a determining role in the behavior of material points.

The concept of the relativity of inertia demands moreover not only that the inertia of a mass A be increased by the accumulation of masses BC... at rest in its neighborhood but also that the increase of the inertial resistance should have no effect if the masses BC... are accelerated together with the mass A. One can also express this as follows: The acceleration of the masses BC... must induce an accelerating force on A that is in the same direction as the acceleration. One sees in this way that the extra accelerating force must overcompensate the increase of the inertia caused by the mere presence of BC..., since in accordance with the relation between energy and inertia of systems the system ABC... as a whole must have a smaller inertia the smaller is its gravitational energy...

Equations (7e′) and (1d) [not reproduced here] show how slowly moving masses act on each other in accordance with the new gravitational theory. To a large degree, the equations correspond to those of electrodynamics: g_{44} corresponds to the scalar potential of electrical masses except for its sign and for the circumstance that the factor 1/2 occurs in the first term on the right-hand side of (1d). **g** [**g** is the vector with components g_{14}, g_{24}, g_{34}] corresponds to the vector potential of electrical currents; the second term on the right-hand side of (1d), which corresponds to an electric field strength that derives from the time variation of the vector potential, yields precisely the induction effect in the same direction as the acceleration that we must expect in accordance with the concept of inertia of energy. The vector **o** [curl **g**] corresponds in electrodynamics to the magnetic field strength (curl of the vector potential), and thus the final term in (1d) corresponds to the Lorentz force.

It should also be recalled that a term of the form $\dot{\mathbf{r}} \times \mathbf{o}$ occurs in the theory of relative motion in mechanics, being known as the Coriolis force. It can be shown on the basis of (7e′) that in the interior of a rotating shell there exists a field of the vector **o**, and this has the consequence that the plane of oscillation of a pendulum that is set up in the interior of the shell is not fixed in space but must, as a consequence of the rotation of the shell, execute a precession in the same sense as the rotation. This result too is to be anticipated – and was anticipated long ago – in the sense of the relativity of inertia. It is noteworthy that in this respect too the theory corresponds to such a conception; unfortunately, the effect that is to be expected is so small that we cannot hope to establish it through terrestrial experiments or in astronomy (Einstein 1913b).

But the confidence that we have in the theory of relativity has still another root. It is difficult to dismiss the following consideration. If K' and K are two coordinate systems that are in uniform motion relative to each other, then from the kinematic standpoint these systems are entirely equivalent. Therefore, we seek in vain for a sufficient reason why one of these systems should be more suitable than the other as a reference system for the formulation of the laws of nature; much rather, we feel forced to postulate the equivalence of the two systems.

However, this argument immediately brings forth a counter argument. Namely, the kinematic equivalence of two coordinate systems is by no means restricted to the case in which the two considered coordinate systems are in *uniform translational motion*. This equivalence from the kinematic standpoint exists, for example, just as well if the systems rotate uniformly relative to each other. One feels forced to the assumption that the theory of relativity as it has hitherto existed should be generalized to a large degree, so that the apparently unjustified preference for uniform translation over relative motions of other kinds disappears. Anyone who has thought about this matter seriously must experience this need for such an extension of the theory.

It does at first appear that such an extension of the theory of relativity is to be rejected on physical grounds. Let K be an allowed coordinate system in the Galilei–Newton sense and K' be a coordinate system that rotates uniformly with respect to K. Then masses at rest in K' are subject to centrifugal forces, which do not act on masses at rest in K. Newton already saw in this a proof that the rotation of K' is to be regarded as 'absolute' and that therefore one cannot treat K' on an equal footing with K as being 'at rest.' But, as E. Mach in particular has noted, this argument is not decisive. We do not need to attribute the existence of the centrifugal forces to the motion of K'; instead, we can just as well attribute them to the average rotational motion of the distant ponderable masses relative to K', this K' now being regarded as 'at rest.' If Newton's laws of mechanics and gravitation do not admit such an interpretation, the reason for this may well be a shortcoming of this theory (Einstein 1914).

In classical mechanics, and no less in the special theory of relativity, there is an inherent epistemological defect which was, perhaps for the first time, clearly pointed out by E. Mach. We will elucidate it by the following example: Two fluid bodies of the same size and nature hover freely in space at so great a distance from each other and from all other masses that only those gravitational forces need be taken into account which arise from the interaction of different parts of the same body. Let the distance between the two bodies be invariable, and in neither of the bodies let there be any relative movements of the parts with respect to one another. But let either mass, as judged by an observer at rest relatively to the other mass, rotate with constant angular velocity about the line joining the masses. This is a verifiable relative motion of the two bodies. Now let us imagine that each of the bodies has been surveyed by means of measuring instruments at rest relatively to itself, and let the surface of S_1 prove to be a sphere, and that of S_2 an ellipsoid of revolution.

Thereupon we put the question - What is the reason for this difference in the two bodies? No answer can be admitted as epistemologically satisfactory, unless the reason given is an *observable fact of experience*. The law of causality has not the significance of a statement as to the world of experience, except when *observable facts* ultimately appear as causes and effects.

Newtonian mechanics does not give a satisfactory answer to this question. It pronounces as follows: The laws of mechanics apply to a space R_1, in respect to which the body S_1 is at rest, but not to a space R_2, in respect to which the body S_2 is at rest. But the privileged space R_1 of Galileo, thus introduced, is a merely *factitious* cause, and not a thing that can be observed. It is therefore clear that Newton's mechanics does not really satisfy the requirement of causality in the case under consideration, but only apparently does so, since it makes the factitious cause R_1 responsible for the observable difference in the bodies S_1 and S_2.

The only satisfactory answer to the question addressed above must be that the physical system consisting of S_1 and S_2 reveals within itself no imaginable cause to which the differing behaviour of S_1 and S_2 can be referred. The cause must therefore lie *outside* this system. We have to take it that the general laws of motion, which in particular determine the shapes of S_1 and S_2, must be such that the mechanical behaviour of S_1 and S_2 is partly conditioned, in quite essential respects, by distant masses which we have not included in the system under consideration. These distant masses (and their motions relative to S_1 and S_2) must then be regarded as the seat of the causes (which must be susceptible to observation) of the different behavior of our two bodies S_1 and S_2. They take over the role of the factitious cause R_1. Of all imaginable spaces R_1, R_2, etc., in any kind of motion relatively to one another, there is none which we may look upon as privileged *a priori* without reviving the above-mentioned epistemological objection. *The laws of physics must be of such a nature that they apply to systems of reference in any kind of motion.* Along this road we arrive at an extension of the postulate of relativity (Einstein 1916).

We therefore reach this result: In the general theory of relativity, space and time cannot be defined in such a way that differences of the spatial co-ordinates can be directly measured by the unit measuring rod, or differences in the time co-ordinate by a standard clock.

The method hitherto employed for laying co-ordinates into the space-time continuum in a definite manner thus breaks down, and there seems to be no other way which would allow us to adapt systems of coordinates to the four-dimensional universe so that we might expect from their application a particularly simple formulation of the laws of nature. So there is nothing for it but to regard all imaginable systems of co-ordinates, on principle, as equally suitable for the description of nature. This comes to requiring that:

The general laws of nature are to be expressed by equations which hold good for all systems of co-ordinates, that is, are co-variant with respect to any substitutions whatsoever (generally co-variant).

It is clear that a physical theory which satisfies this postulate will be suitable for the general postulate of relativity. For the totality of *all* substitutions certainly includes those which correspond to all relative motions of (three-dimensional) systems of co-ordinates. That this requirement of general co-variance, which takes away from space and time the last remnant of physical objectivity, is a natural one, will be seen from the following reflection. All our space-time verifications invariably amount to a determination of space–time coincidences. If, for example, nothing happened in the world but the motion of material points, then ultimately nothing would be observable but the meetings of two or more of these points. Moreover, the results of our measurings are nothing but verifications of such meetings of the material points of our measuring instruments with other material points, coincidences between the hands of a clock and points on the clock dial, and observed point-events happening at the same place at the same time.

The introduction of a system of reference serves no other purpose than to facilitate the description of the totality of such coincidences (Einstein 1916).

In a consistent theory of relativity there can be no inertia *relatively to 'space,'* but only an inertia of masses *relatively to one another*. If, therefore, I remove a mass to a sufficient distance from all other masses in the universe, its inertia must fall to zero (Einstein 1917).

The theory, as it now appears to me, rests on three main points of view, which, however, are by no means independent of each other...:

a) *Relativity principle*: The laws of nature are merely statements about space-time coincidences; they therefore find their only natural expression in generally covariant equations.

b) *Equivalence principle*: Inertia and weight are identical in nature. It follows necessarily from this and from the result of the special theory of relativity that the symmetric 'fundamental tensor' [$g_{\mu\nu}$] determines the metrical properties of space, the inertial behavior of bodies in it, as well as gravitational effects. We shall denote the state of space described by the fundamental tensor as the '*G*-field.'

c) *Mach's Principle*[1]: The G-field is *completely* determined by the masses of the bodies. Since mass and energy are identical in accordance with the results of the special theory of relativity and the energy is described formally by means of the symmetric energy tensor ($T_{\mu\nu}$), this means that the G-field is conditioned and determined by the energy tensor of the matter (Einstein 1918a).

[1]Hitherto I have not distinguished between principles (a) and (c), and this was confusing. I have chosen the name 'Mach's principle' because this principle has the significance of a generalization of Mach's requirement that inertia should be derived from an interaction of bodies.

We want to distinguish more clearly between quantities that belong to a physical system as such (are independent of the choice of the coordinate system) and quantities that depend on the coordinate system. Ones initial reaction would be to require that physics should introduce in its laws only the quantities of the first kind. However, it has been found that this approach cannot be realized in practice, as the development of classical mechanics has already clearly shown. One could, for example, think – and this was actually attempted [Einstein is here presumably referring to the work of Hofmann and Reissner] – of introducing in the laws of classical mechanics only the distances of material points from each other instead of coordinates; *a priori* one could expect that in this manner the aim of the theory of relativity should be most readily achieved. However, the scientific development has not confirmed this conjecture. It cannot dispense with coordinate systems and must therefore make use in the coordinates of quantities that cannot be regarded as the results of definable measurements (Einstein 1918b).

Mach conjectures that in a truly rational theory inertia would have to depend upon the interaction of the masses, precisely as it was true for Newton's other forces, a conception which for a long time I considered as in principle the correct one. It presupposes implicitly, however, that the basic theory should be of the general type of Newton's mechanics: masses and their interaction as the original concepts. The attempt at such a solution does not fit into a consistent field theory, as will be immediately recognized (Einstein 1949).

When not otherwise indicated, translations by Julian B. Barbour

REFERENCES

Einstein, Albert (1907). "Über das Relativitätsprinzip und die aus demselben gezogenen Folgerungen," V. §17. *Jahrbuch der Radioaktivität und Elektronik* 4: 411–462.

Einstein, Albert (1909). Letter to A. Sommerfeld; *Albert Einstein–Arnold Sommerfeld. Briefwechsel.* Armin Hermann, ed. Basel and Stuttgart: Schwabe (1968).

Einstein, Albert (1911). "Über den Einfluss der Schwerkraft auf die Ausbreitung des Lichtes." *Annalen der Physik* 35: 898–908.

Einstein, Albert (1912). "Gibt es eine Gravitationswirkung, die der elektrodynamischen Induktionswirkung analog ist?" *Vierteljahrsschrift für gerichtliche Medizin und öffentliches Sanitätswesen* 44: 37–40.

Einstein, Albert (1913a). "Physikalische Grundlagen einer Gravitationstheorie." *Vierteljahrsschrift der Naturforschender Gesellschaft, Zürich* 58: 284–290.

Einstein, Albert (1913b). "Zum gegenwärtigen Stande des Gravitationsproblems." *Physikalische Zeitschrift* 14: 1249–1266.

Einstein, Albert (1914). "Die formale Grundlage der allgemeinen Relativitätstheorie." *Sitzungsberichte der Preussischen Akademie der Wissenschaften*, pp. 1030–1085.

Einstein, Albert (1916). "Die Grundlage der allgemeinen Relativitätstheorie." *Annalen der Physik* 49: 769–822; quoted with a few modifications from the translation: "The Foundation of the General Theory of Relativity." In *The Principle of Relativity*. Papers by H. A. Lorentz *et al.* New York: Dover (1952), pp. 111–164.

Einstein, Albert (1917). "Kosmologische Betrachtungen zur allgemeinen Relativitätstheorie." *Sitzungsberichte der Preussischen Akademie der Wissenschaften* 142–152; quoted with correction from the translation: "Cosmological Considerations on the General Theory of Relativity." In *The Principle of Relativity*. New York: Dover (1952), pp. 177–188.

Einstein, Albert (1918a). "Prinzipielles zur allgemeinen Relativitätstheorie." *Annalen der Physik* 55: 241–244.

Einstein, Albert (1918b). "Dialog über Einwände gegen die Relativitätstheorie." *Die Naturwissenschaften* 6: No. 48, 697–702.

Einstein, Albert (1949). "Autobiographical Notes." In *Albert Einstein, Philosopher–Scientist*. Paul A. Schilpp, ed. Evanston, Illinois: Library of Living Philosophers.

Wheeler–Einstein–Mach Spacetimes

James Isenberg

1. Introduction

Throughout the history of the natural sciences, physical principles – such as the Atomic Principle, the Cosmological Principle, the Principle of Conservation of Energy, the Equivalence Principle, the Parity Principle, the Quantum Principle, and Mach's Principle – have been formulated and used in addressing fundamental issues in physics such as the nature of matter and its interactions, the nature of space and time, and the typicality in the universe of our local physical laboratory. These principles, which are almost always based on some combination of empirical evidence, philosophical predilection, and theoretical orientation, have played an important role (not always progressive) in the development of theoretical physics.

Often formulated somewhat roughly, at least in its earliest incarnations, a physical principle usually states that certain behavior holds – or *should* hold – in real, observable, physical systems. A theoretical physicist who believes in a given principle may then use that principle as a screening device: Theories at variance with the principle may be tossed out, while those enforcing the behavior described by the principle may be favored. As for a theory which allows the behavior described by the principle but does not enforce it, the physicist may use the principle to screen among the various solutions of the theory,[1] favoring those which behave as prescribed by the principle and disfavoring those which do not. Such a theory draws support from the principle if almost all, if not all, solutions have the prescribed behavior.

While the interaction of physical principle and physical theory has proceeded fairly smoothly in a number of cases (for example, for the Principle of Equivalence and the theory of general relativity), it has also become quite convoluted in others. Mach's Principle (MP) provides a very interesting example of this. Ever since Einstein popularized MP

Einstein Studies, vol. 6: Mach's Principle: From Newton's Bucket to Quantum Gravity, pp. 188–207

and cited it as one of the primary motivations for the particular form he chose for the gravitational field equations in his theory of general relativity (Einstein 1918, 1955), Mach's Principle has hotly - and simultaneously - been used to argue that these equations (and the full theory) need to be modified, need to be abandoned, or should be supported as they are. While the debate has cooled in recent years, Mach's Principle is still used to argue both for and against general relativity and both for and against Einstein's equations in particular.[2]

The convoluted interaction between Mach's Principle and general relativity is largely a result of the difficulties and ambiguities which arise when one attempts to formulate MP precisely. Mach's Principle in its nineteenth-century form could be stated roughly as follows:

> **Mach's Principle** (19th-Century Version): The distribution of matter everywhere in the universe determines the inertial frame at each point in the universe.

Even in the context of the nineteenth-century physics in which Mach discussed his ideas, this statement involves some concepts - like inertia - which are difficult to make theoretically precise. In the context of twentieth-century physics, with vastly changed ideas concerning the nature of space and time and matter, almost every word in this statement is ambiguous. It is easy to see how one can produce greatly varying updated statements of Mach's Principle with consequently contradictory theoretical implications.

Our purpose here is *not* to discuss all of the different interpretations and statements of Mach's Principle which can be made, and debate their respective virtues. Instead, we shall focus on one particular interpretation: the initial value approach, as championed by Wheeler. We shall first state a version of this approach to MP, which we call the Wheeler–Einstein–Mach Principle (WEM Principle), in Sec. 2. Then in Sec. 3, we shall discuss how the WEM Principle relates to the traditional statement of Mach's Principle (as given above), clarifying some of the former's interpretations. Finally in Sec. 4 we shall discuss some theorems and conjectures which are relevant to the WEM Principle and its relationship to Einstein's and other theories of the gravitational field in spacetime.

2. Statement of the Wheeler–Einstein–Mach Principle

The present paradigm[3] (and that of the last 75 years) for mathematically describing a model for the physical universe is that of a *spacetime* (M^4, g, ψ), where M^4 is a smooth, time-orientable, four-dimensional manifold, g is a smooth Lorentz-signature metric field on M^4, and ψ represents a collection of smooth tensor and spinor and connection fields on M^4 (including, perhaps, Maxwell fields, Yang–Mills fields, fluid fields, Vlasov fields, Dirac fields, etc.). The Wheeler–Einstein–Mach Principle works within this spacetime paradigm. It is a statement of certain properties which a spacetime (M^4, g, ψ) should have if it is to be deemed 'Machian' (and therefore physically acceptable) from this WEM point of view.

The first requirement which the WEM Principle makes upon a spacetime is that it be spatially compact. More specifically, the WEM Principle requires that

$$M^4 = \Sigma^3 \times \mathbb{R}, \tag{1}$$

where Σ^3 is a compact (without boundary) orientable three-dimensional manifold, such as S^3, T^3, $S^2 \times S^1$, or the like. Physically, it follows that 'at a given instant,' the universe is finite, with finite volume.

The second WEM Principle requirement is that the spacetime be globally hyperbolic, with no spacetime extensions existing (globally hyperbolic or not). This condition tells us that (M^4, g, ψ) is causal – no closed or almost closed causal paths – and is not part of a bigger spacetime which fails to be causal. Further, this condition tells us that if (M^4, g, ψ) satisfies a set of field equations which has a well-posed Cauchy formulation (relative to g) – for example, the vacuum Einstein, the Einstein–Maxwell, the Einstein–Yang–Mills, the Einstein–Vlasov, or the Brans–Dicke field equations – then it follows that if one knows the values of the fields (g, ψ) and their transverse ('time') derivatives – i.e., the 'initial data' – on a spacelike embedding of Σ^3 in $M^4 = \Sigma^3 \times \mathbb{R}$ – i.e., on a 'Cauchy surface' – one can use the field equations to determine the fields (g, ψ) everywhere on M^4.

The third requirement is that, indeed, (M^4, g, ψ) satisfies a set of field equations having a well-posed Cauchy formulation.

The last requirement that the WEM Principle makes on a spacetime (M^4, g, ψ) is that it satisfy a set of field equations which imposes *constraint equations* on the initial data on any Cauchy surface; and further that one can split the initial data into two sets of fields such that the first set can be freely chosen on the Cauchy surface, and the second set can be determined from the first set using these constraint equations. Typically, the constraint equations take the form of a system of elliptic partial differential equations with coefficients involving the first set of fields, to be solved for the second set.

Note that if a spacetime satisfies all four of the requirements of the WEM Principle (we shall call such a spacetime a 'WEM spacetime'), then it can be reconstructed completely if one knows explicitly the values of the first set of fields on any Cauchy surface in M^4.

3. Are WEM Spacetimes Machian?

We would like to argue that every spacetime that satisfies the requirements listed above (in Sec. 2) for a WEM spacetime is Machian in the traditional sense. From the Wheeler (initial-value) point of view, the key property which makes a spacetime Machian is, in a slightly modified version of the nineteenth century statement of Mach's Principle given in Sec. 1, that *the distribution of matter and field energy–momentum everywhere at a particular moment in the universe determines the inertial frame at each point in the universe.*

This property holds in every WEM spacetime in the following sense: To represent the universe at a particular moment (let us call this moment 't_0'), one chooses a fixed Cauchy surface $\Sigma^3_{t_0}$ in (M^4, g, ψ). This Cauchy surface may be *any* spacelike embedding of Σ^3 in M^4 (Budic, Isenberg, Lindblom, and Yasskin 1978). So, at the moment t_0, "everywhere... in the universe" means everywhere on the Cauchy surface Σ_{t_0}. And the "distribution of matter and field energy–momentum everywhere" means the specification of certain fields on $\Sigma^3_{t_0}$. Which fields? From the WEM Principle point of view, the right fields to know on $\Sigma^3_{t_0}$ are exactly those which make up the 'first set' of the split of the initial data, as discussed in the last WEM Principle requirement (Sec. 2). It follows from this last WEM Principle requirement that, if we know explicitly this first set of fields on $\Sigma^3_{t_0}$ (which we identify as the "distribution of matter and field energy–momentum everywhere at the moment t_0"), then we can use the constraint equations to determine uniquely the full set of initial data on $\Sigma^3_{t_0}$. Then, since the spacetime (M^4, g, ψ) is globally hyperbolic (second WEM Principle requirement) and since the fields (g, ψ) satisfy a set of field equations which have a well-posed Cauchy formulation (third WEM Principle requirement), it follows that we can uniquely determine g and ψ in a spacetime neighborhood of $\Sigma^3_{t_0}$. The inertial frames are determined by the metric, so we have (at least near $\Sigma^3_{t_0}$) determined the inertial frames everywhere at time t_0 from knowledge of the matter and field energy–momentum everywhere at time t_0. In fact, since the spacetime (M^4, g, ψ) is globally hyperbolic and nonextendible (second WEM Principle requirement) we can determine (from knowledge of the matter and field energy–momentum everywhere at time t_0) the inertial frames everywhere in the universe *at all times*.

This claim - that a WEM spacetime is Machian in the traditional sense - and some of the details of the statement of the WEM version of Mach's Principle raise a number of questions, which we shall now address.

a) Why does the Wheeler–Einstein–Mach Principle focus on knowledge of the matter and field energy-momentum *at a specific moment*? Why do we need to choose a Cauchy surface in M^4 to specify a particular moment? Are all choices of a Cauchy surface equally valid, or are some choices better than others?

Whether or not Mach intended the determination of inertial frames to be based on knowledge of the matter distribution at a particular moment, or on knowledge of the matter distribution for all time, is not clear. Regardless of Mach's intentions, it is a much stronger restriction to demand that inertial frames be determined from information at a single moment, rather than from information at all times, and Wheeler has chosen this interpretation. It is the basis of the initial-value approach to Mach's Principle, and it is the basis of the WEM Principle.

One of the key features of general relativity and of our present way of thinking about possible models of the universe in terms of spacetimes (M^4, g, ψ) is that such a model does *not* generally come equipped with a notion of simultaneity. On the other hand, one can *pick* a notion of simultaneity in any given spacetime: One simply assigns certain sets of acausally related points in the spacetime to be simultaneous (labeling them with the same value of time). If such an assignment is made consistently with the causal structure of the spacetime and if it is made as complete and as all-inclusive as possible, then each set of simultaneous points makes up a Cauchy surface. Conversely, every choice of a family of Cauchy surfaces determines a notion of simultaneity. Hence we identify the choice of a notion of simultaneity with the choice of a family of Cauchy surfaces. Note that if the spacetime (M^4, g, ψ) is spatially compact, then every spacelike embedding of Σ^3 in M^4 specifies a Cauchy surface (Budic, Isenberg, Lindblom, and Yasskin 1978), and every Cauchy surface is given by such an embedding. Note also that if one wants to make a choice of time for every point in the spacetime (associating to each point a unique Cauchy surface Σ^3_t), then one simply needs to choose a foliation of (M^4, g, ψ) by Cauchy surfaces.

Now in certain spacetimes there are physically preferred choices of simultaneity which can be made. For example, in Minkowski spacetime, an observer can use various light-bouncing schemes to single out certain 'Lorentz frame' Cauchy surfaces through each point. These are unique up to the action of the Lorentz group. In a Friedman–Robertson–Walker spacetime, the spatial homogeneity of the spacetime picks out a unique Cauchy surface through each spacetime point. However, in a general spacetime with no isometries, there is generally no privileged choice of Cauchy surfaces[4]; hence "everywhere at the particular moment t_0" may be associated to *any* Cauchy surface in the spacetime.

> b) Mach's Principle in its traditional form (see Sec. 1) says that one should be able to determine inertia everywhere based on knowledge of the *matter* distribution everywhere. Exactly which fields does the WEM Principle require one to know (as 'matter and field energy–momentum distribution')? If it requires more than just the matter distribution, why?

In our description of the Wheeler–Einstein–Mach Principle (in Sec. 2), we have been a bit vague concerning the specific fields ('the first set') that one needs to know on $\Sigma^3_{t_0}$ if one wants to use the constraints to determine the rest of the initial data ('the second set'). This vagueness allows us to avoid tying the statement of the WEM Principle to a specific set of field equations. If we examine how this works for spacetimes satisfying, say, the Einstein–Maxwell-perfect-fluid equations, then we can be fairly specific regarding the particular fields we need to know on $\Sigma^3_{t_0}$: (i) the spatial metric, up to conformal factor[5]; (ii) the mean curvature function; (iii) the transverse-traceless part of the extrinsic curvature, up to conformal factor[5]; (iv) the electric and magnetic vector fields, up to conformal factor[5]; and (v) the energy density function and momentum density function of the fluid, up to conformal factor.[5]

This is clearly more than just information regarding the matter distribution. While the specific choice of the needed fields is determined by the mathematics of the field equations under study, it should not surprise us that we do need to know about more than just the 'matter' fields. Physicists have learned in the twentieth century that electromagnetic fields and gravitational fields (as well as most others) carry energy and momentum, and concentrations of these fields can accurately mimic matter concentrations. Indeed, electromagnetic fields can transform into elementary particle matter and vice versa; the same is true of gravitational fields. So knowledge of some parts of the electromagnetic and gravitational (and other) fields is needed, along with knowledge of matter fields directly, to determine inertia. As for which parts of the fields we need to know, we let the mathematics tell us. What it tells us makes sense.

c) What, precisely, does it mean to determine the inertial frames at every point in the universe? Is it *sufficient* to know the metric everywhere? Is it *necessary* to know the metric everywhere?

It has never been easy, either in the context of nineteenth- or twentieth-century physics, to pinpoint precisely where in the structure of space or spacetime it is that the direct information about inertial frames resides. This is one of the reasons why discussions of Mach's Principle have often focused on specific inertial phenomena like frame-dragging by rotating or accelerating matter, rather than on abstract, more complete formulations of the notion of inertial frames. Certainly, however, if we know the full spacetime metric everywhere in a neighborhood of a point, then we know all there is to know about inertial frames at that point. It may not be necessary to know the full spacetime metric – this issue is dodged in the formulation of the WEM Principle – but it is certainly sufficient.

d) Why does the WEM version of Mach's Principle require that a WEM spacetime be spatially compact?

While Mach never said anything about the universe being compact or finite, Einstein argued in his 1917 cosmological paper that if a spacetime is to satisfy the Machian requirement, then it should be spatially compact (Einstein 1917, 1955). Einstein based his argument on the need which he perceived to avoid posing boundary conditions. If a spacetime manifold M^4 is not spatially compact, and if one wishes to solve an initial-value problem for a system of partial differential equations (like the Einstein–Maxwell–fluid equations) on M^4 for the metric g and the fields ψ, then one needs to specify the full metric (and full values of ψ) on the spatial boundary of M^4. (If the spacetime is asymptotically flat, then this means that one needs to specify limits for g and ψ as one approaches spatial infinity.) The need to make such a specification, Einstein argued, is un-Machian because it more or less requires one to choose absolute inertial frames at the spatial boundary, independent of the content of the universe. Since the need to pick boundary conditions for the metric goes away if the spacetime is spatially compact, Einstein required this feature in those spacetimes which he viewed as properly Machian. Wheeler, in his development of the initial-value formulation of Mach's Principle (Wheeler 1964; also Isenberg and Wheeler 1980), also treats spatial compactness as a key feature, and hence we include it in our statement of the Wheeler–Einstein–Mach Principle. We shall see below (in Sec. 4) that spatial compactness also plays a role in the proof of certain results which are useful in considering the extent to which WEM spacetimes are compatible with Einstein's gravitational field equations.

e) Why does the WEM Principle require that a WEM spacetime be inextendible?

A spacetime (M^4, g, ψ) is *smoothly extendible* to a (larger) spacetime $(\tilde{M}^4, \tilde{g}, \tilde{\psi})$ if there exists a smooth diffeomorphism ξ which maps M^4 into an open region $\xi(M^4) \subset \tilde{M}^4$, with $\xi^* \tilde{g}|_{\xi(M^4)} = g$ and $\xi^* \tilde{\psi}|_{\xi(M^4)} = \psi$. The spacetime $(M^4 = T^3 \times (-1,1),\ g = g_{\text{flat}},\ \psi = 0)$ is an example of an extendible spacetime: It extends to $(\tilde{M}^4 = T^3 \times \mathbb{R},\ \tilde{g} = g_{\text{flat}},\ \tilde{\psi} = 0)$ via the natural embedding. The familiar Taub spacetime is also extendible: It extends to Taub–NUT. The standard Big Bang–Big Crunch $(k = +1)$ Friedman–Robertson–Walker spacetime, on the other hand, is not extendible.

Globally hyperbolic spacetimes that only admit extensions to spacetimes which are also globally hyperbolic (the $T^3 \times (-1, +1)$ spacetime mentioned above is an example of such a spacetime) are not a real issue for the WEM Principle. The nonextendibility requirement in the WEM Principle simply replaces such a spacetime by its *maximal* globally hyperbolic extension. As shown by Choquet-Bruhat and Geroch (1969), such maximal globally hyperbolic extensions are unique.

If a globally hyperbolic spacetime admits a *non*globally hyperbolic extension (the Taub spacetime is an example of this), then the nonextendibility requirement does really disqualify the spacetime, and one may ask why this is necessary. Consider the following scenario: A physicist in the original (globally hyperbolic) spacetime (M^4, g, ψ) who is interested in Mach's Principle somehow gathers together the right matter and field energy–momentum information at time t_0 to determine inertial frames. Since (M^4, g, ψ) is globally hyperbolic, he can in fact determine inertial frames for all observers for all time in (M^4, g, ψ). But since (M^4, g, ψ) is extendible, some observers may proceed out of

(M^4, g, ψ) into the extension $(\tilde{M}^4, \tilde{g}, \tilde{\psi})$. Since $(\tilde{M}^4, \tilde{g}, \tilde{\psi})$ is not globally hyperbolic, and since these nonglobally hyperbolic extensions are not generally unique (Chrusciel and Isenberg 1993), the inertial frames of the fugitive observers cannot be determined. Collecting information from $(\tilde{M}^4, \tilde{g}, \tilde{\psi})$ outside (M^4, g, ψ) generally will not help, either.

Our view is that this is un-Machian behavior, so our statement of the Wheeler–Einstein–Mach Principle disallows spacetimes with nonglobally hyperbolic extensions.

One may argue against including this requirement in the WEM Principle, asserting that inertial frames need to be determined only near $\Sigma^3_{t_0}$, not necessarily for every observer for all time. Indeed, I believe that this is Wheeler's point of view, and he does not generally include the nonextendibility requirement in his version of Mach's Principle. However, I do consider nonextendibility to be an important part of the initial value formulation of Mach's Principle, so I include it in the WEM Principle.

f) Do Machian gedanken experiments have the proper outcomes in WEM spacetimes?

As noted above, much of the debate over Mach's Principle has focused on certain gedanken experiments, such as the ones involving rotating matter shells in which one tests for the proper amount of dragging of inertial frames (Dicke 1964). The statement of the WEM Principle does not in any way refer to these experiments; a space is labeled WEM or non-WEM without investigating the expected outcome of any such experiment. One can of course consider various gedanken experiments both in WEM and in non-WEM spacetimes, with Einstein's field equations or any other appropriate set of field equations imposed. Such a study, done systematically, could be interesting. It has not been done.

4. Conjectures and Theorems Relevant to the Wheeler–Einstein–Mach Principle

Do any WEM spacetimes exist? Which field theories do they satisfy? Are there any field theories for which the generic solution on $\Sigma^3 \times \mathbb{R}$ is a WEM spacetime? These questions are not philosophical or theoretical; they are mathematical. We shall discuss some mathematical results and mathematical conjectures which are relevant to questions such as these.

We first need to know that there exist field theories, for the metric g and other fields ψ on a spacetime M^4, which have well-posed Cauchy formulations. A given spacetime field theory for g and ψ is said to have a well-posed Cauchy formulation if one can prove the following: For every choice of the initial data[6] – which consist of a Riemannian 3-metric γ, a symmetric tensor K, and certain other fields θ and π (closely tied to projections of ψ and its derivatives) all on Σ^3, possibly with certain constraint equations and certain differentiability and integrability conditions imposed on (γ, K, θ, π) – there exists a unique globally hyperbolic spacetime which (i) satisfies the field equations of the field theory, (ii) has an embedding $i{:}\Sigma^3 \to M^4 = \Sigma^3 \times \mathbb{R}$ such that the induced metric and induced extrinsic curvature are γ and K, while appropriate projections of ψ and its derivatives on $i(\Sigma^3)$ are θ and π, and (iii) admits no globally hyperbolic extension.

Over thirty years ago, Choquet-Bruhat (Bruhat 1962; Choquet-Bruhat and Geroch 1969) proved that Einstein's vacuum field theory has a well-posed Cauchy formulation; subsequently she and others have shown that the Einstein–Maxwell, Einstein–Yang–Mills, Einstein–Cartan, Brans–Dicke, supergravity, and a number of other theories of interest do as well. Hence, all of these field theories can have WEM spacetimes as solutions. Note, on the other hand, that there are many other field theories for g and possibly ψ that do not admit well-posed Cauchy formulations and hence are incompatible with WEM spacetimes. These include a number of '$R+R^2$' theories, the Horndeski theory for gravity coupled to electromagnetism (Isenberg and Horndeski 1986), and many others.

Next, we wish to consider the nature of the constraint equations that occur in these spacetime field theories and see what we can do with them. Since all of the theories being discussed involve the diffeomorphism group as a gauge freedom, Noether-type considerations show that all of the theories have supermomentum and super-Hamiltonian constraints. If a given field theory is to be compatible with the WEM Principle (and admit WEM spacetimes as solutions) then it is crucial (see Sec. 2) that one be able to split the initial data (γ, K, θ, π) into nontrivial first and second sets, with the second set being obtainable from knowledge of the first set by solving the constraints. This can be done for most of the well-posed theories listed above, and for many others. Rather than discuss how this is done for an abstract, general theory, we shall focus on the specific example of the Einstein–Maxwell–fluid-field theory.

The initial data for this field theory consists of a Riemannian 3-metric γ_{ab} and symmetric tensor K^{cd} for the gravitational field, as noted above, together with a magnetic vector field B^a and an electric vector field E_a for the Maxwell field and an energy density function ρ and momentum density vector field J^a for the fluid.[7] In terms of these quantities, the constraint equations for this theory take the following form:

$$\tilde{\nabla}_a B^a = 0, \tag{2a}$$

$$\tilde{\nabla}_a E^a = 0, \tag{2b}$$

$$\tilde{\nabla}_a K^a_b - \tilde{\nabla}_b (K^c_c) = -8\pi(\varepsilon_{bcd} E^c B^d + J_b), \tag{2c}$$

$$\tilde{R} - K_{ab} K^{ab} + (K^c_c)^2 = 8\pi(E_a E^a + B_a B^a + 2\rho). \tag{2d}$$

Here $\tilde{\nabla}_a$ is the covariant derivative determined by the 3-metric γ_{ab}, $\tilde{R}$ is its covariant derivative, and ε_{bcd} is the skew-symmetric Levi-Civita symbol. Note that certain constants have been set to unity for convenience.

While there may be alternative schemes for splitting the fields into first and second sets, to date the most successful is that which has been developed by Lichnerowicz, Choquet-Bruhat, and York (Choquet-Bruhat and York 1980). One writes out the initial data fields γ, K, B, E, ρ, J as follows:

$$\gamma_{ab} = \phi^4 \lambda_{ab}, \tag{3a}$$

$$K^{cd} = \phi^{-10}(\sigma^{cd} + LW^{cd}) + \frac{1}{3}\phi^{-4}\lambda^{cd}\tau, \tag{3b}$$

$$B^a = \phi^{-6}\beta^a, \tag{3c}$$

$$E^a = \phi^{-6}(\eta^a + \nabla^a \mu), \tag{3d}$$

$$\rho = \phi^{-8} r, \tag{3e}$$

$$J^a = \phi^{-10} j^a. \tag{3f}$$

Here the first set of fields include a Riemannian 3-metric λ_{ab}, a scalar field τ (the mean curvature), a symmetric transverse traceless tensor field σ^{cd} (traceless means that $\lambda_{cd}\sigma^{cd}=0$, while transverse means that $\nabla_c \sigma^{cd}=0$, where ∇_c is the covariant derivative defined by λ_{ab}), a pair of transverse vector fields β^a and η^a (transverse means that $\nabla_a \beta^a=0$ and $\nabla_a \eta^a=0$), a vector field j^a, and a scalar field r. (Note that these fields are really only to be specified up to the joint conformal transformation indicated in Eqs. (3) via the conformal factor ϕ).[8] The second set of fields consists of the scalar field μ (electric potential), the vector field W^a (generating the longitudinal part of K), and the positive definite scalar field ϕ (conformal factor). The notation LW^{ab} indicates the conformal Killing field operator L, which acts on the vector field W^a according to the definition

$$LW^{ab} := \nabla^a W^b + \nabla^b W^a - \frac{2}{3}\lambda^{ab}\nabla_c W^c. \tag{4}$$

Now if one substitutes the field decomposition (3) into the constraint equations (2), one obtains the coupled elliptic system

$$\nabla^2\mu = 0, \tag{5a}$$

$$\nabla_m(LW)^m_b = \frac{2}{3}\phi^6\nabla_b\tau + 8\pi\varepsilon_{bcd}\beta^c(\eta^d + \nabla^d\mu) - 8\pi j_b, \tag{5b}$$

$$\nabla^2\phi = \frac{1}{8}R\phi - \frac{1}{8}(\sigma + LW)^2\phi^{-7} + \frac{1}{12}\tau^2\phi^5 - \pi(\beta_a\beta^a + [\eta + \nabla\mu]^2 + r)\phi^{-3}, \tag{5c}$$

where ∇ and R indicate the covariant derivative and scalar curvature determined by the metric λ. The idea is to solve these equations for (μ, W, ϕ), given (λ, σ, τ, β, η, r, j). Does it work?

This issue has received much attention (Choquet-Bruhat and York 1980; Isenberg 1987; Choquet-Bruhat, Isenberg, and Moncrief 1992). There are two questions involved. The first, *existence*, asks the following: For which choices of the 'free data' (λ, σ, τ, β, η, r, j) can one solve Eqs. (5) for (μ, W, ϕ) and thus construct a new spacetime? The second question, *uniqueness*, asks: If, in a *given* spacetime, one happens to know the information (λ, σ, τ, β, η, r, j) on some Cauchy surface, can one proceed to solve Eqs. (5) uniquely for (μ, W, ϕ), and thence determine the fields and the inertial frames in the given spacetime?

The existence question is the most interesting one mathematically and has received the most attention. Much is known: Existence has been resolved completely for spatially compact Σ^3 with τ=constant [this is the constant-mean-curvature, or CMC case (Isenberg 1987)], and existence is increasingly being understood for non-CMC data as well (Choquet-Bruhat, Isenberg, and Moncrief 1992; Isenberg and Moncrief, unpublished). However, existence is not as important for Machian studies as is uniqueness. Understanding existence is useful for parametrizing and cataloging globally hyperbolic spacetimes, but it does not tell us whether or not these are WEM spacetimes. For this, determining uniqueness is crucial. How much is known about uniqueness?

For compact, constant-mean-curvature hypersurfaces, uniqueness holds, so long as the data (λ, σ, τ, β, η, r, j) do not correspond to a time-symmetric Cauchy surface in a flat spacetime (in which case the solution is unique up to a trivial rescaling constant) (Choquet-Bruhat and York 1980). For non-CMC Cauchy surfaces, the issue of uniqueness (like that of existence) is not yet resolved. However, there has been much progress in verifying uniqueness for non-CMC data in the last three years (Choquet-Bruhat, Isenberg, and Moncrief 1992; Isenberg and Moncrief, unpublished), and thus far there seems to be no indication that uniqueness should ever fail. Indeed, I believe that if efforts were focused on proving uniqueness (rather than examining it only after existence has been verified), then one should be able to obtain comprehensive results - showing that uniqueness always, or almost always, holds - using known techniques. Such results would confirm the following conjecture: If a globally hyperbolic, nonextendible, spatially compact spacetime (M^4, g, ψ) satisfies the Einstein–Maxwell–fluid-field equations, then the conformal split discussed above [Eqs. (3)] for the initial data on any Cauchy surface can always be made, with the data (μ, W, ϕ) determined uniquely by solving Eqs. (5). Hence, such a spacetime is a WEM spacetime.

We further conjecture that the same result is true if one replaces the Einstein–Maxwell–fluid field equations by the Einstein–$\mathscr{F}$ field equations, where $\mathscr{F}$ is any nonderivative coupled theory with a well-posed Cauchy problem. Here we note the useful work of Isenberg and Nester (1977), which shows how to carry out the Lichnerowicz–Choquet-Bruhat–York type conformal decomposition of the initial data for a large collection of such Einstein–$\mathscr{F}$ type field theories.

If this conjecture turns out to be true (it certainly holds if we restrict to constant mean curvature hypersurfaces), then there are very large classes of WEM spacetimes which solve the Einstein vacuum, Einstein–Maxwell, and more generally, Einstein–$\mathscr{F}$ field equations. Are the *generic* solutions of these field equations WEM spacetimes? This question, as stated, does not make much sense, but it does if we refocus it as follows: Fix a compact three-dimensional manifold Σ^3, and consider the space of all smooth initial data (γ, K, θ, π) which satisfy the constraint equations of the chosen field theory. Assuming that the field theory has a well-posed Cauchy problem, then for each choice of this initial data (γ, K, θ, π), there exists a corresponding maximally extended globally hyperbolic spacetime (M, g, ψ). For the generic choice of the data (γ, K, θ, π), is (M, g, ψ) a WEM spacetime?

Assuming that the split of the initial data discussed above can be carried out, we find that this question is exactly equivalent to asking if, for the generic choice of initial data, the spacetime (M, g, ψ) is extendible past a Cauchy horizon to a nonglobally hyperbolic spacetime.

But now, we recall, the Strong Cosmic Censorship Conjecture (SCC) (Penrose 1979) concerns exactly the same issue. This conjecture claims that, indeed, the maximal globally hyperbolic spacetime development of generic initial data cannot be extended. So SCC, if true, would tell us that generic initial data produce a WEM spacetime.

As yet, there is no proof for the Strong Cosmic Censorship Conjecture, even if we restrict attention to the vacuum Einstein equations. Nor is there a serious counter example. There are various results (Isenberg 1992) which support the validity of SCC (all of these consider only spatially compact spacetimes), but the issue remains wide open, with a number of active research efforts currently focused on determining if Strong Cosmic Censorship holds or does not hold.

5. Conclusion

We have not argued that there is one, unique, best way to formulate Mach's Principle in modern physical terms. We have not proven that Einstein's gravitational field equations are 'Machian' or 'un-Machian' in any rigorous sense. Rather, we have focused on defining a class of spacetimes - we call them the Wheeler–Einstein–Mach spacetimes since their definition is based on a succession of ideas developed by Mach, Einstein, and Wheeler - and discussing some of their properties. We believe that these WEM spacetimes, which are traditionally Machian in a certain essential sense (see Sec. 3), are interesting and worth studying from both a mathematical and a physical point of view, whether or not you believe in, or care about, Mach's Principle.

Acknowledgment. We thank K. Kuchař and D. Brill for useful discussions. This work was supported in part by NSF grants PHY93-8117 at the University of Oregon and PHY89-04035 at the Institute for Theoretical Physics in Santa Barbara.

NOTES

[1]In using the terminology 'solutions of the theory' to refer to 'model physical systems compatible with the theory,' we are adopting language which reflects the usual classical situation in which a model physical system is compatible with a given physical theory iff certain model-representative functions satisfy certain theory-representative differential equations.

[2]Just this month (June 1993), I read a paper which relies on Mach's Principle as its chief argument for adding torsion fields to Einstein's theory.

[3]Our discussion here focuses on medium- to large-scale features of the universe, generally far from any singularities, and so we ignore quantum considerations.

[4]In many spacetimes there is a unique foliation by constant-mean-curvature (CMC) Cauchy surfaces (Brill and Flaherty 1978). However, there are some spacetimes (vacuum solutions of Einstein's equations) that do not admit CMC Cauchy surfaces (Eardley and Witt, unpublished; also Bartnik 1988), so this is not a reliable choice of simultaneity in a general spacetime.

[5]The conformal quotienting of all these quantities is coordinated; see (Choquet-Bruhat and York 1980), as well as Sec. 4 below.

[6]We describe here the initial data which are appropriate to a spacetime field theory which is second order; i.e., the spacetime covariant field equations involve second derivatives of the metric (and other fields) and no higher-order derivatives. The generalization to higher-order spacetime field theories is straightforward, but will not be discussed here.

[7]For this particular field theory, (B^a, E^a, ρ, J^a) together make up the '(θ, π)' portion of the initial data.

[8]This is because one can prove, so long as τ is constant, that if we replace (λ, σ, τ, β, η, r, j) by ($\theta^4\lambda$, $\theta^{-10}\sigma$, τ, $\theta^{-6}\beta$, $\theta^{-6}\eta$, $\theta^{-8}r$, $\theta^{-10}j$), then we get the same (constraint-satisfying) initial data (γ, K, B, E, ρ, J) by solving Eqs. (5).

REFERENCES

Bartnik, Robert (1988). "Remarks on Cosmological Spacetimes and Constant Mean Curvature Surfaces." *Communications in Mathematical Physics* 116: 895.

Brill, Dieter and Flaherty, Francis (1978). "Maximizing Properties of Extremal Surfaces in General Relativity." *Annales de l'Institut Henri Poincaré* 28: 335–347.

Bruhat, Yvonne (1962). "The Cauchy Problem," In *Gravitation: An Introduction to Current Research*, L. Witten, ed. New York: Wiley.

Budic, Robert; Isenberg, James; Lindblom, Lee; and Yasskin, Phillip (1978). "On the Determination of Cauchy Surfaces from Intrinsic Properties." *Communications in Mathematical Physics* 61: 87–95.

Choquet-Bruhat, Yvonne and Geroch, Robert (1969). "Global Aspects of the Cauchy Problem." *Communications in Mathematical Physics* 14: 329–335.

Choquet-Bruhat, Yvonne; Isenberg, James; and Moncrief, Vincent (1992). "Solutions of the Constraints for Einstein Equations." *Comptes Rendues de l'Academie Scientifique de Paris* 315: 349–355.

Choquet-Bruhat, Yvonne and York, James (1980). "The Cauchy Problem." In *General Relativity and Gravitation*, Alan Held, ed. New York: Plenum.

Chrusciel, Piotr and Isenberg, James (1993). "Nonisometric Vacuum Extensions of Vacuum Maximal Globally Hyperbolic Spacetimes," *Physical Review D* 48: 1616–1628.

Dicke, Robert (1964). "Experimental Relativity." In *Relativity, Groups and Topology*, Cecile DeWitt and Bryce DeWitt, eds. New York: Gordon and Breach.

Eardley, Douglas and Witt, Don unpublished. See also Bartnik, Robert (1988).

Einstein, Albert (1917). "Kosmologische Betrachtungen zur allgemeinen Relativitätstheorie." *Sitzungsberichte der Preussischen Akademie der Wissenschaften*, 142–152.

Einstein Albert (1955). *The Meaning of Relativity*. Princeton: Princeton University Press.

Isenberg, James (1987). "Parametrization of the Space of Solutions of Einstein's Equations." *Physical Review Letters* 59: 2389–2393.

Isenberg, James (1992). "Progress on Strong Cosmic Censorship." *Contemporary Mathematics* 132: 403–418.

Isenberg, James and Horndeski, Gregory (1986). "Effective Determinism in a Classical Field Theory with Spacelike Characteristics." *Journal of Mathematical Physics* 27: 739–745.

Isenberg, James and Moncrief, Vincent, unpublished.

Isenberg, James and Nester, James (1977). "Extension of the York Field Decomposition to General Gravitationally Coupled Fields." *Annals of Physics* 18: 368–385.

Isenberg, James and Wheeler, John A. (1980). "Inertia Here is Fixed by Mass-Energy There in Every W Model Universe." In *Relativity, Quanta, and Cosmology*, M. Pantaleo and F. DeFinis, eds. New York: Johnson Reprint.

Penrose, Roger (1979). "Singularities and Time Asymmetry." In *General Relativity – an Einstein Centenary Survey*. Stephen W. Hawking and Werner Israel, eds. Cambridge: Cambridge University Press.

Wheeler, John A. (1964). "Geometrodynamics and the Issue of the Final State." In *Relativity, Groups and Topology*. Cecile DeWitt and Bryce DeWitt, eds. New York: Gordon and Breach. See also Isenberg, James and Wheeler, John A. (1980).

Discussion

Ciufolini: I like this approach, but there is one thing I never fully understood. To solve for the spacetime geometry, you give on a Cauchy surface some conditions that are not directly related with the matter distribution such as the trace of the extrinsic curvature, the conformal metric, and the conformal traceless free part of the extrinsic curvature. Probably this last part has some relation to the energy of gravitational waves. My query is that the definition you gave at the beginning was that to determine the spacetime geometry you need the distribution of matter; however, you need to give something else apart from the distribution of matter. Is the conformal traceless free part of the extrinsic curvature related to the energy of gravitational waves?

Isenberg: Yes. Very roughly, this extra gravitational information in the 'first set' includes the quantities which relate to 'gravitational energy' and 'gravitational momentum.' It is hard to be too specific about this – York has talked about it in terms of the Cotton tensor and its conjugate, and Wheeler has as well. But I don't see how to make this relationship exact. Indeed, there is the fact that, in the 'first set,' you're leaving everything conformally undetermined. So the first set data can't really contain quantities *exactly* equivalent to gravitational energy.

While this lack of exactness might bother a serious Machian, I believe that what we should do is let the mathematics guide us, and say "O.K., maybe this isn't exactly what Mach wanted, or what anybody wanted, but it *works* [laughter] in the sense that if you specify this first set of information, then you can determine everything else. Whether it's exactly energy, momentum, or whatever – it's just the stuff that works.

Barbour: You said you'd like to know if Mach formulated Mach's Principle in these terms, talking about initial data at an initial instant. I'm almost certain that he didn't, and of course that's one of the problems with Mach. He had such a distrust of theory he would never pin himself down to any particular theory. I think this is what creates the impression that he was only in favor of a redescription and not something more. However, you will find an absolutely clear, precise statement by Poincaré [pp. 111–112] of the problem in initial-data terms, and it's very close to what Wheeler is arguing for, at least in the original thin-sandwich conjecture.

Isenberg: I don't believe that there is a clear initial-value version in Einstein's work, either.

Barbour: But it's in Poincaré, and that's what I'll be talking about tomorrow [p. 214], which is an approach through the configuration space somewhat different from yours but very much in the same spirit.

Ehlers: I think that what you have presented to us was a very nice way of how to look at a certain class of classical field theories mathematically, and that's fine, but I think there is a big gap with respect to what one would like from the point of view of physics. The big concern – and this is no criticism because I don't know of anybody who could do better than this – is that I think we do not understand what is the relation, even in those cases where one has sorted out the free data; we do not know what is the relation between these free data and quantities which are actually accessible to observation.

Isenberg: Exactly; I agree completely.

Ehlers: Also, the Cauchy initial-value problem is, I think, a very nice tool in order to specify particular solutions within a given theory, and that is its main purpose, but it has little to do with, for example, what the cosmologist wants to do if he has observational data and wants to know what he can infer from these data. I just wanted to say that it is perhaps a little regrettable that people with much knowledge of mathematics work on these questions. They focus their attention more on things which are mathematically nice and avoid those questions which from the point of view of physics are the more urgent ones. This is not a criticism of you.

Isenberg: This is a point well-taken, and I think it's a nice invitation to look at the characteristic initial-value problem, which rather than examining data specified on a spacelike hypersurface, examines data specified on a past causal (null) cone. I agree, my approach here is basically a mathematical one rather than a physical one. As I noted sometime earlier, I'm paid by a math department to come up with and prove theorems. As for questions like "What is the physics of the information in the first set, and how might one go about measuring it?", well, I guess I just don't know. I would like to understand this issue.

Ehlers: Can I make another remark? I think it would be very nice if one could give nearly as clear-cut a description of the framework of the quantum field theory. With respect to such questions we seem to be very far from understanding a similarly clear mathematical structure in quantum field theory.

Isenberg: My approach is completely classical, and I am rather pessimistic regarding the imminence of a clear, consistent theory of the physical gravitational field. I've seen a lot of discussion of quantum gravity, and so far as I can tell none of it yet makes a lot of sense physically.

Goenner: Maybe it's a stupid question, but the most trivial solution I can think of, which I do not expect to be lying in your class, is if you take T^3 cross R.

Isenberg: It's in there. The spacetime you are talking about is basically Minkowski spacetime with a closed spatial topology. You take a cube in ordinary Euclidean space, you identify its opposite faces so that you have the three-dimensional torus T^3, and then cross T^3 with the time axis R^1. This *is* a solution of the Einstein equations, and it *is not* extendible past a Cauchy horizon, so it *does* fit into the set of WEM spacetimes. This is a point where Wheeler and I diverge a bit. Wheeler *hates* these spacetimes [laughter]; and, as I understand it, he includes in his version of Mach's Principle a restriction which throws out this spacetime along with all spacetimes which are spatially T^3. Indeed, his restriction throws out almost all three-dimensional manifolds. It leaves a few, like S^3 and $S^2 \times S^1$ and many other familiar ones; but in a mathematical sense it throws out almost all of them. The restriction he makes is that if a spacetime is to be deemed Machian, then among other things it must be able to stop expanding and then begin to collapse. Such behavior for a solution of Einstein's equations is incompatible with most three-dimensional manifolds. In defining the WEM spacetimes, I do not include this restriction. Consequently, the T^3 Minkowski spacetime solution qualifies. If you think that such a spacetime is un-Machian, then this is something to dislike about my WEM spacetime formulation. I, however, don't find anything drastically un-Machian about the T^3 Minkowski spacetime, so I'm not bothered by its inclusion.

Goenner: Well, I'm very sorry for your class, because test particles in that kind of spacetime would have inertia.

Isenberg: I did think about that issue some years ago, but I ran into some trouble deciding what it *means* to put a test particle into a T^3 Minkowski spacetime. The problem is that these spacetimes are unstable in the sense that if you add a bit of matter to the spacetime, you need to add a corresponding bit of 'gravitational waves.' Then the spacetime becomes dynamic, and then it has other stuff around to help fix inertial frames. I would like to understand this issue better.

King: Just a last point on trying to rule out T^3. At first it sounds like Wheeler was trying to introduce some sort of topological discrimination, which doesn't seem to be terribly fair, but if it's almost impossible to satisfy the constraint equations for a spacetime, then that spacetime is virtually ruled out anyway.

Isenberg: Oh, it's not impossible to satisfy the constraints. There is a whole big fat function space of solutions of the constraints on the three torus T^3; the only problem is that the flat T^3 (Minkowski spacetime) solution is unstable, as I noted before.

King: If you can't put one particle in the spacetime though.

Isenberg: You *can* put a particle in, but *only* if you also add in some gravitational waves at the same time, allow the spacetime to expand (or contract).

Ciufolini: The main reason, as you said, why Wheeler doesn't like these kinds of topologies is because they imply a model universe expanding for ever.

Isenberg: Right, they do. A $T^3 \times R$ spacetime which satisfies Einstein's equations (and is not dead flat) either expands or contracts forever. But while that may bother Wheeler, it doesn't bother me in any Machian sense. What does bother me is if the spacetime extends past a Cauchy horizon. Then, an observer who is doing all kinds of nice experiments may head off into the spacetime region past the Cauchy horizon. Suddenly determinism and causality are lost. Inertial frames are *not* fixed by the distribution of matter or anything. Physics becomes strange, and all bets are off. That's the sort of thing I want to throw out of the class of 'Machian' spacetimes.

Barbour (post-conference comment). Perhaps I may be allowed to make some comments here about the remarks Isenberg makes on p. 204 about the difficulty of being precise about the definition of gravitational stress–energy and about what "works." A little bit of the history might be in order. Although back in the early sixties, when he formulated the thin-sandwich conjecture (see the Wheeler references on p. 231), John Wheeler was not, so far as I know, aware of the remarks of Poincaré (pp. 111–112), which show the intimate connection between Newton's use of absolute space and the formulation of the initial-value problem, the conjecture was nevertheless formulated in a manner that does relate it closely to what Poincaré said. In my contribution (p. 214) and in (Barbour 1994a) cited in it, I show how the parallel can be made very close indeed, especially if the nonexistence of an external time is taken into account.

However, the problem with the thin-sandwich conjecture is that the associated mathematics is still not at all well understood, as Kuchař remarked at Tübingen (unfortunately not recorded and therefore not reproduced here). This was at least partly why, in response to important work done by York at the beginning of the seventies, Wheeler went over to the kind of formalism Isenberg describes. Essentially, York gave up the attempt to solve the initial-value problem in the Lagrangian formulation of general relativity and went over to the same problem in the Hamiltonian ADM formalism. Now for simple dynamical systems, the relationship between the Lagrangian and Hamiltonian formulations is trivial, but in reparametrization-invariant theories like general relativity there are very significant differences. In a remarkable manner, the Hamiltonian formalism seems to be especially well suited to the formulation of the initial-value problem. Having a strong sense of the importance of a dynamical approach for the quantization problem, Wheeler adopted the York formalism, especially since the intuitive transparency of the Lagrangian thin-sandwich formulation had nevertheless not born much fruit. This is all recounted in very readable form in (Isenberg and Wheeler 1980), which also gives references to York's work. Thus, as of now, we have a transparent formalism with almost intractable mathematics and the formalism that "works" described by Isenberg. As Kuchař remarked at Tübingen, further work on the thin-sandwich conjecture is greatly to be desired; recent work in this direction by Bartnik and Fodor (discussed and cited by Giulini, p. 500) is an encouraging sign of a revival of interest in the conjecture.

Comments on Initial-Value Formulation: Response to Isenberg

Dieter R. Brill

1. On Principles

Isenberg's proposal (1995) is remarkable not least because it is intended to cover not one or the other aspect of Machian ideas, but a complete formulation of Mach's Principle. Isenberg gives cogent reasons why the Wheeler–Einstein–Mach (WEM) program expresses the important Machian demands, and it is hard to see how it could be improved as a general program, particularly since Isenberg added the nonextendibility requirement, giving a link between Mach's Principle and cosmic censorship.

Isenberg also considers Mach's Principle in the larger context of principles in physics. In this general context, Mach's Principle is somewhat unusual: It cannot easily be disproved, because we know few if any effects that are unequivocally anti-Machian (for example, Ozsváth and Schücking 1962). By contrast, the most useful principles in physics naturally have a negative or interdictory aspect. For example, the uncertainty principle forbids certain variables from being simultaneously well-defined, the energy principle forbids perpetual motion, the equivalence principle denies distinction between gravity and inertia, the atomic principle excludes infinite divisibility, and so on. Such a formulation is not only heuristically useful (for example, it saves us from useless speculation about impossible situations), but it can also point the way toward progress in the theory: A negative principle implies a challenge, to find the mechanism or rationale behind the prohibition, and can lead to a new theory in which the principle is automatic and no longer needs to be stated explicitly.

At first sight the WEM Principle looks like business as usual (we still do classical general relativity in a way that current lingo might

Einstein Studies, vol. 6: Mach's Principle: From Newton's Bucket to Quantum Gravity, pp. 208–213

associate with STINO[1]), and gives no direct motivation to change the theory. The implication is different if we state it as a negation: No spacetime can fail to satisfy the four WEM requirements. But of course there are solutions of the classical Einstein equations that are not WE Machian. Hence the challenge is to find the mechanism that excludes the offending spacetimes. Thus the WEM Principle also points the way beyond classical general relativity to new and certainly as yet unfinished business.

2. On Inertia

Many, like Einstein, find something fascinating about the idea that in inertia we feel the rest of the universe at work, and they look to Mach's principle for the real origin of inertia. Does the WEM–Isenberg approach finish that business of formulating the principle? Isenberg tells us that if we know the full spacetime metric near a point, we know all there is to know about inertial frames at that point. In Shimony's comparison (1992), you enter Mach's Store looking on the shelves for various useful and fascinating gadgets, many of them somehow connected with inertia. But in the WEM store you find only a general do-it-yourself kit from which you might be able to build your own gadgets. How much more effort is required to build the gadgets we care about out of the WEM kit? Let us consider some of the 'gadgets' that other authors in this volume might hope to find in Mach's Store.

Pfister might care about the inertial frame dragging. Suppose we consider a point inside Pfister's shell. We know the metric there – it is flat. But this knowledge does not tell us all there is to know about the dragging as usually understood (Brill and Cohen 1966, Lindblom and Brill 1974). A true answer about inertia and inertial frames must involve specific frames or coordinates. The WEM Principle, being a child of general relativity, tends to be hostile to picking out a particular frame – the really significant information is considered to be frame-independent. "Frame not included" is written on the packages in the WEM store; but is this not one of the things we expect to get from Mach, not to put into it?

Raine, whose own formulation of Mach's Principle has been questioned concerning the distinction between matter and gravitational waves, might ask of the WEM principle whether there is really a crucial difference between the following two situations: an otherwise closed WEM universe containing either a black hole formed by collapse of matter, or an eternal Kruskal black hole with an asymptotically flat region on the 'other side' of the horizon. The former would be called WE Machian, and the latter would not, because its Σ^3 is not compact. But this distinction is not reasonable: Since the difference can be extremely small between the physical regions on 'this side' of the horizon, and since one cannot look behind a horizon, the Machian nature of a spacetime would be something that could never be ascertained by experiment. Perhaps the attribute *Machian* should apply to regions in spacetimes, for which it does not matter what happens behind horizons.

If we allow this extension of the WEM Principle, we can treat the following situation, which is more amusing than profound. Suppose Narlikar asked the question that has a definite answer in his formulation: What is the smallest number of masses in a WE Machian S^3 universe that is free from other content such as gravitational waves? Suppose we take the absence of wave content to mean that the free data can be chosen to be trivial, and the presence of mass to mean that n asymptotically flat regions behind (apparent) horizons are allowed. Since asymptotically flat regions are conformally equivalent to taking points out of the S^3, an appropriate choice for Isenberg's first set (Σ^3, λ, σ) is ($\mathbb{R}^3$ less $n-1$ points, flat, 0). For $n=1$, the only regular solution for the Lichnerowicz conformal factor ϕ is ϕ=constant, which is flat spacetime without horizon, and hence without Machian region. For $n=2$ the solution is $\phi=1+M/2r$, with r = Euclidean distance in $\mathbb{R}^3$ from the removed point. This is just the single-mass Schwarzschild solution with one horizon, which does not bound a compact Machian region. So one mass is not enough. For $n=3$, we have $\phi=1+M_1/2r_1+M_2r_2$, which is asymptotically flat in three regions, at $r_1\to\infty$, at $r_2\to\infty$, and at (r_1 and r_2)$\to\infty$. For small M_1, M_2 there are only two horizons, not bounding a Machian region. But if M_1 and M_2 are chosen large enough (compared to their Euclidean distance), there can be another apparent horizon surrounding the two (Brill and Lindquist 1963). A Machian region then exists between these three horizons. Thus three masses is the answer by this extended WEM Principle, not unreasonable because three masses usually do define a frame. (Unfortunately in this particular construction they do not, because the solution is rotationally symmetric about an axis through the original M_1, M_2. In this sense the answer is not better than Narlikar's two-mass minimum.)

3. On Details

Examples such as the above suggest that the WEM Principle leaves some room for further refinement. This appears particularly urgent in connection with the distinction between the 'first' and 'second' set of Cauchy data. The first set should contain variables that can be freely chosen; but in Isenberg's examples it consists of a TT tensor σ and *transverse* fields β and η. Because of such transversality requirements these fields are really not free but are themselves subject to constraints. Would it then not be simpler to choose as the first set any constraint-satisfying initial data, so that the second set is empty? If it is allowed to demand transversality of the first set, then why not constraint-satisfaction? Isenberg (1995) suggests that the former condition is linear and does not essentially restrict free choice, whereas the latter condition is nonlinear and implements the Machian determination of the inertial frames. This interpretation itself would, of course, constitute a (small) refinement of the WEM Principle, a refinement motivated by a possible physical meaning of the splitting into first and second sets. Refining the physical meaning of the decomposition of data into the first and second sets seems a promising task. It could have interesting physical significance if a *particular* decomposition were demanded, not just the existence of *some* decomposition (one of possibly many). For example, in the Lichnerowicz-York decomposition, the vector W itself does not appear in the 'Machian' constraints; only LW occurs. Perhaps this (or some other, even more Machian) decomposition can give an appropriate, general definition of the frame dragging by means of a vector like W (which may be related to the shift vector, a coordinate quantity of the type needed really to describe inertia).

It would seem unusual to find that a formulation, one of whose authors is Wheeler, could benefit from greater emphasis on physical meaning, but such are the conclusions to which we are led.

NOTE

[1]STINO, from *stinknormal*, the name of a new popular music phenomenon in Germany, celebrating traditional melodies and folk songs. Perhaps such labels can help us gain recognition in the lay public. (What attention the no-hair theorems might have received if they were identified with skinheads!)

REFERENCES

Brill, Dieter R. and Cohen, Jeffrey (1966). "Rotating Masses and their Effects on Inertial Frames." *Physical Review* 143: 1011–1015.

Brill, Dieter R. and Lindquist, Richard W. (1963). "Interaction Energy in Geometrostatics." *Physical Review* 131: 471–476.

Isenberg, James (1993). Private Communication, Tübingen July 1993.

Isenberg, James (1995). "On the Wheeler–Einstein–Mach Spacetimes." This volume.

Lindblom, Lee and Brill, Dieter R. (1974). "Inertial Effects in the Gravitational Collapse of a Rotating Shell." *Physical Review* D 10: 3151–3155.

Ozsváth, István and Schücking, Engelbert (1962). "An Anti-Mach Metric." in *Recent Developments in General Relativity.* Oxford: Pergamon Press, pp. 339–350.

Shimony, Abner (1992). Private communication, APS Spring Meeting, Washington, DC. Also see Brill, Dieter R. "Comments on Dragging Effects." This volume.

Discussion

Rindler: Just a small point to Dieter's remarks. Mathematically you can continue a static Schwarzschild spacetime region across the horizon either into a collapsing massive object, or into a Kruskal wormhole with vacuum everywhere. So the same external Schwarzschild spacetime with its inertial structure can be 'caused' by matter à la Mach or by pure geometry.

Brill: Well, that was my point. Is it reasonable for it to change from Machian to non-Machian on the basis of something that changes only behind this horizon? It's sort of connected with Bondi's point [p. 88] that distant things somehow shouldn't affect the inertia, which is maybe what we're after very much.

Goenner: You made a remark on only frame-independent quantities being accepted nowadays, but then we can adjourn because inertia is only a concept which is defined relative to a frame.

Brill: Yes, that was my point also – that Mach's Principle may point the way toward giving physical meaning to quantities usually considered frame-dependent.

Goenner: Perhaps the question is whether we should be satisfied to identify coordinate systems and frames, or whether we should come up with some sort of definition of what a frame is, or a rule which selects reference frames from among all the coordinate frames.

Barbour: One definition of the Machian problem is that there do exist locally Minkowskian frames, which seem to be distinguished locally, in which rotation is well defined. The question is: What is their origin? I would say that what one should be doing – and I think this the hidden agenda of what Jim Isenberg was saying this morning – is start from initial data which do not in any sense contain frames in the kinematics, solve the dynamical problem, and construct spacetime. You then find an explanation why you can introduce in the dynamical solution, which is spacetime, local distinguished frames. To repeat, the question we have to ask is: Why do local frames exist in which rotation is so well defined? The answer has to come out of a deeper, fundamental theory in the kinematics of which there are no distinguished frames, but when you solve the problem and you construct spacetime, frames can be then introduced in the resulting spacetime.

Brill: Yes, I'd love to see that spelled out.

Narlikar: I think I would like to add to this and not talk about rotation but just Minkowski frames, one related to another. There is a unique Minkowski frame in which the universe looks isotropic, with respect to the Hubble law.

Bondi: Hear! Hear!

Narlikar: So that if you're moving relative to that with uniform motion you would still notice that. That is something which is a nonrotation effect that is distinguishing.

Bondi: I will talk about that a little on Friday [p. 474].

General Relativity as a Perfectly Machian Theory

Julian B. Barbour

1. Introduction

In this paper, I shall argue that general relativity is as Machian as one could reasonably hope to make any theory. The qualification is to cover an infinite universe, for which there are subtleties (Sec. 5).

The first step to this thesis is to establish just what Mach did want; this I attempt to do in Sec. 2. In contrast to some contributors to this volume, I believe this is clear: an account of inertia containing only relational quantities and different in its observable consequences from Newtonian theory. Moreover, I point out that Mach criticized not only the Newtonian concept of absolute space but also the notion of absolute external time; the need to formulate dynamics without an external time can be regarded as a kind of *Second Mach's Principle* (cf. p. 92 and 102ff).

In Sec. 3, I discuss the relationship between Poincaré's formulation of the problem of absolute *vs* relative motion (p. 111) and Mach's more intuitive statements (p. 109) and argue that Poincaré has provided a *precise criterion of Machianity* of a dynamical theory in at least the nonrelativistic case. The formulation of a second criterion to take into account the nonexistence of an external time is straightforward.

In Sec. 3, I also describe the appropriate kinematic framework for Machian dynamics. The key concept is the *relative configuration space* (RCS) of the universe. The two criteria of Machianity are then formulated in the RCS, and it is noted that the first is met by the theories of Hofmann, Reissner, and Schrödinger translated in Chap. 2 and that the implementation of the second is relatively trivial.

However, these theories are ruled out experimentally by the anisotropic inertia that they predict, and it is fortunate that both Machian

Einstein Studies, vol. 6: Mach's Principle: From Newton's Bucket to Quantum Gravity, pp. 214–236

criteria can be met in theories of a somewhat different form based on a notion that Bertotti and I call the *intrinsic derivative*. This is discussed in Sec. 4. It is important that this notion can be very easily generalized to field theories and theories in which geometry itself is dynamic. In fact, when the Hilbert action of general relativity is appropriately rewritten in terms of the dynamical evolution of 3-geometries, it is found to be based on a natural generalization of the intrinsic derivative. The dynamics is also formulated without an external time. These two results show that general relativity is Machian.

In Sec. 5, I argue that the problem of the so-called anti-Machian solutions of general relativity disappears very largely once it is realized that even the seemingly most anti-Machian spacetime - Minkowski space - still admits interpretation as the outcome of a perfectly Machian evolution of 3-geometries. It is very important here that gravitational degrees of freedom are treated on an equal footing with conventional matter degrees of freedom.

The full relativistic theory of the intrinsic derivative, together with a Machian treatment of time, is given in (Barbour and Bertotti 1982) and (Barbour 1994a, b), in which I also consider the quantum implications. This material also formed the basis of my talk at Tübingen, but it seems to me inappropriate to reproduce this recently published material here, especially since there is a page limit on this volume. I also hope to give a more extended account of the entire matter in Vol. 2 of my study *Absolute or Relative Motion?*, which is in preparation. I therefore ask the interested reader to consult these other works for the detailed elaboration. I should mention that the discussion session at the end of this paper refers to material in my Tübingen talk published in those works and only partly reproduced here.

2. What Was Mach Advocating?

Norton (p. 9) doubts whether Mach really did advocate a new theory of inertia, while von Borzeszkowski and Wahsner (p. 58) say categorically that Mach did not. Mach's writings are sometimes obscure, but if they are considered as a whole I think it is impossible to maintain that Mach did not envisage a quite new account of inertia. Moreover, important issues are at stake. As noted elsewhere (p. 7), around 1670 Newton already recognized the *fundamental problem of motion*, which even now is a central issue of quantum gravity (Barbour 1994a,b). Mach was the first person who saw a way to resolve Newton's problem without invoking absolute space.

I start by referring the reader to Mach's 1872 passage on p. 109 that begins: "Now what share has every mass in the determination of *direction and velocity* in the law of inertia?" (My italics to make the point that by 'inertia' Mach always meant the direction and velocity of inertial motion, not the inertial mass, cf. pp. 91–92.) Given that Newtonian gravitation, the paradigm of a physical force, determines acceleration proportionately to the mass of the attracting body and in inverse proportion to the square of its distance, Mach can hardly have intended here anything other than a physical force that determines direction and velocity in accordance with some definite (but as yet unknown) law containing (in principle) both the mass and the distance of the inertia-determining body.

Such an interpretation accords with the 1912 passage in which Mach (1960, p. 296) says the ideal is an account in which "accelerated and inertial motions result in the *same* way." While granting the strength of this interpretation, Norton (p. 23) claims that the final sentence (not given on p. 109 because of lack of space but quoted on p. 23) in the 1872 passage is a "variant or, possibly, special form" of the possibilities considered in the previous sentence and is "just a redescription of the inertial motion of a collection of noninteracting bodies that avoids mention of 'space,' " However, while the sentence in question does read like that, the statement that Norton makes is not strictly true. The distances between bodies that are moving inertially *do not* vary proportionately to one another (even if they are sufficiently far apart for mutual gravitation and other recognized forces to have negligible effect).

Mach knew this well and mentioned it in the 1883 passage (reproduced in its entirety on p. 110, starting line 5) in the *Mechanik* that Norton analyzes. Mach points out that the distance between two bodies moving purely inertially satisfies the differential equation

$$d^2r/dt^2 = (1/r)[a^2 - (dr/dt)^2], \tag{1}$$

where a is a constant. (This behavior arises because the separation of two bodies that each moves uniformly in absolute space passes once and only once through a minimum $R \geq 0$ but increases linearly at the rate a in the limits $t \to \pm\infty$.) Having noted, among other things, the form of Eq. (1) in the first paragraph of the section, which undoubtedly concerns mere redescription, Mach continues in the second paragraph: "Instead of saying, the direction and velocity of a mass μ in a space remain constant, we may also employ the expression, the mean acceleration of the mass μ with respect to the masses m, m', m'' ... at the distances r, r', r'' ... is $=0$, or

$$d^2(\Sigma mr/\Sigma m)/dt^2 = 0." \qquad (2)$$

Is Mach here proposing a redescription of Newtonian law (as Norton believes) or a specimen of a new relational law?[1]

It is unfortunate that, in either interpretation, Mach's equation is mathematically flawed, since it is a single scalar equation, whereas a vector equation with three components is needed to specify the motion of a particle fully (as Kuchař pointed out in the discussion session at Tübingen). However, if Newtonian mechanics is correct, *any* relational equation like (2) will only hold *exactly* in very exceptional occasions and at most at certain instants. In a universe containing very few particles, it will in general disagree strongly with the Newtonian prediction. Whatever Mach may have intended with (2), as it stands it is a physically distinct law. Moreover, his following remark that Newton's expression and (2) are equivalent "as soon as we take into consideration a sufficient number of sufficiently distant and sufficiently large masses" can hardly be understood in any other way than that he regarded (2) to be a possible exact relational law from which Newton's expression would follow *as a very good approximation* in a universe like ours. Note that, in contrast, Mach's first equation (1) and Newton's expression are always *exactly* equivalent.

While Mach's first equation is obviously mere redescription, the second is unambiguously non-Newtonian because *it contains the masses of the bodies of the universe*. In contrast, the masses play no role at all in (1), and they also play no role in Lange's construction (1885). It is the presence of the masses in (2) that makes it dynamical: Large masses substituted anywhere for some small masses will cause the mass μ to move differently, even if only slightly. This is quite different from Newtonian inertial motion. Schrödinger (p. 148, Note 5) was also in no doubt that (2) was an attempt at a new law.

Mach's comments sometimes seem open to doubt because he was a reluctant (and, in the above example, a somewhat incompetent) theorist who distrusted analytical mechanics (Mach 1960, p. 575). In analytical mechanics, an equation like $\mu d^2(\Sigma mr/\Sigma m)/dt^2=0$ [where I have added to (2) the particle's own mass μ, which cannot be omitted when there are forces] could only be derived from a Lagrange function that contains terms with the mass products μm_i, $=1, 2, \ldots,$ as coefficients, i.e., interaction terms, just as in the electrostatic interaction potential $e_i e_j / r_{ij}$.

There is an even more decisive passage (cited by neither Norton nor von Borzeszkowski and Wahsner) that makes Mach's position clear beyond peradventure. Mach added it to at least the second and third editions (1889, 1897) of the *Mechanik* as his response to Lange's redescription (1885) of Newtonian mechanics in purely relational terms. Following an extremely positive evaluation of Lange's proposal, which is granted to be perfectly possible, Mach says he is still dissatisfied with such an approach and then comments (p. 236, my italics and translation): "It appears *very questionable* whether a fourth force-free material point would follow with respect to one of Lange's 'inertial systems' a straight line (uniformly) *if the heaven of fixed stars were not present*, or were not unvarying, or could not be regarded as unvarying with sufficient accuracy." Here are counterfactuals employed just as Norton requires (p. 15, p. 28). How could Mach say more explicitly that he wanted and expected a law of inertia physically different from Newton's law?[2]

3. Criteria of Machianity

Even among those who recognize that Mach wanted a new law of inertia, there is still much disagreement about precisely what Mach's Principle should be. My view is that there should be no doubt on this score and that virtually all the disagreement has arisen because, ironically, Einstein himself never really sorted out the matter.

In this connection it is illuminating to read (Chap. 2) the six authors Mach, Poincaré, Boltzmann, Benedict Friedlaender, Föppl, and Hofmann,[2] who all wrote *before* Einstein appeared on the scene. They are in essential agreement about what needs to be done - and was done in first tentative steps by the Friedlaenders and Hofmann and then in detail by Reissner and Schrödinger. In contrast, no one can read the selected passages from Einstein on the Machian issue (pp. 180–187) without recognizing that Einstein introduced many different formulations over the years, and *twice* finished up by rejecting his own earlier formulations (rejection of general covariance as a criterion of Machianity in 1918 and rejection of the whole Machian idea in 1949, the latter rejection being made on the basis of an argument - false in my opinion - that requires a mere "moment's reflection"!).

Although what Einstein says is seldom completely divorced from the core of Mach's thought and is often close to it, it is important to get this matter straight, for it goes to the heart of dynamics. Also, it seems to me that some of Einstein's less successful characterizations of the Machian issue have often been the starting points for modern attempts to implement Mach's Principle; for example, the theory of Jordan (1955) and Brans and Dicke (1961), the approach of Hoyle and Narlikar (p. 262), the Green's function approach described by Raine (p. 274), or

Dehnen's Higgs-type mechanism for mass generation (p. 479) all derive from Einstein, not Mach. Interesting as these are, I personally would not call them Machian since they do not start with a radical critique of the kinematic foundations of dynamics. That is the first agendum of a Machian theory.

Lack of space prevents me here from discussing why Einstein took the route he did and why he gave so many different formulations of the Machian idea [however, see (Barbour 1992) and Hoefer, p. 67]; instead, I shall attack that agendum frontally (which, interestingly, Einstein claimed was not possible, p. 187, 1918b), establish unambiguous criteria of Machianity for nonrelativistic particle mechanics, and then see how they should be applied to relativistic geometrodynamics. I think we shall then see that even if Einstein was more often wrong than right in his Machian pronouncements *general relativity itself is perfectly Machian.* I ask the reader to look at what the man did, not what he said.

Mach's arguments against absolute space stem from a gut feeling about the nature of reality that can be traced back several centuries, to Copernicus at least (p. 6). This is that at any instant of time the objects in the universe occupy some definite relative configuration which is changing in some definite relative manner and that this is all that should count in physics. Numerous authors have noted the absurdity of supposing the universe as a whole has any position, orientation, or motion in any sort of space external to the universe. Almost by definition, the universe must be a self-contained whole. Hence Mach's eloquent "The universe is not *twice* given, with an earth at rest and an earth in motion, but only *once*, with its relative motions, alone determinable." Unfortunately, things are seldom as simple as such aphorisms. As Newton correctly sensed, Tait (1883) and Lange (1885) showed, and Mach himself admitted (see above), something very like absolute motions, namely, motions in inertial frames of reference, *can* be deduced from the purely relative motions.

The supremely important point made by Poincaré (pp. 111–112), who shared all Mach's gut convictions, was that, in any one instant, *the complete set of these inertial-frame motions cannot be deduced from the instantaneous relative data* that characterize any particular dynamical system that one may be studying, for example, the solar system. The problem is the overall rotation of the system. For suppose we are told the masses of n material particles and are given the relative configuration of these masses at one instant, i.e., we are given the distances r_{ij} between all pairs of points i and j, and also the rate of change of this relative configuration, i.e., the values of dr_{ij}/dt at the same instant t. Since we know the masses, the instantaneous configuration enables us to calculate its center of mass. However, we can deduce nothing about its overall rotation, which is described by additional nonrelational degrees of freedom. But this means we cannot deduce the angular momentum $\mathbf{M}_0$ of the system about its center of mass. Unfortunately, the vector $\mathbf{M}_0$, which has three components, has a strong influence on the subsequent evolution of the system - a planet without angular momentum will fall straight into the sun. The three nonrelational degrees of freedom are therefore unobservable in the relational initial data but have a dynamical effect and modify the *subsequent relational data*.

Poincaré comments that if all that counted in physics were the purely relative data one would expect such data at any one instant to contain sufficient information to predict the future uniquely. This is simply not the case in Newtonian mechanics. Ultimately, all the unease about absolute space derives from this fact. For subsystems of the universe, there is no way round it. However, Mach's vital hint was that the situation might be different if the entire universe were considered as a dynamical system. His *Mechanik* is full of comments about the need to contemplate the entire universe when formulating dynamics (p. 110). Thus, we may conjecture that the universe as a whole is governed by a purely relational dynamics for which relative initial data do suffice to predict the future uniquely and that the mismatch in a subsystem is due to the ignored influence of the rest of the universe on it.

For decades it has been impossible - and this volume is testimony to the fact - to get scientists to agree on the answer to this question: When is a theory Machian? But the answer suggested by Poincaré's analysis is simple: *when the dynamical evolution of the universe as a whole can be predicted uniquely on the basis of purely relative initial data.* In the case of a universe consisting of point particles, these data will be r_{ij} and dr_{ij}/dt at any one instant.

I shall call this the *First Machian Requirement*.

Let me here recall that the analytical mechanics of n Newtonian point particles is done in a $(3n+1)$-dimensional space, which is formed by the $3n$ dimensions of the ordinary configuration space Q, to which is adjoined the one-dimensional space T of the absolute time t. The coordinates of Q are the positions of the particles *in some inertial frame of reference*.

There is, however, also the configuration space Q_0 of the purely relative variables. I call this the *relative configuration space* (RCS); it has $3n-6$ degrees of freedom. It is the natural arena of Machian dynamics. Machian histories are curves in the RCS. According to

Poincaré's analysis, in a relational theory initial data should be specified in the RCS Q_0, not Q. This cannot be done in standard Newtonian theory because of the problem with the angular momentum $\mathbf{M}_0$.

Before continuing, let me dispose of one matter. Given the purely relative instantaneous state of a system - the *Machian instantaneous state* - it is not only the overall rotation but also the center-of-mass motion that we are unable to determine. However, for an isolated system, such motion has no dynamical consequence because of the Galilean invariance of physics.

What, however, is important is access to an external clock. For the above discussion assumes a measure of time in order to specify the rates of change dr_{ij}/dt of the r_{ij}. In reality, as Mach repeatedly emphasized (p. 92), physicists never have access to *time*. All they can ever do is measure one motion with respect to another or, more generally, one physical change with respect to another. If we consider our universe of n point particles is defined solely by all the r_{ij}, this leaves no variable by means of which we can measure 'time.' We would have to nominate one of the r_{ij}, say r_{12}, to play the role of 'time' and then give the rates of change of the remaining r_{ij}'s with respect to this 'internal time.'

However, it is more satisfactory to note the following. Since we deny the existence of time, we clearly cannot adjoin a one-dimensional 'time space' T to our RCS. Instead, histories of the universe are simply curves in its RCS Q_0. There is no way in which we can say 'how fast' the universe moves along such a curve. That would require an external time. At any point, such a curve has a *direction* in the RCS, but speed along the curve in that direction is meaningless. This is the lesson we must draw from Mach's "It is utterly beyond our power to measure the changes of things by *time*" (Mach 1960, p. 273).

Although time does not exist, in the context of classical (non-quantum) Machian physics we may still assume that the universe in its history occupies a unique continuous sequence of configurations. Each such configuration may be called an *instant*. There are instants, but there is no time. A history is a string of such instants.

We can now extend Poincaré's analysis to include this nonexistence of time. Just as the First Machian Requirement is associated with inability to determine $\mathbf{M}_0$ of a dynamical system from relative initial data, the absence of external time is associated with a *Second Machian Requirement* and an inability to determine kinetic energy - and hence a total energy E - from relative data. Therefore, the initial condition for a fully Machian theory takes the form of specifying an *initial point* of the RCS Q_0 and an *initial direction* in Q_0 at that point. If time existed, we could allow the luxury of specifying both an initial direction in Q_0 and an initial *speed* along that direction. The Second Machian Requirement prohibits specification of such a speed; the direction must suffice.

To formulate a Machian theory, we distinguish different instants by an arbitrary labeling parameter λ that increases monotonically along the Machian histories. In the technical implementation (Barbour 1994a), the action is invariant with respect to arbitrary *reparametrization* of λ, i.e., its replacement by any other monotonic label parameter. It is, however, important that reparametrization invariance by itself does not necessarily implement the Second Machian Requirement. So-called parametrized particle dynamics is reparametrization invariant, but *it is not timeless* (Barbour 1994a). The true criterion of a timeless theory is that its initial conditions require specification of only a direction in a genuine RCS. In parametrized particle dynamics, the configuration space is augmented by a completely heterogeneous 'time space,' which is simply Newton's absolute time in another guise (cf. pp. 103–104).

It should also be said that once dynamical Machian histories have been found in a truly timeless fashion, the very fact that they are obtained as the solution of a Machian variational principle in the RCS makes it possible to introduce along the curve of any such history a uniquely distinguished time *metric*, with respect to which the history of the universe unfolds in a particularly simple manner. In the context of nonrelativistic particle dynamics, this time *metric* is found by exactly the same method as the astronomers used for several decades to determine what they called *ephemeris time*. In fact, its properties are identical to those of Newton's absolute time metric (duration), but it is found operationally.

A very important fact about ephemeris time is that *all the dynamical degrees* of freedom contribute to its determination. Because Mach said "A motion is termed uniform in which equal increments of space described correspond to equal increments of space described by some motion with which we form a comparison, as the rotation of the earth," (Mach 1960, p. 273) he may have misled some people into thinking certain motions can be separated out and used as clocks to measure all the remaining motions. However, in the generic case this is not so (Barbour 1994a). The only satisfactory definition of uniform motion is with respect to the ephemeris time, to which all motions contribute. In the context of geometrodynamics, the global ephemeris time of nonrelativistic theory is generalized to a *local ephemeris time*, which turns out to be identical to Einsteinian local proper time (Barbour 1994a).

In *classical* (nonquantum) physics the consequences of the nonexistence of an external time are relatively minor and amount to little more than the introduction of an operational definition of time and the recognition that any actual classical universe can have only *one* value of its total energy; this means that the history can be described by means of Jacobi's principle as a timeless geodesic in the configuration space (Barbour 1994a). However, in a *quantum* theory of the universe the implications of timelessness are potentially very great (Barbour 1994b); see the final discussion session in this volume.

Let me conclude this section by introducing a new word that was kindly proposed to me at Tübingen by Dieter Brill as an alternative to the somewhat staid *Machianity*. In a play on *machismo*, he proposed *Machismo*. This appropriately matches Ehlers's comment, also made at Tübingen, that you are a Machian, "If you are so courageous as to think that you primarily formulate a theory for the whole universe." To this I would only add that, of course, such a theory must satisfy *both Machian requirements*. That is real Machismo.[3]

4. Implementation of the First Machian Requirement by Means of the Intrinsic Derivative

It is now necessary to discuss the implementation of the First Machian Requirement, first in nonrelativistic physics, and then in geometrodynamics. I do not think anyone can doubt the 'Machianity' of the kinetic energy introduced by Hofmann, Reissner, and Schrödinger in the papers translated and published in Chap. 2 of this book. Since Newtonian potential energies are already Machian, the First Machian Requirement is clearly met by such theories. Moreover, as Bertotti and I (1976) showed, it is easy to translate such theories into a timeless form and so implement the Second Machian Requirement too.

However, most theories of the Hofmann–Reissner–Schrödinger type, in particular those based on the Weber potential (Assis, p. 159), lead to an anisotropic effective mass, in crass disagreement with experiment (Will, Nordtvedt, this volume). It was this fact above all that led Bertotti and me to seek alternative ways of implementing the two Machian requirements in a timeless RCS. This led us to the notion of the *intrinsic derivative* (Barbour and Bertotti 1982, referred to below as BB2; Barbour 1994a), which is found by a procedure in which two complete configurations of the universe that have some small intrinsic difference are compared in a 'best matching' procedure.

One can imagine that one configuration is 'slid around' on top of the other into all possible trial matchings; this has the effect of establishing trial *equilocality pairings* of points in the respective configurations. Given such a trial equilocality relation, one can define rates of change at points provisionally taken to be equilocal and use them to define a trial action by integrating a suitable functional of the dynamical degrees of freedom and their derivatives over space. Extremalization of this action with respect to all trial equilocality relations then leads to an invariantly defined action. As Ehlers notes (p. 460, see also p. 7 and p. 55), in order to define velocity, Newton introduced the notion of absolute space in order to be able to say when some given body is 'at the same place,' i.e., is *equilocal*, at different instants of time. An inertial frame of reference serves the same purpose. The intrinsic derivative is a fully Machian alternative to this use of external frames of reference.

It is interesting that when the equilocality extremalization is carried out explicitly (Lynden-Bell, p. 172), rather than implicitly (BB2), the resulting Machian kinetic energy also has the intuitively Machian form first proposed by Hofmann, i.e., it has the form of an interaction involving *pairs* of masses and their relative separations. However, in contrast to the Weber potential, the BB2–Lynden-Bell action does not lead to mass anisotropy.

The really remarkable and ironic fact is that the relative motions predicted by the BB2–Lynden-Bell model are *identical* to the relative motions in Newtonian mechanics for a system having vanishing center-of-mass angular momentum $\mathbf{M}_0$, $\mathbf{M}_0=0$, and one fixed total energy E. This is how intrinsic dynamics resolves the failure of predictability that Poincaré identified in Newtonian mechanics, in which instantaneous relational data cannot determine $\mathbf{M}_0$ and E. In intrinsic dynamics $\mathbf{M}_0=0$, and E can have only one fixed value.

A further advantage of the intrinsic derivative is that it can be immediately applied as soon as one has chosen an 'ontology' of the world, that is, one has chosen what kind of relative configurations are supposed to embody possible instants. In Newtonian mechanics, the relative configurations are those of mass points in Euclidean space. One could just as easily develop Machian *field theory* (BB2), for which the relative configurations are defined by field intensities. Finally, one can develop *Machian geometrodynamics*, in which the relative configurations are Riemannian 3-geometries (and matter fields defined on them if one wants more than pure geometrodynamics). As shown in BB2 and (Barbour 1994a), formulation of the appropriate Lagrange function leads unambiguously to an action that has the same key basic properties as the

Baierlein–Sharp–Wheeler (1962, BSW) form of the Hilbert action of general relativity, though it is not possible to pin down the precise form without invoking certain additional hypotheses (cf. Kuchař's comments, pp. 454–455). However, this is of no concern to a Machian; the claim is not that general relativity is *the* unique Machian theory but merely that *it is a Machian theory*.

The basic formulas for the intrinsic derivative and ephemeris time in particle dynamics are compared with the corresponding generalizations in geometrodynamics by Goenner (pp. 449–450).

Let me end this section with a response to the difficulties Ehlers finds (p. 466) with extending relational particle mechanics into the context of general relativity. What Ehlers says is perfectly correct if we are talking about relative motion of *particles* in *spacetime*. But in Machian geometrodynamics the intrinsic derivative is used to define *relative motion of fields or geometries* in a context in which spacetime does not yet exist. That is a very different matter. In fact, Bertotti and I developed the notion of the intrinsic derivative precisely in order to overcome, in the context of the modern 'ontology' of fields and dynamic geometry, the very difficulties to which Ehlers refers. It may also be worth mentioning that when we developed the idea we were expecting to create a Machian geometrodynamics with physical predictions different from those of general relativity. We had no idea the notion already existed at the core of Einstein's theory in the form of the BSW action. When Kuchař pointed this out to us in 1980, our initial reaction was one of disappointment. We had lost the chance of finding a new theory!

5. 'Anti-Machian' Solutions and Infinity

Because of program constraints, the material in this section was not presented at Tübingen, and is partly written in answer to remarks made elsewhere in this volume, especially by Hoefer (pp. 82–83, 88), Isenberg (pp. 205–206), Ehlers (pp. 466–468), and Goenner (p. 450). I begin with a brief review of the steps, given in detail in (Barbour 1994a), that lead me to conclude that general relativity is Machian:

1) The natural RCS for general relativity considered as a dynamical theory is *superspace*, the space of all Riemannian 3-geometries, which for the moment we shall assume are compact. If matter fields are present, the RCS is extended accordingly. In pure geometrodynamics, a 3-geometry is characterized formally by three intrinsic degrees of freedom per space point.

2) In the BSW form, the Lagrange function of general relativity

defines a generalized line element in superspace provided the thin-sandwich problem (Wheeler 1964b, 1968) can be solved. The manner in which BSW action is calculated, by variation with respect to a 3-vector field that establishes trial equilocality relations, ensures that the First Machian Requirement is satisfied. The Second Machian Requirement would already be satisfied if the BSW action principle possessed *global* reparametrization invariance. In fact, it possesses *local* reparametrization invariance and therefore satisfies the Second Machian Requirement *a fortiori*. General relativity is more than Machismo – it's *Machissimo*.

To reach this conclusion, we had to shed at least two Einsteinian convictions: 1) Cosmic derivation of the inertial mass is the essence of Mach's Principle; 2) local inertial frames of reference must be exclusively determined by the matter energy–momentum tensor. It was widely accepted at the Tübingen conference (cf. Ehlers's comments, p. 93) that such a formulation of Mach's Principle, as given by Einstein in 1918, is hopelessly flawed from the mathematical point of view. From the physical point of view, there is also no good reason to rule out purely gravitational degrees of freedom as determinants of motion. This too was also widely recognized at Tübingen. See *gravitational degrees of freedom, role in Mach's Principle* in the Index. Moreover, the fact that in the spacetime picture the gravitational degrees of freedom do not possess a generally covariant energy–momentum tensor (Hoefer's objection, pp. 82–83) in no way prevents one from formulating an initial-value problem in which the role of such degrees of freedom is perfectly well defined. This is related to the fact that the intrinsic derivative is defined at a level of the dynamics that is *logically prior* to the appearance of spacetime.

We must now consider those perennial bogey men, the so-called manifestly anti-Machian solutions of general relativity, especially matter-free spacetimes and above all empty Minkowski space. We can go a long way to exorcizing these bogey men if we hold fast to the following principle: Any solution of pure geometrodynamics, i.e., any Ricci-flat spacetime, is not to be analyzed as a matter-free structure in which test particles have inertia or as a structure that has a disconcerting resemblance to Newton's absolute space and time but as a *dynamical history of 3-geometries*. The Machian requirements apply to the structure of such a dynamical history, not to the behavior of test particles within it.

Thus, what we need is a change of perspective – away from the view that flat Minkowski space is the 'natural ground state of the universe' (or, alternatively, a fixed external framework) and to the recognition that it is a highly atypical solution of a sophisticated dynamical theory. It must also be remembered that the worries Einstein struggled with in the 1916–18 period (Hoefer, p. 67) stemmed to a quite considerable degree from his complete misunderstanding of the mathematics of his own field equations (see Ehlers's comments, p. 92). They are not elliptic equations like Poisson's equation, for which one first specifies the matter distribution and then finds the field, but evolution equations with very nontrivial initial-value constraints on *both the geometry and the matter*. The Machismo is all in the constraints.

Let us consider in this light the spatially compactified Minkowski spaces discussed by Isenberg and Goenner (pp. 205–206). To interpret such solutions dynamically, we must foliate them in some manner.

Any foliation of such a spacetime by spacelike hypersurfaces generates a curve in superspace, each point of which represents one leaf of the foliation, i.e., one 3-dimensional hypersurface. If we choose the foliation trivially, i.e., we choose Galilean coordinates in such a spacetime, so that all the 3-dimensional hypersurfaces are flat and identical, the curve that is supposed to represent our history degenerates into a singular point. Nothing happens! However, there are also infinitely many other foliations of the same spacetime, and these correspond to *nontrivial Machian histories* of the 3-geometries. Looked at from the point of view of Machismo, there is nothing wrong with these histories. The 'degeneracy' of general relativity - the fact that foliation invariance can generate so many different histories - in no way diminishes the virility of its Machismo. Quite the opposite - it is evidence of its *potency* (cf. what I said above about general relativity being Machissimo).

It should also be noted that superspace, like all RCSs, has frontiers, or strata. Virtually all exact solutions of general relativity, including the ones that seem so anti-Machian, possess special symmetries and live on these degenerate frontiers, or, rather, some of the histories into which they can be foliated live on the frontiers. This is an important contributory factor that helps to create the impression of anti-Machianity.

It should be noted here that simply because a spacetime like Minkowski space that arises out of Machian geometrodynamics does not seem to contain propagating gravitational waves this does not mean there are no nontrivial geometrical degrees of freedom in the spacetime. Any curved - or even flat - 3-geometry is a *bona fide* state of the geometry even if it happens to fit into a very special, entirely nongeneric 4-geometry.

However, as Isenberg remarks, when you add a particle to such a 4-geometry you simultaneously have to add some rather more nontrivial gravitational degrees of freedom. What this does is shift the solution and all the histories into which it can be foliated away from the degenerate frontiers of superspace and into its generic interior. Then the very superficial similarity between the high-symmetry solutions and Newtonian absolute structures disappears. But – let me emphasize it again – the real difference between the Newtonian absolute structures and all Einsteinian spacetimes is that the former are a rigid kinematic framework for dynamics, whereas the latter arise as a result of Machian dynamical evolution in a situation in which there are no preexisting frameworks at all.

Finally, we must consider the toughest nut – solutions that are not spatially compact, in particular, solutions that are asymptotically flat at infinity. It cannot be denied that here some difficulty does arise. The point about Machian geometrodynamics is that it is formulated solely in terms of 3-geometries. A solution of geometrodynamics is merely a *sequence of 3-geometries*. Now the process of finding such sequences involves solution of the partial differential equations of the thin-sandwich problem (Wheeler 1964b, 1968) for a 3-vector field N_i (subsequently identified with the shift). Once N_i has been found, it can be used in conjunction with the known 3-metric g_{ij} and its derivative with respect to an arbitrary time parameter, both of which are given, to find, by means of purely algebraic relations, the four remaining components g_{00} and g_{0i} of a *four-dimensional* metric tensor. In this way, spacetime is constructed from the sequence of 3-geometries.

Now as Wheeler (1964a,b, 1968) pointed out long ago, if the 3-geometries are spatially compact, the thin-sandwich equations are 'self-contained,' that is boundary conditions do not arise. It is merely (!) necessary to ensure that N_i satisfies the thin-sandwich conditions over the complete compact 3-manifold. If, however, the 3-manifold is not spatially closed, some sort of boundary condition for N_i must be imposed at the frontier of the manifold or at spatial infinity. From the Machian point of view, the imposition of such a condition violates the spirit of 'Machismo'. It is also an affront to the principle of sufficient reason (cf. Einstein's comments on p. 182 and p. 183) and brings in an extraneous element at infinity. It is, however, important to note here the difference between the nonrelativistic form of Mach's Principle as implemented in the BB2–Lynden-Bell model and Einsteinian geometrodynamics. The former is based on a *global* gauge group, the latter on a *local* gauge group. It is for this reason that the thin-sandwich equations are differential, not algebraic, as in the BB2–Lynden-Bell model.

Thus, *everywhere* within *any* Einsteinian spacetime the thin-sandwich equations must be satisfied. *Locally* there is nothing to tell us how this fulfillment of the Machian thin-sandwich conditions is brought about: We cannot tell whether we are in a spatially compact or asymptotically flat spacetime; for we can never get to infinity to put the matter to the test! No matter how far we go in exploration of our spacetime, the thin-sandwich equations will always be satisfied. As we progress further and further, we shall find an ever larger region in which the structure of spacetime can be understood in a perfectly Machian manner. For us, as opposed to mathematicians "paid by their math department" (cf. p. 205) to find Einsteinian solutions, that process can never end. Moreover, even if we could 'get to infinity' all we should find would be some condition on its rim that somehow 'pegs the world down.' All the action within the rim would be totally Machian. Once again, the proper recognition of 3-geometry as a *bona fide* determinant of inertial frames of reference along with ordinary matter fields goes a very long way to defuse the worries that Einstein was expressing in 1916–17 (Hoefer, pp. 74–80). In spacetimes that are nearly flat and nearly free of matter, the inertial frames are *not* determined solely by boundary conditions at infinity but – to a far, far greater extent – by fulfillment of the Machian constraints on the geometry *all the way out to infinity*.

Infinity is always going to be a problem for the human mind. General relativity is wondrously Machian, as perfectly so as any mortal could construct. As Copernicus remarked (1543), we can leave the philosophers to worry about infinity.[4]

Acknowledgments. It was a matter of great regret to me that, because of a family bereavement, Bruno Bertotti was unable to join us at Tübingen. My work with Bruno in which we developed our Machian models of motion was wonderfully stimulating and rewarding, and I am greatly indebted to him. Karel Kuchař is another person to whom I owe a large debt. Lest the reader should get the wrong impression from the various uninhibited exchanges between Karel and me recorded in this volume, we are actually very good friends. It is just we enjoy a good rap. I should also like to thank Herbert Pfister most warmly for the idea of the Tübingen conference and especially for asking me to share with him the planning of the program and the publication of the proceedings.

NOTES

[1]It is certainly true that the opening words of Mach's second paragraph (p. 110) could easily give one the impression that it is simply going to continue the

theme of mere redescription. In correspondence with me, Norton comments: "If he intended the major change of course you propose, one would expect some stronger indication that the enterprise is different." That is a very fair comment, and when I first looked at the passage I initially thought Norton was correct. However, closer examination has persuaded me otherwise. I am very happy to concur with Norton's further proposal in his letter: "We ought now to hand over the decision to the good sense of our readers."

[2]*Note added in proof.* I have just discovered that in the fifth and sixth editions of the *Mechanik* (1908), Mach actually referred to Hofmann (p. 262, my italics): "I have in front of me also a lively, clear text written in a very popular style by W. Hofmann (*Bewegung und Trägheit*, Vienna 1904), who seems unaware of the controversy and who *seeks the solution in almost the same ways as I did* (die Lösung fast auf denselben Wegen sucht, wie ich es seinerzeit gethan habe)." This comment too seems to support my position.

[3]Since *machismo* does have a pejorative overtone, especially in this feminist age, and Dieter Brill's own attitude to Mach's Principle, on the evidence of his contributions to this volume, is somewhat more pragmatic than my robust stance, was there, I just wonder, a playful *subversive* intent behind his coining? Whatever the truth, the happy inspiration was - and is - very gladly accepted.

[4]It will probably not have escaped the reader that only two of my fellow symposiasts at Tübingen joined me in the straw poll (p. 106) in saying that "General relativity is perfectly Machian." However, a majority did agree that "General relativity with appropriate conditions of closure of some kind [is] very Machian." Probably the only substantive difference between my position and that view concerns the delicate issue of the status of conditions at infinity, and that is bound to remain something of a quibble. I hope the new arguments of this final section will win over at least some waverers. I may also mention that one participant told me after the straw poll he completely accepted my argument and had only abstained in the 'perfect Machianity' vote out of a sense of anticlimax - he wanted Mach's Principle to do more than "just be general relativity." Personally, I think much may yet be won from a Machian standpoint, but the dynamic core of general relativity strikes me as a good start. Finally, the quantum aspect needs to be considered too. If, as I - and quite a number of other quantum cosmologists - conjecture (pp. 516–517), quantum gravity simply gives probabilities for 3-geometries, the classical histories dissolve - to be recovered at best in certain regions of superspace in which the wave function of the universe gets into a WKB regime. All this will be determined in a perfectly Machian manner, see p. 478.

REFERENCES

Baierlein, R. F., Sharp, David H., and Wheeler, John A. (1962). "Three-Dimensional Geometry as a Carrier of Information about Time." *Physical Review* 126: 1864–1865.

Barbour, Julian B. and Bertotti, Bruno (1976). "Gravity and Inertia in a Machian Framework." *Nuovo Cimento* 38B: 1–27.

Barbour, Julian B. and Bertotti, Bruno (1982). "Mach's Principle and the Structure of Dynamical Theories." *Proceedings of the Royal Society of London, Series A* 382: 295–306.

Barbour, Julian B. (1992). "Einstein and Mach's Principle." In *Studies in the History of General Relativity*, vol. 3 of *Einstein Studies*. Jean Eisenstaedt and A. J. Kox, eds. Boston: Birkhäuser, pp. 125–153.

Barbour, Julian B. (1994a). "The Timelessness of Quantum Gravity: I. The Evidence from the Classical Theory." *Classical and Quantum Gravity* 11: 2853–2873.

Barbour, Julian B. (1994b). "The Timelessness of Quantum Gravity. II. The Appearance of Dynamics in Static Configurations." *Classical and Quantum Gravity* 11: 2875–2897.

Brans, Carl and Dicke, Robert H. (1961). "Mach's Principle and a Relativistic Theory of Gravitation." *Physical Review* 124: 925–935.

Copernicus, Nicholas (1543). *De Revolutionibus Orbium Coelestium*. Nurenberg.

Jordan, Pascual (1955). *Schwerkraft und Weltall*. Braunschweig: Vieweg.

Lange, Ludwig (1885). "Über das Beharrungsgesetz." *Berichte der Königlichen Sächsischen Gesellschaft der Wissenschaften, Math.-Physik. Klasse* 333–351.

Mach, Ernst (1889). "Die Mechanik in ihrer Entwickelung. Historisch-kritisch dargestellt." Second ed. Leipzig: F. A. Brockhaus.

Mach, Ernst (1960). *The Science of Mechanics: A Critical and Historical Account of Its Development*. LaSalle: Open Court.

Tait, P. G. (1883). "Note on Reference Frames." *Proceedings of the Royal Society of Edinburgh*, Session 1883–84: 743–745.

Wheeler, John A. (1964a). "Mach's Principle as Boundary Condition for Einstein's Equations." In *Gravitation and Relativity*. H-Y Chiu and W. F. Hofmann, eds. New York: Benjamin.

Wheeler, John A. (1964b). "Geometrodynamics and the Issue of the Final State." In *Relativity, Groups, and Topology: 1963 Les Houches Lectures*. Cecile DeWitt and Bryce DeWitt, eds. New York: Gordon and Breach.

Wheeler, John A. (1968). "Superspace and the Nature of Quantum Geometrodynamics." In *Battelle Rencontres: 1967 Lectures in Mathematics and Physics*. Bryce DeWitt and John A. Wheeler, eds. New York: Benjamin.

Discussion

Ehlers: It may be helpful to mention the following: In classical mechanics it was a question how to separate the internal motions of a deformable body, like a planet, from the overall motion. In order to do this Tisserand (1891 *Treatise de Mécanique Céleste*, Vol. 2, Chap. 30, § 1–3, pp. 500–505) has introduced into classical mechanics the concept of a corotating frame of an arbitrary deformable body. He requires to choose that particular reference point, and that system of reference directions, with respect to which the relative kinetic energy is minimized. Then the reference point is the center of mass, and the reference directions are those for which the relative angular momentum vanishes [cf. Frauendiener, pp. 354–356, and Lynden-Bell, pp. 174–175]. And I think your description [of the intrinsic derivative] amounts to something quite similar. Instead of starting with the standard Newtonian description, and attaching such a frame to one particular body at a time, you attach it to the universe as a whole, and then say you should describe the whole situation only with respect to this preferred frame.

Barbour: There are undoubtedly similarities, but I do not think what you describe is exactly the same as Bertotti and I did, since the scheme you describe must presumably work for any rest-frame angular momentum whereas in the Machian scheme one recovers (in relative variables) only those Newtonian motions for which the rest-frame angular momentum is exactly zero.

Lynden-Bell: In your scheme there is length?

Barbour: There is at this stage; to get rid of length, I think you have to solve Riemann's problem of where metrical properties come from. Somehow or other, all of physics works with metrical quantities, intensities, or something that has definite *numerical* values.

Lynden-Bell: You could make this scale invariant?

Barbour: One could certainly make it scale invariant.

Ehlers: A short remark. I think for the conceptual set-up it's essential to realize that in your Q_0 version even [Sec. 3], you accept without criticism two *a priori* structures. Namely, a simultaneity relation and Euclidean distances, and everything else is then made dynamical.

Barbour: That's quite true. And in Machian geometrodynamics I assume there are 3-geometries, described by a spatial metric, just as in the ADM form of general relativity. Each of these 3-geometries will represent instants, or simultaneities. It is important that Einstein never abolished simultaneities. He just abolished the distinguished foliations; but spacelike hypersurfaces, particularly in a Hamiltonian formulation of general relativity, are central concepts. These are what the n-body configurations in Euclidean space become in GR.

Giulini: But this simultaneity is a very abstract one. It is an abstraction really from nonrelativistic Newtonian mechanics. Since there is relativity, special relativity, it's pretty doubtful you can use it. We do not have direct access to positions, even relative positions. Rather, it is the theory which permits us to infer them from observations.

Bondi: As a final word [on particle dynamics], you could say the universe is not only its own clock, it is its own measure of rotation.

Barbour: Yes, and we must find it out from within the universe.

Jones: I don't think that the total angular momentum of the universe can be expressed in the relative configuration space. It seems like an absolute rather than a relative concept.

Barbour: The fact that an *n*-body Newtonian system has a nonvanishing angular momentum certainly does show up in the relative motions, and therefore it can be expressed in relative terms [pp. 111–112].

Nordtvedt: I've probably been thinking about radar and laser ranging too much, but there's this trend in which people more and more reduce so-called distance measurements into effectively local time measurements. In some sense, time is emerging as more fundamental; in my conceptual world anyway, it's more essential and significant than space. However, I got the opposite twist from you, that you saw time disappearing. For me clocks have become more central, at least in experimental physics.

Barbour: I don't think there's any necessary conflict. The important thing is to distinguish between theoretical concepts posited as basic and effective concepts derived from them. This would also be my answer to the remark Giulini just made. After all, we continually use temperature but still look to derive it from microscopic motions of atoms. I don't think there's anything wrong with having a theoretical framework that leads to *effective concepts* which then dominate your thinking. Indeed, this workshop is about recovering the inertial frames of reference, which dominate the way most of physics is done on the earth. They are certainly very good concepts; they are around us here, almost concrete, and confront us all the time. But are they the most fundamental entities? The fact that we use something a lot doesn't mean that we cannot explain where it comes from. My aim is to recover both local time and local inertial frames of reference from the one idea that the history of the universe is an extremal in an RCS.

Bondi: As you quote enough of ancient authors, I'll quote myself [laughter]. I think, quoting from memory, Gold and I in our '48 paper, 45 years ago, wrote "A law of motion is only relevant to describe a large number of different motions. The universe has no law of motion, it has only a motion."

Barbour: I thought a lot about that remark of yours, and it was important to me in clarifying these ideas. In the end, I didn't agree with you. You *can* formulate a law for the whole universe. I believe it's been done here this morning: The unique history of the universe is a geodesic in its RCS. But you can then look at many subsystems within the one universe and describe them by means of effective local inertial frames and an effective time that are both determined by the universe as a whole. You then do indeed have laws of motion that describe a large number of different motions. Our two standpoints are reconciled by noting that the law describing the motion of the universe has a form quite different from the laws that describe subsystems, since they include effective concepts derived from the motion of the complete universe.

Bondi: In the light of your talk, I interpret myself slightly differently [laughter]. In all the subsystems, you use the time defined by the universe, but the universe defines its own time, which is an identical statement.

Barbour: Yes.

Bondi: Which is quite a different thing – relying on somebody else's time and manufacturing it yourself.

Barbour: Well, everything we use must ultimately be extracted from within the world.

Kuchař: Julian, you treated the angular momentum differently than the energy, because you put it equal to zero. You didn't put the energy equal to zero. To proceed symmetrically, you should perform, not a reduction to zero angular momentum, but to a *constant* value of the angular momentum.

Barbour: Yes, this too we have discussed over several years Karel. Rotations are different because the components of the angular momentum do not commute. This is an issue that I will cover elsewhere. [In Vol. 2 of my study *Absolute or Relative Motion?*.]

Will: Do you want to comment on any possible Machian ironies in situations that may arise if atomic time, which replaces ephemeris time, is ultimately replaced by a time based on millisecond pulsars, where then the basis of standard time is rotation? The current millisecond pulsar is at least as stable as all the atomic clocks on earth. We cannot tell which is the stablest, so if we have a collection of millisecond pulsars, like the one we now know, that could then become the operational definition of time.

Barbour: That is a lovely question, because I'm sure you know there's a problem with the binary pulsar and the accuracy with which its emission of gravitational waves is conformed. As Joe Taylor and Thibault Damour note (1991, *Astrophysical Journal* 366: 501), the differential acceleration between the solar system and the binary pulsar in the field of the Galaxy exactly mimics emission of gravitational waves, so to interpret the binary-pulsar data the galactic gravitational field must be modeled. Now certainly the millisecond pulsars that you refer to may well become the most useful actual clocks from which to read out time, but if there are several of them there's no doubt it will be necessary to model the Galaxy - and, in fact, to some extent the universe as a whole because of the role that quasars play in determining frames of reference - in order to extract a meaningful time from them. There is a very nice paper by G.M. Clemence in the *Reviews of Modern Physics* for 1957 in the issue that's got the proceedings of the Chapel Hill relativity conference. It's the very first article in that issue; it's on astronomical time, on ephemeris time, and Clemence asks: What is a clock? His answer is: "A clock is a mechanical device which is continually calibrated against ephemeris time."

Will: Except the ephemeris time is no longer enough.

Barbour: Not the ephemeris time based on the solar system no; I'm talking now about a generalized ephemeris time extracted from observation of the entire universe.

Hoefer: I was wondering whether Newtonian empty absolute space turned out to be a Machian situation with this scheme.

Barbour: No. Here I'm exactly with Hoyle and Narlikar [p. 250 and 262]. In fact they need two particles, but, because I get rid of time, I need at least three particles to get a nontrivial Lagrange function. The simplest nontrivial Machian model must have three particles in it in the nonrelativistic case. In geometrodynamics there is no need for matter since geometry has its own degrees of freedom.

Hoyle: I think it is the case that there are people who go round the world, or were a few years ago, selling time. Their job is to sell it. They carry it in a little suitcase, and I came across one of these chaps some years ago. We got talking, and I said "Well what are you doing?" I mean because he's very careful with this suitcase, you wonder if he's got some sort of explosives in it, and he says "I'm selling time, I'm a salesman for time," and it means he's simply got an exceedingly accurate atomic vibration, and my understanding is that astronomers have for quite some time used this rather than anything connected with the sky.

Barbour: That is, of course, true, but my answer is very much like the one before to Cliff [Will], Fred. My understanding is that time is now determined by a system of about seven such atomic clocks, which are distributed around the world. Just like the earth, atomic clocks have internal jitters, over which the scientists have no control.

Hoyle: There are perturbations on the system.

Barbour: Yes, and to counteract those you have to model the earth and average the clock readings. To extract a time out of the network that Joe Taylor can actually use to test if the binary pulsar's really giving off gravitational waves, you have to model the continental drift, the Chandler wobble of the axis of the earth, and all these things. There is no clock from which you can simply read off time. It does not exist. The present one is a network of such clocks with a model of the motion of the earth.

Hoyle: I think you could use a system in some other place in the universe. It need not be the earth at all.

Barbour: Yes, but you still have got to have a complete dynamical system, and you've got to parametrize the environment in which the clock is read.

Hoyle: If you reduce it to practical time, yes. But it's a system of time in which you count the number of oscillations of a certain transition, and that is going to be the same wherever you are in the universe.

Barbour: With respect, Fred, that's not a clock because an actual atomic clock is a many-body system. An atomic clock corresponds to a complicated many-body problem of solid-state physics. One can never get one's hands on an oscillation of one atom like that; it just isn't there.

Ehlers: I think you talked about proper time, which is defined in terms of atomic systems, and in order to relate different proper times, you have to have a good model of the gravitational potentials and relative motions in order to reduce them to a common time. It is unfortunate that we use just one word. I think for science we need at least two different concepts, which are unfortunately denoted by the same word, namely, we use time in a first sense as a global parameter of events, to order them in a certain sequence, and that is not necessarily the same as what is measured by a good clock. Secondly, we use clocks, and we know that already in special relativity time in the first sense is the coordinate time of some inertial frame, and proper time is something of a different nature. It's idealized by a different mathematical structure and it is different also in its actual scientific use. Would you agree?

Hoyle: Yes, I agree entirely, because clearly there's an infinitely large number of ways in which one can define coordinate time, but my point is that the proper time is unique.

Ehlers: The proper time of one particular clock at one particular place.

Hoyle: At one particular place, yes.

Barbour: I would only add that, nevertheless, in order actually to measure that local proper time one must still in principle model the universe since it is the dynamics of the universe that ultimately 'manufactures' proper time.

A Closed Universe Cannot Rotate

D. H. (Harry) King

1. Introduction

What does it mean for a universe to rotate? In the context of Newtonian physics, a rotating universe is one where there is a net rotation of its contents with respect to absolute space. In other words, a rotating universe has a nonzero total angular momentum with respect to the global inertial frame. Our solar system in otherwise empty space is an example of a Newtonian rotating universe. All the planets orbit the Sun in the same direction, thus contributing to a nonzero total angular momentum for the solar system. This model universe can be said to be rotating because there is nothing to cancel the angular momentum of the solar system.

At this point, I find myself in the somewhat awkward position of having to define Mach's Principle in order to explain why a rotating universe is contrary to this principle. Mach's Principle is dangerous to define because everyone seems to have a different interpretation. Nevertheless, I follow Barbour's lead (Barbour 1989) by saying that, within the context of Newtonian physics, Mach's Principle asserts that only the relative motions of the mass in a universe are significant. According to this principle, absolute space is not a fundamental part of (Newtonian) dynamics, but merely an auxiliary device that is introduced for computational convenience.

It is clear that a rotating model universe is not Machian because the rotation takes place with respect to absolute space instead of with respect to matter. It turns out that the connection between a Machian universe and a nonrotational one is fundamental. Barbour and Bertotti (1982) have shown that the Newtonian dynamics of a model universe can be reinterpreted in terms of a theory of relative motion if, and only if, the total angular momentum of that universe is zero. (See also Lynden-Bell's contribution in this volume, p. 172.)

Einstein Studies, vol. 6: Mach's Principle: From Newton's Bucket to Quantum Gravity, pp. 237–248

In summary, the status of Mach's Principle within Newtonian physics is as follows: 1) Not all Newtonian model universes are Machian, and 2) Machian is equivalent to nonrotating.

Shifting now to general relativity, things become considerably more difficult. Einstein had hoped to create a theory of relative motion (hence the name general relativity) where *all* model universes would be Machian. Relative motion was supposed to be implemented indirectly by making the inertial frame, now local instead of global, a dynamic element of the theory responding to the matter content of the universe. General relativity would indeed be a theory of relative motion if the inertial frame was *completely* determined by the matter content of the universe. However, this is not the case. The inertial frame is a dynamic element of the theory having independent degrees of freedom. Further complications are introduced by the Equivalence Principle, which inextricably combines the gravitational field with the local inertial frame and by the inclusion of continuous fields, which makes the definition of relative motion itself rather unclear. At this point the definition of Mach's Principle becomes open to much discussion, since any definition requires the above complications to be addressed.

Barbour [p. 225], who has interpreted relative motion for continuous fields, has shown that general relativity is a theory of relative motion.

I have taken a different approach to Mach's Principle by avoiding the issue of relative motion altogether and asking the question: "Does general relativity, unlike Newtonian dynamics, automatically exclude rotating model universes?" Because adherents of Mach's Principle would like to answer this question in the affirmative, I will call it the *Mach Question* for general relativity.

The first thing to note about the Mach Question is that it is false for asymptotically flat model universes. Consider a general relativistic model of the solar system in otherwise empty space. Because the gravitational field within the solar system is weak, the general relativistic model is little different from the Newtonian model discussed previously. Like the Newtonian model, the general relativistic model has a nonzero total angular momentum, which qualifies it as a rotating universe. Thus, asymptotically flat general relativistic model universes are no more Machian than Newtonian ones.

The source of this anti-Machian behavior is the need by Einstein's field equations for spatial boundary conditions. Boundary conditions of asymptotic flatness provide an inertial frame at infinity which is completely unaffected by the matter content of the universe. Elsewhere in space, the inertial frame is only modified by the presence of matter. The inertial frame is not completely determined by the universe's matter content. The problem of boundary conditions can be avoided by considering only spatially closed (3-sphere topology) model universes.

The Mach Question for closed model universes has a long history and for the last thirty years has been thought to be false (Ozsváth and Schücking 1962, 1969). The most important conclusion of this paper is that the Mach Question is true for closed model universes when: 1) The angular momentum is measured relative to the average inertial frame for the universe, and 2) the angular momentum of gravitational waves is included in the total for the universe.

As the above two conditions suggest, the Mach Question requires a number of issues to be addressed: 1) Global rotation must be defined for an arbitrary spacetime that lacks rotational symmetry; 2) angular momentum must be defined for a closed universe; and 3) the angular momentum carried by gravitational waves must be taken into account. The key to the solution of all these issues is the introduction in Sec. 2 of an average inertial frame for a given model universe. The average inertial frame permits the definition of total angular momentum in Sec. 3 and permits the definition of the stress-energy of gravitation in Sec. 4. These definitions lead to the proof in Sec. 5 that the total angular momentum of a closed universe is zero.

2. Average Inertial Frame

The lack of a global inertial frame for general relativity makes it much harder to define the rotation of a universe in this theory. The approach adopted in this paper is to restrict the discussion to approximately homogeneous and isotropic universes and to introduce a global frame of reference that is approximately inertial. Despite this restriction, the results of this paper are valid to all orders of the perturbation.

The fact that our own universe is thought to be approximately homogeneous and isotropic helps to motivate the introduction of an approximately inertial frame. Our universe is modeled by astrophysicists as an average Hubble flow plus inhomogeneities corresponding to individual galaxies or clusters of galaxies.

Corresponding to the average Hubble flow is a homogeneous and isotropic background metric tensor, $g^{(B)}{}_{\mu\nu}$, which picks out a preferred coordinate system that is equivalent to the global approximately inertial frame. The background metric is defined to be the average of the real metric,

$$g^{(B)}{}_{\mu\nu} \equiv \langle g_{\mu\nu} \rangle \,. \tag{1}$$

The problem of devising the best way to perform the average in Eq. (1) is as yet unsolved and is called the 'fitting problem' for cosmology (Ellis 1984). This average is difficult to define because the quantity being averaged is a tensor, and in order for the definition to be covariant, the tensor must be parallel transported to a common point in order to sum up the contributions from different points in space.

For the purposes of this paper, it is not necessary to specify a particular solution of the fitting problem. Any reasonable fitting procedure must meet a number of consistency conditions. However, first we must establish some notation. Suppose that we have already selected a background metric. Being homogeneous and isotropic, the background metric has a number of vector fields associated with it. There is the time-like vector field n^μ that is orthogonal to the hypersurfaces of homogeneity. There are the six linearly independent Killing vector fields $\{\xi_{(a)}{}^\mu,\ a=1,\ldots 6\}$ that result from the homogeneity and isotropy of the hypersurfaces. It is convenient to introduce the usual coordinate system $\{x^\mu\}$, where $x^0=t$ is constant on each hypersurface and $\{x^i,\ i=1,\ldots 3\}$ label the points on any given hypersurface. In these coordinates, we have $n^\mu=(1, 0, 0, 0)$, $\xi^\mu=(0, \xi^i)$, and $g^{(B)}{}_{\mu\nu}=\mathrm{diag}(-1,\ g^{(B)}{}_{ij})$, which simplify the consistency conditions.

The consistency conditions result from three types of physical requirements on the background metric:

1) The background metric must measure the same proper time, on average, as the real metric, i.e.,

$$\langle (g_{\mu\nu} - g^{(B)}{}_{\mu\nu}) n^\mu n^\nu \rangle = \langle g_{00} - g^{(B)}{}_{00} \rangle = 0. \tag{2}$$

2) The background metric must measure the same spatial distances, on average, as the real metric, i.e.,

$$\langle (g_{\mu\nu} - g^{(B)}{}_{\mu\nu}) g^{(B)\mu\nu} \rangle = \langle g_k{}^k - g^{(B)}{}_k{}^k \rangle = 0, \tag{3}$$

where Eq. (2) has been used to simplify the equation.

3) The background metric must have no net translation or rotation relative to the real metric, i.e.,

$$\langle (g_{\mu\nu} - g^{(B)}{}_{\mu\nu}) n^\mu \xi_{(a)}{}^\nu \rangle = \langle (g_{0i} - g^{(B)}{}_{0i}) \xi_{(a)}{}^i \rangle = 0. \tag{4}$$

The average performed in each of the above consistency conditions is applied over the entire spatial hypersurface corresponding to a given time. That is, the average of a scalar field A is defined to be

$$\langle A \rangle \equiv \frac{1}{V}\int_V A dV, \tag{5}$$

where V is the volume of the spatial hypersurface at time t, and dV is the volume element for the background metric. In each case, because the hypersurface average is applied to a scalar quantity, there is no difficulty in performing the parallel transport of the quantity from one location to another during the averaging process.

It is always possible to choose a background metric that satisfies these consistency conditions. If the background metric does not satisfy the first condition, it can be adjusted by choosing a new time coordinate $t \to t'$. (Here and in the following procedures for adjusting the background metric, the components of the real metric are transformed to a new coordinate system while the components of the background metric remain the same in the new coordinate system. In this way, the real metric remains the same tensor, but the background metric is changed to a new tensor. This procedure is analogous to the gauge transformation used in perturbation analysis.) If the background metric does not satisfy the second condition, it can be modified by choosing a new scale factor $S(t)$. If the background metric does not satisfy the third conditions, it can be adjusted by the coordinate transformation

$$\begin{aligned} &t \to t' = t, \\ &x^i \to x'^i = x^i + A_{(1)}(t)\,\xi_{(1)}{}^i + \dots + A_{(6)}(t)\,\xi_{(6)}{}^i, \end{aligned} \tag{6}$$

where the functions $A_{(1)}(t) \dots A_{(6)}(t)$ are to be determined by Eq. (4). An elementary way to satisfy the third consistency conditions, but by no means the preferred way, would be to choose synchronous coordinates for both the real and background metrics. In fact, the consistency conditions are extremely unrestrictive, leaving great freedom in the choice of a fitting procedure.

In summary, the background metric:

- gives the best approximation to the geometry of spacetime;
- evolves in response to the evolution of the real metric;
- is completely determined by the real metric (for a given choice of fitting procedure); and
- is coordinate independent (the fitting procedure is covariant).

Although the background metric is certainly not a fundamental element of general relativity, it is nevertheless a valuable tool for the interpretation of a given universe. In the next two sections, I show that the use of the background metric allows: 1) global rotations to be defined, and 2) the stress-energy of the gravitational field to be identified. These resolve the two main outstanding issues in the interpretation of the Mach Question for general relativity.

As a final note, it is important to recognize that the background metric $g^{(B)}{}_{\mu\nu}$ is quite different from the unperturbed metric $g^{(0)}{}_{\mu\nu}$ used in perturbation theory. The unperturbed metric is a solution of the field equations that is completely unaffected by a perturbation, whereas the background metric responds to changes in the real metric as it evolves. Even though both are homogeneous and isotropic, the background and unperturbed metrics differ in the evolution of their scale factors. If, however, a perturbation analysis is carried out only to the first order, then the background metric is equal to the unperturbed metric as long as the consistency conditions are maintained by the choice of gauge. In this case, the selection of a fitting condition completely specifies the gauge.

3. Total Angular Momentum

The homogeneity and isotropy of the background metric introduced in the previous section allows us to give meaning to the total angular momentum. Only when a spacetime is rotationally symmetric about some axis at a point P is it possible to give a coordinate independent definition for total angular momentum. Rotational symmetry implies the existence of a Killing vector field. A Killing vector field ξ^μ is required to define total angular momentum. The total angular momentum L_P of a stress-energy field $T^{\mu\nu}$ is given by

$$L_P(t) \equiv \int_V T^{0i}\xi_i dV, \tag{7}$$

where V is the spatial hypersurface of homogeneity at time t, and dV is the volume element for the hypersurface. This definition reduces to the usual definition of angular momentum for the case of 3-dimensional Euclidean space.

Rewritten in terms of the hypersurface average defined in Eq. (5), the angular momentum is given by:

$$L_P = -V \langle T_{0i}\xi^i \rangle , \tag{8}$$

where $T^{0i} = -T_0{}^i$ holds in the special coordinate system chosen.

4. Stress–Energy for Gravitation

In general, the smoothed background metric does not satisfy Einstein's equations when the source is the stress energy of the Hubble flow because these equations are nonlinear, i.e.,

$$G^{(B)}{}_{\mu\nu} \neq 8\pi \langle T^{(M)}{}_{\mu\nu} \rangle \,, \tag{9}$$

where $G^{(B)}{}_{\mu\nu}$ is the Einstein tensor corresponding to the background metric, and $\langle T^{(M)}{}_{\mu\nu} \rangle$ is the stress–energy tensor for the matter comprising the Hubble flow. The reason for this discrepancy is that the stress–energy for the Hubble flow does not take into account the effective stress–energy of the gravitational field.

It is well known that there is no place for the stress–energy of gravitation in Einstein's equations, nor is there any way to define this stress–energy in a general way. However, the situation changes when Einstein's equations are averaged. To illustrate, the effective total mass for a binary star system is not just the sum of the masses of each star, but this amount *plus* the (negative) potential energy of the gravitational field between the two stars. This potential energy appears only when the two individual stars are modeled as a single entity. When a system is averaged over a nonzero length scale, the stress–energy for gravitational interactions occurring at a smaller scale must be included.

The existence of the background metric allows the effective stress–energy of gravitation relative to this metric to be given a completely rigorous definition. The first step is to define the tensor

$$h_{\mu\nu} \equiv g_{\mu\nu} - g^{(B)}{}_{\mu\nu} \tag{10}$$

and then to expand the Einstein tensor in a power series in $h_{\mu\nu}$, i.e.,

$$G_{\mu\nu} \equiv G^{(B)}{}_{\mu\nu} + G^{(1)}{}_{\mu\nu} + G^{(2)}{}_{\mu\nu} + \cdots, \tag{11}$$

where $G^{(B)}{}_{\mu\nu}$ is the Einstein tensor for the background metric and $G^{(n)}{}_{\mu\nu}$ contains all the terms involving n factors of $h_{\mu\nu}$.

The effective stress–energy for gravitation is defined in terms of the nonlinear portion of the Einstein tensor, i.e.

$$T^{(G)}{}_{\mu\nu} \equiv -\frac{1}{8\pi}(G^{(2)}{}_{\mu\nu} + G^{(3)}{}_{\mu\nu} + \cdots). \tag{12}$$

The background stress–energy is defined in terms of the background portion of the Einstein tensor, i.e.,

$$T^{(B)}{}_{\mu\nu} \equiv \frac{1}{8\pi} G^{(B)}{}_{\mu\nu}. \tag{13}$$

With these definitions, Einstein's equation can be rewritten as

$$G^{(1)}_{\mu\nu} = 8\pi(T^{(\mathrm{M})}_{\mu\nu} + T^{(\mathrm{G})}_{\mu\nu} - T^{(\mathrm{B})}_{\mu\nu}), \tag{14}$$

where $T^{(\mathrm{M})}_{\mu\nu}$ is the stress-energy tensor for matter. When $G^{(1)}_{\mu\nu}$ is expanded, Eq. (14) can be seen to be a linear wave equation for the field $h_{\mu\nu}$ whose source is the difference between the total stress-energy (matter plus gravitation) and the background stress-energy. The unusual feature of this equation is that the field $h_{\mu\nu}$ is a source for itself. The equation shows explicitly that 'gravity gravitates.'

This reformulation of Einstein's equation is called the field theory approach to gravity and has been presented by numerous authors (Gupta 1957; Thirring 1961; Deser 1970; Weinberg 1972; Grishchuk *et al.* 1984), traditionally in terms of a Minkowski background metric. Here, the field theory formulation has been shown in a general form that is valid for any choice of background metric. The appropriate background metric for a model universe similar to our own is the metric for an expanding FRW universe.

5. Total Angular Momentum of a Model Universe

In the previous three sections I have defined the average global inertial frame of reference, the total angular momentum with respect to that frame, and the stress-energy of the gravitational field with respect to the average global inertial frame. I now show that, given these definitions, the total angular momentum of the matter and gravitation in a closed universe is zero.

For a closed universe (3-sphere topology background), the consistency conditions (2), (3), and (4) lead to the result (King 1990)

$$\langle G^{(1)}_{0i}\xi_{(a)}{}^{i}\rangle = 0. \tag{15}$$

By substituting the $0i$-components of the field equations (14) into the above result, we obtain

$$\langle (T^{(\mathrm{M})}_{0i} + T^{(\mathrm{G})}_{0i} - T^{(\mathrm{B})}_{0i})\xi_{(a)}{}^{i}\rangle = 0. \tag{16}$$

The background stress-energy makes no contribution to this equation since $T^{(\mathrm{B})}_{0i} = 0$ in the preferred coordinate system. Therefore we have

$$\langle (T^{(\mathrm{M})}_{0i} + T^{(\mathrm{G})}_{0i})\xi_{(a)}{}^{i}\rangle = 0. \tag{17}$$

Now consider the Killing vector ξ^{μ} corresponding to a rotation about a given axis at some point P. The Killing vector ξ^{μ} is equal to a linear combination of the six linearly independent Killing vector fields $\{\xi_{(a)}{}^{\mu}\}$. Therefore, we have

$$\langle (T^{(M)}{}_{0i} + T^{(G)}{}_{0i})\xi^i \rangle = 0. \tag{18}$$

By Eq. (8), the left-hand side of this equation is proportional to the total angular momentum L_P of the matter and gravitational waves in the universe about the given axis and given point P; therefore,

$$L_P = 0. \tag{19}$$

This proves that a closed universe cannot rotate.

6. Conclusions

One intuitively expects a closed universe to be nonrotating. The proof that closed universes cannot rotate has been sought since Einstein introduced closed model universes to general relativity. The difficulty has been in accounting for the energy and momentum carried by the gravitational field and in interpreting solutions of Einstein's equations. The need to include gravitational waves in the analysis of rotation had been established as early as the 1960's (Dehnen and Hönl 1962, Wheeler 1964). The problem was not the identification of the theorem, but the formulation of the required definitions to include the momentum carried by gravitational waves.

Admittedly, one can argue that the introduction of the average inertial frame is unnecessary for general relativity and therefore should not be done. However, the situation is exactly the same for nonrotating (Machian) Newtonian model universes since the global inertial frame can be dispensed with in these models just as the average inertial frame can be dispensed with in general relativity. In both cases, these inertial frames are introduced as an auxiliary device to aid in the interpretation of these models. The widespread use of this technique by astrophysicists indicates its great practical value in cosmology.

The Ozsváth-Schücking model universe (and other Bianchi type-IX closed model universes) were thought previously to contradict Mach's Principle. In fact, they are a special case of the theorem. The angular momentum of rotating matter in the Bianchi type-IX model universes is exactly canceled by the angular momentum of a longest-wavelength gravitational wave rotating in the opposite direction to the matter (King 1991). Because the Bianchi IX model universes are homogeneous, the momentum of the matter is exactly canceled by the momentum of the gravitational wave at each point in space.

In conclusion, every closed model universe is nonrotating in that the total angular momentum of matter and gravitation about any point is zero. This result holds for every 3-sphere topology model universe that is approximately homogeneous and isotropic and is valid to all orders of perturbation from homogeneity and isotropy.

REFERENCES

Barbour, Julian B. and Bertotti, Bruno (1982). "Mach's Principle and the Structure of Dynamical Theories." *Proceedings of the Royal Society of London* A382: 295–306.

Barbour, Julian B. (1989). *Absolute or Relative Motion?*, Vol. 1: *The Discovery of Dynamics*. Cambridge: Cambridge University Press.

Dehnen, Heinz and Hönl, Helmut (1962). "Finite Universe and Mach's Principle." *Nature* 196: 362–363.

Deser, Stanley (1970). "Self-Interaction and Gauge Invariance." *General Relativity and Gravitation*. 1: 9–18.

Ellis, George F.R. (1984). "Relativistic Cosmology: Its Nature, Aims and Problems." In *General Relativity and Gravitation*, Bruno Bertotti, ed. Dordrecht: D. Reidel Publishing, pp. 215–288.

Grishchuk, Leonid P., Petrov, A.N., and Popova, A.D. (1984). "Exact Theory of the (Einstein) Gravitational Field in an Arbitrary Background Space-Time." *Communications in Mathematical Physics* 94: 379–396.

Gupta, Suraj N. (1957). "Einstein's and Other Theories of Gravitation." *Reviews of Modern Physics* 29: 334.

King, D.H. (Harry) (1990). "Mach's Principle and Rotating Universes." Ph.D. thesis, University of Texas at Austin.

King, D.H. (Harry) (1991). "Gravity-Wave Insights to the Bianchi Type-IX universes." *Physical Review* D 44: 2356–2368.

Ozsváth, I. and Schücking, Englebert (1962). "Finite Rotating Universe." *Nature* 193: 1168–1169.

Ozsváth, I. and Schücking, Englebert (1969). "The Finite Rotating Universe." *Annals of Physics (New York)* 55: 166–204.

Thirring, Walter E. (1961). "An Alternative Approach to the Theory of Gravitation." *Annals of Physics (New York)* 16: 96–117.

Weinberg, Steven (1972). "Photons and Gravitons in Perturbation Theory: Derivation of Maxwell's and Einstein's Equations." *Physical Review* B138: 988–1002.

Wheeler, John A. (1964). "Mach's Principle as Boundary Condition for Einstein's Equations." In *Gravitation and General Relativity*. H.-Y. Chiu and W.F. Hoffmann, eds. New York: W.A. Benjamin, Inc.

Discussion

Ehlers: I didn't quite get your proposal concerning the averaging. Imagine that you have a realistic, complicated metric; how do you then identify first of all your preferred hypersurfaces and then do the construction of the best-fit Robertson–Walker metric?
King: That's a very difficult problem that nobody has solved.
Ehlers: It seems to me you go the other way round. You assume that you have already some background, and then you formulate a condition for what you consider as permitted deviations from the flat background.
King: That's right. It's a consistency requirement for the background. For a selected background to be valid, it must satisfy that condition. Now that condition does in no way uniquely select a background metric. A very large set of background metrics will satisfy that condition. The condition is very weak indeed, but that's the only requirement you need to get this theorem out.
Ehlers: I think for cosmology this is really a serious question. What does this mean when we usually say in cosmology on a sufficiently large scale the metric is Robertson–Walker? This is, I think, so far always done in some rather rough intuitive way, and it seems to me, considering the hierarchical structure which seems to be observed, that it becomes more and more urgent to give a really precise meaning to this splitting into what we call a large-scale average of the metric and the actual metric which is supposed to obey Einstein's field equations. Usually, if one would have such an averaging procedure, one could not expect that the average metric itself should satisfy Einstein's equations because of various nonlinearities.
King: That's in fact exactly where this gravitational stress-energy tensor comes in. It is the gravitational stress-energy tensor that is taking care of these nonlinearities.
Ehlers: In order just to see how one actually makes use of the properties of the averaging, I have looked in which way one assumes properties of the averaging in statistical gravitational lens theory. There one does actually make assumptions concerning the relation between the average Robertson–Walker metric and the actual inhomogeneous metric. There it is essential to make two assumptions; namely, if we consider our past light cone, which is, after all, where we really observe, and go to a particular value of the red shift in the actual inhomogeneous universe and then compare it with what we call the background universe, then the areas of the z=constant

sections are supposed to have the same magnitude in the average and in the actual metric which is used for statistical purposes; the second assumption is that the relationship between the red shift z and the affine parameter is essentially the same in the background model as in the inhomogeneous one. If you would not make these two assumptions, the statistical arguments in gravitational lens theory simply would not work, and therefore the purpose of my remark is only to show I think one is not at liberty to make any assumptions about the averaging which are convenient in a particular setting. It is essential to find what, if one compares the theory with the observations, is needed there for the averaging process.

Isenberg: I don't see what this averaging has to do with the observations one would make as a physicist.

King: Well, the trouble is we are not able to observe enough of the universe to be able to tell whether Mach's Principle is right.

Isenberg: Right, so it's not clear to me that this averaging is relevant to what we actually see.

King: Well, it only tells you something about what we see if you believe the universe is closed and that is very problematic.

Isenberg: Even if the universe is spatially closed, I don't see what this averaging calculation over the whole 3-sphere gives us, particularly since, as Jürgen [Ehlers] was noting, we only see information propagating along our past causal cone. How does this have anything to do with averaging over three-spheres?

Nojarov: You assume the existence of these gravitational waves which compensate the angular momentum, but are there more fundamental principles to claim this for the universe?

King: There is a difference between conservation of angular momentum and a theorem which shows that the total angular momentum is zero. For example, for an open universe the angular momentum is conserved but it can easily be nonzero. For a closed universe the angular momentum is constrained to be zero. But, you know, there's an interpretational framework put on this model universe and that framework is necessary to get an intuitive feel for what is going on in this model. The background metric is not real, it's a useful interpretational tool. That has to be kept in mind.

4. Other Formulations of Mach's Principle

Introduction

This chapter illustrates the strikingly different ways in which different people have attempted to implement Mach's Principle. Like Brans and Dicke, Hoyle and Narlikar set out to realize in a systematic manner Einstein's contention (p. 180) that in a Machian approach the *inertial mass* of any body must be determined by a kind of interaction of that body with all the other masses in the universe. Narlikar gives a particularly clear rationale for such an approach at the beginning of his paper.

In contrast, Raine takes as his point of departure Einstein's brief 1918 paper (pp. 185–186) in which he actually coined the expression Mach's Principle and gave a formal definition of it: The metric tensor in a Machian solution of general relativity must be completely determined by the energy-momentum tensor of *matter*, understood in the narrow sense (i.e., gravitational waves are not to contribute). The development of the Green's function approach (or integral formulation) as a way to give rigorous mathematical expression to this idea must represent one of the most remarkable examples of simultaneous discovery in science – it was developed independently by Al'tshuler, Lynden-Bell, Sciama and Waylen, and Gilman, as Raine recounts.

Finally, Bleyer and Liebscher's paper is the most radical attempt in this volume to relate the distribution of matter in the universe as a whole to the deep structure of local physics, in this case the causal (light-cone) structure of Minkowski space. This is work in the spirit of Dicke's 'generalized Mach's Principle,' in accordance with which one seeks systematically for ways in which the universe at large might influence local physics (cf. the remarks of Brill and Brans, pp. 333 and 337).

J.B.B.

Einstein Studies, vol. 6: Mach's Principle: From Newton's Bucket to Quantum Gravity, p. 249

Direct Particle Formulation of Mach's Principle

Jayant V. Narlikar

1. Introduction

There are two ways of measuring the Earth's spin about its polar axis. By observing the rising and setting of stars the astronomer can determine the period of one revolution of the Earth around its axis: the period of $23^h 56^m 4^s.1$. The second method employs a Foucault pendulum whose plane gradually rotates around a vertical axis as the pendulum swings. Knowing the latitude of the place of the pendulum, it is possible to calculate the Earth's spin period. The two methods give the same answer.

At first sight this does not seem surprising. Closer examination, however, reveals why the result is nontrivial. The first method measures the Earth's spin period against a background of distant stars, while the second employs the standard Newtonian mechanics in a spinning frame of reference. In the latter case, we take note of how Newton's laws of motion get modified when their consequences are measured in a frame of reference spinning relative to the 'absolute space' in which these laws were first stated by Newton.

Thus, implicit in the assumption that equates the two methods is the coincidence of absolute space with the background of distant stars. It was Ernst Mach in the last century who pointed out that this coincidence is nontrivial. He read something deeper in it, arguing that the postulate of absolute space that allows one to write down the laws of motion and arrive at the concept of inertia is somehow intimately related to the background of distant parts of the universe. This argument is known as 'Mach's Principle,' and we will analyze it further.

When expressed in the framework of the absolute space, Newton's second law of motion takes the familar form

Einstein Studies, vol. 6: Mach's Principle: From Newton's Bucket to Quantum Gravity, pp. 250–261

$$\mathbf{P} = m\mathbf{f}. \tag{1}$$

This law states that a body of mass m subjected to an external force $\mathbf{P}$ experiences an acceleration $\mathbf{f}$. Let us denote by S the coordinate system in which $\mathbf{P}$ and $\mathbf{f}$ are measured.

Newton was well aware that his second law has the simple form (1) only with respect to S and those frames that are in uniform motion relative to S. If we choose another frame S' that has an acceleration $\mathbf{a}$ relative to S, the law of motion measured in S' becomes

$$\mathbf{P}' \equiv \mathbf{P} - m\mathbf{a} = m\mathbf{f}'. \tag{2}$$

Although (2) outwardly looks the same as (1), with $\mathbf{f}'$ being the acceleration of the body in S', something new has entered into the force term. This is the term $m\mathbf{a}$, which has nothing to do with the external force but depends solely on the mass m of the body and the acceleration $\mathbf{a}$ of the reference frame relative to the absolute space. Realizing this aspect of the additional force in (2), Newton termed it 'inertial force.' As this name implies, the additional force is proportional to the inertial mass of the body.

According to Mach, the Newtonian discussion was incomplete in the sense that the existence of the absolute space was postulated arbitrarily and in an abstract manner. Why does S have a special status in that it does not require the inertial force? How can one physically identify S without recourse to the second law of motion, which is based on it?

To Mach the answers to these questions were contained in the observation of the distant parts of the universe. It is the universe that provides a background reference frame that can be identified with Newton's frame S. Instead of saying that it is an accident that Earth's rotation velocity relative to S agrees with that relative to the distant parts of the universe, Mach took it as proof that the distant parts of the universe somehow enter into the formulation of local laws of mechanics.

One way this could happen is by a direct connection between the property of inertia and the existence of the universal background. To see this point of view, imagine a single body in an otherwise empty universe. In the absence of any forces, (1) becomes

$$m\mathbf{f} = 0.$$

What does this equation imply? Following Newton we would conclude that $\mathbf{f}=0$, that is, that the body moves with uniform velocity. But we now no longer have a background against which to measure velocities. Thus $\mathbf{f}=0$ has no operational significance. Rather, $\mathbf{f}$ should be completely indeterminate. And it is not difficult to see that such a conclusion follows naturally provided we argue that

$$m = 0. \tag{3}$$

In other words, the measure of inertia depends on the existence of the background in such a way that in the absence of the background the measure vanishes! This aspect introduces a new feature into mechanics not considered by Newton. The Newtonian view that inertia is the property of matter has to be augmented to the statement that inertia is the property of matter as well as of the background provided by the rest of the universe.

Einstein, an avid reader of Mach, was impressed by this chain of reasoning and hoped that his theory of gravity would turn out to incorporate Mach's principle. This hope was not realized in the end. There are several anti-Machian solutions in general relativity.

For example, there are empty space solutions that are nontrivially different from the flat spacetime of special relativity. In these solutions $R_{ik}=0$ but $R_{iklm}\neq 0$. What do the timelike geodesics in such spacetime mean? With no 'background' of matter why are these trajectories of 'particles under no force' singled out?

On a second count there are cosmological solutions of Einstein's equations wherein the distant background *rotates* with respect to the local inertial frame. Ironically, the classic paper of Kurt Gödel (1949) which produced one such model appeared in the 70th birthday festschrift for Einstein. By then, however, Einstein himself had lost his enthusiasm for Mach's Principle. In his autobiographical notes he writes (Einstein 1949):

> Mach conjectures that in a truly rational theory inertia would have to depend upon the interaction of the masses, precisely as was true for Newton's other forces, a conception which for a long time I considered as in principle the correct one. It presupposes implicitly, however, that the basic theory should be of the general type of Newton's mechanics: masses and their interaction as the original concepts. The attempt at such a solution does not fit into a consistent field theory, as will be immediately recognized.

Although Einstein himself moved away from Mach's Principle, there were others who felt its impact and sought to give expression to it in quantitative theories of gravity. For example, Dennis Sciama (1953) and Carl Brans and Robert Dicke (1961), among others, proposed alternative theories of gravity. However, these were field theories, since the general belief (shared by these authors with Einstein) was that field theories alone provide a proper description of physics.

Nevertheless, action-at-a-distance theories can also foot the bill if they are properly formulated and applied with the right cosmological boundary conditions. I will discuss this possibility here.

2. The Hoyle-Narlikar Formulation

In 1964, Fred Hoyle and I proposed an action-at-a-distance theory of inertia that directly incorporated Mach's principle. In this theory the inertial mass of ath particle ($a=1, 2, \ldots$) at world point X was given by

$$m_a(X) = \lambda_a \sum_{b \neq a} \lambda_b \int G(X,B)\, ds_b, \tag{4}$$

where ds_b is the element of proper time on the worldline of particle b and λ_b a coupling constant. The action at a distance is through the two-point scalar propagator G satisfying the relation

$$\Box G(X,B) + \frac{1}{6} R G(X,B) = \frac{\delta_4(X,B)}{\sqrt{-g(X)}}, \tag{5}$$

and we define

$$m^{(b)}(X) = \lambda_b \int G(X,B)\, ds_b. \tag{6}$$

[$\Box$ and R in (5) are evaluated at X.]

The propagator G is symmetric with respect to its two points:

$$G(X,B) = G(B,X). \tag{7}$$

The rationale for these formulas will be considered next.

First we notice that the interaction conveys the property of inertia from one particle to another. Next, from (7) we also learn that the interaction works symmetrically between pairs of particles. Finally, the wave equation (5) ensures that the mass interaction propagates with the speed of light.

3. A Digression into Electromagnetic Theory

What are these functions $m^{(b)}(X)$? That they communicate the property of inertia from particles b to any particle placed at the spacetime point X is clear from the context. To arrive at a suitable form for them we take hints from action-at-a-distance electromagnetism, in which it is usual to introduce electromagnetic disturbances that arise specifically from sources, that is, from moving electrical charges. Accordingly, we introduce the 4-potential $A_i^{(b)}(X)$ as denoting the electromagnetic effect at X from the electric charge b. The $A_i^{(b)}(X)$ satisfies the wave equation

$$\Box A_i^{(b)} + R_i^k A_k^{(b)} = 4\pi J_i^{(b)}, \tag{8}$$

where $J_i^{(b)}$ is the 4-current generated by the charge b. The solution of (8) may be written in the integral form

$$A_i^{(b)}(X) = 4\pi \int e_b G_{ik}(X, B)\, db^k, \tag{9}$$

where $G_{ik}(X, B)$ is a Green's function of the wave operator $(g_i^k \Box + R_i^k)$. The well-known Coulomb potential is a special case of (8).

The Green's function is not uniquely fixed from the form of the wave operator alone. Boundary conditions must also be specified. The customary boundary condition is imposed by causality; that is, the influence from B to X must vanish if X lies outside the future light cone of B. The Green's function satisfying this condition is called the *retarded Green's function*. We will denote such a Green's function with a superscript R. Similarly, a Green's function confined to the past light cone of B is called the *advanced Green's function* and is denoted with a superscript A.

These Green's functions have played a key role in action-at-a-distance theories. It was originally believed that action at a distance must be instantaneous and hence inconsistent with the framework of special relativity. However, Schwarzschild (1903), Tetrode (1922), and Fokker (1929a,b; 1932) demonstrated during the first three decades of this century that a relativistically consistent action-at-a-distance theory can indeed be formulated. If we consider two spacetime points A and B with s_{AB}^2 as the invariant square of the relativistic distance between them, then $\delta(s_{AB}^2)$, where δ is the Dirac delta function, is a convenient function for transmitting physical influences between A and B. For, this function acts only when A and B are connectable by a light ray (that is, when $s_{AB}^2=0$). This delta function therefore necessarily occurs as the main component in any Green's function in the action-at-a-distance theory. The action principle, which is the basis of the electromagnetic theory in Riemannian spacetime, is described below. We start with the action

$$\boldsymbol{A} = -\sum_a \sum_{<b} 4\pi e_a e_b \int\int \bar{G}_{ik}\, da^i db^k \tag{10}$$

where $\bar{G}_{ik}$ is the *symmetric Green's function* given by

$$\bar{G}_{ik}(A, B) \equiv \frac{1}{2}[G_{ik}^R(A, B) + G_{ik}^A(A, B)]. \tag{11}$$

Thus $\bar{G}_{ik}(A, B) = \bar{G}_{ik}(B, A)$ and each term in the action is completely symmetric between each pair of particles. The electromagnetic potential given by (9) is a symmetric half-advanced plus half-retarded combination, rather than the more familiar pure retarded one. However, the action (10) together with suitable cosmological boundary conditions reproduces all the electromagnetic effects of the standard Maxwell field theory. The key issue recognized first by Wheeler and Feynman (1945, 1949) is that no charge is isolated. The motion of a typical charge a invokes a reaction from all other charges in the universe, which we may term the *response of the universe*.

What is the response of the universe? It was shown by Dirac (1938) that when an electric charge a accelerates, it suffers a force of radiative damping, and that this force can be calculated by evaluating half the difference of the retarded and the advanced fields F of the charge *on its worldline*:

$$Q(a) = \frac{1}{2}[F^R(a) - F^A(a)]. \tag{12}$$

In the Maxwell field theory Dirac's result had remained just a curiosity without a proper reasoning as to why the radiative reaction must be determined by the above formula. In the Wheeler–Feynman theory the 'correct' response from the universe to the motion of a is precisely this!

Moreover, if we add (12) to the basic time-symmetric direct particle field of a, *viz.*

$$F(a) = \frac{1}{2}[F^R(a) + F^A(a)] \tag{13}$$

we get the total effect in the neighborhood of a to be a pure retarded one. A correct response therefore eliminates all advanced effects except those present in the radiation reaction. However, all this works provided we have the correct cosmological boundary conditions, which are spelled out below.

In 1945, Wheeler and Feynman had shown that to get the correct response the universe has to be a perfect absorber. Their work was carried out within the framework of a static universe. When Hogarth (1962) repeated the calculation in an expanding universe, he found that the correct response (12) is possible in a universe that is a perfect absorber in the future but not in the past. The steady-state cosmology fulfills this condition, but all known Friedman models fail to meet it. In 1963, Hoyle and I arrived at the same conclusion with somewhat more general assumptions (Hoyle and Narlikar 1963). Because of the crucial requirement of perfect absorption, this theory is sometimes called the 'absorber theory of radiation.'

4. Inertia and Gravity

Our purpose in the above digression into electromagnetism was to show that a similar approach to inertia leads us to a Machian theory of gravity. In the case of inertia, we note that the functions $m^{(b)}(X)$ are scalars, and so we have to deal with scalar Green's functions. Thus we wrote (6) in analogy to (9), and (7) in analogy to (11), while the inertial action in analogy to (10) becomes

$$\mathcal{A} = -\sum_{a}\sum_{<b}\int\int \lambda_a \lambda_b \tilde{G}(A, B)\, ds_a\, ds_b. \tag{14}$$

The analogy continues further. The wave equation (5) is conformally invariant and gives us a conformally invariant Machian theory just as (10) gives us a conformally invariant electromagnetic theory.

The action of HN theory is given by (14), and with the help of definition (6) we may write it as

$$\mathcal{A} = -\sum_{a}\int m_a\, ds_a. \tag{15}$$

Written in this form, this action appears to have only the inertial term of Newtonian mechanics. How can such an action yield any gravitational equations?

The answer to this question lies in the fact that the m_a's in (15) are not constants but depend on spacetime coordinates *as well as on spacetime geometry*. For they are defined with the help of Green's functions, which in turn are defined in terms of spacetime geometry. Thus if we make a small variation

$$g_{ik} \rightarrow g_{ik} + \delta g_{ik},$$

the wave equation (5) will change and so will its solution. Thus we will have

$$\tilde{G}(A,B) \rightarrow \tilde{G}(A,B) + \delta\tilde{G}(A,B)$$

and hence $\mathcal{A} \rightarrow \mathcal{A} + \delta\mathcal{A}$. We therefore have a nontrivial problem whose solution may be expressed in the following way. To simplify matters we will take all λ_a to be equal to unity.

Define the following functions:

$$m(X) = \sum_{a} m^{(a)}(X) = \frac{1}{2}[m^R(X) + m^A(X)], \tag{16}$$

$$\phi(X) = m^R(X) m^A(X), \quad m_k \equiv m_{,k}, \ldots, \tag{17}$$

$$N(X) = \sum_{a}\int \delta_4(X, A)[-g(X)]^{-1/2} ds_a. \tag{18}$$

As in the electromagnetic case, we have chosen the symmetric (half $R+$

half A) Green's function. The gravitational equations then become

$$R_{ik}-\frac{1}{2}g_{ik}R=-6\phi[T_{ik}-\frac{1}{6}(g_{ik}\Box\phi-\phi_{;ik})$$
$$-\frac{1}{2}(\overset{R}{m_i}\overset{A}{m_k}+\overset{R}{m_k}\overset{A}{m_i}-g_{ik}g^{pq}\overset{R}{m_p}\overset{A}{m_q})], \tag{19}$$

together with the 'source' equation for $m(X)$

$$\Box m+\frac{1}{6}Rm=N. \tag{20}$$

The derivation leading to the final set of equations of the theory may appear somewhat long-winded to anybody unfamiliar with the techniques of direct interparticle action. We have followed here the method used by Hoyle and the author, who arrived at this theory via their earlier work on electromagnetism. As in the electromagnetic case, the universe responds to a local event. To ensure causality and to eliminate advanced effects, the correct response should be given by

$$\sum_a m^{(a)A}(X)=\sum_a m^{(a)R}(X)=m(X). \tag{21}$$

Under these conditions the equations (19) further simplify to

$$R_{ik}-\frac{1}{2}g_{ik}R=-\frac{6}{m^2}\left[T_{ik}-\frac{1}{6}(g_{ik}\Box m^2-m^2_{;ik})\right.$$
$$\left.-\left\{m_i m_k-\frac{1}{2}g_{ik}m^l m_l\right\}\right]. \tag{22}$$

Had we accepted the standard field theoretical approach and introduced a scalar inertia field $m(X)$, we could have arrived at (20) and (22) from an action given by

$$\boldsymbol{A}=\int\left(\frac{1}{12}Rm^2-m^i m_i\right)\sqrt{-g}d^4x-\sum_a\int m ds_a. \tag{23}$$

The action-at-a-distance approach, although unfamiliar to a typical theoretical physicist, is useful in that it gives a more direct expression to Mach's Principle. The physical interpretation of the field theoretical term (23) is not so easy to see. For this reason, we have discussed the former approach at some length.

Notice that in the former approach our action (15) contained only the last term of (23), but there m was made up of nonlocal two-point functions. Here m is a straightforward field with sources whose dynamical properties are defined through the first term in the above action.

Since the property of conformal invariance was used in the formulation of the theory, we expect the final equations (20) and (22) to exhibit conformal invariance. This expectation is borne out. If (g_{ik}, m) are a solution of these equations, then so are

$$\bar{g}_{ik} = \Omega^2 g_{ik}, \quad \bar{m} = \Omega^{-1} m. \tag{24}$$

Thus, apart from coordinate invariance of general relativity, this theory also shows conformal invariance.

The symmetry of conformal invariance of the action leads to a vanishing of trace of the field equations. It may be easily verified that the trace of (22) vanishes in view of (20). The vanishing of trace represents the fact that the problem is underdetermined. Just as the vanishing of $T^{ik}{}_{;k}$ in general relativity shows that more solutions can be generated from any given solution by coordinate transformations, so we can generate more solutions through (24). All these solutions are physically equivalent provided we stick to the rule that Ω does not vanish or become infinite.

5. The Transition to General Relativity

Suppose we are allowed to choose an Ω in the above range that ensures that

$$\bar{m} = \Omega^{-1} m = \text{constant} = m_0. \tag{25}$$

This choice of Ω is possible provided m does not vanish or become infinite. This conformal frame is called the *Einstein frame*, in which we get a simplified form for (22):

$$R_{ik} - \frac{1}{2} g_{ik} R = -\kappa T_{ik}, \tag{26}$$

with the constant κ given by

$$\kappa = \frac{6}{m_0^2}. \tag{27}$$

Thus we have arrived at Einstein's equations! At first sight we don't seem to have gained anything. We have no new theory and hence no new predictions, as in the Brans–Dicke theory. Closer examination, however, reveals several ways in which this theory goes beyond relativity.

1. Our starting point was based on Mach's Principle. It is only in the many-particle approximation, when the response condition (21) is satisfied, that we arrive at the final Einstein-like field equations. An empty universe in relativity is given by

$$R_{ik} = 0,$$

which can have well-defined spacetimes as solutions. Test particles in such spacetimes will have well-defined trajectories. Such trajectories would not make any sense according to Mach, since we no longer have a material background against which to measure the motion of these particles. These solutions in fact correspond to the $\mathbf{f}=0$ solutions of Newtonian theory. In the HN theory, an empty universe corresponds to

$$m=0, \text{ indeterminate } g_{ik},$$

in accord with the Machian $m=0$ solution of (3).

2. The sign of κ is fixed arbitrarily in general relativity. Neither in the heuristic derivation of Einstein nor in the Hilbert action principle is κ required to be positive. It is only when κ is determined by reference to Newtonian gravity in the weak-field approximation that we conclude that $\kappa>0$. In the HN theory (27) shows that κ must necessarily be positive. (This conclusion does not depend on our assumption of $\lambda_a=1$; the result follows whatever sign the λ_a are given.)

3. In the direct interparticle approach, it is not possible to accommodate the λ-term of cosmic repulsion without making the wave equation (5) nonlinear. Thus Occam's razor automatically comes into play. In relativity, the λ-term is still possible.

4. The transition from (22) to (26) is possible provided $0<\Omega<\infty$. What happens if we break this rule? Suppose in the solution of (22) we had a hypersurface on which $m=0$. If we insist on the transformation (25) in a region that contains such a hypersurface, we have to pay the price of $\Omega\to 0$, which in turn produces spacetime singularities. The work of Kembhavi (1978) showed that the well-known cases of spacetime singularities of relativity arise because of the occurrence of zero-mass hypersurfaces in the solution of the equations (22). For a simple example of this conclusion, let us look at the standard Big-Bang singularity of relativity.

Consider the Minkowski line element (with $c=1$)

$$ds^2=d\tau^2-dx^2-dy^2-dz^2 \tag{28}$$

as a solution of (22). It is easily verified that the mass function satisfying both (20) and (22) for a uniform number density N of particles is

$$m\propto\tau^2. \tag{29}$$

This is the simplest possible cosmological solution in this theory.

If we now insist on going over to a frame with constant mass $\bar{m}$, then from (24) we see that the appropriate Ω must be given by

$$\Omega\propto\tau^2. \tag{30}$$

However, Ω vanishes on the hypersurface $m=0$. The transformation to the Einstein conformal frame is 'illegal.' The price paid for insisting that $\bar{m}$=constant is that the resulting model has a geometrical singularity at $\tau=0$. In fact, it is easily verified that the new model is none other than the singular Einstein–de Sitter model.

5. It is instructive to see how the phenomenon of Hubble redshift is explained in the flat spacetime model of (28) and (29). Clearly, a photon traveling in Minkowski spacetime does not undergo redshift. Consider, however, what happens to a photon arriving at the observer at the present epoch τ_0 from a galaxy at a distance r. This photon originated in an atomic (or molecular) transition at time τ_0-r.

From atomic physics, the wavelength of a photon so transmitted varies inversely as the mass of the electron (making the atomic transition). From (29) we see that if λ is the wavelength of this photon and λ_0 is the wavelength of a photon emitted in a similar transition at τ_0 at the observer, then

$$1+z \equiv \frac{\lambda}{\lambda_0} = \frac{m(\tau_0)}{m(\tau_0-r)} = \frac{\tau_0^2}{(\tau_0-r)^2}. \tag{31}$$

Thus the redshift in the above HN cosmology arises from the variation of particle masses.

6. Concluding Remarks

This basic theory therefore resembles general relativity in the Einstein frame but has more general implications in the sense that unlike the relativity theory it is conformally invariant. It has the advantage that it starts with the Machian notion of inertia of a particle arising from other particles in the universe.

Further work along these lines has opened up the possibilities of a variable gravitational constant (Hoyle and Narlikar 1974), anomalous redshifts (Arp and Narlikar 1993), and creation of matter (Hoyle, Burbidge, and Narlikar 1993). These investigations lead to observationally testable results, thus bringing the theory scientific respectability.

REFERENCES

Arp, Halton C. and Narlikar, Jayant V. (1993). "Flat Spacetime Cosmology: A Unified Framework for Extragalactic Redshifts." *Astrophysical Journal* 405: 51–56.

Brans, Carl and Dicke, Robert H. (1961). "Mach's Principle and a Relativistic Theory of Gravitation." *Physical Review* 124: 925–935.

Dirac, Paul A. M. (1938). "Classical Theory of Radiating Electrons." *Proceed-*

ings of the Royal Society, Ser. A 167: 148–169.

Einstein, Albert (1949). "Autobiographical Notes." In *Albert Einstein, Philosopher–Scientist*, ed. P. A. Schilpp, Evanston, Illinois: The Library of Living Philosophers, Inc., Illinois, p. 29.

Fokker, Adrian D. (1929a). "Ein invarianter Variationssatz für die Bewegung mehrerer elektrischer Massenteilchen." *Zeitschrift für Physik* 58: 386–393.

Fokker, Adrian D. (1929b). *Physica (Eindhoven)* 9: 33.

Fokker, Adrian D. (1932). *Physica (Eindhoven)* 12: 145.

Gödel Kurt (1949). "An Example of a New Type of Cosmological Solutions of Einstein's Field Equations of Gravitation." *Reviews of Modern Physics* 21: 447–450.

Hogarth, Jack E. (1962). "Cosmological Considerations of the Absorber Theory of Radiation." *Proceedings of the Royal Society, Ser. A* 267: 365–383.

Hoyle, Fred and Narlikar, Jayant V. (1964). "A New Theory of Gravitation." *Proceedings of the Royal Society, Ser. A* 282: 191–207.

Hoyle, Fred and Narlikar, Jayant V. (1963). "Time Symmetric Electrodynamics and the Arrow of Time in Cosmology." *Proceedings of the Royal Society. Ser. A* 277: 1–23.

Hoyle, Fred and Narlikar, Jayant V. (1974). *Action at a Distance in Physics and Cosmology*, New York: W. H. Freeman.

Hoyle, Fred; Burbidge, Geoffrey; and Narlikar, Jayant V. (1993). "A Quasi-steady Cosmological Model with Creation of Matter." *Astrophysical Journal* 410: 437–457.

Kembhavi, Ajit K. (1978). "Zero Mass Surfaces and Cosmological Singularities." *Monthly Notices of the Royal Astronomical Society* 185: 807–822.

Schwarzschild, Karl (1903). "Zur Elektrodynamik." *Nachrichten der Gesellschaft der Wissenschaften zu Göttingen* No. 3, 126–131.

Sciama, Dennis W. (1953). "The Origin of Inertia." *Monthly Notices of the Royal Astronomical Society* 113:34–42.

Tetrode, H. (1922). "Über den Wirkungszusammenhang der Welt. Eine Erweiterung der klassischen Dynamik." *Zeitschrift für Physik* 10: 317–328.

Wheeler, John A. and Feynman, Richard P. (1945). "Interaction with the Absorber as the Mechanism of Radiation." *Reviews of Modern Physics* 17: 157–181.

Wheeler, John A. and Feynman, Richard P. (1949). "Classical Electrodynamics in Terms of Direct Interparticle Action." *Reviews of Modern Physics* 21: 424–433.

The discussion for this paper followed Hoyle's talk and is on p. 272ff.

Mach's Principle and the Creation of Matter

Sir Fred Hoyle

1. What is Creation of Matter?

Narlikar has shown that a Machian approach to the problem of inertia nevertheless leads to Einstein's equations, and therefore to the problem of what to do about the so-called Big-Bang in which the entire universe is supposed to have originated at a particular moment of time. Choosing this moment at the zero of the time t, Narlikar's expression (15) for the action of a set of particles a, b, ... is truncated to

$$\boldsymbol{A} = -\sum_a \int_{t=0} m_a da, \tag{1}$$

time being with respect to a particular choice of coordinates in which the metric takes the well-known Robertson–Walker form

$$ds^2 = dt^2 - S^2(t)\left[\frac{dr^2}{1-kr^2} + r^2(d\theta^2 + \sin^2\theta d\phi^2)\right], \tag{2}$$

where k can be 0 or ± 1.

Now unless we are willing to forego everything we are supposed to have learned from 20th-century physics, this simply will not do. A basic physical step as in (1) cannot be made in such a grossly noninvariant way with respect to a particular choice of coordinates.

We have tried two ways of attempting to remedy this situation. One is to switch the signs of the coupling parameters λ_a, λ_b, ... at the moment $t=0$ (Hoyle and Narlikar 1972a, b). This leads to a situation that is symmetric about $t=0$. There is a physical existence before the Big-Bang, and it is just like the Big-Bang but with respect to t decreasing rather than t increasing. There were objections to this procedure. It still involves a noninvariant physical requirement, the switching of the mass couplings at a particular moment in a special coordinate system. To some extent this objection could be met by arguing for a kind of quantum transition of the universe, or at any rate for the portion of the universe

Einstein Studies, vol. 6: Mach's Principle: From Newton's Bucket to Quantum Gravity, pp. 262–273 © 1995 *Birkhäuser Boston, Inc. Printed in the United States.*

we observe. Perhaps worse for this approach, however, was that we were never able to get much profit out of it. There were no really interesting consequences that we could relate to observation, a shortcoming which does not apply to the second remedy, as we have recently been able to show in collaboration with Geoffrey Burbidge (Hoyle, Burbidge, and Narlikar, 1993, 1994).

The second remedy accepts the Big-Bang as a demonstration that matter must originate, but it does not accept the noninvariant form of (1). Instead of requiring all matter to originate at a particular time in a particular coordinate system, it argues that a material particle can originate at any spacetime point X provided appropriate energy and momentum conservation conditions are satisfied. To give expression to such conditions, a scalar field $C(X)$ is introduced and the action formula for a set of particles a, b, ... is upgraded from (1) to

$$\boldsymbol{A} = -\int_{A_0} m_a da - \int_{B_0} m_b db - \cdots - C(A_0) - C(B_0) - \cdots \tag{3}$$

The points of origin A_0, B_0, ... of the particles are then determined by applying the principle of stationary action to (3), the variations of $\boldsymbol{A}$ being calculated for small shifts of the points A_0, B_0, ... The results are

$$\begin{aligned} m_a \frac{da^i}{da} &= g^{ik} C_k(A) \text{ at } A = A_0, \\ m_b \frac{db^i}{db} &= g^{ik} C_k(B) \text{ at } B = B_0, \\ &\cdots \end{aligned} \tag{4}$$

The introduction of a scalar field has some resemblance to the situation in inflationary cosmological models. But whereas the scalar field plays the role of a *deus ex machina* in inflationary cosmology, being introduced to suit the investigator from some unspecified source, here the field $C(X)$ satisfies an explicit wave equation with sources at the points A_0, B_0, ..., *viz.*

$$\left[\Box + \frac{1}{6}R\right] C(X) = f^{-1} \sum_{A_0} \frac{\delta_4(X, A_0)}{\sqrt{-g(A_0)}}, \tag{5}$$

where f^{-1} is a coupling constant with dimensionality (length)^2, like the gravitational constant.

At any rate, this was the position as described in the last cited paper. Here I want to go further, starting by noticing that, apart from the coupling f^{-1}, Eq. (5) is the same as Narlikar's Eq. (5), which raises the question of whether there should be two scalar fields, one determining particle masses, $m(X)$, and the other, $C(X)$, determining particle creation, both related to the same wave equation. Or should Narlikar's $m(X)$ and the above $C(X)$ both be manifestations of the same scalar field? And if so, how?

2. Quantum Mechanics and Gravitation

With the dimensionalities introduced above and in Narlikar's contribution, the action $\boldsymbol{A}$ is a dimensionless number. By introducing the dimensionless couplings

$$\lambda = \lambda_a = \lambda_b = \ldots, \tag{6}$$

put equal to unity in (3), $\boldsymbol{A}$ is multiplied by λ^2. This makes no difference at all within a purely gravitational theory, since Narlikar's gravitational equations (19) follow just as well from $\lambda^2 \delta \boldsymbol{A} = 0$ applied for $g \to g_{ik} + \delta g_{ik}$ as they do from $\delta \boldsymbol{A} = 0$. But when the action is given a quantum interpretation, it matters whether exp $i\boldsymbol{A}$ is the amplitude for a particular metric or exp $i\lambda^2 \boldsymbol{A}$ is the correct amplitude. It is usual to write the correct amplitude in the form exp $i\boldsymbol{A}/\hbar$ so that λ^{-2} plays the role of the Planck scaling constant $\hbar$.

A Planck particle is defined as a particle whose gravitational radius is $3/2\pi$ times its Compton wavelength, which condition determines the mass of a Planck particle as

$$\left[\frac{3\hbar}{4\pi G}\right]^{1/2}, \tag{7}$$

the units being such that the velocity of light is unity, as it has been assumed throughout both this and Narlikar's contributions. In conventional units the mass is about 10^{-5} g, about $6 \cdot 10^{18}$ times the proton mass. In relation to Narlikar's Eq. (27), which required particle masses m_0 in the gravitational theory to be given by

$$m_0 = \left[\frac{3}{4\pi G}\right]^{1/2}, \tag{8}$$

we now have

$$m_0 = \left[\frac{\text{Planck mass}}{\hbar^{1/2}}\right] = \lambda(\text{Planck mass}). \tag{9}$$

The interpretation of the coupling constant λ is therefore that it determines particle masses with respect to the Planck mass. Thus, for m_0 to relate to a proton or neutron, the coupling λ must be $\sim (6 \cdot 10^{18})^{-1}$. It is this smallness of the gravitational coupling that explains why gravitation, interpreted with respect to nucleonic particles, is an extremely weak force.

But now let us ask if this is a reasonable conclusion. Is it reasonable to introduce so tiny a coupling into what is claimed, at any rate by implication, to be a fundamental theory? The answer, I believe, is that it is not. The correct value for λ in a purely gravitational theory should be unity, or else some dimensionless number of order unity. The particles in a purely gravitational theory should, therefore, be Planck particles. Starting from Mach and arguing perhaps a little remorselessly, this is the conclusion one reaches.

When interactions other than gravitation are included, Planck particles decay quickly, probably in a moderate multiple of 10^{-43} seconds. Ultimately they decay into baryons and radiation in a moderate multiple of 10^{-24} seconds. Conservation requires the coupling λ to change during this decay. If $\lambda=1$ for the Planck particles, it is to be expected that $\lambda \simeq N^{-1}$ after decay into N similar particles, or less than this to the extent that radiation is produced. The energy of radiation goes largely into the kinetic explosive motions of the resulting fireball of baryons, which can be $\sim 5 \cdot 10^{18}$ in number. It is thus the decay of Planck particles into a vast number of secondaries which explains the small value of λ appropriate for the particles of our everyday world. A number of interesting conclusions and avenues to be explored open up immediately:

(1) We see where the so-called large numbers of physics and cosmology come from, from λ^{-1}. Such numbers simply reflect the circumstance that Planck particles decay into a very large number of secondaries.

(2) The physical conditions in an expanding Planck fireball are analogous to the conditions in the very early universe of Big-Bang cosmology. We would say the attractions of Big-Bang cosmology for particle physicists are misplaced. The attractions belong to Planck fireballs, not to cosmology. Among the attractions are eventual nuclear reactions analogous to those which are supposed to produce helium and lithium in the Big-Bang but which also produce beryllium and boron in the case of Planck fireballs.

(3) The kinetic energies acquired by the nucleons emerging from Planck fireballs provide a power source for a wide range of astrophysical processes - active galaxies, radio sources, and quasars - and the situation is directly observable in the modern universe, not relegated as in Big-Bang cosmology to the remote past in a way that is inherently unobservable.

(4) In the treatment of mass as discussed by Narlikar, the total mass

field can be divided into the sum of two components, written as $m(X)+c(X)$, say. Here $m(X)$ is the contribution to the mass field from stable particles, and it satisfies the wave equation

$$\left[\Box+\frac{1}{6}R\right]m(X)=\sum_{a}\lambda_{a}\int\frac{\delta_{4}(X,A)}{\sqrt{-g(A)}}da, \tag{10}$$

in which the coupling constants λ_a, λ_b, ... are small with values of order $(6\cdot 10^{18})^{-1}$ appropriate to nucleons. The field $c(X)$ satisfies the same wave equation but with $\lambda=1$ and with the line integrals extending only over the brief moment of existence of the Planck particles,

$$\left[\Box+\frac{1}{6}R\right]c(X)=\sum_{A_0}\int_{A_0}^{A_0+\delta A_0}\frac{\delta_{4}(X,A)}{\sqrt{-g(A)}}da, \tag{11}$$

in which for simplicity the Planck particle a is considered to decay into stable particles all in one go, a particular particle being created at A_0 and decaying at $A_0+\delta A_0$.

The theory is no different in principle from that discussed by Narlikar. It differs in detail, however, in that everywhere in the equations given by Narlikar when $m(X)$ appears, there should now be $m(X)+c(X)$, in effect because it was implicit in Narlikar's treatment that the particles under discussion were unchanging, whereas here we have the dramatic change of Planck particle decay to cope with as well. In particular, the gravitational equations are given by replacing m by $m+c$ in Narlikar's Eq. (22). At first sight the situation looks very complicated, but there are simplifying features. To start with, the magnitude of m at a typical cosmological spacetime point X is vastly greater than the magnitude of c, so that replacing the coefficient $-6/m^2$ in Narlikar's Eq. (22) by $-6/(m+c)^2$ has negligible effect. Indeed it is not hard to see that the magnitude of c is smaller than that of m by the Hubble constant H_0 multiplied by the decay time τ, say, of the Planck particles, and for $\tau\approx 10^{-43}$ seconds this is a minute factor of $\sim 10^{-60}$. So on a quick judgement we might feel inclined simply to strike out all the terms involving $c(X)$ and its derivatives. But this would be wrong because although $c(X)$ is tiny in magnitude compared to $m(X)$, the derivatives of $c(X)$ are not similarly small, as can be seen by considering the behavior of the solution of the wave equation (11) applied to a particular Planck particle.

The field $c(X)$ propagates outwards from its source in a region of spacetime contained between two very closely spaced light cones, one with its vertex at A_0, the other with vertex at $A_0+\delta A_0$. Along a timelike line cutting these two light cones $c(X)$ rises from zero as the line crosses the light cone from A_0, attains some maximum value, and then falls back to zero as the line crosses the second light cone from $A_0+\delta A_0$. Although $c(X)$ may be small its time derivative thus involves the reciprocal of the time difference between the two light cones, i.e., τ^{-1}. And again it is not hard to see that the derivatives of $c(X)$ gain in the region between the light cones on those of $m(X)$ by $(H_0\tau)^{-1}$, the same order as the factor by which m exceeded c, whence we see that the derivatives of $c(X)$ are of the same general order as those of $m(X)$. However, the derivatives of $c(X)$ can as well be negative as positive, so that when averages over many particles are taken, linear terms in the derivatives of $c(X)$ do disappear. Considering such features, the gravitational equations become

$$R_{ik}-\frac{1}{2}g_{ik}R=\frac{6}{m^2}\left[-T_{ik}+\frac{1}{6}(g_{ik}\Box m^2-m_{;ik})\right.$$
$$\left.+\left\{m_i m_k-\frac{1}{2}g_{ik}m_l m^l\right\}+\frac{2}{3}\left\{c_i c_k-\frac{1}{4}g_{ik}c_l c^l\right\}\right]. \tag{12}$$

Since the same wave equation is being used for $c(X)$ as for $m(X)$, the theory remains conformally invariant. Hence the procedure used by Narlikar to reduce his gravitational equations (22) to the Einstein form can also be used here, with the result

$$8\pi G=6/m_0^2,\ m_0=\text{constant}, \tag{13}$$

$$R_{ik}-\frac{1}{2}g_{ik}R=-8\pi G\left[T_{ik}-\frac{2}{3}\left\{c_i c_k-\frac{1}{4}g_{ik}c_l c^l\right\}\right], \tag{14}$$

not Einstein's equations on this occasion, because the conformal transformation which removes the derivatives of $m(X)$ does not remove those of $c(X)$.

3. Discussion

Applied cosmologically with only time derivatives of $c(X)$ retained, and using the metric (2), the dynamical equation for $\ddot{S}$ is

$$\frac{\ddot{S}}{S}=\frac{4\pi G}{3}(\overline{\dot{c}^2}-\bar{\rho}), \tag{15}$$

in which $\overline{\dot{c}^2}$ is the cosmologically averaged value of $\overline{\dot{c}^2}$ and $\bar{\rho}$ is the cosmologically averaged value of the mass density ρ. Thus the c-field acts to accelerate the universe, which is the sense of a *negative* pressure, the sense in which scalar fields are used in all forms of inflationary cosmology, and in the recent models studied by Burbidge, Narlikar, and myself (1993, 1994). It is not my purpose here to discuss the many applications of the latter to astrophysics and to observational cosmology. My purpose is to point out that the theory has passed a test of sign, without any possibility of its being adjusted to do so. Just as Narlikar's discussion passed a test of sign (in that there was no means of arbitrarily adjusting the gravitational constant G to be positive so as to make gravity an attraction rather than a repulsion – the relation (13) required $G>0$), so the automatic positive sign of $\overline{\dot{c}^2}$ requires the effect of $c(X)$ be to produce an expansion of the universe. One can conceive that the theory might have led to the wrong sign, but it doesn't. Whereas in Big-Bang cosmology exceedingly fine tuning is needed to produce the observed present-day expansion of the universe – a fine-tuning of about 1 part in 10^{60} in some models and as extreme as 1 part in 10^{100} in others – there is no tuning here at all. The universe simply expands at whatever rate the present degree of creation of matter dictates, and it does so through the positive $\overline{\dot{c}^2}$ term in (15).

By now I have indicated many directions in which the theory can go in its relation to practical observational issues, and evidently the scope of my present contribution cannot go further into details. What remains to end with is to relate the explicit wave equation (11) for $c(X)$ to the wave equation (5) for $C(X)$ written previously as a capital rather than as a small letter. With the creation point A_0 of particle a close to its decay point $A_0+\delta A_0$, Eq. (11) can be approximated somewhat crudely by

$$\left[\Box+\frac{1}{6}R\right]c(X)=\tau\sum_{A_0}\frac{\delta_4(X,A_0)}{\sqrt{-g(A_0)}}, \tag{16}$$

where τ is again the decay time of the Planck particle, which decay time determines the separation of A_0 and $A_0+\delta A_0$. Also when the c-field acts to promote the creation of some particle b, say, it does so through the action integral

$$\int_{B_0} c(B)db, \tag{17}$$

which can be approximated to $-\tau c(B_0)$. This we previously wrote as $C(B_0)$ in (3), whence $C(X)$ was just an approximation of the ends of the line integrals, i.e.,

$$C(X)=\tau c(X), \tag{18}$$

and using (16), the result for $C(X)$ is as in (5),

$$\left[\Box+\frac{1}{6}R\right]C(X)=\tau^2\sum_{A_0}\frac{\delta_4(X,A_0)}{\sqrt{-g(A_0)}}, \tag{19}$$

but with the previous f^{-1} in (5) now seen to be related to the square of the decay time of the Planck particles. The form (5) and hence (19) was an initial exploratory approach to the problem of the creation of particles, whereas (11) and (17) use additional physical ideas concerning the decay of the Planck particles.

One may wonder why a conformal transformation could not be used to make $m(X)+c(X)$ constant with respect to X. If this were done the conformal function Ω in Narlikar's treatment would have exceedingly fine-scale ripples in it, which would introduce corresponding ripples into the metric tensor. While these would be very small in amplitude their derivatives, arising in evaluating the tensor R_{ik}, would be greatly enhanced by the inverse time factor τ^{-1}, making the problem of obtaining R_{ik} exceedingly awkward. It is better, therefore, to use the device of a conformal transformation only with respect to the mass field $m(X)$. Strictly speaking, the latter does have fine-scale ripples in it occasioned by the appearance of baryons as the Planck particles decay, but the effect by such ripples is smaller for a baryon than for a Planck particle by a factor of order $(6 \cdot 10^{18})^{-1}$. Once the baryons have appeared and are well separated on a scale $\sim 10^{-14}$ cm, rather than the Planck particle scale of $\sim 10^{-33}$ cm, they can be considered to produce an essentially smooth mass field.

I think it was Hermann Bondi who once said that physics is such a consistent and connected logical structure that if one starts to investigate it at any point and if one pursues correctly every issue that branches away from one's starting point, in the outcome one will be led to understand the whole of physics. With Mach's Principle it seems something like that.

REFERENCES

Hoyle, Fred and Narlikar, Jayant V. (1972a). "Cosmological Models in a Conformally Invariant Gravitational Theory – I. The Friedmann Models." *Monthly Notices of the Royal Astronomical Society* 155: 305–321.

Hoyle, Fred and Narlikar, Jayant V. (1972b). "Cosmolgoical Models in a Conformally Invariant Gravitational Theory – II. A New Model." *Monthly Notices of the Royal Astronomical Society* 155: 323–335.

Hoyle, Fred; Burbidge, Geoffrey R.; and Narlikar Jayant V. (1993). "A Quasi-Steady Cosmological Model with Creation of Matter." *Astrophysical Journal* 410: 437–457.

Hoyle, Fred; Burbidge, Geoffrey R.; and Narlikar, Jayant V. (1994). "Astrophysical Deductions from the Quasi-Steady-State Cosmology." *Monthly Notices of the Royal Astronomical Society* 267: 1007–1019.

Discussion (to both Narlikar's and Hoyle's paper)

Ehlers: I would first like to come back to your [Narlikar's] starting point when you formulated the basis of the theory. Your action functional depends on two types of independent objects, namely, a metric and a collection of worldlines. If that is so, then if you evaluate the principle of stationary action, you should get one equation from the variation of g_{ik}, and that's in fact the equation which you wrote down, but you should also get another set of equations from the variation of the particle worldlines, which I did not see.
Narlikar: Yes, in fact you get something that is equivalent to the geodesic equations in relativity, but you get additional terms which are derivatives of masses, the mass functions, which come too. I didn't write them but they are there.
Ehlers: If then you find out that the field equation is a differential equation between the metric and the particle worldlines, because of the source terms, and you assume in the beginning that the metric and the particle worldlines could be varied independently, but then the equations that you get out of the variational principle tell you that they are not independent, I have difficulties seeing whether that's all consistent. It's a different situation from an ordinary Lagrangian-based field theory.
Narlikar: I think if you took the field equations which I wrote down and took their divergence and then evaluate them as limiting quantities on the worldlines of particles, because they are singular lines, with the delta function, then you will find that they've reduced to the equations of motion which you get out by the variation of worldlines, so this whole thing is mutually self-consistent.
Lynden-Bell: I think the real answer to your question is, in part, the argument of Eddington: That you always vary in such a manner as to violate the field equations. You compare the universe that obeys the field equations with an imaginary universe that violates the field equations, and I think they're doing it just as consistently as any other person who varies field equations.
Bondi: A fairly minor question, but in our discussion in the last day or so we have very often said that, in the Cauchy problem of relativity, space in the form of gravitational waves has an independent degree of freedom in addition to the matter. Now is that not so in your theory?
Narlikar: To the extent that the geometry is described by the metric g_{ik}, and we are varying this metric to derive the field equations, this means we are investing them with some degrees of freedom. So to that extent, supposing you have a particle which you wobbled like this; that would

produce ripples in the spacetime geometry because of the field equations, and so gravitational radiation would be there.

Bondi: It's a little different from in the discussion yesterday. It was very much stressed by various people, by you [Barbour] in particular, that the space has independent degrees of freedom, that if you prescribe on a Cauchy surface the right conditions for the future it was not obligatory to be able to trace back the gravitational wave content to particle motions. I'm not saying it should or should not be; I'm just trying to clarify.

Narlikar: I think in our framework we need to look carefully at what is the formulation of the initial-value problem because, since we have worldlines, it is different from the standard way of specifying data on a Cauchy surface or spacelike hypersurface. So we shall do that at some stage.

Hoyle: Any freedom that exists in general relativity should be present here because what we're doing is saying that you can have an arbitrary configuration, specification of the worldlines, and then by varying the geometry you get the field equations; since in Narlikar's form it reduces to Einstein, it should be the same.

Barbour: I'm not at all expert in this, but am I right in thinking that the whole question of the initial-value problem is also unclarified in the Wheeler–Feynman theory?

Narlikar: In the 1949 paper of Wheeler–Feynman, they have looked at models and how things propagate from one spacelike hypersurface to another, taking account of how the light cones cross, and it's quite a difficult paper to read, but they have done their best under the circumstances, describing what can be done.

Barbour: Surely in your action, you only have an action if there are particles present, is that not right?

Narlikar: Yes.

Barbour: So surely it's not quite the same, Fred [Hoyle], as in general relativity, because in general relativity you can have an action with no particles there at all, just pure gravity, whereas in this case you've got to have the particles. So in some sense it's not quite so free surely?

Narlikar: Yes, in that sense this theory is strongly Machian, which general relativity is not. For example, if there were only two particles, I'm sure the problem would be quite complicated, and you wouldn't be able to use the relativistic approximation.

Raine: Once you have a large number of particles though, the free degrees of freedom are smuggled in by specifying what the worldlines of the particles are. So you have some free initial data, as one would say from the Cauchy point of view; what you say equivalently is that you have a wiggle in the worldlines later on.

Ehlers: I would first like to apologize. I think my question was indeed nonsensical, as was pointed out by Lynden-Bell. But still if I am to understand the starting point of your theory, where you have a collection of particles and you have not yet gone to the smoothed-out limit, then if you take your particles seriously, isn't it so that your propagator diverges at the positions of the particles? I wonder whether you think that the original field equations have any solutions at all mathematically. I would very strongly doubt that they have any solutions. Maybe the theory should be taken seriously only after you have gone to the fluid average.

Narlikar: We had one solution which is equivalent to the standard Schwarzschild solution in general relativity, in which we took one particle and examined how the equations look as we approach that particle, taking account of the fact that it has a singularity.

Ehlers: Yes, but the singularity in the Schwarzschild field is not a timelike line, as follows from the Kruskal extension; you never meet a singular particle worldline.

Narlikar: The geometry turns out to be different, more like Reissner–Nordström.

Raine: I understand how the theory works if you've only got one type of particle, but suppose you set the length scale of the theory by the mass of the electron. Then you discover the proton's made actually of three particles. You are then rescaling the mass of the proton, but you're not rescaling the length of the theory, so the separate terms that come into the fundamental equations will be different if you think the proton is three particles or if you think it's one particle. In other words, your theory actually differs from general relativity as you go back through the evolution of the universe depending on what the universe is actually made of. Is that wrong?

Hoyle: The couplings of particles to the mass field are in proportion to the particle masses. Or you can put it the other way round: The different couplings of different particles determine their mass ratios.

Xu: Have you calculated for a neutron star what the particle creation rate is and can this be compared with binary pulsar data now?

Narlikar: I cannot answer the question within the framework of the two lectures which were given today, but I can give you an additional reference; there is a paper in the *Astrophysical Journal* of June 20th of this year (1993) [410: 437–457 "A Quasi-Steady State Cosmological Model with Creation of Matter"] in which we discuss the creation process near massive objects, collapsed massive objects; if we use those ideas given in the paper, we can apply the theory Professor Hoyle was describing. Then you find that you need to have very much more collapsed objects than neutron stars; they're not collapsed enough to produce fresh creation; it's $1-2\text{G}m/c^2r$ needs to be very close to zero, and in neutron stars it is about 1/3 or of that order.

Bondi: Can I elucidate this question: If the creation rate is roughly such as to produce enough in a period of the order of the age of the universe, then in the period of the binary pulsar its mass should be increased by about one part in 10^9.

Hoyle: No, because the conditions for creation then have to be very close to the black-hole condition. The bosons of the *C*-field as they exist everywhere in space have much too little energy to promote creation by themselves; they have to fall into a very strong gravitational field.

Bondi: I see, a neutron star isn't good enough.

Hoyle: That's right. You have to go very close to the black-hole condition. That's discussed in the paper that Jayant [Narlikar] referred to in the recent *Astrophysical Journal*.

Will: Does that mean you get the same rate around a solar-mass black hole as around a supermassive black hole? In the latter case the gravitational field is weak.

Hoyle: Near any event horizon, the creation rate per unit proper volume should in general be the same, except perhaps insofar as the process has feedback in itself; so if it is in a big volume the possibility for feedback is greater. That is to say, once it gets started, it modifies its own local gravitational fields. It is a feature of the Planck particle that it has a gravitational radius of the order of its own Compton wavelength, so that as one starts to produce them, they do modify their local gravitational field quite markedly.

Will: Then it's just the strength, if you say the density in a neutron star is not strong enough.

Hoyle: But as the explosion starts it could well be that it matters whether it's a small explosion or a large one. But this is, as you can appreciate, an awful calculation to carry out. The best one can hope to do is to see how it starts, and once the amplitude has become big for the explosion then it's hard to know how it will go. But if we then turn to observation as our guide, rather than calculation, the implication is that when one gets up to big objects the effects are very large.

Nordtvedt: Do you have any estimate at this point as to how much mass is being created per year in the universe?

Hoyle: On average, it's about 100 solar masses per major galaxy per year.

The Integral Formulation of Mach's Principle

Derek J. Raine

1. Introduction

In this paper, we begin with a brief sketch of the concepts surrounding Mach's Principle which will lead to the idea of inertial induction. Complete inertial induction in the context of general relativity can be expressed as a condition on an integral representation of the Weyl tensor as a linear function of the 'acceleration currents' of matter. However, this does not fully express the idea that the matter content of the universe should determine the local inertial frame; for example, it allows flat and asymptotically flat metrics. A complete statement requires a further condition on how the metric is determined by the curvature. By means of these conditions, Mach's Principle is implemented as a selection rule in general relativity. A simpler formulation can be obtained via a direct integral representation of the metric tensor as a function of the 'velocity currents' of matter (i.e., the stress tensor). A heuristic derivation of this is given that shows how matter and its associated gravitational energy contributes to the determination of the local inertial frame in those cosmologies which satisfy the conditions of the representation. This integral representation therefore provides us with a rule for selecting Machian universes. An outline of how this selection rule can be applied in practice is presented and the known results summarized. The relation of these results to the notion of an isotropic singularity and gravitational entropy is noted. Finally we discuss problems with the view of Mach's Principle as a selection rule and with the integral equation approach and speculate on possibilities for incorporating Mach's Principle as a deduction from quantum gravity.

Einstein Studies, vol. 6: Mach's Principle: From Newton's Bucket to Quantum Gravity, pp. 274–292

2. A Short 'History' of 'Mach's' Principle

We start with Aristotle: In this picture the spacetime is a product $R^3 \times T$. This structure is associated with the first law of motion that bodies remain at rest unless forced to move (Raine and Heller 1981). At least in one version of the theory, forces produce velocities according to a second law of motion of the form $F=mv$. A minimal component of Mach's Principle would be that the velocities in this law are those measured relative to the matter in the universe. This condition is satisfied in Aristotelian physics if by distant matter we mean the earth, which defines the fundamental rest frame. In a frame in motion with relative velocity u, the law of motion becomes $F+mu = mv$, where, according to later interpretations of Mach, the 'mu' should be induced by motion relative to matter. If we regard this stronger condition as an expression of Mach's Principle, then Aristotelian physics is non-Machian because it provides no mechanism for this inertial induction. Equivalently, $mu=0$ in the preferred frame is not predicted by the theory. This is a reflection of the existence of an absolute element in the theory: Spacetime is endowed with a *preferred vector field on spacetime* specifying the state of rest with respect to the earth.

A similar analysis of Newtonian physics is more difficult at least in part because it is hard to make it consistent without changing it. Newton himself argued forcibly for absolute *acceleration*, but in the context of the Aristotelian spacetime structure which depends on absolute *velocities*. The correct spacetime structure for Newtonian dynamics, in the absence of gravity, is (1) that the spacetime of events has a slicing by absolute time, and (2) that the bundle of frames over spacetime is a product. This latter statement means that we can identify (objectively) the same frame at different events - by nonrotating transport along any piecewise-differentiable inertial path between the events. Mach's contribution was to point out that by nonrotating here we mean nonrotating relative to the fixed stars. In *The Science of Mechanics*, Mach explicitly states that this is a deduction from observation at the limited accuracy then available: He does not imply it is a necessary truth. (In fact, he suggests that further observation may reveal discrepancies.) In Einstein's hands it was elevated to a matter of principle: Thus, in a frame subject to an acceleration a_0 relative to distant matter, Newton's law of motion becomes $F+ma_0=ma$ and the force ma_0 should be induced by the stars. An equivalent result should also hold for rotation, so the observed nonrotation of the stars would be a prediction of the theory. Of course, as we know, applying these ideas *locally* leads to the equivalence principle and undermines the Newtonian spacetime structure: In the presence of gravity the transport of frames is path-dependent. This is well known for general relativity, but it is true also for the *spacetime* of Newtonian gravity (for example, Misner, Thorne, and Wheeler 1973). Thus, we arrive at the Einstein–Cartan structure for the spacetime of Newtonian physics: Gravity and inertia contribute to a local connection which allows us to specify the same (zero) accelerations at different events. Of course, this preferred field, which incorporates gravity as spacetime curvature, is supposed to be determined by the matter content of the universe.

The motion of one body relative to another is given by the deviation equation for the connecting vector ε^μ:

$$\frac{d^2\varepsilon^\mu}{dt^2}+\Gamma^\mu_{\nu\rho}\frac{dx^\nu}{dt}\frac{d\varepsilon^\rho}{dt}+R^\mu{}_{\nu\rho\sigma}\frac{dx^\nu}{dt}\varepsilon^\rho\frac{dx^\sigma}{dt}=0,$$

where Greek indices range over 0,1,2,3, t is the Newtonian time, and Γ and R are the Newtonian connection and spacetime curvature given in an inertial frame by the nonzero components:

$$\Gamma^i_{00}=\phi_{,i} \tag{1}$$

$$R^0{}_{i0j}=\phi_{,ij}. \tag{2}$$

with $i, j=1,2,3$ and ϕ the Newtonian potential. Mach's Principle therefore comes twice into Newtonian gravity: Once through the fact that the connection must be fully determined by the matter content of the universe, and also through the determination of the curvature by matter.

Finally, Einstein spacetime differs only in the replacement of the time-slicing of spacetime by a metric structure by virtue of special relativity. This leads to the amazing simplification that this metric also determines the connection and the curvature. Thus if the metric is generated entirely by matter, then so will the connection and the curvature.

It is the purpose of what follows to define what it means for matter to *determine* the metric, connection, and curvature. Within the context of general relativity, we shall find that these definitions select certain spacetimes, which we can call 'Machian,' from among all possible solutions. This selection is made without prejudice as to whether we find the outcome, in terms of which solutions are selected, pleasing or not. Once we know what the consequences of this (Einstein's) interpretation of Mach are, we can decide whether to abandon the principle or redefine it. I suspect the majority view will be in favor of redefinition, but with little agreement as to how. It should be noted though that, once we accept a form of physical law that treats interactions between matter by the gauge principle, Mach's Principle inevitably appears in the context of boundary conditions for the gauge field and hence as a selection rule.

3. Inertial Induction

It is well known that Einstein's early attempt (Einstein 1922) to calculate the effect of inertial induction on the mass of a body is a coordinate-dependent artefact, but Thirring's famous result (Thirring 1918), although not the last word on the subject, does indeed show how a rotating shell drags the inertial frame located at its center: If the shell has mass M, radius R, and angular velocity ω the induced angular velocity of a Foucault pendulum at the center of the shell is

$$\Omega = \frac{4}{3}\frac{GM}{c^2R}\omega .$$

Mach's Principle is satisfied in a universe of such shells if frames are maximally dragged, $\Omega=\omega$. Summing over the universe, we get $\Omega=\omega$ if $\Sigma GM/c^2R \sim 1$.

Of course, this argument is only approximate because the gravitational effect of a shell depends on the geometry in which it is embedded, i.e., on the effects of all the other shells. (See Pfister, this volume, for a review.) It is also complicated by the expansion of the universe and retardation effects. Nevertheless, to understand the problem in a straightforward way, we shall construct a Newtonian model.

In his 'toy model' of inertia, Sciama posits that the induction of inertia by an acceleration current of the matter in the universe takes place through an additional component of the gravitational interaction proportional to the acceleration of a body and to $1/r$. In this picture the linear frame dragging is complete again if $\Sigma GM/c^2R \sim 1$.

This idea can be incorporated into a 'post-Newtonian' picture by making use of the Bianchi identities to describe the effect of the acceleration currents of distant matter in generating curvature. Specifically, we shall be concerned with the Weyl curvature given in general by

$$C^{\mu\ \nu}_{\ \rho\ \sigma} = R^{\mu\ \nu}_{\ \rho\ \sigma} - 2g^{[\mu}_{[\rho}R^{\nu]}_{\sigma]} + \frac{1}{3}Rg^{[\mu}_{[\rho}g^{\nu]}_{\sigma]},$$

and in terms of the Newtonian potential in 3-space, by

$$C^{0}_{\ i0j} = E_{ij} = \phi_{,ij} - \frac{1}{3}\delta_{ij}\nabla^2\phi .$$

3.1. Newtonian inertial induction. Suppose a body B of mass m has acceleration a along the z-axis in a uniform isotropic universe of density ρ. In the 'post-Newtonian' picture we propose to investigate here, we take the spacetime curvature to be 'painted' on to a flat background. The net Newtonian gravity of the universe at B vanishes, so the gravitational potential in this model carries only the effects of inertial induction. From the point of view of B, the universe has an acceleration along the z-axis; this will be shown to induce an inertial force on B of magnitude ma provided the universe has of order the critical density.

We assume the acceleration is small so that the velocity of the universe relative to B remains nonrelativistic. Then the four-velocity of the universe is $u^\mu=(1,0,0,at)$ with $u^\mu u_\mu \sim -1$, where we have set $c=1$ and we are neglecting the terms of order v^2/c^2. The energy–momentum tensor of the universe, assumed to consist of pressureless fluid, is

$$T_{\mu\nu}=\rho u_\mu u_\nu,$$

from which the Einstein field equations yield

$$(R_{\mu\nu})=\kappa\rho\begin{bmatrix} \frac{1}{2} & 0 & 0 & -at \\ 0 & \frac{1}{2} & 0 & 0 \\ 0 & 0 & \frac{1}{2} & 0 \\ -at & 0 & 0 & \frac{1}{2} \end{bmatrix}.$$

The Bianchi identities $R_{\mu\nu[\rho\sigma;\tau]}=0$ can be written in terms of the Weyl tensor $C_{\mu\nu\rho\sigma}$ in the flat background space as

$$C_{\mu\nu\rho}{}^{\sigma}{}_{;\sigma}=R_{\rho[\mu;\nu]}-\frac{1}{6}\eta_{\rho[\mu}R_{;\nu]}. \tag{3}$$

Note that if we substitute for $R_{\mu\nu}$ on the right of (3) from the field equations, then (3) contains an expression of the induction of curvature by 'acceleration currents' of matter (i.e., by gradients of $T_{\mu\nu}$). For the simple example we are considering, the only nonzero component of (3) is

$$C_{0z0}{}^{i}{}_{,i}=\frac{1}{2}\kappa\rho a,\ i=1,2,3. \tag{4}$$

Now, putting

$$C_{0i0j}=E_{ij}=\frac{\partial^2\Phi}{\partial x^i\partial x^j}-\frac{1}{3}\delta_{ij}\nabla^2\Phi,$$

(4) becomes

$$\nabla\nabla^2\Phi=\nabla^2\nabla\Phi=\frac{3}{4}\kappa\rho a\hat{\mathbf{z}}.$$

The force on B is $\nabla\Phi$, which is clearly the sum of contributions from the matter in the universe proportional to a/r. There is also, of course, the possibility of adding a homogeneous solution of (4) to Φ, which would represent an inertial induction force not dependent on the acceleration of matter in the universe. In such a case the local inertial frame would not coincide with the frame of the fixed stars.

This calculation is obviously heuristic. But it serves to illustrate the fact that inertial induction is to be looked for in general relativity in the Bianchi equations. Mach's Principle should be expressed through the fact that the solutions of these equations contain no free modes not attributable to the acceleration currents of matter. In summary, the velocity currents of matter generate Ricci curvature, hence the usual contribution to the Newtonian gravitational potential; acceleration currents generate an additional Weyl curvature contribution to the potential, through which they induce inertial effects.

3.2. Integral representation of the Weyl curvature. We now show how this Machian program can be carried over to the relativistic theory. The details are rather complicated [they are given in (Raine 1975)], so I shall restrict myself here to explaining the general structure.

The Bianchi identities can be written in terms of the electric (E_{ij}) and magnetic (H_{ij}) parts of the Weyl tensor: $E_{ij}=C_{0i0j}$ and $H_i^j=\frac{1}{2}\eta_{0ikl}C^{0jkl}$. The 'constraint' equations take the form

$$E^{ij}{}_{;j}=J^i, \tag{5}$$

$$H^{ij}{}_{;j}=\bar{J}^i. \tag{6}$$

This form is deceptively simple: When written out explicitly, the 4-space covariant derivative introduces magnetic terms into the first equation and electric terms into the second. Nevertheless, the symmetry of the system allows us to combine the equations into a single system for a complex field

$$\Psi^{ij}=E^{ij}+\mathrm{i}H^{ij}$$

of the form

$$\boldsymbol{D}_j\Psi^{ij}=\boldsymbol{J}^i$$

involving only a covariant derivative in the three-spaces. Formally, $\boldsymbol{D}$ behaves as a 'covariant' derivative (except that the three-space metric, h_{ij}, is not covariantly constant) so we can use the standard decomposition of a symmetric tensor (Deser 1967) to solve the constraints.

We begin by writing $\Psi_{ij}=\Psi^{\mathrm{T}}_{ij}+\Psi^{\mathrm{L}}_{ij}$, where $\boldsymbol{D}^j\Psi^{\mathrm{T}}_{ij}=0$ and

$$\Psi^{\mathrm{L}}_{ij}=\boldsymbol{D}_i\xi_j+\boldsymbol{D}_j\xi_i-\frac{2}{3}h_{ij}\boldsymbol{D}^m\xi_m$$

When this form is substituted into the constraints (4), we obtain an elliptic system for ξ_i which can be solved in terms of a Green function which goes to zero sufficiently rapidly at infinity (if the three-spaces are open). This gives us the contribution to Ψ^{L}_{ij} that represents data on the initial surface due to gravitating matter at earlier times or to the instantaneous 'Coulombic' part of the gravitational field from matter on the initial surface, but ignoring any free component.

We then turn to the evolution equations for the Weyl curvature. These have the general form

$$\dot{\Psi}_{ij} + L_{ij}^{kl}\Psi_{kl} = J_{ij},$$

where L_{ij}^{kl} is a linear differential operator in three-space. Formally the solution is

$$\Psi_{ij}(t) = \int_{t_0}^{t} (e^{-L(t-t')})_{ij}^{kl} J_{kl}(t')dt' + \Psi^{\mathrm{L}}_{ij}(t_0),$$

where the initial conditions have been set by $\Psi \rightarrow \Psi^{\mathrm{L}}$ as $t \rightarrow t_0$, i.e., there is no free Weyl tensor. A spacetime will be Machian if as $t_0 \rightarrow 0$ this Ψ yields the Weyl tensor of the spacetime.

3.3. Results. The detailed implementation of this selection rule for Machian spacetimes is somewhat complicated; most of the known results are obtained indirectly by showing that the known Weyl tensor cannot satisfy a representation of this form.

We shall refer to the non-FRW spatially homogeneous cosmologies as Bianchi cosmologies. In these solutions the Weyl tensor constraints reduce to a set of algebraic equations for the off-diagonal elements of Ψ_{ij} in a natural frame in terms of the diagonal elements Ψ_i. The essence of the proof of the non-Machian character of the Bianchi cosmologies is to compare the behavior of the known Ψ_i with the Machian solution of the evolution equations, which are here autonomous ordinary differential equations in cosmic time. Specifically we consider a rescaling of the matter density $\mu \rightarrow \lambda\mu$. The field equations give the behavior of the coefficients L_{ij}^{kl} as a function of λ. The Machian solution has an expansion in a power series in λ which begins with a term of order λ. The burden of the proof is then to show that the known Weyl tensor has a series expansion which begins with a (nonzero) term of order λ^0.

A second class of models that can be subjected to a Machian analysis in this way are the so-called Bondi models (Bondi 1947). I shall treat these in somewhat more detail since although the results in the literature are stated correctly (Raine 1975) the derivations are flawed.

We follow the treatment of Eardley *et al.* (1972). The metric of these models has the form

$$ds^2 = -dt^2 + A^2(r, t)d\Omega^2 + B^2(r, t)dr^2, \; d\Omega^2 = d\theta^2 + \sin^2\theta d\phi^2.$$

With a stress-energy of pressure-free perfect fluid, the field equations yield:

$$A' = [1 + \beta^2(r)]^{1/2} B,$$

$$2A\ddot{A} + \dot{A}^2 - \beta = 0$$

for some function $\beta(r)$, where A' denotes $\partial A/\partial r$ and $\dot{A}$ denotes $\partial A/\partial t$. The second equation can be integrated to give

$$A(\dot{A}^2 - \beta^2) = \lambda(r),$$

where, from the Bianchi identities,

$$\lambda' = \mu_0(1 + \beta^2)^{1/2}$$

with

$$\mu_0 = A^2 B\mu.$$

To investigate the nature of the singularity, Eardley *et al.* expand the metric as a function of a cosmic time $t - {}_0t(r)$, relative to a shifted origin $t = {}_0t(r)$ on each worldline. Solutions which behave like FRW to leading order as we approach the singularity are characterized by ${}_0t'(r) = 0$, the Heckmann–Schücking(HS)-like behavior by ${}_0t'(r) \neq 0$.

We discuss the Machian character of these solutions in the case $\beta = 0$; the general case follows similarly. We shall need an expression for the nonzero component of the Weyl tensor, which in our notation is

$$\Psi_{11} = -\frac{1}{3}(\ddot{B}/B - \ddot{A}/A).$$

In standard notation $\Psi_{11}/2$ is just the Newman–Penrose (NP) function ψ_2, which we shall denote by χ. The Bianchi constraint for χ is

$$\chi' + 3(A'/A)\chi = \frac{1}{6}\mu',$$

and the evolution equation is

$$\dot{\chi} + 3(\dot{A}/A)\chi = \frac{\mu}{6}(\dot{A}/A - \dot{B}/B).$$

A solution of the constraint will automatically be propagated by the evolution equation, so we can restrict our attention to that.

The Machian solution on the initial surface $t = t_0$ is

$$\chi = \frac{\mu}{6} - \frac{\lambda(r)}{2A^3} - \frac{A^3(0, t_0)\mu(0, t_0)}{6A^3} + \frac{\lambda(0)}{2A^3}.$$

For a FRW singularity the leading terms of A and $B=A'$ near the singularity satisfy $B \propto A$, i.e., they have the same time dependence. This means we can arrange $\lambda(0)=\mu(0,t_0)A^3(0,t_0)/3$ for all t_0 and the Machian solution is the known NP function. However, for the HS singularity, A and B have a different leading time behavior [proportional to $(t-{}_0t(r))$ to the powers 2/3 and $-1/3$, respectively], so we cannot arrange for these terms to cancel for all times. Therefore, we must remove them by adding a solution of the homogeneous equation, constant/A^3. Thus, the Bondi models with HS singularities are non-Machian.

3.4. The first Mach condition. Unfortunately, imposing a selection rule on the Weyl tensor – that there should be no homogeneous contribution (which would represent free gravitational waves) in the integral representation – does not rule out all obviously non-Machian solutions. Minkowski spacetime is the simplest counterexample: The constraint is trivially satisfied but the metric is obviously non-Machian. Asymptotically flat spacetimes would also violate such a constraint. Although some of these solutions satisfy the constraint that the curvature be determined by matter, they violate the obvious additional Machian condition that the curvature determine completely the local inertial frame, i.e., that the curvature determine the metric. In fact, it is precisely those solutions representing plane source- free gravitational waves that form the simplest examples of inequivalent metrics that have the same Riemann tensor and are precisely the sort of solution we want Mach's Principle to rid us of.

In (Raine 1975), I showed how one could set up an integral representation of the metric in terms of the Riemann tensor using a generalized inverse to take care of the gauge freedom. The requirement that this representation should be a linear (homogeneous) relation between the curvature and the metric is then called the first Mach condition, and the requirement on the Weyl tensor becomes the second Mach condition. The first Mach condition provides, I believe, a valid, but very labor-intensive way of ruling out asymptotically flat and plane-wave spacetimes. Is there a simpler way of incorporating complete inertial induction into relativistic cosmological models?

4. The Integral Representation of the Metric

The ideas of this section go back (independently) to Al'tshuler (1967), to Lynden-Bell (1967), Sciama and Waylen, and to Gilman (Sciama, Waylen, and Gilman 1969). They showed how one could represent the metric in a curved spacetime as the 'sum' of contributions from the matter (and boundary conditions) in a meaningful way (i.e., despite the nonlinearity of the Einstein field equations). Rather than just repeat the derivation here, I will try to explain what lies behind it.

We begin by recalling how the relativistic field equations can be 'derived.' Start from a spin-two theory for a symmetric tensor field; this is essentially the linearized Einstein theory for a potential $\phi_{\mu\nu}$:

$$[(\eta_{\mu\alpha}\eta_{\nu\beta}-\eta_{\mu\nu}\eta_{\alpha\beta})\Box+\eta_{\mu\nu}\partial^2_{\alpha\beta}+\eta_{\alpha\beta}\partial^2_{\mu\nu}-\eta_{\mu\alpha}\partial^2_{\nu\beta}-\eta_{\nu\beta}\partial^2_{\mu\alpha}]\phi_{\alpha\beta}=\kappa T_{\mu\nu}. \quad (7)$$

This is inconsistent because the sources on the right side should include the energy of the $\phi_{\mu\nu}$ field, $\tau_{\mu\nu}(\phi)$, say. We therefore add this on the right of the field equations. But this change to the field equations modifies the field energy and so on. The process must converge to the Einstein equations $G_{\mu\nu}(\eta_{\mu\nu}+\phi_{\mu\nu})=\kappa T_{\mu\nu}$ (because expanding these equations gives back the infinite series). Deser (1970) gives a proof that stops after the second iteration by making use of a first-order form for the Lagrangian and a particular choice of field variable. In the final theory only the full metric $g_{\mu\nu}$ appears, and all trace of the auxiliary background metric $\eta_{\mu\nu}$ is lost.

Now imagine we carry out the same procedure, but starting from an integral representation of $\phi_{\mu\nu}=\eta_{\mu\nu}+h_{\mu\nu}$. We will end up with a representation of the form:

$$\phi_{\mu\nu}=\int G^{\rho\sigma}_{\mu\nu}(K_{\rho\sigma}+\tau_{\rho\sigma})+\text{ a surface integral.}$$

What are G and τ? Recall that the final equation, in differential form, is obtained from a Lagrangian $\boldsymbol{L}(\eta+h)$ by variation with repect to h; but that this is the final form of the equation, i.e., that the iteration process is complete, is signalled by $\delta\boldsymbol{L}/\delta\eta$ is identically zero because this is the correction to the stress–energy of the h-field. We conclude that (i) $\tau_{\mu\nu}=0$ and (ii) $\delta\boldsymbol{L}/\delta h$ is identically zero, i.e., the variation of the action of the field operator is zero, and consequently that $\delta\int GK$ is zero. This last condition is equivalently

$$g+\delta g=\int G(K+\delta K)+\text{ surface terms}$$

with no contribution from a δG term. We refer to this by saying that the Green function is stable with respect to perturbations of the metric. If we call L the differential operator inverse to G, then we have $L(g+\delta g)=K+\delta K$; so $Lg=K$ must be the Einstein field equations, $R_{\mu\nu}=K_{\mu\nu}$, and the operator L can be obtained by variation of these.

Note that the stability of the Green function (or its inverse) is a condition here on the convergence of the iteration scheme to the Einstein field equations. Alternatively we may interpret it as giving a meaning to the superposition of the influence on the gravitational field of each element of matter: This influence is propagated linearly over the self-consistent final spacetime. Note also that we start from an integral expression for the field in terms of the stress–energy of matter and the field itself; in the limit the explicit contribution from $\tau_{\mu\nu}$ vanishes, but the field energy is included implicitly in the Green function.

Some care is required in the choice of index positioning and the use of tensors or tensor densities to enable this procedure to work (i.e, to yield a stable operator), but for each form of the Einstein equations there is an appropriate field variable and operator. By varying the mixed form of the field equations $R^{\mu}_{\nu}=T^{\mu}_{\nu}-\frac{1}{2}\delta^{\mu}_{\nu}T=K^{\mu}_{\nu}$, using the contravariant metric as fundamental variable, $\phi^{\mu\nu}=g^{\mu\nu}+\delta g^{\mu\nu}$, Sciama, Waylen, and Gilman (1969) obtain

$$\Box\phi^{\mu\nu}-2R^{\mu}{}_{\rho}{}^{\nu}{}_{\sigma}\phi^{\rho\sigma}+2\nabla^{(\mu}\psi^{\nu)}=2\kappa K^{\mu\nu},$$

where $\psi^{\mu}=\nabla_{\nu}(\phi^{\mu\nu}-\frac{1}{2}g^{\mu\nu}\phi^{\rho}_{\rho})$ (see also Al'tshuler 1967, Gilman 1970).

In the gauge $\psi^{\mu}=0$ this becomes

$$\Box\phi_{\mu\nu}-2R^{\rho}{}_{\mu}{}^{\sigma}{}_{\nu}\phi_{\rho\sigma}=K_{\mu\nu}+\delta K_{\mu\nu}. \tag{8}$$

It can be shown that (to first order) the gauge conditions are preserved by the field equations.

By varying the covariant field equations using the mixed components of the metric variation, $\phi_{\mu\nu}=\delta_{\mu\nu}+g_{(\mu\rho}\rho\delta g^{\rho}_{\nu)}$, Al'tshuler obtains, instead of (8),

$$[-(\delta^{\mu}_{(\rho}\delta^{\nu}_{\sigma)}-\frac{1}{2}g_{\rho\sigma}g^{\mu\nu})\Box-R_{\rho}{}^{\mu}{}_{\sigma}{}^{\nu}+g_{\rho\sigma}R^{\mu\nu}+g^{\mu\nu}R_{\rho\sigma}-\frac{1}{2}g_{\rho\sigma}g^{\mu\nu}R]\phi_{\mu\nu}=2\kappa K_{\rho\sigma} \tag{9}$$

in the gauge $\psi^{\mu}=0$.

5. Applications of the Integral Representation

5.1. The Mach Green Function. We obtain a Machian selection rule from an integral representation of the metric by imposing conditions on the surface terms. We shall assume that the spacetimes we are dealing with are globally hyperbolic, so have the form $\Sigma\times T$, with the initial surface (possibly singular) labeled by $t=0$. If the operator equations are self-adjoint, then the volume integral is a particular integral of the system showing the dependence of the metric on the matter interior to the volume. The simplest suggestion would then be that for a Machian spacetime the surface terms vanish when the integral is over the whole of spacetime. This presents a *prima facie* problem in spacetimes with singularities because the retarded Green function is not well defined; physically one can see this because the surface terms may contain information from *matter* outside the volume, which, in a spacetime with boundaries, can be a nonzero term, in addition to a genuinely 'non-Machian' source-free contribution. The scheme can, however, be implemented *provided* we are careful in the choice of Green function. Specifically, we choose a Green function which satisfies the constraint equations $G^{\alpha'\beta'}_{\mu\nu;\alpha'}=0$ *on the initial surface*. The Green function will not satisfy this condition off the surface, but the volume integral as a whole will (because the gauge conditions are preserved). We write a Green function satisfying this condition as $M^{\alpha'\beta'}_{\mu\nu}$. The surface term in the integral representation is

$$I=\int M^{\alpha'\beta'}_{\mu\nu;\gamma'}g_{\alpha'\beta'}n^{\gamma'}dS,$$

where the integral is over an initial 3-surface. Mach's Principle is therefore satisfied if $I\to 0$ as the surface tends to the past singularity, or equivalently, if

$$\int M^{\alpha'\beta'}_{\mu\nu}K_{\alpha'\beta'}dV\to g_{\mu\nu}.$$

5.2. Results. The direct application of this criterion is again somewhat involved. However, in spatially homogeneous models it can be greatly simplified by what amounts to integrating out the spatial variables before testing the selection criterion. Specifically, we look at a representation of $\eta_{ab}=g_{\mu\nu}e^{\mu}_{a}e^{\mu}_{b}$, where e^{μ}_{a} are a vierbein and $a,b,\ldots=0,1,2,3$ are vierbein indices, by considering a purely time-dependent perturbation $\delta\eta_{ab}(t)$. This leads to a set of ordinary differential equations in t for the diagonal components $\phi_a=diag(\eta_{ab}+\delta\eta_{ab})$.

In FRW spacetimes the symmetry reduces the number of components to two, ϕ_0 and ϕ_1 say. The Sciama–Waylen–Gilman (SWG) equations become

$$\begin{aligned}
&-\ddot{\phi}_0-\theta\dot{\phi}_0+\frac{2}{3}\theta^2\phi_0-2\dot{\theta}\phi_1=2K_0,\\
&-\ddot{\phi}_1-\theta\dot{\phi}_1+\frac{2}{3}\theta^2\phi_1-\frac{2}{3}\dot{\theta}\phi_0=2K_1,
\end{aligned}\tag{10}$$

where dots denote differentiation with respect to cosmic time, t, and $\theta=3\dot{R}/R$, with R the FRW scale factor. The source terms for a perfect fluid with equation of state $p=(\gamma-1)\mu$ are

$$K_0=\frac{3}{2}(\mu+p),\tag{11}$$

$$K_1 = \frac{1}{2}(\mu - p), \tag{12}$$

and $K_i \propto R^{-3\gamma}$. The Einstein field equations give

$$3\ddot{R}/R = -\frac{\mu_0}{2}(3\gamma - 2)R^{-3\gamma},$$

and near the singularity this integrates to $\dot{R}^2 \sim (\mu_0/3)R^{-3\gamma+2}$. Rewriting (10) in terms of derivatives with respect to R instead of t, we have, near the initial singularity,

$$\begin{aligned} -R\phi_0'' + \left[\frac{3\gamma}{2} - 4\right]\phi_0' + \frac{6\phi_0}{R} + \frac{(6\gamma+2)\phi_1}{R} &= \frac{3\kappa}{2}\frac{(3\gamma-2)\mu_0}{R}, \\ -R\phi_1'' + \left[\frac{3\gamma}{2} - 4\right]\phi_1' + \frac{6\phi_1}{R} + \frac{(2\gamma+2/3)\phi_0}{R} &= \frac{1\kappa}{2}\frac{\gamma\mu_0}{R}. \end{aligned} \tag{13}$$

The Machian solution of these equations is $\phi_0 = -1$, $\phi_1 = 1$, as required.

One can proceed on similar lines to discuss simple anisotropic spatially homogeneous cosmological models. In these cases the shear terms dominate the left side of the equations near the singularity, and $\phi_0 = -1$, $\phi_1 = 1$ is a solution of the homogeneous system. Thus the matter serves only to modify the source-free metric and the solutions are non-Machian. The details are given in (Raine 1981).

These examples are straightforward because the spatial homogeneity guarantees that the constraints are algebraic and so can be eliminated explicitly, while the simplicity of the anisotropy provides explicit forms for the metric. It is not clear to what extent one can use this criterion to investigate the Machian character of solutions which are given only implicitly, i.e., for which the explicit form of the metric is unknown, although the above considerations suggest that it should be sufficient to know the behavior near the initial singularity.

6. Isotropic Singularities

We imagine the evolution back in time towards the initial cosmological singularity of a small quantity of cosmological fluid while the metric is rescaled so that the volume of the fluid blob remains constant and the density finite. In this rescaled spacetime, the Ricci tensor remains finite as we approach the singularity. The distortion the blob undergoes may be infinite in some directions and finite in others, or finite in all directions. In the latter case we say the singularity is isotropic (because up to finite corrections the distortion is the same in all directions). In the isotropic case the rescaled Weyl tensor remains finite, so the only singularity is in the Ricci tensor, i.e., in the stress tensor. Such spacetimes are candidates for Machian cosmologies (Tod 1987, 1993).

One might guess that only models with zero Weyl tensor would be Machian: This would uniquely select the FRW models. We have already seen that this is not the case, presumably because the putative contribution from a finite initial Weyl tensor is redshifted away at finite times. There is indeed general agreement (although not perfect identity) between the cosmologies with isotropic singularities and those that satisfy our Mach conditions. In particular, the FRW solutions are Machian and isotropic, models with rotation are nonisotropic (Goode 1987) and non-Machian, and of the Bondi models those with FRW-like (hence isotropic) singularities are Machian and those with Heckmann–Schücking like (nonisotropic) singularities are not.

7. Problems and Speculations

Does this version of the Machian program work? We shall look at some problems: the flatness problem, the alternative Green function problem, the problem of inhomogeneous universes, and the relation to quantum gravity. A problem of a different nature is: How do we know if it is working? If the only prediction of Mach's Principle is that the universe is as it is, how can we test it? This can only be addressed in the context of a theory in which Mach's Principle becomes a derivable result.

7.1. The flatness problem. This is just the standard problem of the age of the universe in a different guise. Mach's Principle as originally conceived was supposed to supply a reason for the existence of the matter in the universe, in the quantity found, through the relation $G\rho\tau^2=1$ (i.e., an $\Omega=1$ universe). In fact, we find that any FRW universe fulfills the conditions, so we need a different explanation for the longevity of the universe. Most of the competing explanations (inflation, anthropy) also require the universe to be reasonably isotropic. This leaves precious little for Mach's Principle to explain! In this case Mach's Principle is simply true by accident.

7.2. Alternative formulations. Insufficient attention has been paid to the possibility that there is not one Mach selection criterion but as many criteria as there are different Green function formulations, and these have not been shown to be equivalent. [Al'tshuler (1982) suggests they are not.] The hint that Mach's Principle may be related to the existence of isotropic singularities ameliorates this somewhat but does not resolve it (because there may be other candidate classes of singularities).

7.3. Inhomogeneous universes. An isotropic singularity is not restricted to have small inhomogeneity; Mach's Principle therefore provides no reason for believing the universe is close to homogeneous. Again, if we have to invoke an entirely different explanation for this, Mach's Principle is a *post hoc* accident except in as much as it might secure only a small anisotropy for a small inhomogeneity.

7.4. Unified physics. One aspect of the problem is that Mach's Principle is a piece of classical physics which is being applied in a manifestly quantum regime. Even worse, from the point of view of some unified (higher-dimensional) theories, there is little fundamental distinction between gravity and other fields; so why single out the free *gravitational* modes to be unexcited? Another version of this dilemma asks why a universe which starts off as a gravity wave which is then transformed into matter should be non-Machian, whereas a matter universe that evolves into gravitational radiation can be Machian. In fact, everyone seems to have their favorite example of a solution that is manifestly Machian but ruled out by the selection criterion of the integral formulation. It is important to recall therefore that the selection rule is an attempt to give a meaning to the requirement that inertial motions be determined by matter. Once we know what that statement means, we can decide whether we like its consequences. I make no claims about what we should or should not like: Only that if we do not now wish to accept the consequences, then we must abandon either Einstein's form of Mach's Principle or general relativity itself.

In fact, the integral formulation itself gives rise to an alternative version of Mach's Principle. As far as the integral formulation is concerned, there is no fundamental distinction between spatially closed and open universes. In both cases the initial conditions in the 3-surface integral represent the influence of source-free gravitational energy in some sense. Thus, if we allow all gravitational energy (source-free or not) to contribute to inertia, then the integral representation shows how the compass of inertia is determined by the distribution of matter and of gravitational energy in general relativity: In this sense all solutions of general relativity are Machian!

If this problem does have an answer, it is not to be found within general relativity itself, but only when Mach's Principle arises as a result of some more all-embracing considerations.

7.5. Matter and Quantum Decoherence. One such more all-embracing consideration is quantum gravity. Let us retrace our steps a bit. To understand why the initial conditions of the universe do not allow large inhomogeneities, one can appeal to the notion of a low-entropy initial state. The gravitational entropy in this state will be measured by something like the magnitude of the Weyl tensor (Penrose 1979). A small Weyl tensor accords with low entropy, and hence with an isotropic singularity and consequently with Mach's Principle. This should emerge as a requirement of a time-asymmetric quantum gravity.

We can go further and ask why such a universe should contain matter. One possible answer (Raine 1993) is that a certain matter content is required for the decoherence of the wave function which signals the emergence of a classical universe from a quantum state. A (very rough) estimate of the required matter content leads to $G\rho\tau^2 \sim 1$. On this view, Mach's Principle would become a necessary condition for the existence of a classical universe (Raine 1986); states which do not satisfy such a condition would 'evolve' to no universe at all.

REFERENCES

Al'tshuler, Boris (1967). "Integral Formulation of Einstein's Equations and a Covariant Formulation of Mach's Principle" *Soviet Physics JETP* 24: 766–769.

Al'tshuler, Boris (1982). *International Journal of Theoretical Physics* 24: 99.

Bondi, Hermann (1947). "Spherically Symmetric Models in General Relativity." *Monthly Notices of the Royal Astronomical Society* 107: 410–425.

Deser, Stanley (1967). "Covariant Decomposition of Symmetric Tensors and Gravitational Cauchy Problems." *Annales de l'Institut Henri Poincaré* 7: 149–188.

Deser, Stanley (1970). "Self-Interaction and Gauge Invariance." *General Relativity and Gravitation* 1: 9–18.

Eardley, Doug; Liang, ed; and Sachs, Ray (1972). "Velocity Dominated Singularities in Irrotational Dust Cosmologies." *Journal of Mathematical Physics* 13: 99–106.

Einstein, Albert (1922). *The Meaning of Relativity*. London: Methuen.

Gilman, Robert C. (1970), "Machian Theory of Inertia and Gravitation." *Physical Review* D2: 1400–1410.

Goode S. W. (1987). "Vorticity and Isotropic Singularities." *General Relativity and Gravitation* 19: 1075–1082.

Lynden-Bell, Donald (1967). "On the Origins of Spacetime and Inertia." *Monthly Notices of the Royal Astronomical Society* 135: 453–467.

Misner, Charles W., Thorne, Kip S., and Wheeler, John Archibald (1973).

Gravitation. San Francisco: Freeman.

Penrose, Roger (1979). In *General Relativity, An Einstein Centenary Volume*. Werner Israel and Stephen W. Hawking, eds. Cambridge: Cambridge University Press.

Raine, Derek J. (1975), "Mach's Principle in General Relativity." *Monthly Notices of the Royal Astronomical Society* 171: 509–528.

Raine, Derek J. (1981), "Mach's Principle and Space–Time Structure." *Reports on Progress in Physics* 44: 1151–1195.

Raine, Derek J. (1986). "The Quasi-Isotropic Universe." in *Gravitation in Astrophysics, Cargese 1986*. James Hartle and Brandon Carter, eds. New York: Plenum.

Raine, Derek J. (1993), "Decoherence of the Cluttered Quantum Vacuum." In *A Survey to Celebrate the 65th Birthday of Dennis Sciama. The Renaissance of General Relativity and Cosmology*. George Ellis, Antonio Lanza, and John Miller, eds. Cambridge: Cambridge University Press.

Raine, Derek J. and Heller, Michael (1981). *The Science of Space–Time*, Tucson: Pachart.

Sciama, Dennis W., Waylen, Peter C., and Gilman, Robert C. (1969), "Generally Covariant Integral Formulation of Einstein's Field Equations." *Physical Review* 187: 1762–1766.

Thirring, Hans (1918), "Über die Wirkung rotierender ferner Massen in der Einsteinschen Gravitationstheorie." *Physikalische Zeitschrift* 19: 33–39.

Tod, K. Paul (1987), "Quasi-Local Mass and Cosmological Singularities." *Classical and Quantum Gravity* 4: 1457–1468.

Tod, K. Paul (1993). "Mach's Principle and Isotropic Singularities." In the volume cited in Raine 1993 above.

Tod, K. Paul (1994), "Mach's Principle Revisited." *General Relativity and Gravitation* 26: 103–111.

Discussion

Goenner: Could you tell us something about vacuum solutions? Are they non-Machian?

Raine: Yes, and in fact you know when to stop deriving a criterion at some level when it definitely eliminates vacuum solutions.

Goenner: So you lose all black holes? You're unhappy with that?

Raine: I lose all black holes in empty universes, and I think all Machians will be happy with that.

Goenner: Well, but not all relativists of course.

Isenberg: You just said all vacuums, but what happened to Minkowski? It still seems that the integral equation is satisfied.

Raine: No, the Minkowski metric arises from the bit that you want to throw away. It doesn't arise from the integral over the matter; it arises from the homogeneous solution.

Isenberg: I thought that what you were writing is more or less a minimal relation between the metric and the matter; and so if matter is gone, then the metric is just basically flat Minkowski.

Lynden-Bell: No! The metric's gone.

Raine: It's zero. You must in some sense have selection criteria. You have your solution first and then ask whether it's Machian. It's not a way of generating solutions to Einstein's equations; it's not using it like that. You take your solution and ask how it's made up, and the Minkowski metric is made up of zero from the matter plus something that's been put in by hand.

Barbour: Just a thought about time reversal. If I'm given a solution of Einstein's equations, it seems to me entirely arbitrary to say one end of it is the beginning of time and one is the end of time. How does this selection criterion work out if the integral is calculated in the other direction? Is that an issue that's been discussed?

Raine: I believe the Machian character will turn out to depend on which way you take time. I think you can see this from the anisotropic Bianchi cosmologies which are Robertson–Walker at late times.

Barbour: That's very interesting and mysterious, it seems to me, because if Einstein's dynamics is time symmetric it seems strange that the Machianity is time-direction dependent.

Raine: Selection criteria do often select reasonable initial conditions which would not be reasonable final conditions.

Goenner: I'm not yet convinced that your selection rule is very powerful. Nobody doubts that Friedman solutions are Machian, but you pushed quite a lot under the rug when you explained to us this integral representation of the metric. The Green's function depends on the metric as well, so the representation really is an integral equation, and thus a bit more complicated than you wanted to make it. If you wish to get exact solutions from your formulation, you have to work very hard. Do you expect to find these Green's functions?

Raine: I do not believe you can use this as a way of generating solutions of Einstein's equations, but given solutions of Einstein's equations and assuming you want to give expression to the fact that the metric depends on the matter and not on free boundary data, this provides a legitimate sense that you can make of that.

Goenner: Without numerical relativity you have about 1,500 solutions to Einstein's equations or more.

Raine: Some of them will turn out to be Machian and some of them will turn out to be non-Machian. That's all you do, and that's all I can do.

This provides a physically justifiable way of imposing a Machian initial condition in GR.

Brill: Just to understand your scheme: Is it true that it depends very much on knowing exactly whether the source of gravitational fields is matter or black holes? If I have a star that is about to collapse into a black hole it might be Machian just before but not afterwards; how does that go?

Raine: Essentially, it depends upon how you set the problem up, not what happens later, so this is a condition on the metric integrated over the past of the point, so it essentially depends on what's happening initially. So if you put a lot of black holes into the universe, it will presumably turn out to be non-Machian. If you put a lot of stars that are about to collapse in at the beginning of the universe, then you can do that in such a way as to make the solution Machian. This is the problem that I wouldn't hide; it's the legitimacy problem as to why you regard the gravitational modes as so much more restricted than the modes of all the other matter fields, which in a unified theory will have the same status.

King: What happens with model universes that don't have singularities? What happens there when you try to integrate back? Do you need to put the start of the integration back infinitely far to the past?

Raine: Yes.

King: And will that converge?

Raine: Oh, it will depend on the solution. You have to get the metric. You know you've got a solution, so the sum of the volume integral and the surface integral will give you the metric. The point is whether the volume integral over the matter alone will give you the metric or not.

King: But does that mean somewhere in the past all the gravitational waves had to have disappeared?

Raine: That's right, yes, exactly yes.

Bondi: I have a comment about the legitimacy question. I mean, to me what Bishop Berkeley and Mach worried about was that there was no independently assessable source of inertia. If the source of inertia is measurable, I don't care two hoots whether it's black holes or solid bodies or gravitational waves. They are independently measurable, in principle, and I think that answers the legitimacy question.

Raine: But why do we worry then that Minkowski space is non-Machian?

Bondi: Because there's no legitimate rule for selecting a particular Minkowski space rather than another. We cannot bring it back to something observable that anchors it.

Mach's Principle and Local Causal Structure

Ulrich Bleyer and Dierck-E. Liebscher

1. Introduction

The question of a possible influence of global structure of the universe on local physical laws is one of the most fundamental problems of natural science. The topicality of this question comes from the geometrization of all interactions by modern gauge field theories following Einstein's theory of gravitation as well as the consideration of energy regions in elementary particle physics which could be realized only in early stages of the evolution of the universe (Bleyer and Liebscher 1988). Corresponding to Poincaré's epistemological sum, stating that the physical content of a theory is defined by geometry plus dynamics, we might handle the interconnection between physics at small and large distances in two different ways. In unified field theories, dynamics is based on a geometry of the space-time manifold in which the global existence of a causal structure is assumed *a priori* and the local laws determine everything else. In the opposite case, the connection between local motion and global structure will be given by the Mach-Einstein postulate of the induction of inertial properties of matter mediated by the joint gravitational influence of cosmic masses. We argue for a realization by pregeometric models in which the metrical structure, and therefore the causal properties, are induced dynamically due to dynamical principles (Treder 1974a).

In the framework of mechanics, a consistent realization of the Mach-Einstein principle was given in an analytical description of inertia-free mechanics by Schrödinger (1925) with the help of the Weber potential, by Treder with the help of the Riemann potential (Treder 1972, 1974), and by Barbour in his forceless or relational mechanics (Barbour 1975; Barbour and Bertotti 1977). In these constructions, inertia is replaced by interaction in that the kinetic terms are replaced by (velocity-dependent) potentials. The integral of these interactions over the

Einstein Studies, vol. 6: Mach's Principle: From Newton's Bucket to Quantum Gravity, pp. 293–307 © 1995 *Birkhäuser Boston, Inc. Printed in the United States.*

surrounding universe leads for small subsystems to kinetic terms, which can be interpreted as induced inertia. The Galilei invariance arises locally as the remaining part of a larger 'telescopic' (Planck 1913; Neumann 1870) symmetry of the total universe, the latter being broken in the restriction to a subsystem by the interaction with the universe in its actual state. That is, the laws ruling the universe are assumed to be more symmetric than the locally observed ones, but the state of the universe in detail may be much less symmetric and produces local symmetry only by the averaging effect of integrals (see also the contributions in this volume of Ehlers and Nordtvedt). Measurable effects of such an induction scheme should appear as small deviations from locally Galilei-invariant laws due to comparatively short-range inhomogeneities and the expansion of the surrounding universe.

However, in the conventional approach to relational mechanics, the definition of rotation requires a definition of simultaneity, and the unbroken part of the telescopic group can never be the Lorentz group. At most, we may expect to get the Galilei group. Of course, one may try a bimetric theory and construct Minkowski-signature metrics in an effective space on background Galilei symmetry. This last, together with the preexisting definition of simultaneity in the background, would be hidden in the equations of motion of nongravitational fields (Liebscher 1981). However, the *a priori* simultaneity shows up in the gravitational interaction, in particular in the post-Newtonian approximation. For any bimetric construction of the kind mentioned, the absolute simultaneity in the background shows up in the coefficient (Kasper and Liebscher 1974)

$$\alpha_2 = \left(\frac{v_l}{v_g}\right)^2 - 1, \tag{1}$$

where v_l denotes the velocity of light propagation and v_g the velocity of gravitation propagation. In the case of relational mechanics, the problem becomes that of getting Poincaré or Lorentz invariance as unbroken residue of the telescopic symmetry.

Full Mach–Einstein constructions should be consistent with special relativity theory (SRT). If the classical symmetry breakdown (as yet constructed only in mechanics) is the essence of Mach–Einstein constructions, we have now to construct an induction formalism for local Lorentz invariance (LLI) (Liebscher and Yourgrau 1979). If this succeeds, the LLI, i.e., the Minkowski metric of tangent spaces, which currently appears as an absolute element of GRT, will be shown to be dynamically induced (Ehlers, *loc. cit.*). We believe this to be necessary, because if one accepts the existence of local Lorentz frames *a priori*, the influence of the cosmos is washed out (see the contribution of Bondi in this volume). Generalizations of the mechanical induction schemes for inertia have to end up with LLI for a perfectly symmetric universe. Consequently, Mach–Einstein effects will arise as perturbations of LLI due to the potentials of short-range cosmological inhomogeneities. By analogy with the induction schemes for mechanics, deviations from Lorentz invariance have to be expected in the kinetic terms, i.e., in the differential operators (Liebscher 1985).

In the first half of this paper, we consider a framework in which breakdown of a large telescopic group to an approximate Lorentz group might occur. In the second part (which was mentioned only briefly at the workshop), we consider a model that illustrates the possible experimental effects of the reduction to only approximate Lorentz invariance.

2. Mach's Principle and Local Causal Structure

As we have seen, a relativistic Mach–Einstein principle should be realized in a pregeometric theory that starts without the assumption of LLI. Based on other considerations, this has been stated implicitly by Heller (1975a). He started with Mach's Principle in the following formulation: *The local inertial frames are entirely determined by the distribution and motion of all matter present in the Universe* (Bondi 1960; McCrea 1971). Under Heller's assumption, the local inertial compass and the local light compass must coincide due to the dynamical properties of matter (Pirani 1956; Goenner 1970; Reinhardt 1973). Consequently, a principle like the one stated cannot be realized in a physical theory which fulfils the following assumptions normally used to introduce General Relativity:

1. Spacetime is a four-dimensional, connected, orientable, paracompact and Hausdorff C^r $(r \geq 1)$ manifold without boundary.
2. An affine connection is given together with:
3. A Lorentz metric related by Ricci's Lemma to the connection: $g_{kl;m}=0$.

In such a spacetime, it is possible to introduce a continuous system of linear frames induced by the tangent space at every point of the manifold. Local inertial frames are linked to these linear frames in a manner not referring to any matter fields. Therefore, local Lorentz–Minkowski structure exists independently of dynamical properties of matter. This contradicts Mach's Principle in the formulation given above.

It is also possible to introduce a cosmological time referring only to topology and causal structure of the spacetime manifold (Heller 1975).

This is another argument for demanding the realization of Mach's Principle in a theory that starts without local Lorentz structure, i.e., with pregeometric models in which the local causal structure has to be induced dynamically.

One might consider at this point a conformal theory which is reduced to Lorentz invariance by some (presumably scalar) mass-generating field. There are two reasons why we do not want to follow this route. First, we believe it to be more interesting to have a larger extension of Lorentz invariance, and to ask for a dynamical explanation of causality. Second, there are a lot of theories that begin with conformal invariance which are purely local, i.e., in which the local (quantum) vacuum instead of the universe mediates the symmetry breakdown. We would consider this case to be an anti-Machian option. Hence, a purely affine theory should provide the simplest nontrivial scheme for our purpose.

3. Affine Symmetry and Its Breaking

Equations of motion or field equations can be formulated only on differentiable manifolds or locally trivial fiber bundles, on which an appropriately introduced topology allows free choice of reference frames. In mechanics, we start from a manifold of events. A system of axioms may ensure that the topology permit a C^2 differential structure such that any trajectory of a particle is a one-dimensional C^2 manifold denoted as the worldline of the particle. The physical equations restrict the configurations of the manifold of events to the physically possible states. Further axioms are needed in order to define the invariance properties of the physical laws and, as a consequence, the geometry of the manifold of events. One possible axiom is connected with the law of inertia: At every point of the spacetime manifold M there exists a Riemannian coordinate system $\{x^i\}$, so that we have for every worldline of a non-interacting particle with an appropriately chosen parameter s the equation

$$\frac{dx^i}{ds}\frac{d^2x^j}{ds^2} - \frac{d^2x^i}{ds^2}\frac{dx^j}{ds} = 0. \tag{2}$$

This expression is invariant with respect to the group of affine transformations. If we restrict ourself to four-dimensional spacetime, we find the physical geometry given by the Klein geometry $(M, A(4))$ (Treder and Bleyer 1988).

In axiomatic foundations of mechanics, the affine group is restricted *ad hoc* or with the help of axioms to special subgroups, the Galilei group or the Poincaré group. In a Mach–Einstein program, this reduction should be a symmetry breakdown by the actual state of the universe. This is the reason to try first an affine invariance as telescopic symmetry in our approach, and to expect the reduction to local Poincaré invariance by the state of the universe.

We now consider some general aspects of fields in affine space.

The existence of a unique pseudo-Riemannian metric, together with general covariance, implies LLI. Therefore, the dynamical induction of LLI means dynamical induction of the existence of such a metric.

For a procedure inducing the metric of macroscopic motion as a consequence of the dynamical equations of auxiliary fields, the metric tensor has to be eliminated from the usual microscopic Lagrangian. This was tried by Terazawa *et al.* in their approach to pregeometry (Akama and Terazawa 1983). But in order to get scalar-density Lagrangians out of vectors or spinors, one needs some tensor to form scalars, and they use the Levi-Civita symbol. In the scalar case, the action reads

$$S = \int d^4x \Big(\det[\sum_A \Phi^A{}_{,k}\Phi^A{}_{,l}]\Big)^{1/2} F[\Phi], \tag{3}$$

with some scalar function $F[\Phi]$. In such a way, the Lorentz group of the principal bundle is apparently replaced by the centroaffine group if we consider a chosen field on the background of the others. However, to construct the pregeometric Lagrangian, a nondynamical Lorentz metric is already used, and from this point of view the proposed pregeometry is just a special kind of a bimetric theory. The sum over the scalar fields hides a pseudo-Euclidean metric in the space of Φ (Liebscher 1985). This construction shows that the existence of a Minkowski metric in the tangent space has to be *a posteriori* in the proposed Mach–Einstein induction scheme too. The only measure that the Lagrangian of the multicomponent fields Φ^A can have is the metric of the affine group, i.e., the Levi-Civita symbol. A Lagrangian that avoids the *a priori* existence of a pseudo-Euclidean metric has to consist of terms that use only this symbol apart from the fields Φ^A and their derivatives.

Another important point has to be made in connection with the gauge field theory based on the affine group given by Ne'eman and Sijacki (1988). Here, too, a metric is hidden from the very beginning in the assumption of the existence of a 'flat gauge' as a Lorentz-subgroup invariant. The corresponding matter coupling forms the symmetry breakdown to the Poincaré symmetry beforehand. That is, the reduction to LLI is formed by the assumed coupling to matter and not by the actual state of the matter distribution. In addition, the procedure of getting at LLI by a local symmetry breakdown produced by the state of the local vacuum is an entirely anti-Machian procedure. From this point of view, the question of Mach's program is whether it is the local vacuum or the state of the universe which is responsible for LLI.

What will field theory without metric look like? If we expect a wave equation for some multicomponent field quantity Φ, the effective coefficients g^{kl} in the wave operator $\Box = g^{kl}(\partial^2/\partial x^k \partial x^l)$ have to be constructed from these field quantities themselves. Therefore, in manifolds without *a priori* metric tensor field, the effective metric has to be an integral, i.e., a nonlocal quantity. This is the technical aspect of the epistemological expectation that inertial (in relativity: metric, or causal) properties are to be determined by the global distribution of the fields in the manifold. The possibility of constructing an effective wave equation from an affinely invariant action lies in higher-order spacetime integrals.

Second-order field equations for the multicomponent field Φ^A take in general the form

$$C_{AB}{}^{nl}\,\Phi^B{}_{,nl} = \text{first derivatives and source terms.} \tag{4}$$

If for some field configuration Φ^A the quantity $C_{AB}{}^{nl} = \partial^2 L/\partial\Phi^A{}_{,n}\partial\Phi^A{}_{,l}$ decomposes into a product $a_{AB}g^{nl}$, we get the factor g^{nl} as the effective (contravariant) metric induced by the field itself. The factor a_{AB} might mix the field components in the chosen representation.

Despite the fact that there is no constructive example of such an induction scheme, the formal construction shows in which direction deviations from the usual picture of relativistic field theories are to be expected. An *a posteriori* recovery of wave equations implies that the wave equation is only approximately separable for the different components of the field, and the finiteness of the potentials of the matter distribution in the universe can be expected to give rise to small deviations from the usual wave operator. These deviations should be at least of order 10^{-40} (Dirac's number), at most of order 10^{-6} (Newtonian potential of the Galaxy).

4. Matter Field Equations for Generalized Causal Structure

Before considering experimental consequences, we want to note the relation of premetric constructions to the axiomatic approach to spacetime structure. The particle concept of quantum field theory suggests the derivation of the spacetime structure from the basic exigencies of field theory (Liebscher 1985a). The procedure is to discover and to describe the geometrical structure of spacetime by means of the behavior of appropriately selected physical systems (called primitive objects), in particular physical effects taken as basic experiences (Lämmerzahl 1990). Extending the axiomatics of Ehlers, Pirani, and Schild (1972) based on light rays and test particles to the concept of free matter waves as primitive elements, Audretsch and Lämmerzahl (1990) gave a complete axiomatics leading to Riemann–Cartan spacetime. The basic experiences refer essentially to interference experiments. Subsequently, Audretsch and Lämmerzahl (1991) improved this approach by considering plane matter waves as a particular limiting case of wave mechanics defined by a general field equation in a manifold with a conformal structure. As field equation for the vector-valued complex field, the most general linear system of partial differential equations of arbitrary order was considered. This procedure was physically justified for the description of matter in a further paper (Audretsch and Lämmerzahl 1991a).

Constructive axiomatics do not include Lorentz invariance from the beginning. But they are usually restricted to Lorentz-invariant structures. In addition to fundamental assumptions such as a deterministic and local evolution of fields and the validity of a superposition principle, the demand of LLI is one of the assumptions in constructive axiomatics (Audretsch and Lämmerzahl 1991). Not demanding Lorentz invariance in advance raises the necessity of independent tests leading to upper limits for possible deviations from LLI. The general interest in such tests meets our interest in testing the effects of a scheme that we try to design.

Relational mechanics produces the Galilei group as unbroken residue of a telescopic group. Local inhomogeneities in the universe lead to effects in the kinetic part of the theory (for instance, to anisotropic mass). In analogy, we have to expect effects in the kinetic part of a field theory, which exist in our local spacetime. These kinetic effects show that the LLI is only approximate, like the approximate Galilei invariance in relational mechanics. Only effects in the kinetic terms, i.e., the leading degree in the field equation, can be characteristic for an only approximate LLI. Additions to the lower-degree terms cannot be expected to differ qualitatively from other ordinary fields coupled to the field in question. Therefore, we want to model just the effects in the kinetic terms in order to see what might be expected to be testable. It turns out that the central point is a kind of spin-dependent propagation of signals. Different components of a multicomponent physical field follow different propagation cones (Bleyer and Liebscher 1988; Treder and Bleyer 1988; Bleyer 1991). The mutual configuration of these cones, such as a common time axis and spatial isotropy, can be used to define special reference frames.

We consider first an arbitrary second-order Euler–Lagrange equation of a multicomponent field. The highest (second) derivatives with respect to the field functions are

$$C_{AB}{}^{ik} = \frac{\partial^2 \ldots}{\partial \Phi^A{}_{,i} \partial \Phi^B{}_{,k}}. \tag{5}$$

Using an ansatz for a shock wave front on the surface $z=0$ given by

$$\Phi^A{}_{z>0} = \Phi^A{}_{z<0} + \phi^A z^2, \tag{6}$$

we find for the jump function the equation [for a mathematically more explicit treatment see (Audretsch, Bleyer, and Lämmerzahl 1993)]

$$C_{AB}{}^{ik} z_{,i} z_{,k} \phi^B = 0. \tag{7}$$

The existence condition for jumps reads

$$\det(C_{AB}{}^{ik} z_{,i} z_{,k}) = 0. \tag{8}$$

In the case of N components of the field Φ^A, this is an equation of order $2N$ in $z_{,i}$.

In SRT, shock fronts are possible only on the light cone. The existence condition for jumps degenerates to

$$8(g^{ik} z_{,i} z_{,k})^N = 0, \tag{9}$$

where N denotes the number of components of the Φ-field. Therefore, Lorentz invariance is ensured by the factorization of the coefficients of the field equations

$$C_{AB}{}^{ik} = a_{AB} g^{ik}. \tag{10}$$

As a consequence, all field components fulfill the wave equation separately, and all field components follow a common light cone.

If there are deviations from the factorization condition (10), then

$$C_{AB}{}^{ik} = a_{AB} g^{ik} + \epsilon_{AB}{}^{ik}. \tag{11}$$

In this case, the different field components no longer satisfy the wave equation separately; they are mixed. To first order in the perturbation $\epsilon_{AB}{}^{ik}$, we find in an appropriate field representation

$$(g^{ik} z_{,i} z_{,k})^{N-1} (g^{lm} + a^{AB} \epsilon_{AB}{}^{lm}) z_{,l} z_{,m} = 0. \tag{12}$$

So we have the product of two different 2-surfaces, the first for $N-1$ field components and the second for the last one. This means that in an appropriately chosen field representation, one field component follows a propagation cone different from the common light cone. In the general case, one more field component leaves the common light cone in each higher approximation. The light cone is replaced by a surface of order $2N$, which may be constructed from N different propagation cones. In this way, we find a general field theoretical model for a component-dependent propagation behavior. If we can connect the different field components with spin projections or polarizations, we can speak of a spin or polarization dependent propagation. Some analogy to this situation is known from Maxwell's theory of birefringent media. This component-dependent propagation for one multicomponent field is a generalization of non-LLI model theories in which different fields are assumed to follow different propagation cones (see the contribution of Will in this volume).

One can show that the Dirac equation

$$i\gamma^{k}{}_{A}{}^{B}\partial_k\Psi_B = M_A{}^B\Psi_B \tag{13}$$

can be generalized like the wave equation to provide an analogous model theory for non-LLI. This will be a generalization of the Dirac matrices, which will no longer satisfy the usual anticommutation relations, but

$$(C^{ik})^B_A = \tfrac{1}{2}\left((\gamma^i)^C_A(\gamma^k)^B_C + (\gamma^k)^C_A(\gamma^i)^B_C\right). \tag{14}$$

If we restrict the perturbations by physically meaningful conditions like spatial isotropy in the preferred frame and helicity conservation (Audretsch, Bleyer, and Lämmerzahl 1993), we can write

$$\gamma^k = \begin{cases} \tilde{\gamma}^0 + \epsilon_1\tilde{\gamma}^5\tilde{\gamma}^0 \\ \tilde{\gamma}^\mu + \epsilon_2\tilde{\gamma}^5\tilde{\gamma}^\mu \end{cases} \tag{15}$$

In this case, the dispersion relations read

$$E = \frac{1\pm\epsilon_2}{1\pm\epsilon_1}\,|\bar{\mathrm{p}}|. \tag{16}$$

The effective parameter describing deviations from LLI is given by $\epsilon = \epsilon_1 - \epsilon_2$. The choice of perturbations of the form (15) avoids at least lower-order anisotropy problems.

5. Testing Possible Machian Effects

The first effect of the explained generalization of the Dirac equation (GDE) should be an additional hyperfine splitting of the energy levels of the hydrogen atom given by Bleyer (1993):

$$E_n = m\left[1 + \frac{\alpha^2}{(n+s)^2}\right]^{-1/2}, \tag{17}$$

with $\epsilon_2 = 0$

$$s=\left[k^2-\alpha^2(1\pm\epsilon_1)\right]^{1/2}. \tag{18}$$

On the other hand, these effects can be made arbitrarily small by limitations on the perturbation parameters ϵ.

Experiments give upper limits on the numerical values of the effective perturbation $\epsilon_1-\epsilon_2$. In the case of the hydrogen atom, we find for the fine structure splitting

$$E_{n,k}=m\left[1+\frac{\alpha^2}{(n+k)^2}\right]^{-1/2}$$

$$-\frac{m}{2}\left[1+\frac{\alpha^2}{(n+k)^2}\right]^{-3/2}\frac{\alpha^4(1\pm\epsilon_1)^2}{k(n+k)^3}+\ldots, \tag{19}$$

and we get the bound

$$\epsilon_1<10^{-8}. \tag{20}$$

The change to μ-mesic atoms does not give stronger limitations. This shows that the hydrogen atom is not such a sensitive indicator of deviations from the Lorentz invariant Dirac theory as is widely believed.

These results show the GDE to be meaningful in order to look for further experimental consequences that give us the possible order of magnitude of the perturbations. For this problem, it is important to notice that the GDE can be connected to other model theories for the breaking of Lorentz invariance (Nielsen and Picek 1983; Froggatt and Nielsen 1991).

Up to now, the most restrictive experimental limit on the perturbation parameters in the GDE is given by the so-called Phillips experiment (Phillips and Woolum 1969; see also Froggatt and Nielsen 1991). This experiment determines the daily variation of the torque acting on a ferromagnet hanging on a string. In this way, one can examine the existence of a preferred reference frame, in which the velocity of the earth $\boldsymbol{v}$ is connected with the spin $\mathbf{S}$ of the electrons via a coupling term (we use $c=\hbar=1$)

$$H_{\text{int}}=bm_e\boldsymbol{v}\cdot\mathbf{S}. \tag{21}$$

The experiment limited the expected splitting of the two different spin states

$$\Delta E=H_{\text{int}}(S=1/2)-H_{\text{int}}(S=-1/2)=bm_e v. \tag{22}$$

The same coupling term occurs for the GDE. This can be seen in the Pauli approximation up to the first order in the perturbation parameters. We find with $\Psi=\binom{\varphi}{\chi}$

$$\mathrm{i}\frac{\partial\varphi}{\partial t} = \frac{p^2}{2m}\varphi + \epsilon \mathbf{S}\cdot\mathbf{p}\varphi. \tag{23}$$

If we substitute in (22) the values for the electron mass and the velocity of the earth on its orbit around the sun ($v=30$ km/s), we find

$$cm_e v = 10^{-17}\ \mathrm{J}. \tag{24}$$

The experimental result gives

$$\Delta E \leq 7\cdot 10^{-35}\ \mathrm{J}, \tag{25}$$

and, using the above result, we find

$$|\epsilon| < 10^{-18}. \tag{26}$$

This is the upper limit for the ϵ perturbations in the order v/c. But the disadvantage of this experiment is that we have to put in a velocity of the laboratory frame with respect to an assumed global reference frame.

This will be not the case for atomic or neutron interferometers, where the only assumption will be that we have nonrelativistic velocities and can neglect terms of the order v^2/c^2. We use (23), which can be also written as

$$\mathrm{i}\frac{\partial\varphi}{\partial t} = H\varphi, \tag{27}$$

with

$$H = \frac{p^2}{2m} + H_{\mathrm{int}},\ H_{\mathrm{int}} = \epsilon\mathbf{S}\cdot\mathbf{p}. \tag{28}$$

For the interferometer experiment, the incoming beam of particles with definite helicity state will be split into two beams, which after some traveling along different paths will be recombined. In one of these two paths, a spin flip will be performed along a definite distance l corresponding to a time of flight Δt. This leads with H_{int} from (28) to a phase shift (Audretsch, Bleyer, and Lämmerzahl 1993),

$$\Delta\phi = \oint p_0\,\mathrm{d}t = \oint H_{\mathrm{int}}\,\mathrm{d}t = 2\epsilon p\Delta t = 2\epsilon\frac{l}{\lambda_c}, \tag{29}$$

with the Compton wavelength $\lambda_c := \hbar/mc$ of the particles used.

For the neutron interferometry, we find with $\lambda_c = 10^{-15}$ m and $l = 10^{-1}$ m

$$\delta\Phi \approx 10^{14}\epsilon. \tag{30}$$

Together with the accuracy 10^{-3} of the neutron interferometer, this gives us for the perturbations a bound of the order of magnitude

$$|\epsilon| < 10^{-17}. \tag{31}$$

For an atomic interferometer, this value can be improved by at least two orders of magnitude. We find for the helium atom $\lambda=0.2\cdot 10^{-15}$ m, and the measuring device has an effective length of $l=1.3$ m. So finally we have finally the most restrictive limitation expected from future measurements (Audretsch, Bleyer, and Lämmerzahl 1993):

$$|\epsilon| < 10^{-19}. \tag{32}$$

6. Conclusions

The Lorentz group defines the causal structure of the Minkowski space-time, the light cone, the mass shell, and so on. Every theory producing the Lorentz group has to explain the existence of a light cone or the Lorentz–Minkowski causality. This is also the demand on theories realizing Mach's Principle constructively. For such theories, Mach–Einstein effects appear as a perturbation of LLI. Disturbance of Lorentz invariance means that this symmetry is broken and the field equations are no longer Lorentz-invariant, but their deviation from Lorentz invariant equations is small in a reference system chosen appropriately. This can be realized in model theories like the GDE.

A model theory based on a generalization of the Dirac equation represents a simple violation of local Lorentz invariance (LLI). This violation of LLI is related to the fact that the generalized Dirac matrices do not fulfill any Clifford algebra. Using physically meaningful requirements like conservation of helicity and isotropy of the null cones, we reduce the problem to the general violation of LLI in a minimal non-trivial model. In the nonrelativistic limit, the result is a special spin–momentum coupling leading to a splitting of the mass shells and consequently of the null cones.

This spin–momentum coupling can be most suitably tested with atomic beam interferometry using spin flip devices. Our model would lead to a phase shift proportional to the parameter ϵ, which characterizes the splitting of the null cone. Assuming a negative outcome of atomic beam interference experiments and taking into consideration the accuracy of the respective apparatus, we obtain upper limits for the parameter characterizing the violation of LLI. The great and increasing accuracy of atomic beam interferometers makes it very desirable to perform such experiments, because this would lead to improved limitations of LLI violations: $|\epsilon| < 10^{-19}$.

Two points have to be stated again. The minor one is the remark that the null result of the experiments discussed may only prove that the fields constitutive for the full system in question are not the first ones to leave the common light cone, Eq. (12). To find the components that split off in the first or second place might be a difficult task. The second remark concerns the far more difficult question of the status of the local symmetry breakdown, i.e., the question whether the actual state of the universe or the actual state of the quantum vacuum is responsible for the symmetry breakdown to LLI. In our understanding, only the first variant should be labeled Machian.

REFERENCES

Akama, Kei-Ichi and Terazawa, Hideki (1983). "Pregeometric Origin of the Big Bang." *General Relativity and Gravitation* 15: 201–207.

Audretsch, Jürgen and Lämmerzahl, Claus (1990). "Establishing Spacetime Geometry by Means of Light Rays and Particle Interferometry." *Wissenschaftliche Zeitschrift der Universität Jena, Naturwissenschaftliche Reihe* Heft 1: 9–12.

Audretsch, Jürgen and Lämmerzahl, Claus (1991). "Reasons for a Physical Field to Obey Partial Differential Equations." *Journal of Mathematical Physics* 32: 1354–1359.

Audretsch, Jürgen and Lämmerzahl, Claus (1991a). "Establishing the Riemannian Structure by Means of Light Rays and Free Matter Waves." *Journal of Mathematical Physics* 32: 2099–2105.

Audretsch, Jürgen; Bleyer, Ulrich; and Lämmerzahl, Claus (1993). "Testing Lorentz Invariance with Atomic Beam Interferometry." *Physical Review A* 47: 4632–4640.

Barbour, Julian B. (1975). "Forceless Machian Dynamics." *Nuovo Cimento* 26B: 16–22.

Barbour, Julian B. and Bertotti, Bruno (1977). "Gravity and Inertia in a Machian Framework." *Nuovo Cimento* 38B: 1–27.

Bleyer, Ulrich (1991). "On a Lorentz Noninvariant Generalization of the Dirac Equation." *Annalen der Physik (Leipzig)* 48: 489–496.

Bleyer, Ulrich (1993). "Energy Levels of the Hydrogen Atom Due to a Generalized Dirac Equation." *Foundations of Physics* 23: 1025–1048.

Bleyer, Ulrich and Liebscher, Dierck-E. (1988). "Quantum Mechanical Consequences of a Pregeometry." In *Proceedings of the IVth Seminar on Quantum Gravity.* Moissey A. Markov, Valeri M. Frolov, Valeri A. Berezin, eds. Singapore: World Scientific, 1988, pp. 282–288.

Bondi, Hermann (1960). *Cosmology.* Cambridge: Cambridge University Press.

Ehlers, Jürgen, Pirani, Felix A.E., and Schild, Alfred (1972). "The Geometry of Free Fall and Light Propagation." In *General Relativity, Papers in Honour of J.L. Synge.* L. O'Raifeartaigh, ed. Oxford: Clarendon Press.

Froggatt, C.D. and Nielsen, H.P. (1991). *Origin of Symmetries.* Singapore: World Scientific.

Goenner, Hubert (1970). "Mach's Principle and Einstein's Theory of Gravitation." *Boston Studies in the Philosophy of Science* 4: 200–215.

Heller, Michael (1975). "Some Remarks About the Concept of Cosmic Time." *Acta Cosmologica* 3: 41–44.

Heller, Michael (1975a). "Space–Time Manifolds and the Inertial Frames." *Acta Cosmologica* 3: 45–48.

Kasper, Uwe and Liebscher, Dierck-E. (1974). "On the Post-Newtonian Approximation of Theories of Gravity." *Astronomische Nachrichten* 295: 11–17.

Lämmerzahl, Claus (1990). "The Geometry of Matter Fields." In *Quantum Mechanics in Curved Spacetime*. Jürgen Audretsch and Venzo de Sabbata, eds. New York: Plenum Press.

Liebscher, Dierck–E. (1981). "Inertia-Free Mechanics and the Bi-Metric Procedure." *Astronomische Nachrichten* 302: 127–132.

Liebscher, Dierck–E. (1985). "Classical and Quantum Pregeometry." In *Proceedings of the Third Seminar on Quantum Gravity*. Moissey Markov A., Valeri M. Frolov, Valeri A. Berezin, eds. Singapore: World Scientific, pp. 223–227.

Liebscher, Dierck–E. (1985a). "The Geometry of the Dirac Equation." *Annalen der Physik (Leipzig)* 42: 35–40.

Liebscher, Dierck–E. and Yourgrau, Wolfgang (1979). "Classical Breakdown of Symmetry and the Induction of Inertia." *Annalen der Physik (Leipzig)* 36: 20–24.

McCrea, William H. (1971). "Doubts about Mach's Principle." *Nature* 230: 95–97.

Ne'eman, Yuval and Sijacki, Djordje (1988). "Gravity from Symmetry Breakdown of a Gauge Affine Theory." *Physics Letters* B200: 489–494.

Neumann, Carl (1870). *Ueber die Principien der Galilei–Newtonschen Theorie.* Leipzig: Teubner.

Nielsen, H.P. and Picek, I. (1983). "Lorentz Noninvariance." *Nuclear Physics* B211: 269–296.

Phillips, P.R. and Woolum, D. (1969). "A Test of Lorentz Invariance Using a Torsion Pendulum." *Nuovo Cimento* 64B: 28–33.

Pirani, Felix A. E. (1956). "On the Definition of Inertial Systems in General Relativity." *Helvetica Physica Acta, Suppl.*, IV: 198–203.

Planck, Max (1913). *Das Prinzip der Erhaltung der Energie.* Leipzig: Teubner.

Reinhardt, Michael (1973). "Mach's Principle - A Critical Review." *Zeitschrift für Naturforschung* 28a: 529–537.

Schrödinger, Erwin (1925). "Die Erfüllbarkeit der Relativitätsforderungen der klassischen Mechanik." *Annalen der Physik (Leipzig)* 77: 325–336. (Translation in this volume, p. 147.)

Treder, Hans–J. (1972). *Die Relativität der Trägheit.* Berlin: Akademie Verlag.

Treder, Hans–J. (1974). *Über die Prinzipien der Dynamik von Einstein, Hertz, Mach und Poincaré.* Berlin: Akademie Verlag.

Treder, Hans-J. (1974a). *Philosophische Probleme des Physikalischen Raumes.* Berlin: Akademie Verlag.

Treder, Hans-J. and Bleyer, Ulrich (1988). "Quantum Gravodynamics and Pregeometry." In *Proceedings of the Conference on Fundamental Problems of Quantum Theory.* L. Kostro, A. Posiewnik, J. Pykacz, and M. Zukowski, eds. Singapore: World Scientific, pp. 832–836.

Discussion

Bondi: I want to explore your concept of causality *a posteriori.* Do you mean by that a Cauchy surface with the right conditions prescribes later developments perfectly with no influence traveling outside the light cone? Is that a definition of your theory?

Liebscher: At the point when I was trying to do this in a mechanical model, yes. The action will be just summing up terms which describe interaction between total worldlines. There is no causality *a priori.* It might be, and that, of course, is the hope, that for a local subsystem in a well-behaved universe you get *a posteriori* just the terms in an order which you have to interpret as a wave equation and as a hyperbolic problem. But that is only expectation. There is no theoretical model so far that really works.

Bondi: If it leads to that, you will say causality emerges.

Liebscher: It has to; it's intended to be. What I'm looking for is a model which breaks down to the Lorentz group, of course.

Goenner: Could I make a comment which may be helpful? He doesn't have a causal structure *a priori.* He just has a projective structure. He wants to derive the causal structure from deeper.

Liebscher: Yes, I have often been asked why not to start with a conformally invariant theory (instead of the more difficult affine or even projective constructions). Conformally invariant theories exist in different forms, but they all suppose (*unterstellen*) the existence of a causal structure *a priori.* The really interesting point is to think about a light-cone structure *a posteriori.* If such a scheme will work, it will give the reason for the existence of a causal structure.

5. Frame Dragging

Introduction

In the realm of the elusive Mach's Principle, the phenomenon of dragging of inertial frames due to accelerated masses seems to be the least elusive effect, and there is quite general agreement on its realization in nature. Already in Mach's *Mechanik* (1883) there is a relatively clear conjecture towards this phenomenon when he asks "how the experiment [Newton's bucket] would turn out if the sides of the vessel increased in thickness and mass till they were ultimately several leagues thick," and some of his contemporaries deduced therefrom clear suggestions for interesting though unsuccessful dragging experiments (Friedlaender, p. 309, Föppl, p. 312). The dragging phenomenon and other Machian ideas were then a strong stimulus and heuristic principle for Einstein in his search for a generalized theory of gravitation, and indeed dragging showed up in all his predecessor-theories of general relativity, beginning in 1912. Curiously, Einstein left it to Thirring and Lense to calculate the concrete dragging result in the definitive form of general relativity for a simple model system. The article by Pfister (p. 315) tries to follow the path through most other model systems in which dragging was considered within general relativity. Since dragging is an essentially global effect (Brill, p. 332), one has to address the question whether it can be reduced to causal retarded interactions or whether it has to be seen as an instantaneous phenomenon. Worked out examples (for example, Lindblom and Brill) favor the second alternative. The fact that several different plausible but incompatible definitions of dragging appear in the literature (see, for example, Frauendiener) is also connected with the globality. In recent times, dragging could also be studied for relativistic angular velocities of specific matter models (Meinel and Kleinwächter p. 339), and there are indications that dragging might lead to observable effects in active galactic nuclei (Karas and Lanza p. 347). H.P.

Einstein Studies, vol. 6: Mach's Principle: From Newton's Bucket to Quantum Gravity, p. 308

Absolute or Relative Motion?[1]

Immanuel Friedlaender

Without knowing that Mach had already done this, I have already for many years doubted the completeness of these [Newtonian] foundations of mechanics; in particular, I have gained the conviction that, given the correct mechanical understanding, the phenomenon of centrifugal force should be explicable solely on the basis of the relative motions of the considered system and without recourse to absolute motion. It was however clear to me that the mere statement of this doubt would not amount to much, and that either it would be necessary to find a new formulation of the expression for the *vis viva* [*lebendige Kraft*, i.e., kinetic energy] of a moving mass and thus an improved form of the law of inertia or that the inadequacy of the present conception would have to be demonstrated experimentally. Precisely the phenomena of centrifugal force appeared to me to be suited to such an experimental resolution of the problem; for if the centrifugal force that occurs in a flywheel is to be explained solely by its relative motion, it must be possible to derive the force under the assumption that the flywheel is at rest while the earth rotates about the axle of the flywheel with the same angular velocity in the opposite sense. Then just as the centrifugal force arises on the flywheel at rest as a consequence of the rotation of the massive earth together with the universe, there should also arise, I believed, a centrifugal force effect – on a correspondingly smaller scale – in fixed bodies near to heavy moving flywheels. If this effect could be demonstrated, a stimulus would be given for the reformulation of mechanics; simultaneously a deeper insight into the nature of gravitation would have been gained, since in the case of gravity it can only be a question of influences[1)] of masses at a distance, specifically a question of the dependence of these influences on relative rotations.

[1)]It makes no difference whether, as the writer believes, or not these are transmitted by a medium.

Einstein Studies, vol. 6: Mach's Principle: From Newton's Bucket to Quantum Gravity, pp. 309–311 © 1995 *Birkhäuser Boston, Inc. Printed in the United States.*

However, given the smallness of the masses with which we can experiment, I had little hope of an experimental solution until in fall 1894 I came across an experimental arrangement that promises success. This arrangement consists of setting up the most sensitive of all physical instruments, a torsion balance, in the extension of the axis of a heavy mass that rotates as rapidly as possible, namely, a large flywheel, for example, in a rolling mill. If the arm of the balance, having at its ends two balls, is not parallel to the (vertical) plane of the flywheel at rest but inclined at about 45° to it, our theory predicts that motive forces will arise that strive to move the balls from the extended axis, i.e., they will tend to move the torsion balance to the parallel position. However, a sensitive torsion balance is a tricky instrument, and a rolling mill is not the most comfortable and favorable location for precision measurements, and thus my experiments, which I began already in November 1894 at the rolling mill in Peine - with the most friendly support on the part of the directors and the engineers - have not yet, as a consequence of the many error sources, yielded indisputable results that I would be willing to present to the public even though a deflection in the expected sense was observed at the beginning and end of the motion. ... [description of various disturbances] ... Although reliable results do not yet exist, the constant occupation with the issue and the frequent discussions with my brother Dr Benedict Friedlaender have led us to the conclusion that the matter is of sufficient importance to publish our thoughts already. My brother drew my attention to Mach right at the start, and in collaboration we have drawn many of the consequences that should flow from our conception. My brother has put together the results of these considerations, together with some views that I cannot completely share, in the second part of this tract[2] and has also made an attempt to express the law of inertia in a different manner, so that the relative nature, and thus the invertibility, of the centrifugal force can be deduced. However, it seems to me that the correct form of the law of inertia will only then have been found when *relative inertia* as an effect of masses on each other and *gravitation*, which is also an effect of masses on each other, have been derived on the basis of a *unified law*.2) The challenge to theoreticians

2)In this connection it is greatly to be desired that the question of whether Weber's law is to be applied to gravitation and also the question of the propagation velocity of gravitation should be resolved. For the second issue, one could use an instrument that makes it possible to measure statically the diurnal variations of the earth's gravity as a function of the position of the heavenly bodies.

and calculators to attempt this will only be crowned with success when the invertibility of centrifugal force has been successfully demonstrated.

Berlin, New Year 1896

NOTES

[1]Translated by Julian B. Barbour from: Friedlaender, Benedict and Friedlaender, Immanuel (1896). *Absolute oder Relative Bewegung?* Teil I: *Die Frage nach der Wirklichkeit einer absoluten Bewegung und ein Weg zur experimentellen Lösung*, pp. 14–17. Berlin: Leonhard Simion.

[2]Partial translation in Chap. 2 of this volume, p. 114.

Commentaries on the Friedlaenders' work can be found after the partial translation of the part written by Benedict Friedlaender (p. 114) and in the papers by Norton (pp. 30–31) and Pfister (p. 321).

In the editions of the *Mechanik* of 1897, 1901, 1904, and 1908 Mach makes the following brief comment: "P. and J. Friedländer [sic] (*Absolute und relative Bewegung* [sic]) attempt to settle the question by an experiment along the lines [*nach dem Schema*] that I mention [in the paragraph at the bottom of p. 109]; my only worry is that its accuracy will not suffice." In the 1912 edition, this reference is replaced by the even shorter mention on p. 111 (the 1912 citation at least gets the initials of the brothers right but still not their title). The brevity of these references is in striking contrast to Mach's responses to accredited academics and perhaps suggests a touch of professional snobbery and embarrassment at some of Benedict's wilder ideas. However, Mach did discuss the Friedlaender experiment seriously and at some length in 1904 in a letter to Petzoldt (p. 25).

Citations from Mach can be especially confusing on account of the numerous German and English editions of his works, especially the *Mechanik.* Most Mach quotations in this volume are from *Die Geschichte und die Wurzel des Satzes von der Erhaltung der Arbeit* (1872, 1909) with English translation *History and Root of the Principle of the Conservation of Energy* first published in 1911 and from *Die Mechanik in ihrer Entwickelung. Historisch-kritisch dagestelt* with editions published in Mach's lifetime in 1883, 1889 (preface dated 1888), 1897, 1901, 1904, 1908, and 1912. Whereas the first work was reprinted unchanged except for a page of supplementary notes, most of the lifetime revisions of the *Mechanics* contain significant changes relating to the issues surrounding Mach's Principle. Many of these are not contained in the 1912 German edition and in the frequently cited 1960 English edition *The Science of Mechanics: A Critical and Historical Account of Its Development.* Besides the one above, such quotations not contained in the 1912 and 1960 editions can be found on p. 120, 218, 230 (Note 2). See also the Notes on p. 127.

On a Gyroscope Experiment to Measure the Rotation Velocity of the Earth[1]

August Föppl

By rotation of the earth, we understand here the rotation that it makes with respect to a space for which the law of inertia holds and, indeed, under the explicit assumption that this rotation is to be established through observation of processes involving motion that take place in the vicinity of the surface of the earth. In principle, it is by no means impossible that terrestrial phenomena involving motions are themselves influenced by the rotation of the earth, so that for them the rotation of the earth is not identical to the rotation relative to the heaven of fixed stars. Only an experiment can establish whether this is the case or not. Now in fact the experiments already made with this in mind have made the existence of such a deviation improbable. Specifically, the Foucault pendulum experiment, which among all the experimental arrangements of this kind has so far given the most accurate results, indicates that for terrestrial motions the law of inertia holds for a space that does not rotate relative to the heaven of the fixed stars.

However, it should be pointed out first that the Foucault pendulum experiment has error sources of a magnitude such that even in the case of the most careful execution the accuracy leaves something to be desired. Moreover, it is still possible that some special influence of the earth's rotation that one would like to discover through this experiment is cancelled out by the forward and backward oscillatory motions of a pendulum, whereas in the case of a gyroscope that always rotates in the same direction an effect could be observed. Even if the accuracy of the Foucault pendulum experiment left nothing to be desired, a complementary experiment with a gyroscope would therefore be in no way made redundant.

It is true that such gyroscope experiments were already made by Foucault himself and have often been repeated. One can find a bibliography of the corresponding literature on this topic in Winkelmann's

Einstein Studies, vol. 6: Mach's Principle: From Newton's Bucket to Quantum Gravity, pp. 312–314

Handbuch der Physik, Vol. 1, Breslau (1891), p. 187. In this connection, it is well worth reading the discussion of the experiments of this kind that have hitherto been made together with the criticism of the accuracy achieved that is given in the final section of the recently published third part of the well-known book of Klein and Sommerfeld *Ueber die Theorie des Kreisels*, Leipzig (1903). According to this account, the accuracy leaves much to be desired and falls far short of that achieved by the Foucault pendulum experiment.

It was therefore by no means a superfluous undertaking to make a new gyroscope experiment with greatly improved resources, in order to establish the rotation velocity of the earth in the sense discussed above. It will be seen that I have succeeded in carrying out this experiment with an accuracy that significantly improves on the accuracy of even the Foucault pendulum experiment.

My original hope to achieve here a new result, namely, to find a clear difference between the rotation velocity of the earth deduced from accurate measurements of terrestrial motions as compared with the rotation relative to the heaven of the fixed stars, was not however fulfilled. Nevertheless, the establishment that such a difference, should it after all exist, can only be a small fraction of the magnitude of each of these two quantities, is not without value.[2]

....[Translation of pp. 7–22 omitted]....

As general result we can state that the difference between the angular velocity of the earth's rotation deduced from the observation of terrestrial motions and the astronomical angular velocity cannot, if such a difference should after all exist, be more than about 2 parts in a hundred.

....[Translation of pp. 24–28 omitted]....

NOTES

[1]First published in *Sitzungsberichte der Bayerischen Akademie der Wissenschaften, Mathematisch-Physikalische Klasse* (1904) 34: 5–28 (submitted February 6, 1904). Translated by Julian B. Barbour.

[2]Föppl's account of the experiment contains many details of no modern relevance, so instead we give the following summary (prepared by Herbert Pfister):

The gyroscope consisted of two flywheels of ingot iron with diameters of 50 cm and a weight of 30 kg each. They were fixed to the two ends of a horizontal spindle which was spun by an electric motor with ω up to 2300 rpm. The whole apparatus hung on three steel wires from the ceiling of a high room. The flywheels rotated within narrow metal boxes in order to eliminate disturbance by turbulent air. Mechanical resonances were damped by connecting the relevant parts of the apparatus with wings which were immersed in oil. If θ (=26.7 kg·cm·sec^2) is the moment of inertia of the gyroscope, φ the geographic latitude of the laboratory, u the sidereal angular velocity of the earth, and ψ the angle of the spindle axis with the east–west direction, then, according to the theory of the spinning top (and disregarding any direct-dragging effects of the rotating earth), the torque acting on the gyroscope was $M=\theta\omega u \cos\varphi \cos\psi$. After the electric motor had been started, this torque deflected ψ from the initial position ψ_i in such a way, that, after an initial buildup of oscillations ψ executed damped oscillations with typical periods of 6 to 8 min (determined by the torsion properties of the steel wires) around a new mean position ψ_f. Each run of the experiment was stopped at the latest after 30 min, because thereafter the heating of the motor changed the conditions. After some 'null test runs,' for example, with $\psi_i=90°$, Föppl performed 10 real experiments with $\psi_i=0°$ and 180°, with ω values in the range 1500–2300 rpm, and with results for $|\psi_f-\psi_i|$ in the range 5.60°–8.23°. From the proportionality of $|\psi_f-\psi_i|$ to ω, and from the formula $\theta\omega u \cos\varphi|\cos\psi|=M=c|\psi_f-\psi_i|$, a value for u was extracted which coincided to within the experimental accuracy of 2% with the sidereal angular velocity of the earth. Therefore, the 'final position' ψ_f of the gyroscope axis stayed fixed with respect to the distant stars, and no direct effect of the rotating earth could be measured. (Today we know, mainly from the Lense–Thirring result, that a measurement of the dragging effects due to the rotating earth affords an experimental accuracy better than 10^{-9}, which seems to be completely out of reach for a Föppl type gyroscope, even with present technology.)

Dragging Effects Near Rotating Bodies and in Cosmological Models

Herbert Pfister

1. Historical Remarks

To my knowledge, the first hint to dragging effects near rotating bodies was given by Ernst Mach (1883, p. 216 f.) in his *Mechanik*:

> Newton's experiment with the rotating bucket teaches only that the rotation of the water relative to the *sides of the vessel* does not induce noticeable centrifugal forces, but that such forces are induced by its rotation relative to the mass of the earth and the other celestial bodies. Nobody can say how the experiment would turn out if the sides of the vessel increased in thickness and mass till they were ultimately several leagues thick.

Equally important in Mach's *Mechanik* are remarks to the effect that a purely relative description of all physical phenomena should only be expected in a realistic cosmological context:

> The universe is not *twice* given, with an earth at rest and an earth in motion, but only *once* with its relative motions only determinable. Therefore we cannot say what would happen if the earth did not rotate. We may interpret the one case that is given us in different ways. If, however, we so interpret it that we come into conflict with experience, our interpretation is simply wrong. The principles of mechanics can presumably be so conceived that centrifugal forces arise also for relative rotations.

The last sentence formulates an ambitious program that has only partly been realized in the intervening 110 years. (Compare the contributions by Assis, Barbour, and Lynden-Bell to this volume.)

Einstein Studies, vol. 6: Mach's Principle: From Newton's Bucket to Quantum Gravity, pp. 315–331

The next really interesting contribution to the problems of inertia and dragging with important new ideas, and the first experimental attempt comes from Benedict and Immanuel Friedlaender (1896, pp. 10–33). Immanuel Friedlaender says:

> Let us suppose we take the point of view that there are only relative motions; then we cannot explain those phenomena if we assume the earth as fixed; why the sun, the moon, and the stars should then drag free axes or the plane of the Foucault pendulum with their daily revolution around the earth would not be explicable mechanically.... It was however clear to me that the mere statement of this doubt would not amount to much, and that either it would be necessary to find a new formulation of the expression for the *vis viva* of a moving mass and thus an improved form of the law of inertia or that the inadequacy of the present conception would have to be demonstrated experimentally.

The concrete experiment of Immanuel Friedlaender consisted of a rapidly rotating, heavy fly-wheel with a torsion-balance in line with its axis, about which he said:

> Then just as the centrifugal force arises on the flywheel at rest as a consequence of the rotation of the massive earth together with the universe, there should also arise, I believed, a centrifugal force effect – on a correspondingly smaller scale – in fixed bodies near to heavy moving flywheels. If this effect could be demonstrated, a stimulus would be given for the reformulation of mechanics; simultaneously a deeper insight into the nature of gravitation would have been gained, since in the case of gravity it can only be a question of influences of masses at a distance, specifically a question of the dependence of these influences on relative rotations.... It seems to me that the correct form of the law of inertia will only then have been found when *relative inertia* as an effect of masses on each other and gravitation, which is also an effect of masses on each other, have been derived on the basis of a *unified law*.

Added is here a prophetic footnote: "In this connection it is greatly to be desired that the question of whether Weber's law is to be applied to gravitation and also the question of the propagation velocity of gravitation should be resolved." At the end of the book, Benedict Friedlaender even vaguely anticipates the incorporation of inertia and gravity into the properties of space and time:

> It is also readily seen that in accordance with our conception the motions of the bodies of the solar system can be regarded as pure inertial motions,

> whereas in accordance with the usual conception the inertial motion, or rather its gravitationally continually modified tendency, strives to produce a rectilinear tangential motion.

In 1904, August Föppl (1904a) asked the question whether the rotation of the earth could not induce a dragging of the local inertial frames relative to the fixed stars, and he tried to answer this question by a gyroscope experiment, a primitive forerunner of the Stanford gyroscope. Unfortunately, his experiment had only an accuracy of 2% of the angular velocity of the earth, whereas 10^{-9} would have been necessary for a positive effect. Föppl (1904b) also conjectured the existence of 'velocity forces,' emanating from moving (for example, rotating) masses and producing Coriolis-type effects on gyroscopes and Foucault pendulums.

Coming to Einstein, an early very interesting and future-pointing remark is made in a letter to Sommerfeld (Einstein 1909):

> The treatment of the uniformly rotating rigid body seems to me to be of great importance because of an extension of the relativity principle to uniformly rotating systems along similar lines of thought as I have tried to carry out in the last section of my paper in the *Zeitschrift für Radioaktivität* [*sic*] for uniformly accelerated translation.

Three years later, Einstein (1912) produced the first definite result that a relativistic gravitation theory can – in contrast to Newton's theory – produce a new type of 'force,' analogous to electromagnetic induction, that leads to dragging. In this paper he introduces also for the first time the model of an infinitely thin spherical mass shell, which turned out to be a very useful testbed for fundamental questions in general relativity even today. By ingenious gedanken experiments concerning the mutual influence of translational accelerations Γ of this shell of mass M and radius R, and γ of a point mass m at the center of the shell (see Fig. 1), Einstein arrives, within a preliminary relativistic scalar gravitation theory, at the following results:

a) The presence of the mass shell M increases the inertial mass of the point mass m to $m+k(mM/Rc^2)$ (k = Newtonian gravitational constant).

b) An acceleration Γ of the mass shell M induces an acceleration, a dragging, $\gamma=(3/2)k(M/Rc^2)\Gamma$ of the point mass m.

From our present knowledge of final general relativity we must conclude that the increase of the inertial mass through nearby masses is an illusion – an untestable coordinate effect – as was convincingly

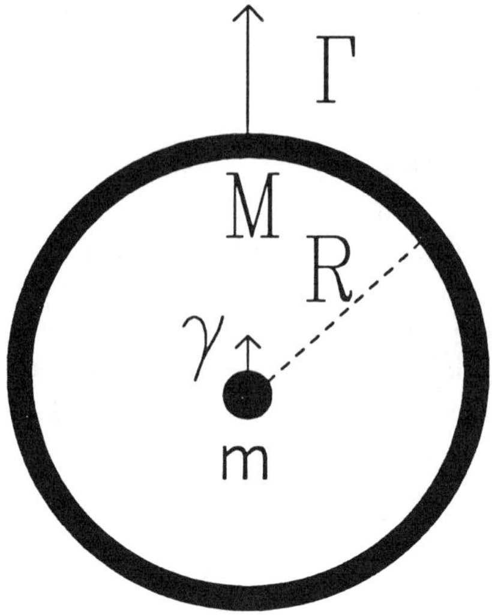

Fig. 1

shown, after numerous controversial claims, not earlier than by Brans (1962). In contrast, the dragging of test masses, or of inertial frames, by accelerated masses is a real and conceptually important effect of general relativity, presumably the single most direct realization of Machian ideas in general relativity. One year after his calculation of a translational dragging effect, Einstein (1913) showed, now in the tensorial Entwurf-theory, that a rotating mass shell induces a Coriolis-type dragging force on the inertial frames in its interior.

2. Dragging Effects Near Rotating Bodies

All of us know that within general relativity H. Thirring deduced the first classic dragging result. It is a curiosity of history, and a nice example for the relations between experiment and theory, that Thirring in 1917 first planned to perform a dragging experiment with a rotating hollow cylinder (Thirring 1966). But since he could not organize and finance the equipment, he sat down to calculate the expected effect (Thirring 1918). He did this in Einstein's model of the rotating mass shell, and in Einstein's scheme of perturbation theory of general relativity in first order in the shell mass M, in second order in the angular velocity ω, and for points $\mathbf{r}$ in the mass shell with $|\mathbf{r}| \ll R$. After surprisingly complicated calculations (compared to modern calculations, which are moreover exact in M and $\mathbf{r}$), Thirring reaches essentially two results: In first order in ω, there results a Coriolis-type force

$$\mathbf{K}_C = -\frac{8Mm}{3R}(\omega \times \mathbf{v}), \tag{2.1}$$

whose order of magnitude must have made Thirring thankful that he did not try the experiment. In second order, an additional force showed up:

$$\mathbf{K}_z = -\frac{4Mm}{15R}[\omega \times (\omega \times r) + 2(\omega \cdot r)\omega]. \tag{2.2}$$

It was treated by Thirring as a centrifugal force although it also has an axial component and cannot be made zero in the same rotating frame in which $\mathbf{K}_C$ vanishes. Equations (2.1) and (2.2) show that inertial effects behave, at least in this model and in this approximation, as R^{-1}. If the universe is regarded as a series of concentric mass shells of equal density and thickness (Cohen and Brill (1968)), for which then $M \sim R^2$, we see that the most distant cosmological regions should have the greatest effect on the local compass of inertia. In the sequel of Thirring (1918), Lense and Thirring (1918) did a similar calculation for the exterior far field ($|\mathbf{r}| \gg R$) of a slowly rotating solid sphere, this time only in first order in ω, and reached the result that is the basis for all presently possible experiments (Ciufolini 1995; Will 1995), that again a Coriolis force is induced with a dragging factor

$$A_{LT} = \frac{4MR^2}{3r^3}. \tag{2.3}$$

Lanczos (1923) noticed that Thirring's paper suffers from an inconsistency because his solutions violate the local energy–momentum conservation law $T^{\mu}_{\nu;\mu} = 0$ in order ω^2, since he had neglected any stresses in the rotating mass shell. Lanczos also made some general and, as I would judge, overcritical remarks concerning Mach's Principle in general relativity. A full analysis and correction of Thirring's error was finally carried out, independently by Bass and Pirani (1955) and by Hönl and Maue (1956), but it did not produce a correct centrifugal force inside the rotating mass shell, even if one allowed for a latitude-dependent mass density on the shell. Further doubts about the interpretation of (2.2) as a centrifugal force came from the argument of Soergel-Fabricius (1961) that a gravitationally induced centrifugal force should not be of order $M\omega^2$ but of order $M^2\omega^2$, thereby calling for a treatment of the rotating mass shell in higher orders or even exactly in M.

This was then first carried out in a very important and well-known paper by Brill and Cohen (1966) by basing a series expansion in the angular velocity ω not on Minkowski spacetime but on the Schwarzschild spacetime of a spherical mass shell of arbitrary mass M. With a view to later extensions to higher orders in ω, we use a metric form that differs slightly from the Brill–Cohen form:

$$ds^2 = -U^2dt^2 + U^{-2}[K^2(dr^2 + r^2d\vartheta^2) + W^2(d\varphi - \omega A dt)^2]. \quad (2.4)$$

Then the Schwarzschild mass shell is given (with $\alpha = M/2$) by

$$\overset{0}{U} = \begin{cases} (r-\alpha)/(r+\alpha) & \text{for } r>R \\ (R-\alpha)/(R+\alpha) & \text{for } r\leq R \end{cases}$$

$$\overset{0}{K} = \begin{cases} (r^2-\alpha^2)/r^2 & \text{for } r>R \\ (R^2-\alpha^2)/R^2 & \text{for } r\leq R \end{cases} \quad (2.5)$$

$$\overset{0}{W} = \overset{0}{K} r \sin\vartheta; \; \overset{0}{A} = 0.$$

The discontinuities in the radial derivatives of these functions at $r=R$ produce through the field equations a δ-type energy–momentum tensor $T^\mu_\nu = (1/8\pi)U^2K^{-2}\tau^\mu{}_\nu\delta(r-R)$ with $(\{x^\mu, \mu=0, 1, 2, 3\} = \{t, r, \vartheta, \varphi\})$:

$$\overset{00}{\tau_0} = -4\alpha/R(R+\alpha), \; \overset{02}{\tau_2} = \overset{03}{\tau_3} = 2\alpha^2/R(R^2-\alpha^2), \; \overset{01}{\tau_1} = \overset{01}{\tau_2} = 0. \quad (2.6)$$

where the singularities of the surface stresses at $R=\alpha$ indicate the limit of stability against collapse.

In first order in ω, only the field equation for $\overset{0}{A}$ has to be solved, which reads in the dimensionless and inverse Schwarzschild coordinate $y = 4\alpha/r(1+\alpha/r)^2$: $(d^2/dy^2 - 2y^{-1}d/dy)\overset{0}{A} = 0$, with the continuous and singularity-free solution [with $Y = y(r=R)$]:

$$\overset{0}{A} = \lambda \begin{cases} y^3 & \text{for } r>R \\ Y^3 & \text{for } r\leq R \end{cases} \quad (2.7)$$

The integration constant λ is fixed by demanding that the energy–momentum tensor $T^\mu{}_\nu$ really represents a rigidly rotating body, i.e., that the timelike eigenvector of $T^\mu{}_\nu u^\nu = -\rho u^\mu$ has the form $u^\mu = u^0(1, 0, 0, \omega)$ with constant ω, ρ being the invariant mass density. Then the dragging coefficient inside the shell takes the form

$$\overset{0}{A}(r\leq R) = \frac{4\alpha(2R-\alpha)}{(R+\alpha)(3R-\alpha)}. \quad (2.8)$$

In the limit $R \gg \alpha$, (2.8) coincides with Thirring's result $\overset{0}{A} = 4M/3R$ [cf. (2.1)]. For $R \to \alpha$, Brill and Cohen get $\overset{0}{A} = 1$, and therefore total dragging of the inertial frames, or total screening of the asymptotic Minkowski frame, by a compact mass shell. Brill and Cohen say: "In this sense our result explains why the 'fixed stars' are indeed fixed in our inertial frame." [An indication of this remarkable result was already obtained earlier by Hönl and Dehnen (1962, 1964) by considering the slowly rotating interior Schwarzschild metric.] The 'total dragging' defined by $\overset{0}{A}=1$ is as seen by an observer at infinity, and this does not automatically mean that the (shell) matter is at rest for a nearby locally nonrotating observer. The locally measured angular velocity of the shell $\tilde{\Omega}=\omega(1-\overset{0}{A})\overset{0}{U}{}^{-1}$ gives for the Brill–Cohen shell with (2.5) and (2.8) $\tilde{\Omega}=3\omega(R-\alpha)(3R-\alpha)^{-1}$, and therefore indeed $\tilde{\Omega}\rightarrow 0$ for $R\rightarrow\alpha$. It is however mainly the diverging surface stresses $\overset{0}{\tau}{}^{2}{}_{2}=\overset{0}{\tau}{}^{3}{}_{3}$ that are responsible for this result. [According to (2.6), the dominant energy condition $\overset{0}{\tau}{}^{0}{}_{0}+\overset{0}{\tau}{}^{2}{}_{2}\leq 0$ is violated for $R<3\alpha/2$, so that the stationary mass shell has to be considered as unphysical at least for $R<3\alpha/2$.] For comparison, Cohen (1968) and Lindblom and Brill (1974) considered expanding and recollapsing shells of dust at the point of maximum expansion and during the process of collapse, and they found that for these more physical models $\tilde{\Omega}$ stays nonzero for $R\rightarrow\alpha$. The same is true (even in the extreme relativistic limit) for the rotating disks of dust considered by Bardeen and Wagoner (1971) and Meinel and Kleinwächter (1995). In these cases, the asymptotically observed total dragging has therefore to be judged as an illusion due to infinite time dilatation. It is also worthwhile to notice that, while the metric (2.4) is continuous across the mass shell according to (2.5) and (2.7), the Christoffel symbols and therefore the dragging effects are discontinuous, except at the poles. For a test particle in the equatorial plane and outside the mass shell, the induced precession is even antiparallel to ω, which might be considered as counterintuitive or even anti-Machian. It was however indicated by Schiff (1960a,b) and analyzed in detail by Thorne (1971) that this behavior is easily comprehensible due to the radial fall-off of the gravitational field, and in analogy to the dragging of little rods in a rotating viscous fluid. [In this connection see also Cohen (1968).]

An extension of the Brill–Cohen results to higher orders in ω, especially the long-standing problem of induction of a correct centrifugal force by rotating masses, had to await another 20 years for a solution (Pfister and Braun 1985). In retrospect, one sees two main obstacles to an earlier solution: On one hand it had to be realized that the necessary and sufficient condition for correct Coriolis and centrifugal forces and no other inertial forces is the flatness of the metric, i.e., quasi-Newtonian conditions, inside the mass shell, which requires for our metric and coordinate choice (2.4) under stationary conditions U, K, $A=$const, and $W=Kr\sin\vartheta$. Furthermore, it had to be understood mathematically that this flat interior metric can be continuously connected to a 'rotating Schwarzschild metric' in the exterior in orders ω^2 and higher only for a nonspherical mass shell. Indeed, $\overset{0}{W} \sim \sin\vartheta$ acts as a source term for all field equations of order ω^{2n}, and therefore requires an expansion of $\overset{2n}{U}$, $\overset{2n}{K}$, $\overset{2n}{A}$, and $\overset{2n}{W}/\sin\vartheta$ in, for example, even Legendre polynomials up to order $2n$. In the stationary and axisymmetric case the expansion coefficients depend only on r and satisfy second-order differential equations. Existence theorems for asymptotically decreasing solutions of these equations are known, and therefore the continuity conditions to a flat interior metric at the mass shell can be explicitly analyzed, with the following results (Pfister and Braun 1986):

a) From the equations for $\overset{2}{U}$, $\overset{2}{K}$, and $\overset{2}{W}$, it follows that the shell position is given by

$$r_s = R(1 + \omega^2 R^2 f \sin^2\vartheta) \text{ with } f < 0, \tag{2.9}$$

i.e., surprisingly the shape is prolate. Analogous corrections up to $\sin^{2n}\vartheta$ are obtained in order ω^{2n}.

b) The discontinuities in the shell-orthogonal derivatives of the metric functions lead then automatically to ϑ-dependent mass and stress densities.

c) From the equation for $\overset{2}{A}$, it follows that a mass shell with flat interior cannot rotate rigidly, i.e., that the angular velocity has the form

$$\omega_s = \omega(1 + \omega^2 R^2 e \sin^2\vartheta) \text{ with } e > 0. \tag{2.10}$$

d) For fixed total mass M and radius R, the solution for a rotating mass shell with flat interior is unique in all orders of ω.

e) Spherical symmetry both of the shell geometry and of the mass distribution and rigid rotation are only restored in the collapse limit $R \to \alpha$, in which case the exterior metric is Kerr, as was already observed (up to order ω^3) by de la Cruz and Israel (1968) and Orwig (1978).

For arbitrary and not only small angular velocity ω, the existence of a rotating mass shell with flat interior and therefore correct inertial forces obviously reduces to the question whether the Dirichlet problem for the elliptic system of the stationary and axially symmetric Einstein equations with flat boundary data at the shell has a solution. Although a general answer to this question seems not to be available at present, it is interesting to observe (Pfister 1989, 1990) that the special nonlinearity of Einstein's equations (Ricci tensor quadratic in the Christoffel symbols!) puts them just on the borderline between mathematical existence and nonexistence theorems. Furthermore there are indications (Schaudt 1992) that some additional mathematical conditions on the magnitude of the boundary data in relation to the size of the region, which guarantee at least regularity of Dirichlet solutions, are connected with physical stability criteria for rotating bodies. Concerning dragging effects near rapidly rotating bodies, it is encouraging that recently some detailed calculations have been possible for special, although quite idealized, sytems (Meinel and Kleinwächter 1995; Karas and Lanza 1995; Bičák and Ledvinka 1993). The formation of ergospheres in such models can be seen as one of the most striking Machian effects in general relativity.

The success of the work of Thirring, Brill, and Cohen and others in showing that general relativity realizes at least in these models the hopes of Mach and Einstein concerning dragging effects and relativity of rotation provokes the question whether general relativity realizes dragging effects also for other types of acceleration. To my knowledge no attempts in this direction have been made hitherto, but I suspect that a model of linear acceleration, taking up the first attempt of Einstein (1912), but using, for example, a cylindrical mass distribution, should be workable, although a stationary situation as in rotating systems seems to be out of reach. Meanwhile, I should like to advance the following conjecture of some type of quasiglobal equivalence between acceleration fields and gravitational fields in general relativity (Pfister and Braun (1985)): If some large but finite laboratory is in arbitrarily accelerated motion relative to the distant masses in the universe, then all motions of free particles and all physical laws, measured from laboratory axes, are modified by inertial forces. It is argued that exactly the same modified motions and laws can be induced (at least for some time) at all places of the laboratory by suitable and suitably moving masses outside the laboratory. Mathematically, this implies the conjecture that there exist continuous solutions of Einstein's field equations (with matter) with the following boundary conditions: flatness in a finite region of spacetime, and asymptotic flatness, but with nearly arbitrary acceleration between the asymptotic and the 'interior' inertial frames.

Some vague ideas, which go in the direction of this hypothesis, may already be read off a letter (June 29, 1912) by P. Ehrenfest to A. Einstein (Klein *et al.* (1993), Document 411), a letter (August 14, 1913) by A. Einstein to H. A. Lorentz (Klein *et al.* (1993), Document 467), and a discussion remark of G. Mie following the Vienna lecture of Einstein (1913).

Coming back to worked-out examples of dragging by rotating matter, we have to mention some papers on rotating, in most cases infinitely long and infinitely thin, hollow cylinders. Since in these models there is only one nontrivial (radial) coordinate, Einstein's equations reduce to ordinary differential equations, and all mathematical problems are much easier. For example, it could be proven that the metric inside such a cylinder is always flat (Davies and Caplan 1971), so that automatically the correct Coriolis and centrifugal forces are induced. Also in the exterior, Einstein's equations can be solved exactly (Frehland 1972; Embacher 1983), and all details of the models can be worked out explicitly, even for relativistic rotation velocities. On the other hand, the cylindrical symmetry and the missing asymptotic flatness of these models impede the physical interpretation of the mathematical results, especially their significance for a cosmological context.

Concerning the transferability of the classical dragging results for mass shells to more realistic situations, important progress has been accomplished by Brill and coworkers in the years 1966–1974: First they proved (Brill and Cohen 1966) that the essential dragging results are also valid (at least in first order in ω) for an expanding and recollapsing ball of dust at the time of momentary stationarity. In this model one can then, at least formally, go to the limit where the coordinate radius of the ball (of fixed mass) goes to zero, with the consequence that the ball separates from the asymptotically flat manifold, forms a closed universe with complete dragging, and in this way "is consistent with the conjecture that Mach's Principle is satisfied in a closed Friedman universe." In a subsequent paper, Cohen and Brill (1968) calculated similar dragging effects for a slowly rotating incompressible fluid sphere, thereby confirming and extending earlier results by Hönl and Dehnen (1962, 1964). Especially interesting and important to me is the work of Lindblom and Brill (1974), which extends the dragging phenomena to a nonstationary situation, namely the free collapse of a slowly rotating spherical shell of dust, and which therefore should be especially significant for our real dynamic universe. In particular, it could be shown in this work that the dragging effects result from Einstein's constraint equations, and that "there are no retardation effects between the shell and the inertia of a gyroscope at its center." It would be desirable to extend this work to at least the second order of the angular velocity of the shell. [For a recent extension of the Lindblom–Brill analysis in other directions see Lynden-Bell, Katz, and Bičák (1995).]

Before coming to dragging effects in cosmological models in Sec. 3, let me mention that there have been attempts to clarify whether the dragging phenomena inside mass shells and their Machian interpretation extend also to physical properties other than inertial ones, especially to electromagnetic systems. On the one hand, there are historic papers by Alexandrow (1921a, b) on electrodynamics in weak (stationary and nonstationary) gravitational fields; on the other hand, there are more recent and more special calculations on charged point sources, shells, and spheres within a slowly rotating mass shell (Hofmann 1962; Cohen 1966; Ehlers and Rindler 1971). Quite generally, it can be said that naive Machian expectations are fulfilled, for example, the rotational induction of a dipolar magnetic field on top of an electrostatic field. However, in detail there are difficulties to interpret all results in accordance with Machian views, as is particularly stressed by Ehlers and Rindler (1971).

3. Dragging Effects in Cosmological Models

As mentioned in Sec. 1, Ernst Mach stressed that a description of nature with only relative concepts is to be expected only in a cosmological context. Also, the most important observational impetus for such a description is cosmological in nature, namely the remarkable fact that the local inertial compass coincides with the frame of the most distant galaxies and quasars within the present measurement accuracy of $2{\cdot}5{\cdot}10^{-4}$ arcsec/year. In this respect, the models with isolated rotating bodies, covered in Sec. 2, have often been criticized as being very imperfect models, since their asymptotic Minkowski frame functions as an absolute element.

So far as I know, the first detailed studies of dragging effects in cosmological models were performed by Soergel-Fabricius (1960) and by Hönl and Soergel-Fabricius (1961). However, they did not base their considerations on the most realistic cosmological model but on the mathematically simplest model, the static and spatially closed Einstein cosmos. Superimposed on this solution of general relativity (with positive cosmological constant Λ) they considered a rotational perturbation induced by an infinitesimal amount $\delta\rho/\rho$ of cosmic matter, distributed either uniformly or as a spherical shell. The Coriolis-type effect of $\delta\rho$ on the geodesics is qualitatively in accordance with Machian views and with Thirring's result, for example, the Coriolis field vanishes precisely in the rotating coordinate system in which the total angular momentum of the cosmic matter vanishes. The results for the 'centrifugal force' are, as is to be expected, divided. The remark of Hönl and Soergel-Fabricius that "the fulfilment of Mach's Principle and the topological demand for a spatially closed and finite universe are altogether identical" is not confirmed by later work. Lausberg (1969) extended the work of Soergel-Fabricius and Hönl to a rotating mass shell of finite thickness and matter content and studied (in first order in ω) the interesting and qualitatively Machian dependence of the dragging factor on the radius and thickness of the shell. The result that a 'mass shell' that covers the whole universe leads to total dragging is however trivial in this case, in contrast to the result of Brill and Cohen.

More realistic cosmological models within the class of stationary and axially symmetric solutions with ideal fluid on a spatially compact manifold do not seem to exist, because Frauendiener (1987) was able to show that general relativity does not allow such solutions if the 'effective pressure' $p-\Lambda/8\pi$ is nonnegative. Taking for granted that our present universe has positive pressure and a very small or zero cosmological constant, this result may be seen as a hint that realistic Machian dragging effects in cosmological models show up only within a class of noncompact and/or dynamic (nonstationary) solutions of general relativity.

This brings us finally to investigations of dragging effects on the basis of rotational perturbations of expanding Friedman universes, which should represent the most realistic models for our present universe. The first important initiative in this direction was taken by Lewis (1980), who realized that a flat 'interior' Minkowski region can be continuously connected via a coexpanding spherical mass shell to a spatially closed $k=1$ Friedman dust solution. Axially symmetric perturbations superimposed on this model were calculated in first order in the angular velocity, leading to a dragging factor for the interior inertial frames which depends on the properties of the shell, on the cosmic matter and on the cosmic time and "tends to confirm the spirit of Mach's Principle," although the interpretation of the mathematical results has to be thought of more carefully than in the asymptotically flat models.

Recently, Klein (1993a) extended these investigations considerably, with many new and interesting results: He took into consideration also the spatially open Friedman models with $k=0$, -1, and allowed for a nonvanishing pressure p of the cosmic fluid. He showed that the energy–momentum tensor of the shell fulfils the standard energy conditions if the cosmic fluid satisfies $|p| \leq 2\mu/3$, and that the total mass of the shell is equivalent to the mass cut out of the Friedman universe. He assumed that the rotational perturbation is induced primarily by the rotation of the mass shell and goes to zero with increasing cosmic distance from the shell. The dragging factor for the first-order Coriolis force on the interior inertial frames is then analyzed in detail. Under physically reasonable conditions, it always lies between 0 and 1, decreases monotonically with the cosmic time τ, and increases with the (initial) radius of the mass shell. Complete dragging is only reached for the $k=0$ universe for which a very big part is replaced by the mass shell with flat interior. Finally, second-order perturbations of these models are considered (Klein 1993b, 1994) for which the metric form (2.4) has to be supplemented by a term $2\omega^2 W^2 D(t, r) dt dr$. In analogy to (Pfister and Braun 1985), the mass shell acquires a centrifugally deformed shape. In combination with the dynamics of the system, this leads to a time-dependent quadrupole moment and therefore to the production of gravitational waves. However, perhaps surprisingly, the deformation and mass distribution of the shell can be adjusted in such a way that there are no gravitational waves in the interior of the shell - the shell acting like a total reflector for this type of gravitational waves - so that flatness and therefore quasi-Newtonian concepts are still valid in the interior.

Let me summarize the effort of more than 100 years of research on these dragging effects in the following way: Although Einstein's theory of gravity does not, despite its name 'general relativity,' yet fulfil Mach's postulate of a description of nature with only relative concepts, it is quite successful in providing an intimate connection between inertial properties and matter, at least in a class of not too unrealistic models for our universe. Perhaps against majority expectation, this connection is instantaneous in nature. Furthermore, general relativity has brought us nearer to an understanding of the observational fact that the local inertial compass is fixed relative to the most distant cosmic objects, but there is surely desire for still deeper understanding.

REFERENCES

Alexandrow, W. (1921a). "Über das Verhalten der ruhenden Elektrizität im nicht-statischen (veränderlichen oder stationären) Gravitationsfelde." *Annalen der Physik* 65: 675–696.

Alexandrow, W. (1921b). "Die Elektrodynamik im schwachen Gravitationsfelde." *Physikalische Zeitschrift* 22: 593–601.

Bardeen, James M. and Wagoner, Robert V. (1971). "Relativistic Disks. I. Uniform Rotation." *The Astrophysical Journal* 167: 359–423.

Bass, L. and Pirani, Felix A.E. (1955). "On the Gravitational Effects of Distant Rotating Masses." *Philosophical Magazine* 46: 850–856.

Bičák, Jiri and Ledvinka, Tomáš (1993). "Relativistic Disks as Sources of the Kerr Metric." *Physical Review Letters* 71: 1669–1672.

Brans, Carl H. (1962). "Mach's Principle and the Locally Measured Gravitational Constant in General Relativity." *Physical Review* 125: 388–396.

Brill, Dieter R. and Cohen, Jeffrey M. (1966). "Rotating Masses and their Effect on Inertial Frames." *Physical Review* 143: 1011–1015.

Ciufolini, Ignazio (1995). "Dragging of Inertial Frames, Gravitomagnetism, and Mach's Principle." This volume.

Cohen, Jeffrey M. (1966). "Electromagnetic Fields and Rotating Masses." *Physical Review* 148: 1264–1268.

Cohen, Jeffrey M. (1968). "Gravitational Collapse of Rotating Bodies." *Physical Review* 173: 1258–1263.

Cohen, Jeffrey M. and Brill, Dieter R. (1968). "Further Examples of 'Machian' Effects of Rotating Bodies in General Relativity." *Nuovo Cimento* 56B: 209–219.

Davies, H. and Caplan, T. (1971). "The Spacetime Metric Inside a Rotating Cylinder." *Proceedings of the Cambridge Philosophical Society* 69: 325–327.

De La Cruz, Vicente and Israel, Werner (1968). "Spinning Shell as a Source of the Kerr Metric." *Physical Review* 170: 1187–1192.

Ehlers, Jürgen and Rindler, Wolfgang (1971). "An Electromagnetic Thirring Problem." *Physical Review* D4: 3543–3552.

Einstein, Albert (1909). Letter to A. Sommerfeld. *Albert Einstein - Arnold Sommerfeld. Briefwechsel.* Armin Hermann, ed. Basel and Stuttgart: Schwabe (1968).

Einstein, Albert (1912). "Gibt es eine Gravitationswirkung, die der elektrodynamischen Induktionswirkung analog ist?" *Vierteljahrschrift für gerichtliche Medizin und öffentliches Sanitätswesen* 44: 37–40.

Einstein, Albert (1913). "Zum gegenwärtigen Stande des Gravitationsproblems." *Physikalische Zeitschrift* 14: 1249–1266.

Embacher, Franz (1983). "Rotating Hollow Cylinders: General Solution and Machian Effects." *Journal of Mathematical Physics* 24: 1182–1186.

Föppl, August (1904a). "Über einen Kreiselversuch zur Messung der Umdrehungsgeschwindigkeit der Erde." *Akademie der Wissenschaften, München, Mathematisch-Naturwissenschaftliche Klasse, Sitzungsberichte*, Sitzung vom 6. Februar, 1904, pp. 5–28. (Partial translation in this volume, p. 317.)

Föppl, August (1904b). "Über absolute und relative Bewegung." *Akademie der Wissenschaften, München, Mathematisch-Naturwissenschaftliche Klasse, Sitzungsberichte*, Sitzung vom 5. November, 1904, pp. 383–395. (Partial translation in this volume, p. 120.)

Frauendiener, Jörg (1987). "Nonexistence of Stationary, Axisymmetric Dust Solutions of Einstein's Equations on Spatially Compact Manifolds." *Physics Letters* A120: 119–123.

Frehland, Eckart (1972). "Exact Gravitational Field of the Infinitely Long Rotating Hollow Cylinder." *Communications of Mathematical Physics* 26: 307–320.

Friedlaender, Benedict and Friedlaender, Immanuel (1896). *Absolute oder relative Bewegung?* Berlin: Leonhard Simion. (Partial translation in this volume, p. 114 and p. 314.)

Hofmann, Klaus D. (1962). "Über Wechselwirkung von Gravitation und elektromagnetischem Feld gemäss der allgemeinen Relativitätstheorie."

Zeitschrift für Physik 166: 567–576.

Hönl, Helmut and Maue, August W. (1956). "Über das Gravitationsfeld rotierender Massen." *Zeitschrift für Physik* 144: 152–167.

Hönl, Helmut and Soergel-Fabricius, Charlotte (1961). "Coriolis-Kräfte im Einstein-Kosmos und das Machsche Prinzip." *Zeitschrift für Physik* 163: 571–581.

Hönl, Helmut and Dehnen, Heinz (1962). "Zur Frage der Relativität der Rotationsbewegungen." *Zeitschrift für Physik* 166: 544–558.

Hönl, Helmut and Dehnen, Heinz (1964). "Über Machsche und anti-Machsche Lösungen der Feldgleichungen der Gravitation. II." *Annalen der Physik* 14: 271–295.

Karas, Vladimir and Lanza, Antonio (1995). "On the Dragging Effects and the Theory of Active Galactic Nuclei." This volume.

Klein, Christian (1993a). "Rotational Perturbations and Frame Dragging in a Friedman Universe." *Classical and Quantum Gravity* 10: 1619–1631.

Klein, Christian (1993b). "Rotative Störungen und Mitführungseffekte in einem Friedman-Universum." Dissertation, University of Tübingen.

Klein, Christian (1994). "Second-Order Effects of Rotational Perturbations of a Friedman-Universe." *Classical and Quantum Gravity* 11: 1539–1553.

Klein, Martin J., Kox, A. J., and Schulman, Robert, eds. (1993). *The Collected Papers of Albert Einstein*, Vol. 5. Princeton: Princeton University Press.

Lanczos, Kornel (1923). "Zum Rotationsproblem der allgemeinen Relativitätstheorie." *Zeitschrift für Physik* 14: 204–219.

Lausberg, A. (1969). "Coriolis Effects in the Einstein Universe." *Astronomy and Astrophysics* 3: 150–155.

Lense, Josef and Thirring, Hans (1918). "Über den Einfluss der Eigenrotation der Zentralkörper auf die Bewegung der Planeten und Monde nach der Einsteinschen Gravitationstheorie." *Physikalische Zeitschrift* 19: 156–163.

Lewis, Steve M. (1980). "Machian Effects in Nonasymptotically Flat Spacetimes." *General Relativity and Gravitation* 12: 917–924.

Lindblom, Lee and Brill, Dieter R. (1974). "Inertial Effects in the Gravitational Collapse of a Rotating Shell." *Physical Review* D10: 3151–3155.

Lynden-Bell, Donald, Katz, Joseph, and Bičák, Jiri (1995). "Mach's Principle from the Relativistic Constraint Equations." *Monthly Notices of the Royal Astronomical Society* 272: 150–160.

Mach, Ernst (1883). *Die Mechanik in ihrer Entwicklung. Historisch-kritisch dargestellt.* Leipzig: F. A. Brockhaus.

Meinel, Reinhard and Kleinwächter, Andreas (1995). "Dragging Effects near a Rigidly Rotating Disk of Dust." This volume.

Orwig, Lawrence P. (1978). "Machian Effects in Compact, Rapidly Spinning Shells." *Physical Review* D18: 1757–1763.

Pfister, Herbert (1989). "Rotating Mass Shells with Flat Interiors." *Classical and Quantum Gravity* 6: 487–503.

Pfister, Herbert (1990). "Zur Frage nach globalen Lösungen der Einsteinschen

Feldgleichungen für rotierende Sterne." *Wissenschaftliche Zeitschrift der Universität Jena* 39: 152–160.

Pfister, Herbert and Braun, Karlheinz (1985). "Induction of Correct Centrifugal Force in a Rotating Mass Shell." *Classical and Quantum Gravity* 2: 909–918.

Pfister, Herbert and Braun, Karlheinz (1986). "A Mass Shell with Flat Interior Cannot Rotate Rigidly." *Classical and Quantum Gravity* 3: 335–345.

Schaudt, Urs (1992). "Zur Dirichlet-Analyse für die stationären axialsymmetrischen Einsteinschen Feldgleichungen." Diploma Thesis, University of Tübingen.

Schiff, Leonard I. (1960a). "Possible New Experimental Test of General Relativity Theory." *Physical Review Letters* 4: 215–217.

Schiff, Leonard I. (1960b). "Motion of a Gyroscope According to Einstein's Theory of Gravitation." *Proceedings of the National Academy of Sciences* (Washington, DC) 46: 871–882.

Soergel-Fabricius, Charlotte (1960). "Thirring-Effekt im Einsteinkosmos." *Zeitschrift für Physik* 159: 541–553.

Soergel-Fabricius, Charlotte (1961). "Über den Ursprung von Coriolis- und Zentrifugalkräften in stationären Räumen." *Zeitschrift für Physik* 161: 392–403.

Thirring, Hans (1918). "Über die Wirkung rotierender, ferner Massen in der Einsteinschen Gravitationstheorie." *Physikalische Zeitschrift* 19: 33–39; errata in *Physikalische Zeitschrift* 22: 29–30 (1921).

Thirring, Hans (1966). "Ernst Mach als Physiker." *Almanach der Österreichischen Akademie der Wissenschaften, Wien* 116: 361–372.

Thorne, Kip S. (1971). "Relativistic Stars, Black Holes, and Gravitational Waves." In *General Relativity and Cosmology*. R. K. Sachs, ed. New York and London: Academic Press, pp. 237–283.

Will, Clifford M. (1995). "Testing Machian Effects in Laboratory and Space Experiments." This volume.

Discussion

Nordtvedt: There is one form of theoretical and empirical evidence for the accelerative dragging in general relativity that you were talking about, and that occurs when you calculate the inertial and gravitational mass of a self-gravitating body like the Earth. In order for the gravitational self-energy of the Earth to modify the total inertial mass of the Earth, you need the Mach-like accelerative induced acceleration of the mass elements of the Earth by the other mass elements of the Earth, and since we experimentally know that the Earth's ratio of gravitational to inertial mass ratio is one, to very good precision, it would be very difficult to get that result without the proper presence of that

acceleration-induced effect.
Pfister: Yes I agree.
Ciufolini: May I comment on the present potential accuracy, with the present technology, for measuring the frame-dragging effect. It's not one order away from the Lense–Thirring effect but with LAGEOS III it's just 3%, so in other words when we will launch this laser-ranged satellite in three years we will have a 3% experiment. In 1989, there was a NASA committee that concluded one could have a 10% experiment, but since then other laser-ranging satellites have been launched and the knowledge of the potential of the Earth has been improved, so now we have calculated LAGEOS III and it is a 3% experiment with the present technology.
Pfister: But it has not yet worked and ...
Will: You're right. It's an order of magnitude away in terms of funding.
Isenberg: You said that with your student you looked at frame dragging in $k=0$ or $k=-1$. You can take those universes and spatially compactify them. I was wondering if you checked if that had any effect on the amount of frame dragging that occurs.
Pfister: We didn't do that. We did the calculations in the $k=1$, 0, and -1 case, so we had the one compact case included, but you are thinking of a different topological situation?
Isenberg: Right, does that affect anything else?
Pfister: We have not looked into that but I am quite sure one could.
Bondi: Just in relation to the last remark my guess is that it is wholly irrelevant what the value of k is or whether the universe is open or closed. I mean the fact of the matter is that in an expanding universe with a high recession velocity at great distances, the effects from very far away are greatly devalued by the Doppler shift, and I would expect all effects to be negligible. So whether at distances a multiple of the Hubble distance the thing closes up or not I think is a very minor matter.
Pfister: Yes, this is in a way confirmed by the calculations that you get at least qualitatively the same or similar results in all three types of cosmological models; in some details there are differences, of course.

Comments on Dragging Effects: Response to Pfister

Dieter R. Brill

1. Preliminaries and Frivolities

The 'Machian' effects of rotation on inertial frames have a distinguished history, as pointed out in the previous paper by Pfister (1995). But beyond that, the appeal of Mach's ideas has also been used to motivate points of view that often are well-founded in inverse relation to their attraction for the general public. For example, in a recent 'popular' scientific book (Fahr 1993), Mach occurs as early as page 55, and there is a whole chapter devoted to matters Machian, with the somewhat ominous title[1] *The view in the large–doomed to failure?* Even if the answers of this particular book may leave much to be desired, it reminds us that these questions are not esoteric considerations of a select few, but of direct and immediate interest to a large number of intellectuals. One can only wish that the present volume may have some fraction of the impact that such 'popular' books have on the general public!

Pfister's emphasis is not so much on Mach as an abstract principle but on definite and calculable effects suggested by the Machian line of thought. This is a very useful and practical approach; it is likely that Mach's ideas have brought more progress by suggesting calculations of such effects than by any general formulation of his principle. This view has been expressed by Prof. Shimony (1992), who compares Mach to a Yankee storekeeper with many useful items on his shelves, connected to each other only loosely if at all. Happily, in Vermont Prof. Shimony has actually found Machs' General Store (Fig. 1), and characterizes it thus: "What we have is on the shelves; what you don't see don't exist; we give no credit."

The dragging effects are probably the most prominent items in Mach's store, and Pfister gives their interesting history. One could

Einstein Studies, vol. 6: Mach's Principle: From Newton's Bucket to Quantum Gravity, pp. 332–338

Fig. 1: Machs' General Store, Pawlet, Vermont
Reprinted with permission of Frank L. Forward

supplement this with the history of other Machian effects, because one use of a good principle is to make simple and quick qualitative or semi-quantitative suggestions for a variety of different physical situations. The most fruitful and imaginative use of Mach's Principle in this fashion occurred in the 50's and 60's, largely inspired by Dicke. Some of the resulting ideas have appeared in the literature (for example, Dicke 1961, 1963; Peebles and Dicke 1962), but they found their most creative expression in Dicke's informal 'brainstorming' sessions for the Princeton relativists. Here all sorts of ways that the universe might influence local physics were explored, and the magnitudes of possible experimental observations were determined by liberal use of Mach's Principle.

2. Dragging of inertial frames

Of all the predictions that follow from, or have been read out of, Mach's Principle, the dragging of inertial frames by rotating bodies is certainly the most definite and least controversial. If one measures this dragging by the Coriolis forces - that is, one calculates in the first order of the angular velocity - then the answer of general relativity is unambiguous, and can be derived simply and elegantly (Pfister 1995). This answer can be interpreted to agree with the Machian expectations, at least if one is willing to refine them suitably and to some extent after the fact. Mach himself, in his famous passage (Mach 1883), simply informs us that the dragging might be measured by centrifugal forces, but it is reasonable to include Coriolis forces (and other effects!) with respect to Mach's Principle. After all, no one is competent to say how Mach's Principle would have turned out if the background of its author increased in scope and style till it ultimately encompassed several modern alternative theories of gravity.

It is the particular merit of Pfister and his students that they did a detailed study of the Machian centrifugal force terms. By the 'correct' centrifugal force Pfister means the one that corresponds to the spacetime region being *flat*. As he remarks, to have a centrifugally correct region, flat to second order in the angular velocity, necessarily demands a rather special situation: Nonlinearity will generally cause second-order forces other than pure centrifugal ones. Thus he finds that the correct forces inside a shell require that the shell have special properties, such as a prolate shape, differential rotation, etc.

But given these special properties one *can* induce the 'correct' inertial forces for the case of rotation; in fact, Pfister conjectures that the effects of *arbitrary* acceleration can be induced in a laboratory at rest with respect to the fixed stars by suitably moving exterior masses. This is a very interesting conjecture, not least because it forces us to think about the proper definition of global measurements, such as acceleration with respect to the fixed stars, and about other global Machian quantities. For example, naively one might reject the conjecture because the inertial forces are 'coordinate effects,' whereas the effect of exterior masses is presumably a curvature effect; but this would neglect the global nature of the acceleration that is at issue here.

In the general situation, when the exterior masses are not moving 'suitably,' Pfister shows that it is difficult to distinguish between Machian dragging and gravitational radiation. If we say that it is in fact impossible, we come very close to the type of minimalistic view of Mach's Principle advocated, for example, in the Mach–Einstein–Wheeler formulation (Isenberg 1995). But just because the Machian dragging is a (global) 'coordinate effect,' whereas gravitational radiation is a (local) 'curvature effect,' we can equally well view the difficulty as a challenge, to define the Machian dragging in a general and useful way so that it *can* be distinguished from gravitational radiation.

3. Alternative Theories

This is not the place to review the many ways in which Mach's Principle has suggested alternative theories. In the case of inertial dragging, Mach of course makes no explicit prediction that might confirm or contradict the general relativity result; but in the limit when the rotating mass is the whole universe, the usual expectation is that of 'complete dragging.' It may be that in this limit the small-rotation results will be exact (because complete dragging presumably means that the rotation with respect to the inertial frame is vanishingly small). As a first step toward a proof of such conjectures, nothing would be more helpful than a clear, general definition of 'complete dragging.'

If this cosmological Machian expectation is not satisfied, it may be the fault of the cosmological model. The task would then be to find a more restricted class of models that are Machian, as advocated in a number of formulations (such as the Einstein–Wheeler–Mach version). Or it may be the fault of the theory, in which case one would look for a more Machian theory of gravity; this is one of the claims of the Brans–Dicke theory. Calculations of dragging effects in alternative theories are appropriate not only to determine this 'degree of Machismo,' but also because these effects will eventually become measurable experimentally, and it is interesting to know how accurately one will have to measure in order to distinguish alternative theories on the basis of dragging. This question has been answered (to lowest order) in the PPN formalism, but beyond that the literature appears limited.

Two alternative theories are particularly interesting in this context: Brans–Dicke theory (because it takes Mach's Principle as a motivational basis) and low-energy string theory (because string theory is supposed to be more fundamental than general relativity). Both of these incorporate a scalar field ('dilaton') as part of the gravitational force. That the scalar part of gravity should not contribute to dragging is suggested by symmetry considerations: To lowest order, rotational perturbations will affect only the g_{0i} components of the metric, and *not* the one component of the scalar field. Therefore in Brans–Dicke theory, for example, the dragging is reduced compared to general relativity (Brill 1962) because the scalar field contributes positively to the gravitational force but 'does not drag.' One can also show that this theory is, in a sense, *more* Machian than general relativity: It leads to complete dragging in the limit when the rotating shell is the only matter in the universe, independent of the shell's mass; the gravitational 'constant' will automatically adjust so that the shell is at its Schwarzschild radius.

Dragging effects appear not to have been discussed in the literature on low-energy string theory, which has no natural place for phenomenologically described matter. The long-range fields in this theory are the gravitational, scalar (dilaton), electromagnetic, and axion fields. However, one does have solutions to this theory for rotating black holes (Sen 1992). For uncharged black holes, these show no difference from general relativity and appear to suggest that the dragging is not reduced as in Brans–Dicke theory. This may be surprising if one expects the rotating black hole solutions to reproduce, at least to lowest order, the exterior dragging effects of other rotating configurations, but the reason is easily explained.

The black hole solutions of low-energy string theory are the simplest solutions to the *sourceless* equations corresponding to black holes. In these equations the electromagnetic field generates the scalar field; if the former vanishes (zero charge), the scalar field is constant. In the Brans–Dicke theory, on the other hand, one has in mind matter sources, and the scalar field is generated by the trace of the matter's stress–energy tensor. Thus the difference already occurs in the description of an unperturbed mass center: Brans–Dicke theory attributes a part of the gravitational force from matter to the scalar field, whereas low-energy string theory considers the scalar field to be constant for uncharged black holes. In fact, collapse to a black hole in Brans–Dicke theory also leads to black holes with a constant scalar field, which are the same as those of general relativity (Hawking 1972). Thus, there is no reason to suppose that the role of the scalar field in dragging effects is different in these theories. (Whether there are also other types of black holes in Brans–Dicke theory is the subject of current research, Campanelli and Lousto 1993.)

NOTE

[1]Chapter 6, "Der Blick ins Große – zum Scheitern verurteilt?"

REFERENCES

Brans, Carl H. and Dicke, Robert H. (1961). "Mach's Principle and a Relativistic Theory of Gravitation." *Physical Review* 124: 925–935.

Brill, Dieter R. (1962). "Erweiterte Gravitationstheorie, Machsches Prinzip und rotierende Massen." *Zeitschrift für Naturforschung* 22a: 1336–1341.

Campanelli, M. and Lousto, C. O. (1993). "Are Black Holes in Brans–Dicke

Theory Precisely the Same as in General Relativity?" *International Journal of Modern Physics* D4: 451.
Dicke, Robert H. (1961). "Experimental Tests of Mach's Principle." *Physical Review Letters* 7: 359–360.
Dicke, Robert H. (1963). "Cosmology, Mach's Principle and Relativity." *American Journal of Physics* 31: 500–509.
Dicke, Robert H. (1962). "The Earth and Cosmology." *Science* 138: 653–664.
Fahr, Hans Jörg (1993). *Der Urknall kommt zum Fall.* Frankh-Kosmos.
Hawking, Stephen W. (1972). "Black Holes in the Brans–Dicke Theory of Gravitation." *Communications in Mathematical Physics* 25: 167–171.
Isenberg, James (1993). "On the Wheeler–Einstein–Mach Spacetimes." This volume, p. 188.
Mach, Ernst (1883). *Die Mechanik in ihrer Entwickelung. Historisch–kritisch Dargestellt.* Leipzig: F. A. Brockhaus.
Peebles, P. James and Dicke, Robert H. (1962). "Significance of Spatial Isotropy." *Physical Review* 127: 629–631.
Pfister, Herbert (1995). "Dragging Effects Near Rotating Bodies and in Cosmological Models." Previous paper, this volume.
Sen, Ashoke (1992). "Rotating Charged Black Hole Solution in Heterotic String Theory." *Physical Review Letters* 69: 1006.
Shimony, Abner (1992). Private communication, APS Spring Meeting, Washington, DC.

Discussion

Brans: Dieter, I'm glad you brought up Bob Dicke's name and that you recall his hours of sparring with theoretical physicists back in the 50s. Bob certainly did produce a lot of ideas and lots of suggestions for experiments, but I recall one time being one-on-one with Bob and his telling me his idea of Mach's Principle and the inertial effects in rather direct terms for an aspiring theorist. He said "Well, Carl, my idea is that some theoretical physicsts should get a real intuition for what inertial reaction is all about, perhaps by giving them a good swift kick up the pants" [laughter].
Unknown: Did it work?
Brans: I got out of there very quick ...
Kuchař: You were quoted by Herbert [Pfister] as calculating the dragging effect in an expanding ball of dust which stops and recontracts and has some rotation. This is the situation in which one expects gravitational wave production under the outgoing-radiation condition, but presumably you have also some waves coming in that cancel the waves that are going out. Are you able to separate the effects which are due to

the waves from those which are due to the rotation?
Brill: It's just too low an order. It's a spherically symmetrical situation, and then just a rotational perturbation. To see gravitational waves you need higher order multipoles.
Pfister (post-conference comment): Work on higher order perturbations of collapsing rotating dust shells is in progress. There are indications (but not yet a proof) that the gravitational waves can be confined to the exterior of the shell.
Isenberg: I recall your paper with Lee Lindblom that you were very careful in following the paths of photons to measure dragging. I was wondering if, based on that experience, you think that's really important whenever you're looking at dragging effects, or can you just read things off from metric coefficients as is often done?
Brill: The reason we followed photons was of course so that we'd get something coordinate independent, measure something invariant; now if one had some theorem that told you these metric coefficients in the situation always measure the outcome of such and such experiment, then of course that would be nice. I'm not aware of any such thing and so one should calculate some kind of invariant, whether it be photons or carrier pigeons....
Ehlers: Yes, I just wanted to say that one can consider also the effect of a rotating mass distribution on light rays, and this is perhaps not quite as far away from realistic situations as other things, namely, you know that gravitational lensing has been observed now in various cases, and one may ask whether, if you have a more or less rapidly rotating object, such as a galaxy where at least the inner parts might be rotating rapidly, whether this would produce a type of gravitational lensing which has a signature that could distinguish it from lensing of an object without rotation. Now unfortunately, calculations which have been done on that show that in principle, yes, there is a difference between the nonrotating and rotating object as a gravitational lens, but the effect is so small that there is no chance in the foreseeable future to actually see it.
Brill: Presumably it also has an effect on the plane of polarization of the light.
Ehlers: Yes.
Ciufolini: I think that would be a useful and interesting exercise to find a theory that is in agreement with all the past experiments but that doesn't have any dragging effects. It should be possible.
Nordtvedt: I'll bet you an Italian dinner it cannot.

Dragging Effects Near a Rigidly Rotating Disk of Dust

Reinhard Meinel and Andreas Kleinwächter

1. Introduction

Dragging effects near rotating bodies, which appear in Einstein's theory of gravitation, confirm without doubt some of the ideas of Ernst Mach (cf. the review given by Pfister 1995). For a detailed discussion of these dragging effects, exact solutions of Einstein's field equations describing rotating objects are desirable. Beside the solution of Kerr (1963), representing the gravitational field of a rotating black hole, almost no global solutions of that kind are available to date.

The exact solution of the general relativistic problem of a rigidly rotating disk of dust recently found by Neugebauer and Meinel (1993, 1994) allows the explicit calculation of dragging effects. In the present paper we investigate the dragging of the locally nonrotating frames of reference, as well as direct and retrograde circular geodesic orbits just at the rim of the disk. It turns out that retrograde geodesic orbits of test particles are impossible for $z > 0.28511\ldots$, where z is the relative redshift of photons from the center of the disk measured at infinity. The disk solution exists for $0 < z < \infty$ with $z \ll 1$ being the Newtonian limit (the classical zero-pressure Maclaurin disk) and $z \to \infty$ leading to the extreme Kerr solution. For $z > 1.88867\ldots$, no retrograde particle motion (geodesic and nongeodesic) is possible at all since the rim of the disk belongs to the ergosphere in this case. Inside the ergosphere all particles or observers are forced to rotate in the same direction as the disk (seen from an inertial frame at infinity). This is one of the most striking Machian effects in general relativity.

Einstein Studies, vol. 6: Mach's Principle: From Newton's Bucket to Quantum Gravity, pp. 339–346

2. The Rigidly Rotating Disk of Dust

The general relativistic problem of a rigidly rotating disk of dust has been solved approximately by Bardeen and Wagoner (1969, 1971). The exact solution found by Neugebauer and Meinel (1993) is given in terms of two linear integral equations (a 'big' and a 'small' one).

The line element can be written in the Weyl–Lewis–Papapetrou form

$$ds^2 = e^{-2U}[e^{2k}(d\rho^2 + d\zeta^2) + \rho^2 d\varphi^2] - e^{2U}(dt + a\,d\varphi)^2, \tag{1}$$

where the functions U, k, and a depend on ρ and ζ only. The energy–momentum tensor is

$$\begin{gathered} T^{ij} = \epsilon u^i u^j, \text{ with } \epsilon = \delta(\zeta) e^{2(U-k)} \sigma(\rho), \\ u^i = e^{-V}(\xi^i + \Omega \eta^i), \; u^i u_i = -1. \end{gathered} \tag{2}$$

Here $\sigma(\rho)$ is a surface mass density (which cannot be prescribed but has to be calculated from the solution), Ω is the constant angular velocity of the disk, and $\xi^i = \delta^i_4$ and $\eta^i = \delta^i_3$ are the Killing vectors corresponding to stationarity and axisymmetry, respectively. (Note that we use units where the velocity of light as well as Newton's gravitational constant are equal to 1.) The disk is characterized by $\zeta = 0$, $\rho \leq \rho_0$ with the coordinate radius ρ_0. The asymptotically flat solution of the outlined problem depends on two parameters: Ω and z. The redshift parameter z is related to $V_0 \equiv U(\rho = 0, \zeta = 0)$:

$$z = e^{-V_0} - 1. \tag{3}$$

All other parameters characterizing the solution are functions of Ω and z, for example, the total gravitational mass $M = M(\Omega, z)$ and the total angular momentum $J = J(\Omega, z)$. It turns out that $\Omega\rho_0$ depends on z alone. Therefore, instead of z, another parameter μ can be used:

$$\mu = 2\Omega^2 \rho_0^2 (1 + z)^2. \tag{4}$$

This parameter enters the small integral equation for some function $\hat{\beta}(x, \mu)$:

$$\hat{\beta} = x(1 - x^2) + \mu^2 \mathbf{D}\hat{\beta} \tag{5}$$

with

$$\mathbf{D}[f(x)] = -(1 - x^2)^2 f(x) - (1 - x^2)\mathbf{C}[(1 - x^2)\mathbf{C}f(x)], \tag{6}$$

and

$$\mathbf{C}f(x) = \frac{1}{\pi} \oint_{-1}^{1} \frac{f(x')\,dx'}{x' - x}, \tag{7}$$

where $\oint$ denotes Cauchy's principal value. From the solution $\hat{\beta}$ of this integral equation one immediately gets the metric coefficients on the symmetry axis and on the disk, as well as many interesting parameter relations. For instance, $V_0(\mu)$ is given by

$$V_0(\mu) = -\frac{1}{2}\operatorname{arsinh}\left\{\mu + \frac{\mu^3}{\pi^2}\left[\int_{-1}^{1}\frac{\beta(x,\mu)dx}{x}\right]^2\right\}. \tag{8}$$

Together with Eq. (3), this provides us with the function $z(\mu)$. The parameter range $0<z<\infty$ corresponds to $0<\mu<\mu_0$, where μ_0 is related to the first nontrivial eigenvalue of the homogeneous equation $\beta_0=\mu_0^2\mathbf{D}\beta_0$:

$$\mu_0 = 4.62966184347\ldots\ . \tag{9}$$

Using the Neumann series representation of the solution of Eq. (5), one can derive a series expansion of the function $z(\mu)$:

$$z(\mu)=\frac{1}{2}\mu+\frac{1}{8}\mu^2+\left[-\frac{1}{16}+\frac{8}{9\pi^2}\right]\mu^3+\left[-\frac{5}{128}+\frac{4}{9\pi^2}\right]\mu^4$$
$$+\left[\frac{7}{256}-\frac{9}{35\pi^2}\right]\mu^5+\left[\frac{21}{1024}-\frac{151}{630\pi^2}+\frac{32}{81\pi^4}\right]\mu^6+\ldots \tag{10}$$

where the coefficients are polynomials in $1/\pi^2$ which can be calculated recursively. This series converges for $0\le\mu<\mu_0$. Figure 1 shows $z/(1+z)$ as a function of μ.

For a calculation of the metric coefficients at arbitrary ρ, ζ, the solution of the big integral equation is necessary. However, at the rim of the disk, explicit algebraic formulae for e^{2U} and a in terms of Ω, μ, and $z(\mu)$ can be derived. The same holds for the derivatives with respect to ρ.

3. Dragging Effects at the Rim of the Disk

For $\zeta=0$, $\rho=\rho_0$ the metric coefficients $g_{\varphi\varphi}$, $g_{\varphi t}$, and g_{tt} and the corresponding ρ derivatives are given by

$$\Omega^2 g_{\varphi\varphi}=\frac{\mu}{2}-\left[\frac{z}{1+z}\right]^2,\quad \Omega g_{\varphi t}=\frac{z}{1+z}-\frac{\mu}{2},\quad g_{tt}=\frac{\mu}{2}-1, \tag{11}$$

$$\Omega^2\rho_0 g_{\varphi\varphi,\rho}=\mu\frac{1-z}{1+z},\quad \Omega\rho_0 g_{\varphi t,\rho}=\mu\frac{z}{1+z},\quad \rho_0 g_{tt,\rho}=-\mu. \tag{12}$$

See also Fig. 2.

We immediately obtain the angular velocity Ω_{nr} of the so-called locally nonrotating observers as seen from infinity:

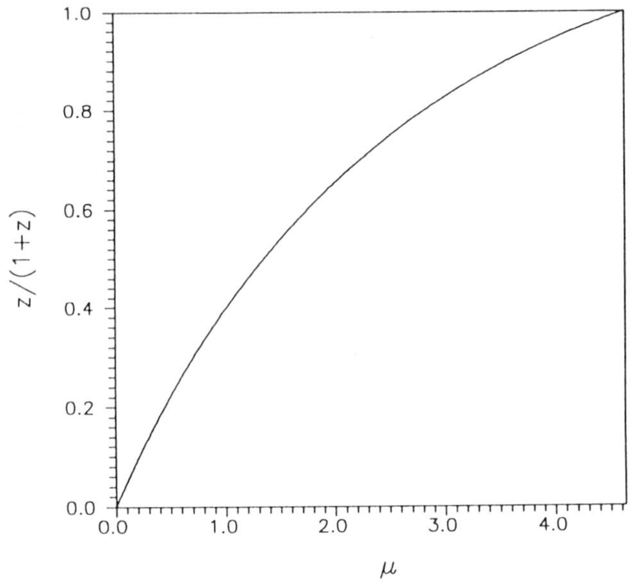

Fig. 1. The relation between the redshift parameter z and the parameter μ.

$$\Omega_{nr} = -\frac{g_{\varphi t}}{g_{\varphi\varphi}} = \Omega\frac{\mu(1+z)^2-2z(1+z)}{\mu(1+z)^2-2z^2}. \tag{13}$$

Ω_{nr} approaches exactly Ω in the 'ultrarelativistic limit' $\mu \to \mu_0$ (i.e., $z \to \infty$). The latter statement holds not only at the rim of the disk but throughout the whole disk. Thus, viewed from infinity, perfect dragging is reached. However, as already pointed out by Bardeen and Wagoner (1971), locally no perfect dragging occurs. The explanation of this apparent contradiction lies in the infinite time dilation between the disk and infinity in this limit.

Furthermore, Eqs. (11) and (12) allow us to calculate circular geodesic orbits of test particles at the rim of the disk. There are two solutions in general: direct orbits with an angular velocity $\Omega_+ > 0$ and retrograde orbits with $\Omega_- < 0$. (We assume without loss of generality that the angular velocity of the disk Ω is positive; all angular velocities are coordinate angular velocities, i.e., those angular velocities seen from infinity.) Of course, $\Omega_+ = \Omega$, since the dust particles of the disk move on circular geodesics themselves. For Ω_- we obtain

$$\Omega_- = -\Omega\frac{1+z}{1-z}. \tag{14}$$

However, the retrograde geodesic motion is only possible as long as

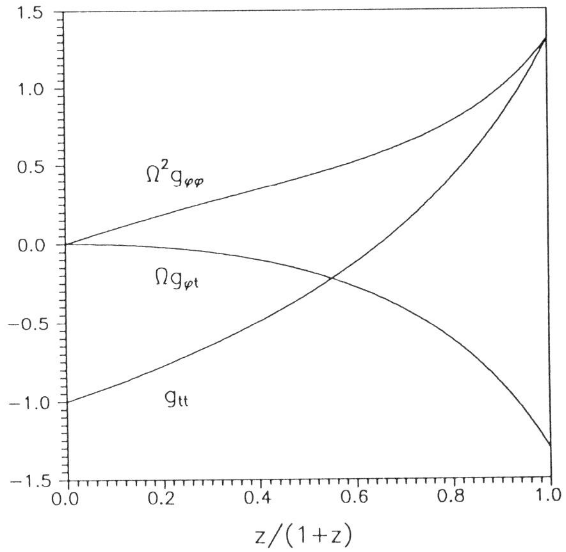

Fig. 2. The metric coefficients $g_{\varphi\varphi}$, $g_{\varphi t}$, and g_{tt} at the rim of the disk in dependence on the redshift parameter z.

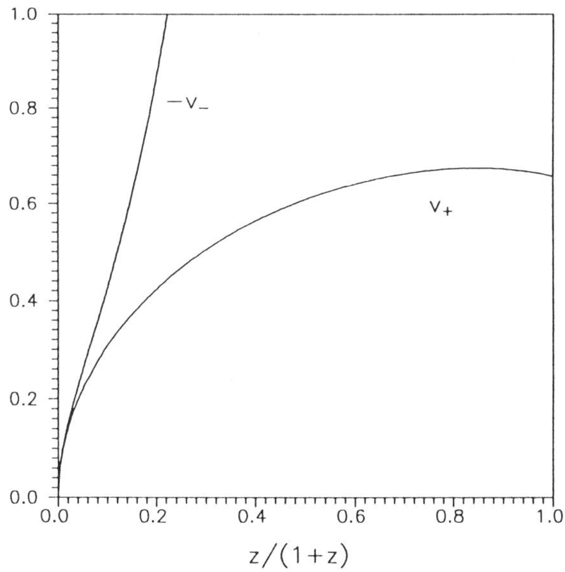

Fig. 3. The linear velocities of rotation $v_\pm$ of test particles moving on circular geodesic orbits at the rim of the disk measured in the locally nonrotating frame of reference. The retrograde motion (v_-) is possible for $z<0.28511\ldots$ only.

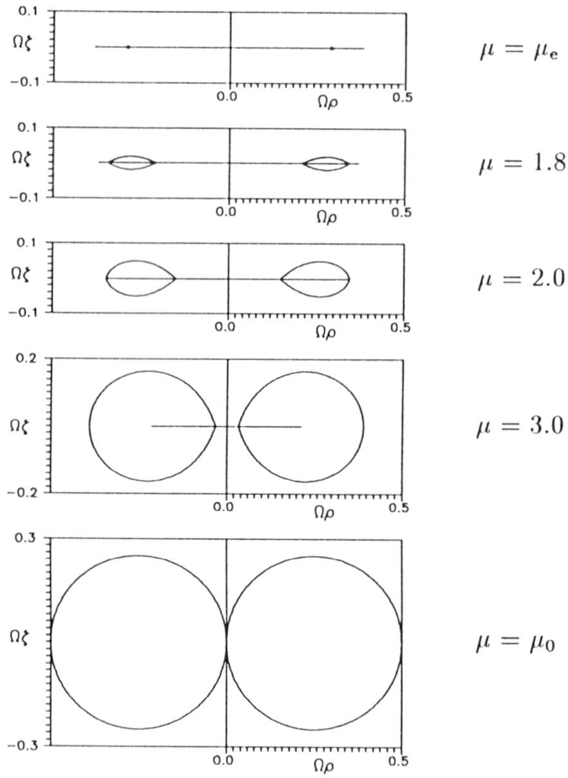

Fig. 4. The ergosphere. It appears for $\mu > \mu_e = 1.68849\ldots$ and reaches the rim of the disk at $\mu = 2$. For $\mu \to \mu_0$, the ergosphere of the extreme Kerr solution (with $M = 1/2\Omega$) occurs. The horizontal lines represent the disk. Its coordinate radius ρ_0 vanishes in the limit $\mu \to \mu_0$.

$\mu < ½$. (For $\mu \geq ½$, no timelike retrograde circular geodesics exist.) The specific energy $E_\pm = -\xi_i u^i_\pm$ and the specific angular momentum $L_\pm = \eta_i u^i_\pm$ ($u^i_\pm$ means the four-velocity) are given as follows:

$$E_+ = 1, \quad L_+ = \frac{1}{\Omega}\frac{z}{1+z}, \tag{15}$$

$$E_- = \frac{1-\mu}{\sqrt{1-2\mu}}, \quad L_- = -\frac{1}{\Omega}\frac{\mu(1+z)-z}{(1+z)\sqrt{1-2\mu}}. \tag{16}$$

Another interesting quantity is the linear velocity of rotation $v_\pm$ measured in the locally nonrotating frame of reference:

$$\frac{v_\pm}{\sqrt{1-v_\pm^2}}=\frac{\eta_i u_\pm^i}{\sqrt{\eta_k\eta^k}}. \tag{17}$$

We get

$$v_+=\sqrt{2/\mu}\frac{z}{1+z} \text{ and } v_-=-\sqrt{2/\mu}\frac{\mu(1+z)-z}{1-z}. \tag{18}$$

It can easily be seen that at the limit $\mu=½$ of the retrograde motion, E_- and L_- become infinite and $|v_-|$ approaches 1, i.e., the velocity of light. Figure 3 shows v_+ and v_- in dependence on $z/(1+z)$.

4. The Ergosphere

For sufficiently large values of the relativity parameter z, the rigidly rotating disk of dust possesses an ergosphere, i.e., a region where no static observer (seen from infinity) is possible. Within this region $d\varphi/dt>0$ must hold for any timelike worldline.

The ergosphere is characterized by

$$\xi_i\xi^i=g_{tt}>0, \tag{19}$$

meaning that the Killing vector ξ_i of stationarity (normalized by $\xi_i\xi^i=-1$ at infinity) becomes spacelike there. From the last Eq. (11) we find that the rim of the disk belongs to the ergosphere for $\mu>2$. Thus, whereas for $\mu\geq½$ no *geodesic* retrograde motion is possible, for $\mu\geq2$ no motion of a real body or observer in retrograde direction is possible at all at the rim of the disk. The torus-like shape of the ergosphere in dependence on z may be found by a numerical solution of the big integral equation; see Fig. 4. Bardeen and Wagoner (1971) obtained similar shapes.

For $\mu\to\mu_0$ the 'exterior' solution ($\rho^2+\zeta^2\neq0$) approaches exactly the extreme ($J=M^2$) Kerr solution; see (Neugebauer and Meinel 1993). Note that ρ and ζ are Weyl's canonical coordinates.

REFERENCES

Bardeen, James M. and Wagoner, Robert V. (1969). "Uniformly Rotating Disks in General Relativity." *Astrophysical Journal* 158: L65–L69.

Bardeen, James M. and Wagoner, Robert V. (1971). "Relativistic Disks. I. Uniform Rotation." *Astrophysical Journal* 167: 359–423.

Kerr, Roy P. (1963). "Gravitational Field of a Spinning Mass as an Example of Algebraically Special Metrics." *Physical Review Letters* 11: 237–238.

Neugebauer, Gernot and Meinel, Reinhard (1993). "The Einsteinian Gravitational Field of the Rigidly Rotating Disk of Dust." *Astrophysical Journal* 414: L97–L99.

Neugebauer, Gernot and Meinel, Reinhard (1994). "General Relativistic Gravitational Field of a Rigidly Rotating Disk of Dust: Axis Potential, Disk Metric, and Surface Mass Density." *Physical Review Letters* 73: 2166–2168.

Pfister, Herbert (1995). "Dragging Effects Near Rotating Bodies and in Cosmological Models." This volume, p. 315.

Dragging Effects and the Theory of Active Galactic Nuclei

Vladimír Karas and Antonio Lanza

1. Introduction

The dragging of inertial frames is considered a typical Machian effect that partially incorporates Mach's ideas about inertial forces in general relativity. A lot has been said and written about the conceptual importance of the dragging effects; we direct the interested reader to the articles in the recently published *Proceedings of the Conference on Ernst Mach and the Development of Physics* (Prosser and Folta 1991) and to Professor Pfister's contribution at this conference [p. 320], where a number of earlier references can be found. Although dragging is considered to be a firm consequence of the general theory of relativity, it has not yet been confirmed by experiment. Several approaches were proposed to verify the predicted value of the dragging in the field of the rotating Earth (Everitt 1971; Braginskij, Polnarev and Thorne 1984; Ciufolini 1986 and 1993; Reasenberg 1987). In the strong gravitational field of a rotating compact object, the dragging becomes more pronounced and it can result in specific effects. In our present contribution we want to emphasize that these effects are important in the theory of certain astronomical objects and that observations could reveal them in the near future.

The Penrose process (Penrose 1969) and its electromagnetic analog, the Blandford–Znajek process (Blandford and Znajek 1977; Wagh and Dadhich 1989), have been proposed as the mechanisms for extracting the rotational energy from the black hole. In their standard formulation, these processes can act if a rotating black hole is surrounded by particles or electromagnetic fields. As such, they are a definite example of Machian effects in general relativity. Although these processes are very important from the theoretical point of view, astronomers have not yet

Einstein Studies, vol. 6: Mach's Principle: From Newton's Bucket to Quantum Gravity, pp. 347–352

found a way to detect them by direct observation. Here, we discuss the dragging effects in a slightly different context, namely, we consider the specific precession motion of the orbit of a solar-mass star orbiting a supermassive [$M \approx (10^6 - 10^{11}) M_\odot$] rotating black hole, and the dragging in a fully relativistic model of a self-gravitating disk. The above-mentioned effects are relevant for the theory of active galactic nuclei (AGNs), the most luminous objects in the universe.

2. Gravitomagnetic Precession of the Orbit of a Star Near a Supermassive Rotating Black Hole

According to the so-called standard scenario, which many astrophysicists adopt, AGNs harbor a supermassive black hole and a plasma accretion disk (Shlosman, Begelman, and Frank 1990). Neither the black hole nor the disk itself has been directly confirmed observationally, but it seems that the standard model is the only one which is capable of explaining diverse properties of different types of AGNs. Unfortunately, the strong relativistic effects which are expected near the black hole horizon cannot easily be detected. They are largely masked by violent plasma processes in the surrounding material, which is gravitationally attracted towards the black hole. Also, present-day techniques do not possess enough resolution to resolve the innermost parts of these objects. Calculations show that the matter in the inner regions becomes extremely hot, and the radiation we observe is in the X-band (Novikov and Thorne 1973). A large amount of data has been collected from astronomical satellites which show that the sources are very often variable on all observed timescales, ranging from minutes to months. The variability is either featureless or, in some cases, semiperiodic. However, very little information about physical processes in AGNs has been gained by studying their variability. Here, we speculate about possible consequences of the AGN variability.

Several authors proposed that a solar-mass star can be captured in an eccentric orbit around the black hole in the core of an AGN. The star may originate from a dense star cluster that is assumed to surround the nucleus. The process of the capture is not particularly well understood. It has been proposed that the effects of tidal distortion and related dissipation of energy or disruption of a binary system play a major role (Hills 1988; Rees 1988). Alternatively, cumulative effects of the collisions of the star with the accretion disk can change the orbital parameters and result in a capture of an originally unbound or a weakly bound star (Syer, Clarke, and Rees 1992; Vokrouhlický and Karas 1993). The orbit is then gradually circularized. The star periodically intersecting the disk will modulate the signal from the source and give us information about the location and time of the star–disk collisions (Rees 1993). If the central black hole rotates, and this turns out to be very likely, the orbital plane of the star will be dragged by the Lense–Thirring precession (Lense and Thirring 1918). Correspondingly, the places of intersection with the disk will also be dragged, and this will result in specific periodicities in the observed X-ray flux (Karas and Vokrouhlický 1993).

At present there are no 'hot' candidates for where the relevant periodicities could be discovered. Data from *ROSAT* will be of particular relevance in the search for suitable objects. Detection of the Lense–Thirring precession would provide us with independent evidence of black holes in the nuclei of active galaxies.

Self-consistent analysis of the accretion disk theory (Wiita 1982) has shown that in order to match the energetical output of quasars, the ratio m/M of the disk mass m to the black hole mass should be greater than 0.1. For such a high mass ratio, the self-gravity of the disk will not be negligible. We discuss this subject briefly in the following section.

3. Self-Gravitating Disks around Rotating Black Holes

Solutions of Einstein equations for self-gravitating disks or rings with a central black hole (Will 1974; Lanza 1992) or without it (Bardeen and Wagoner 1971; Neugebauer and Meinel 1993) have been found in the past. Self-gravitating disks around relativistic spheroidal configurations have also been considered (Nishida, Eriguchi, and Lanza 1992). The exact solutions of Einstein's equations describing the self-gravitating disks or rings are interesting on their own (Chakrabarti 1988; Turakulov 1990; Bičák, Lynden-Bell, and Katz 1993; Bičák and Ledvinka 1993). Specific relativistic instabilities may considerably change our current understanding of the accretion process (Lanza *et al.* 1993). It has been recently proposed that massive disks around neutron stars or stellar-mass black holes could be relevant for the models of γ-ray bursts (Narayan, Paczyński, and Piran 1992). We are mainly interested in the disks around rapidly rotating black holes (Lanza 1992), where the dragging effects are most apparent.

In view of the nonlinearities of the equations describing the equilibrium of self-gravitating disks around black holes, one can solve them only numerically (Nishida, Eriguchi, and Lanza 1992). Both the

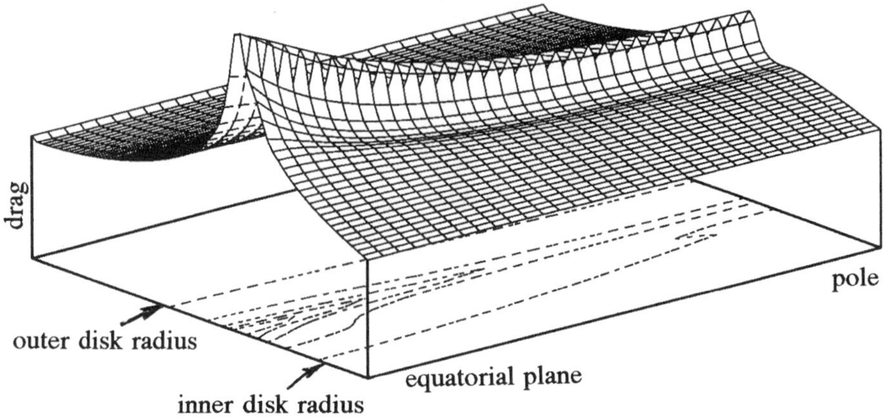

Fig. 1. Dragging due to a self-gravitating thin disk near a rotating black hole. This figure illustrates that the dragging effect reaches its maximum at the center of the disk rather than at the black hole horizon. Here, the ratio of the disk mass to the black hole mass is 0.81; for details see (Lanza 1992).

angular momentum of the black hole and the angular momentum of the disk contribute to the total dragging. (Indeed, it is possible to reach a balance between the two contributions if the positive angular momentum of the disk is exactly compensated by the negative angular momentum of the black hole.) If the black hole rotates slowly and the disk has sufficient mass, the maximum of the dragging effect is located close to the center of the disk rather than at the horizon. One possible observational consequence is that the light rays near a self-gravitating disk are significantly distorted, which results in a change of the spectrum compared to the case when the disk self-gravity is ignored. Another consequence is that trajectories of massive bodies near the disk are attracted to the disk. This makes the scenario outlined in the previous section more complicated because relevant precession frequencies must be corrected for the contributions due to the disk. It appears that even with a disk of relatively moderate mass the change can be important in some cases, in particular for orbits with a low inclination, but this problem still remains to be solved.

To conclude, the dragging effects near a rotating compact object can be of considerable astrophysical importance for the models of black holes surrounded by an accretion disk. It turns out that specific precession frequencies induced by the frame dragging can, in principle, be detected in future observations of active galactic nuclei. This would give us additional evidence about the presence of a rotating black hole and support the standard model of these objects.

REFERENCES

Bardeen, James and Wagoner, Robert W. (1971). "Relativistic Disks I. Uniform Rotation." *Astrophysical Journal* 167: 359–423.

Bičák, Jiří; Lynden-Bell, Donald; and Katz, Joseph (1993). "Relativistic Disks as Sources of Static Vacuum Spacetimes." *Physical Review D* 47: 4334–4343.

Bičák, Jiří and Ledvinka, Tomáš (1993). "Sources for Stationary Axisymmetric Gravitational Fields." *Physical Review D*, submitted.

Blandford, Roger D. and Znajek, Roman L. (1977). "Electromagnetic Extraction of Energy from Kerr Black Holes." *Monthly Notices of the Royal Astronomical Society* 179: 433–456.

Braginsky, Vladimir B., Polnarev, Aleksander G., and Thorne, Kip. S. (1984). "Foucault Pendulum at the South Pole: Proposal for an Experiment to Detect the Earth's General Relativistic Gravitomagnetic Field." *Physical Review Letters* 53: 863–866.

Chakrabarti, Sandip K. (1988). "Spacetime with Self-Gravitating Thick Disk." *Astrophysics and Astronomy* 9: 49–61.

Ciufolini, Ignazio (1986). "Measurement of the Lense–Thirring Drag on High-Altitude Artificial Satellites." *Physical Review Letters* 56: 278–281.

Ciufolini, Ignazio (1993). "Gravitomagnetism, Dragging of Inertial Frames and their Measurement," contribution at this Conference.

Everitt, C. W. Francis (1971). "The Gyroscope Experiment." in: *Experimental Gravitation, Proceedings of the Course 56 of International School of Physics E. Fermi*. Bruno Bertotti, ed. New York: Academic Press, pp. 331–358.

Hills, J. G. (1988). "Hyper-Velocity and Tidal Stars from Binaries Disrupted by a Massive Galactic Black Hole." *Nature* 331: 687–689.

Karas, Vladimír and Vokrouhlický, David (1994). "Relativistic Precession of the Orbit of a Star near a Supermassive Rotating Black Hole." *Astrophysical Journal* 422: 208–218.

Lanza, Antonio (1992). "Multigrid in General Relativity: II. Kerr Spacetime." *Classical and Quantum Gravity* 9: 677–696; "Self-Gravitating Thin Disks around Rapidly Rotating Black Holes." *Astrophysical Journal* 389: 141–156.

Lanza, Antonio; Nishida, Shogo; Eriguchi, Yoshiharu; and Abramowicz, Marek A. (1995). In preparation.

Lense, Josef and Thirring, Hans (1918). "Über den Einfluss der Eigenrotation der Zentralkörper auf die Bewegung der Planeten und Monde nach der Einsteinschen Gravitationstheorie." *Physikalische Zeitschrift* 19: 156–163.

Narayan, Ramesh; Paczyński, Bohdan; and Piran, Tsvi (1992). "Gamma-Ray Bursts as the Death Throes of Massive Binary Stars." *Astrophysical Journal*

395: L83–L86.

Neugebauer, Gernot and Meinel, Reinhard (1993). "The Einsteinian Gravitational Field of the Rigidly Rotating Disk of Dust." *Astrophysical Journal* 414: L97–L99.

Nishida, Shogo; Eriguchi, Yoshiharu; and Lanza, Antonio (1992). "General Relativistic Structure of Star-Toroid Systems." *Astrophysical Journal* 401: 618–627.

Novikov, Igor D. and Thorne, Kip S. (1973). "Black Hole Astrophysics." in: *Black Holes*. Cecile DeWitt and Bryce DeWitt, eds. New York: Gordon and Breach, 343–450.

Penrose, Roger (1969). "Gravitational Collapse: The Role of General Relativity." *Rivista Nuovo Cimento* 1: 252–276.

Prosser, Václav and Folta, Jiří, eds. (1991). *Ernst Mach and the Development of Physics*. Prague: Charles University.

Reasenberg, Robert D. (1987). "Astronomical and Space Experiments." in: *General Relativity and Gravitation, Proceedings of the 11th International Conference on General Relativity and Gravitation*. Malcolm A. H. MacCallum, ed. Cambridge: Cambridge University Press, pp. 276–281.

Rees, Martin J. (1988). "Tidal Disruption of Stars by Black Holes of 10^6–10^8 Solar Masses in nearby Galaxies." *Nature* 333: 523–528.

Rees, Martin J. (1993). "Quasars: Progress and Prospects." in: *The Renaissance of General Relativity and Cosmology: A Survey Meeting to Celebrate the 65th Birthday of Dennis Sciama*. George F. R. Ellis, Antonio Lanza, and John C. Miller, eds. Cambridge: Cambridge University Press, 175–190.

Syer, D., Clarke, Christopher J., and Rees, Martin J. (1992). "Star-Disc Interactions near a Massive Black Hole." *Monthly Notices of the Royal Astronomical Society* 250: 505–512.

Shlosman, Isaac, Begelman, Mitchell C., and Frank, Julian (1990). "The Fuelling of Active Galactic Nuclei." *Nature* 345: 679–686.

Turakulov, Z. Ya. (1990). "Relativistic Rotating Disk: The General Solution." *Modern Physics Letters A* 5: 725–731.

Vokrouhlický, David and Karas, Vladimir (1993). "A Star Orbiting around a Supermassive Rotating Black Hole: Free Motion and Corrections due to Star–Disc Collisions." *Monthly Notices of the Royal Astronomical Society,* 265: 365–378.

Wagh, Sanjay M. and Dadhich, Naresh (1989). "The Energetics of Black Holes in Electromagnetic Fields by the Penrose Process." *Physics Reports* 183: 137–192.

Wiita, Paul J. (1982). "Physical Properties of Thick Supercritical Accretion Disks." *Monthly Notices of the Royal Astronomical Society* 256: 666–680.

Will, Clifford W. (1974). "Perturbation of a Slowly Rotating Black Hole by a Stationary Axisymmetric Ring of Matter. I. Equilibrium Configurations." *Astrophysical Journal* 191: 521–531.

On the Interpretation of Dragging Effects in Rotating Mass Shells

Jörg Frauendiener

1. Introduction

It is generally accepted that dragging effects are the single most direct manifestation of Machian ideas in general relativity (Pfister 1995). The dragging of test bodies and inertial frames by accelerated masses is a real and conceptually important effect of the gravitational interaction. The 'gravimagnetism' influences the relative motion of the test bodies and the surrounding matter.

Up to now the dragging effects have been exhibited mostly in asymptotically flat spacetimes where the matter content is modeled as an infinitely thin shell. Additionally, it was assumed that these spacetimes were stationary and axially symmetric. The magnitude of the dragging effect is measured by the dragging coefficient, i.e., as the ratio of the angular velocities of the mass shell and the inertial frames in the interior. Both angular velocities are measured by an observer at infinity. It is this reference to the preferred Minkowski frame at infinity that has often been criticised as being 'anti-Machian,' as "reintroducing absolute space through the back door."

A way to determine this dragging coefficient is the following: In the asymptotically flat mass shell models, one can cover the whole spacetime by one single global coordinate system which is asymptotically Minkowskian. In this system, the metric in the interior has the form of a flat metric in a coordinate system that rotates with respect to the global system with an angular velocity depending on the radius and the mass of the shell. The dragging coefficient is the ratio between this angular velocity and some representative angular velocity of the shell that depends on the matter model.

In this contribution we want to propose a different definition for the

Einstein Studies, vol. 6: Mach's Principle: From Newton's Bucket to Quantum Gravity, pp. 353–363 © 1995 *Birkhäuser Boston, Inc. Printed in the United States.*

magnitude of the dragging effect which does not refer to the infinitely distant observer, is applicable in principle to a whole range of situations – even spacetimes which are not asymptotically flat – and which can be applied explicitly in the present situation of mass shells with a flat interior. The drawback of this definition is that due to its inherent nonlocal character it is nearly impossible to apply it to any 'realistic' situation. Also, it emerges that one has to consider a whole one-parameter family of spacetimes in order to achieve the independence from an observer.

The idea is to define a reference frame that is uniquely determined by the motion of the matter content of the spacetime. The motion of test bodies or inertial frames is described with respect to that frame. This relative motion then defines the dragging coefficient by reference to a 'standard situation.' The matter frame is 'comoving' with the matter in a sense to be described later. The basic ingredient is the definition of relativistic multipole moments according to Dixon. In this consideration it is not important that the spacetime be asymptotically flat. In fact, the matter frame can be defined (at least in principle) for cosmological situations just as well.

The plan of the paper is as follows: In Sec. 2 we want to describe the Newtonian analog of the situation. In Sec. 3 we will briefly explain the essentials of Dixon's theory of multipole moments in general relativity which will be used in Sec. 4. to compute the dragging effect for mass shells with flat interior. A brief discussion concludes the paper.

2. The Newtonian Analog

Let us suppose we want to describe the motion of a compact body in Euclidean 3-space. At each instant of time we pick an origin O and an orthonormal basis centered at O which we use to define Euclidean coordinates $(x^1, x^2, x^3, t)=(\mathbf{r}, t)$. We define this reference frame as an inertial frame. Then the body is described by its mass distribution $\rho(\mathbf{r}, t)$ and velocity distribution $\mathbf{v}(\mathbf{r}, t)$. Our aim is to define a reference frame in such a way that the body is 'as much at rest as possible' in that frame.

First, we define the mass dipole moment $\mathbf{m}(t)=\int\rho(\mathbf{r}, t)\mathbf{r}d^3x$. Under a change of origin $O\to O+\mathbf{a}(t)$, the mass dipole moment changes

$$\mathbf{m}(t)\to\mathbf{m}(t)+\mathbf{a}(t)\int\rho(\mathbf{r},t)d^3x=\mathbf{m}(t)+M\mathbf{a}(t),$$

with M being the mass of the body. Therefore, we can always assume that the origin has been chosen such that the mass dipole moment $\mathbf{m}(t)$

vanishes for all times t. The origin is then called the center of mass, and the line traced out by it is the center-of-mass line. When there are no external forces present (which we will assume), it is a consequence of the Newtonian equations of motion that the reference frame obtained in that way is still an inertial frame.

Next, we want to fix the freedom in the choice of the triad at O. This being an orthonormal triad, we only have time dependent rotations at our disposal. Consider the kinetic energy of the body

$$E(t) = \frac{1}{2}\int \rho(\mathbf{r},t)\mathbf{v}(\mathbf{r},t)^2 d^3x.$$

Under infinitesimal time-dependent rotations the velocity changes,

$$\mathbf{v}(\mathbf{r},t) \rightarrow \mathbf{v}(\mathbf{r},t) - \mathbf{\Omega}(t) \times \mathbf{r},$$

with $\Omega(t)$ being the instantaneous angular velocity. The kinetic energy changes according to

$$E(t) \rightarrow E'(t) \equiv E(t) - \int \rho(\mathbf{r},t)\mathbf{\Omega}(t) \cdot (\mathbf{v}(\mathbf{r},t) \times \mathbf{r}) d^3x$$

$$+ \frac{1}{2}\int \rho(\mathbf{r},t)(\mathbf{\Omega}(t) \times \mathbf{r})^2 d^3x.$$

At each instant of time we now seek $\mathbf{\Omega}(t)$ such that $E'(t)$ is minimal as a function of Ω. This results in the condition

$$\int \rho(\mathbf{r},t)(\mathbf{r} \times \mathbf{v}(\mathbf{r},t)) d^3x$$

$$= \int \rho(\mathbf{r},t)\{\mathbf{\Omega}(t)\mathbf{r}^2 - (\mathbf{\Omega}(t) \cdot \mathbf{r})\mathbf{r}\} d^3x$$

or

$$\mathbf{S} = \mathbf{I}\mathbf{\Omega},$$

where $\mathbf{I} = (I_{ij}) = \int \rho(\mathbf{r}, t)\{\delta_{ij}\mathbf{r}^2 - x_i x_j\} d^3x$ is the tensor of inertia and $\mathbf{S}$ is the angular momentum of the body at time t. In a frame that rotates with angular velocity $\mathbf{\Omega}(t)$ with respect to the inertial frame, the body is as much at rest as possible in the sense that its kinetic energy is as small as possible. This, uniquely defined, frame is called the 'comoving frame' of the body. [Cf. Ehlers's comments on pp. 231–232.] Note that the kinetic energy need not be zero. This will only be the case if the body has been in rigid motion from the start, i.e., for a rigid body. For then the velocity distribution of the body is given by $\mathbf{v} = \boldsymbol{\omega} \times \mathbf{r}$, and the above criterion for choosing $\mathbf{\Omega}$ gives $\mathbf{\Omega}(t) = \boldsymbol{\omega}(t)$ for all t such that the comoving frame coincides with the rest frame of the body.

In order to define the comoving frame, we had to be able to locate the center-of-mass line, and we needed the angular momentum of the body and its mass quadrupole moment (the tensor of inertia is the negative trace reversal of the quadrupole moment). Fortunately, exactly these notions can be rigorously defined within general relativity. Therefore, it is possible to define a comoving frame in general relativity as well.

3. The Comoving Frame in General Relativity

Since the Dixon theory of relativistic multipole moments is very elaborate and not suited to be presented in the context of this conference we want to describe the relevant definitions and calculations as briefly as possible. For a detailed account of the Dixon theory, see (Dixon 1979) and references therein.

We will assume from the outset that the spacetime is asymptotically flat, stationary, and axisymmetric and contains only a mass shell with flat interior. In addition, we will assume the existence of a discrete equatorial reflection symmetry (this is probably a consequence of stationarity and axisymmetry as in Newtonian theory). Important quantities in the Dixon theory are the world function σ and the so-called Jacobi propagators H and K. In general, these are very difficult to come by because one needs to know the general solution of the geodesic and the Jacobi equations in order to determine them. However, here we need to know them only in the flat interior of the mass shell, so there is no problem in this case.

In general, the world function is defined as follows. Let O and O' be two points in M, and let γ be the geodesic with affine parameter u connecting $O=\gamma(u_0)$ and $O'=\gamma(u_1)$; then

$$\sigma(O,O') = \frac{1}{2}(u_1-u_0)\int_{u_0}^{u_1} g_{ab}(\gamma(u))\dot{\gamma}^a(u)\dot{\gamma}^b(u)du.$$

Here, σ is a two-point function, and hence its derivative at (O,O') is a map from $T_OM\times T_{O'}M$ to the reals. In order to distinguish between vectors at O and O', we will denote vectors (and tensors) at O (O') with unprimed (primed) Latin letters. Then $\partial_a\sigma=:\sigma_a$ is the partial derivative of σ with respect to the first (unprimed) argument, and $\sigma_{aa'}:=\partial_a\partial_{a'}\sigma$ is the mixed part of the second derivative of σ. One can show that $\sigma^a(O, O')=(u_1-u_0)\dot{\gamma}^a(u_0)$ and $\sigma^{a'}(O, O')=-(u_1-u_0)\dot{\gamma}^{a'}(u_1)$.

In flat space, using Cartesian coordinates (t,x,y,z), we have

$$\sigma(O,O') = \frac{1}{2}\{-(t-t')^2+(x-x')^2+(y-y')^2+(z-z')^2\}$$

and

$$\sigma_a(O,O') = (t-t')t_a(O) + (x-x')x_a(O) + (y-y')y_a(O) + (z-z')z_a(O),$$
$$\sigma_{aa'}(O,O') = t_a(O)t_{a'}(O') - x_a(O)x_{a'}(O') - y_a(O)y_{a'}(O') - z_a(O)z_{a'}(O'),$$
$$\sigma_{ab}(O,O') = g_{ab}(O), \quad \sigma_{a'b'}(O,O') = g_{a'b'}(O'),$$

where we define $t^a = \partial^a_t$, etc. Next, we introduce the Jacobi propagator H. This is most conveniently done by the relation:

$$\frac{\partial}{\partial X^a}(\exp_{(O)}X)^{a'} = H^{a'}{}_a(O,O'),$$

for all $X \in T_O M$, where $O' = \exp_{(O)}(X)$. This defines the propagator H via the derivative of the exponential map. The value of H at (O,O') is a map from $T_O M$ to $T_{O'}M$, and therefore it allows the transport of vectors from $T_O M$ to $T_{O'}M$. Note that this transport is different from the parallel transport in general. Having defined H, the second Jacobi propagator is defined as $K^{a'}{}_a(O,O') = H^{a'}{}_b(O,O')\sigma^b{}_a$. The propagators are called Jacobi propagators because they map the initial values at O of solutions to the Jacobi equation into the values at O'; see (Schattner and Trümper 1981). In flat space, one identifies the tangent space at any point with the space itself, so that the exponential map is the identity. Therefore, we obtain

$$H_{aa'}(O,O') = K_{aa'}(O,O') =$$
$$-t_a(O)t_{a'}(O') + x_a(O')x_{a'}(O') + y_a(O)y_{a'}(O') + z_a(O)z_{a'}(O').$$

The main result of Dixon's theory is the equivalence of two descriptions for extended isolated bodies in general relativity. One way to describe a system of extended bodies is by means of a symmetric tensor field T^{ab}, the stress–energy tensor, whose support is spatially compact and which is divergence free: $\partial_a T^{ab} = 0$. The information contained in T^{ab} can be coded into a collection of tensors P^a, S^{ab}, $J^{e_1 \cdots e_n abcd}$ with $n \geq 0$, which are defined along a timelike line ℓ in the neighborhood of the support of T^{ab}. They depend on ℓ and a timelike covector n_a along ℓ. For each choice of (ℓ, n_a), there exists such a collection. The vector P^a is called the 4-momentum, S^{ab} is the angular momentum and $J^{e_1 \cdots e_n abcd}$ is the sequence of higher multipoles. Dixon has given an integral representation of these tensors that we will use below. In general, P^a and n^a are different and also different from the tangent vector u^a to ℓ. The multipole moments have the symmetry and orthogonality properties

$$J^{e_1, \ldots e_n abcd} = J^{(e_1, \ldots e_n)[ab][cd]},$$
$$J^{e_1 \ldots e_n a[bcd]} = 0, \qquad J^{e_1 \ldots e_{n-1}[e_n ab]cd} = 0,$$
$$n_{e_1} J^{e_1 \ldots e_n abcd} = 0.$$

This collection of tensors along ℓ satisfies a system of 10 ordinary differential equations which determines the evolution of the momentum and angular momentum along ℓ in terms of the curvature of spacetime and the higher multipoles.

We now proceed to introduce the center-of-mass line within the Dixon theory. To this end we define the mass dipole moment m^a along ℓ by $m^a = n_b S^{ab}$. It has been shown (Schattner 1979) that there exists exactly one pair (ℓ_0, n_a) such that

$$n_{[a}P_{b]} = 0, \text{ and } m^a = 0.$$

This line ℓ_0 is the center-of-mass line. The parameter s along ℓ_0 is not fixed by this construction. Therefore, we will assume a normalization of the tangent vector by the condition $n_a u^a = -1$. We will assume henceforth that the multipole moments are given in the center-of-mass description. Finally, we introduce the spin vector $S^a = -(1/2)\epsilon^{ab}{}_{cd} n_b S^{cd}$, which is orthogonal to n_a by construction.

Following (Ehlers and Rudolph 1977), we now introduce the co-moving frame. This is done by means of the mass quadrupole moment

$$m^{ab} = \frac{4}{3} J^{acbd} n_c n_d,$$

which serves to define the tensor of inertia

$$I^{ab} = m^c{}_c h^{ab} - m^{ab}.$$

Here, we have used $h_{ab} = g_{ab} + n_a n_b$, the metric in the 3-space orthogonal to n_a. Both m^{ab} and I^{ab} are purely spatial with respect to n_a. Now the angular velocity vector Ω^a is defined in complete analogy to the Newtonian theory by the conditions

$$S^a = I^{ab}\Omega_b, \quad n_a \Omega^a = 0.$$

To end this section we will show how momentum and angular momentum are expressed as integrals over the mass distribution. Given a pair (ℓ, n_a), we define the spacelike hypersurface $\Sigma(s)$ as the image of (part of) the subspace of $T_O M$ orthogonal to n_a at $O = \ell(s)$. So $\Sigma(s)$ is a surface of simultaneity with respect to n_a. Let $\Sigma_{b'}$ be the oriented hypersurface element of $\Sigma(s)$. Then

$$P^a(O,n) = \int_{\Sigma(s)} K^a_{a'}(O,O') T^{a'b'}(O') \Sigma_{b'},$$

$$S^{ab}(O,n) = 2\int_{\Sigma(s)} H^{[a}_{a'}(O,O') \sigma^{b]}(O,O') T^{a'b'}(O') \Sigma_{b'}.$$

The integral expressions for the higher multipole moments are given explicitly in (Dixon 1974). However, they are so complicated in general that we will only show how to compute the mass quadrupole moment in the present special case.

4. The Dragging Coefficient in Mass Shells with Flat Interior

Our first task now is to obtain the center-of-mass line. In our situation of a stationary and axisymmetric system, it is obvious and has, in fact, been proven rigorously by Schattner and Streubel (1981) that the center-of-mass line has to be an integral curve of the timelike Killing vector that is contained entirely in the two-dimensional axis, the fixed point set of the rotation symmetry. Under our assumption of a reflection symmetry, there is a unique line that is invariant under reflections. This must be the center-of-mass line. We now introduce Cartesian coordinates (t,x,y,z) such that the axial Killing vector is tangent to the (xy)-plane and vanishes on the z-axis for all values of t. In addition, we take $z \mapsto -z$ as the reflection symmetry. Then ℓ consists of all points with $x=y=z=0$. The unit normal vector field to $\Sigma(s)$ has constant coefficients with respect to the Cartesian frame (t^a,x^a,y^a,z^a) and coincides with the given n_a at O. So $\Sigma_{b'} = -n_{b'}d^3S$ with the volume element d^3S of $\Sigma(s)$.

The support of the energy–momentum tensor is distributional in our case. Therefore, the volume integrals for computing the various moments degenerate into surface integrals over the intersection $S(s) = \Sigma(s) \cap \mathrm{supp}(T^{ab})$. Now $S(s)$ is a two-dimensional surface with sphere topology that can be described by an equation $F(r,z)=0$, where $r=(x^2+y^2)^{1/2}$ is the distance from the axis. In fact, in the approximation that will be considered later, the surface is a sphere. Note that the components of T^{ab} orthogonal to $S(s)$ vanish. Also, because of the symmetry, $T^{ab}t_a z_b=0$.

It is clear that even though the support of the energy–momentum tensor is concentrated on a shell, there should be contributions to the world function. However, it is a relatively simple matter to convince oneself that these cannot be seen on the level of the first and second derivatives; they are continuous across the shell. Therefore, it suffices to consider only the world function for the flat interior region.

By symmetry, the frame with vanishing mass dipole moment will be the one with $n_a=t_a$; indeed, for this choice and using the expression for $K_{aa'}$ from the previous section, we obtain

$$P_a(s) = t_a \int_{S(s)} t_{a'} T^{a'b'} t_{b'} d^2S = M_D t_a,$$

thus defining the Dixon-mass M_D. Here, d^2S is the surface element of the shell $S(s)$. Similarly, the angular momentum should be proportional to $x^{[a}y^{b]}$. Define the one-form $\phi_a = xy_a - yx_a$ to obtain

$$S^{ab} = 2x^{[a}y^{b]} \int_{S(s)} \phi_{a'} T^{a'b'} t_{b'} d^2S.$$

Hence, the spin vector is $S^c = Sz^c$, where S is the integral above.

After specializing the general integral representation in Dixon (1974) to the present case, we find for the mass quadrupole moment

$$m^{ab}(s) = \int_{\Sigma(s)} \sigma^a \sigma^b t_{a'} t_{b'} T^{a'b'} \Sigma,$$

which in turn can be written as

$$m^{ab}(s) = \frac{1}{2} Q_r \{x^a x^b + y^a y^b\} + Q_z z^a z^b,$$

$$Q_r = \int_{S(s)} r^2 t_{a'} t_{b'} T^{a'b'} d^2S, \quad Q_z = \int_{S(s)} z^2 t_{a'} t_{b'} T^{a'b'} d^2S.$$

And so we get for the tensor of inertia the expression

$$I^{ab} = (Q_z + Q_r/2)\{x^a x^b + y^a y^b\} + Q_r z^a z^b.$$

This expression has the same algebraic appearance as in the Newtonian theory. It is now a simple matter to compute the angular velocity vector Ω^a, which on symmetry grounds has to be of the form $\Omega^a = \Omega z^a$. We get $\Omega = S/Q_r$.

We now want to apply the above to the mass shell of Brill and Cohen (1966) as given in (Pfister and Braun 1985). Apart from a trivial coordinate change, the metric is

$$ds^2 = -e^{2U} dT^2 + e^{2K-2U}(d\rho^2 + dZ^2) + W^2 e^{-2U}(d\Phi - \omega A dT)^2$$

with functions U, K, W, and A, depending on ρ and Z only, which are constant inside the shell. To make the connection with the formulae above, we transform to Cartesian coordinates in the interior: $t = e^U T$, $x = e^{U-K} \rho \cos(\Phi - \omega A T)$, $y = e^{U-K} \rho \sin(\Phi - \omega A T)$, $z = e^{U-K} Z$. To first order in the rotation parameter ω, the shell is spherical with radius R. In this approximation the components of the energy–momentum tensor appearing in the integrals can be determined. We obtain $T^{ab} t_a t_b = -(T^0{}_0 + \omega A T^0{}_3)$ and $T^{ab} \phi_a t_b = -T^0{}_3$. The components $T^0{}_0$ and $T^0{}_3$ can be taken directly from Pfister and Braun (1985). We see that $T^0{}_3$ is proportional to ω, so that the second term in $T^{ab} t_a t_b$ can be neglected in the first-order approximation. Inserting in the integrals, we find the angular velocity of the comoving frame with respect to the inertial frame in the interior

$$\Omega = \omega \frac{R - M/4}{R - M/6} \leq \omega.$$

Here, M is the ADM-mass of the spacetime.

5. Discussion

As was explained in Sec. 3, Ω is the angular velocity of the comoving frame with respect to the inertial frame attached to the center-of-mass of the shell. As such, we may also regard it as the angular velocity of the shell with respect to that inertial frame. To first order in ω, the shell rotates rigidly, so the Newtonian theory would predict $\Omega=\omega$, since ω corresponds to the angular velocity of the shell in the Newtonian limit. However, here we have $\Omega<\omega$ for $M>0$. This means that, in fact, the shell rotates slower than in Newtonian theory as seen from the inertial frame inside; or, the other way round, the inertial frame is dragged around by the rotating mass. We define the dragging coefficient to be

$$q=\frac{\omega-\Omega}{\omega}.$$

Defined in this way $q=(M/2)(6R-M)^{-1}$ is a measure of the dragging of the frames with respect to the *Newtonian* situation. The dragging q vanishes in the Newtonian limit and increases with increasing M/R, which is qualitatively what one would expect. In the second-order approximation the shell is no longer spherical and the components of the energy–momentum tensor become quite complicated. It might be interesting to find the dragging coefficient also in that approximation.

In contrast to the other description of the dragging effect, we do not get a total dragging ($q=1$) of the inertial frames in the limiting case of a compact shell where $M=2R$ (note that we are not using the usual Schwarzschild coordinates). In fact, we get $q\to 1/4$ in that limit. Another discrepancy is the fact that we cannot reproduce the classical result of Thirring (1918). There the dragging coefficient was calculated in a weak field approximation, resulting in $q=4M/3R$, while we obtain $q=M/12R$ in the limit $R\gg M$. There are two explanations for these discrepancies. The first is a physical one: The observations at infinity of the angular velocities of the shell and the inner inertial frames are subject to time dilations. These become infinite for the compact shell because then a horizon appears. However, the ratio of the angular velocities remains finite and approaches unity.

However, the main reason for the discrepancy is that the two ways of defining the dragging coefficient are conceptually different. The previous description assumes one given spacetime and depends on a chosen observer, usually taken to be at infinity. The present definition, however, assumes a one-parameter family of spacetimes and compares the relative angular velocities at a given member of the family with a reference value, usually taken as the Newtonian value. Therefore, the first definition is an observer-dependent property of one spacetime; it describes how the effect is perceived by an observer. The second definition is not observer dependent but only at the expense of referring to a whole family of spacetimes. It measures how the effect changes within the family. Usually, this means that it gives a way of stating the 'distance' from the Newtonian situation.

From the way it was defined, it is obvious that one can always (under weak assumptions on the curvature) find the center-of-mass line and the comoving frame for a mass distribution. Hence, it is always possible to determine a dragging coefficient like we did in the previous section. The advantage of the present definition is its independence of coordinates and the fact that it does not make any reference to a preferred observer. This makes it applicable in principle to any 'reasonable' situation, where we may call 'reasonable' a situation which has a Newtonian counterpart. However, one has to be aware of the fact that then the definition refers to some 'standard' situation and is therefore not an operational one. It is just as obvious that the definition is of only a very limited practical value because, in order to be able to do anything, one has to know the world function and the Jacobi propagators. As long as one considers mass shells with a flat interior, one is in a relatively good position. Another possible situation to consider is a rotating mass shell as a perturbation in an expanding Friedman universe (Klein 1993). There it might be possible to get the general solution of the Jacobi equation because the unperturbed background is conformally flat.

Acknowledgments. I would like to thank H. Pfister and C. Klein for critical discussions and a careful reading of the manuscript. This work was supported by the SCIENCE program of the European Community.

REFERENCES

Brill, Dieter and Cohen, Jeffrey (1966). "Rotating Masses and their Effects on Inertial Frames." *Physical Review* 143: 1011–1015.

Dixon, William G. (1974). "Dynamics of Extended Bodies in General Relativity; III. Equations of Motion." *Transactions of the Royal Society A* 277: 59–119.

Dixon, William G. (1979). "Dynamics of Extended Bodies in General Relativity." In *Course 67 of the Enrico Fermi International School of Physics*. Jürgen Ehlers, ed. Amsterdam: North Holland, pp. 156–219.

Ehlers, Jürgen and Rudolph, Ekkart (1977). "Dynamics of Extended Bodies in General Relativity; Center-of-Mass Description and Quasirigidity." *General*

Relativity and Gravitation 8: 197–217.

Klein, Christian (1993). "Rotational Perturbations and Frame Dragging in a Friedmann Universe." *Classical and Quantum Gravity* 10: 1619–1631.

Pfister, Herbert (1995). "Dragging Effects near Rotating Bodies and in Cosmological Models." This volume, p. 315.

Pfister, Herbert and Braun, Karl-Heinz (1985). "Induction of Correct Centrifugal Forces in a Rotating Mass Shell." *Classical and Quantum Gravity* 2: 909–918.

Schattner, Ruprecht (1979). "The Center of Mass in General Relativity." *General Relativity and Gravitation* 10: 395–399.

Schattner, Ruprecht and Streubel, Michael (1981). "Properties of Extended Bodies in Spacetimes Admitting Isometries." *Annales de l'Institut Henri Poincaré* 34: 117–143.

Schattner, Ruprecht and Trümper, Manfred (1981). "World Vectors, Jacobi Vectors and Jacobi One-forms on a Manifold with a Linear Symmetric Connection." *Journal of Physics A* 14: 2345–2352.

Thirring, Hans (1918). "Über die Wirkung rotierender ferner Massen in der Einsteinschen Gravitationstheorie." *Physikalische Zeitschrift* 19: 33–38.

6. Experimental Status

Introduction

The final judge of any physical principle or theory is experiment. But any tests of effects connected with gravity that go beyond Newton's law are evidently quite difficult because these are tiny effects under all 'normal' circumstances. It has therefore to be admitted that presently there exists no unequivocal direct experimental confirmation of any 'Machian effect.' There is however legitimate optimism that the most prominent Machian effect – dragging – will be observed within the next decade (Will p. 365, Ciufolini p. 386), and there already exist a number of indirect confirmations of at least some Machian effects made in the framework of the PPN-formalism (Nordtvedt p. 422, see also p. 330, Will p. 365). Moreover, the observational fact that lies at the root of most 'Machian discussions,' namely, the nonrotation of the local inertial compass with respect to the most distant quasars and galaxies, is now confirmed with an accuracy of 0.00025 arcsec/year. One can go even further and say that all curvature effects due to masses are in truth Machian effects. On the other hand, it has been shown that some (possibly too naive) Machian expectations are at least below present experimental accuracy, for example, cosmologically induced anisotropies in local physics (Nordtvedt p. 422), and time variations of fundamental physical 'constants' (Sisterna and Vucetich p. 403). Following the conjectures of Mach (p. 110–111) we may however argue that "we are not able to observe enough of the universe to be able to tell whether Mach's Principle is right" (King, p. 248). So we may look forward to future, even more interesting 'Machian experiments,' when we succeed in looking deeper into our universe.

H.P.

Einstein Studies, vol. 6: Mach's Principle: From Newton's Bucket to Quantum Gravity, pp. 364

Testing Machian Effects in Laboratory and Space Experiments

Clifford M. Will

1. Introduction

To the gravitational experimentalist or to the theorist with close ties to experiment, Mach's Principle is a strange thing. It is said to be a principle that gets to the very foundation of our ideas of space and time, yet it apparently has few, if any, observable consequences. In some statements of the principle, local dynamics is to be defined in a purely relational manner, that is, with direct reference only to the distribution of distant matter. Yet in general relativity, the theory of gravitation supposedly inspired by Mach's Principle (at least according to Einstein's early pronouncements), there are almost no observable effects that shed light on how such a reference to distant matter is embodied in the theory. Indeed, this apparent 'effacement' of Machian effects in general relativity has been seen as a long-standing conundrum by some investigators. It led Dicke, among others, to propose alternative theories of gravity.

Indeed, given that gravity is a long-range force, it is a miracle that local dynamics is not actually *dominated* by the effects of distant matter. Moreover, there could be other, weakly interacting, long-range forces in nature, that are too weak to be detected by direct local measurements, but that have strong effects generated by distant matter. If one takes such Machian ideas seriously, then, it is natural to ask "Are Machian effects actually observed?" and "If not, why not?"

This paper will address these questions. For our purposes, 'Machian effects' will be defined to be potentially observable effects in local, freely-falling, nongravitational or gravitational experiments, that depend upon the location or velocity of the local frame relative to distant matter. This discussion excludes such naturally occurring local effects of distant matter as tidal gravitational effects, by assuming either that the scale over

Einstein Studies, vol. 6: Mach's Principle: From Newton's Bucket to Quantum Gravity, pp. 365–385

which any relevant observation is made is small compared to the scale of inhomogeneities in external fields, or that those inhomogeneities are simply irrelevant to the observation in question. We will then describe a number of high-precision experimental constraints on Machian effects provided by laboratory and space experiments.

It is useful to divide Machian effects into three groups: Machian effects in violations of the Einstein Equivalence Principle (EEP) (Sec. 2), Machian effects in violations of the Strong Equivalence Principle (SEP) (Sec. 3), and Machian effects related to gravitomagnetism and the dragging of inertial frames (Sec. 4). Concluding remarks will be made in Sec. 5. For a comprehensive review of many of the theoretical and experimental issues treated in this paper, the reader is referred to Will (1993); for an abbreviated review, see Will (1992).

2. Machian Effects and the Einstein Equivalence Principle

2.1. The Einstein Equivalence Principle. The Einstein Equivalence Principle is the foundation for all metric theories of gravity, such as general relativity, Brans–Dicke theory, and many others.[1] It states, roughly, that all test bodies fall in a gravitational field with the same acceleration (Weak Equivalence Principle), and that in local, freely falling or inertial frames, the outcomes of nongravitational experiments are independent of the velocity of the frame (Local Lorentz Invariance) and the location of the frame (Local Position Invariance). A consequence of this principle is that the nongravitational interactions must couple *only* to the symmetric spacetime metric $g_{\mu\nu}$, which locally has the Minkowski form $\eta_{\mu\nu}$ of special relativity. Because of this local interaction only with $\eta_{\mu\nu}$, local nongravitational physics is immune from the influence of distant matter, apart from tidal effects. Local physics is Lorentz invariant (because $\eta_{\mu\nu}$ is), and position invariant (because $\eta_{\mu\nu}$ is constant in space and time).

How could violations of EEP arise? From the viewpoint of field theory, violations of EEP would generically be caused by other long-range fields additional to $g_{\mu\nu}$ which also couple to matter, such as scalar, vector and tensor fields. Such theories are called nonmetric theories. A simple example of a nonmetric theory is one in which the matter action for charged particles is given by

$$I = -\sum_a m_a \int (g_{\mu\nu} v_a^\mu v_a^\nu)^{1/2} dt + \sum_a e_a \int A_\mu(x_a^\nu) v_a^\mu \, dt$$
$$-(16\pi)^{-1} \int \sqrt{-h} h^{\mu\alpha} h^{\nu\beta} F_{\mu\nu} F_{\alpha\beta} \, d^4x, \tag{1}$$

where m_a, e_a, x_a^ν, and $v_a^\nu = dx_a^\nu/dt$ are the mass, charge, worldline, and ordinary velocity, respectively, of the a-th body, A_μ and $F_{\mu\nu}$ are the electromagnetic vector potential and Maxwell field, $g_{\mu\nu}$ is the metric, and $h_{\mu\nu}$ is a second, second-rank tensor field. Locally one can always find coordinates (local inertial frame) in which $g \rightarrow \eta$, but in general $h \not\rightarrow \eta$, instead, $h \rightarrow h_0$, where h_0 is a tensor whose values are determined by the cosmological or nearby matter distribution. In the rest frame of the distant matter distribution, h_0 will have specific values, and there is no reason *a priori* why those should correspond to the Minkowski metric (unless $h_{\mu\nu}$ were identical to $g_{\mu\nu}$ in the first place, in which case one would have a metric theory). The value of h_0 could also vary with the location of the local frame in space or time relative to the distant matter. This can lead to violations of Lorentz invariance or position invariance in the local physics of electromagnetic systems.

A number of explicit theoretical frameworks have been developed to treat a broad range of nonmetric theories, of which this was just one example.[2] They include the $TH\epsilon\mu$ framework of Lightman and Lee (1973), the $\chi - g$ framework of Ni (1977), the c^2 framework of Haugan and coworkers (Haugan and Will 1987; Gabriel and Haugan 1990), and the extended $TH\epsilon\mu$ framework of Vucetich and colleagues (Horvath *et al.* 1988).

2.2. Local Lorentz Invariance. Tests of Local Lorentz Invariance are most profitably discussed using the c^2 framework. This is a special case of the $TH\epsilon\mu$ formalism, adapted to situations in which one can ignore the variation with space and time of the external fields that couple to matter, and instead focus on their dependence on the velocity of the local frame. It assumes a class of nonmetric theories in which the matter part of the action of Eq. (1) can be put into the local special relativistic form, using units in which the limiting speed of neutral test bodies is unity, and in which the sole effect of any nonmetric fields coupling to electrodynamics is to alter the effective speed of light. The result is the action

$$I = -\sum_a m_a \int \sqrt{1 - v_a^2}\, dt + \sum_a e_a \int A_\mu v_a^\mu dt + (8\pi)^{-1} \int (E^2 - c^2 B^2) d^4x, \tag{2}$$

where E and B are the usual electric and magnetic fields defined using components of $F_{\mu\nu}$. Because the action is explicitly non-Lorentz invariant if $c^2 \neq 1$, it must be defined in a preferred universal rest frame (presumably that of the 3K microwave background); in this frame, the value of c^2 is then determined by the cosmological values of the nonmetric field. Even if the nonmetric field coupling to electrodynamics is a tensor field, the homogeneity and isotropy of the background cosmology in the preferred frame is likely to collapse its effects to that of the single parameter c^2. Because this action violates Lorentz invariance, systems moving through the universe will exhibit explicit effects dependent upon the velocity of motion. Detailed calculations of a variety of experimental situations show that those effects depend on the magnitude of the velocity through the preferred frame ($\sim$300 km/sec), and on the parameter $\delta \equiv c^{-2} - 1$. In any metric theory or theory with local Lorentz invariance, $\delta = 0$.

One can then set observable upper bounds on δ using a variety of experiments. Modest bounds on δ can be set by the 'standard' tests of special relativity, such as the Michelson–Morley experiment and its descendants (Miller 1933; Shankland *et al.* 1955), the Brillet–Hall (1979) interferometry experiment, a test of time-dilation using radionuclides on centrifuges (Champeney, Isaak, and Khan 1963), tests of the relativistic Doppler shift formula using two-photon absorption (TPA) (Riis *et al.* 1988), and a test of the isotropy of the speed of light using one-way propagation of light between hydrogen maser atomic clocks at the Jet Propulsion Laboratory (JPL) (Krisher *et al.* 1990).

Very stringent bounds $|\delta| < 10^{-21}$ have been set by 'mass isotropy' experiments of a kind pioneered by Hughes and Drever (Hughes, Robinson, and Beltran-Lopez 1960; Drever 1961). The idea is simple: In a frame moving relative to the preferred frame, the non-Lorentz-invariant electromagnetic action of Eq. (2) becomes anisotropic, dependent on the direction of the velocity **V**. Those anisotropies then are reflected in the energy levels of electromagnetically bound atoms and nuclei (for nuclei, we consider only the electromagnetic contributions). For example, the three sublevels of an $l=1$ atomic wavefunction in an otherwise spheri-cally symmetric atom can be split in energy, because the anisotropic perturbations arising from the electromagnetic action affect the energy of each substate differently. One can study such energy anisotropies by first splitting the sublevels slightly using a magnetic field, and then monitoring the resulting Zeeman splitting as the rotation of the Earth causes the laboratory **B**-field (and hence the quantization axis) to rotate relative to **V**, causing the relative energies of the sublevels to vary among themselves diurnally. Using nuclear magnetic resonance techniques, the original Hughes–Drever experiments placed a bound of about 10^{-16} eV on such variations. This is about 10^{-22} of the electromagnetic energy of the nuclei used. Since the magnitude of the predicted

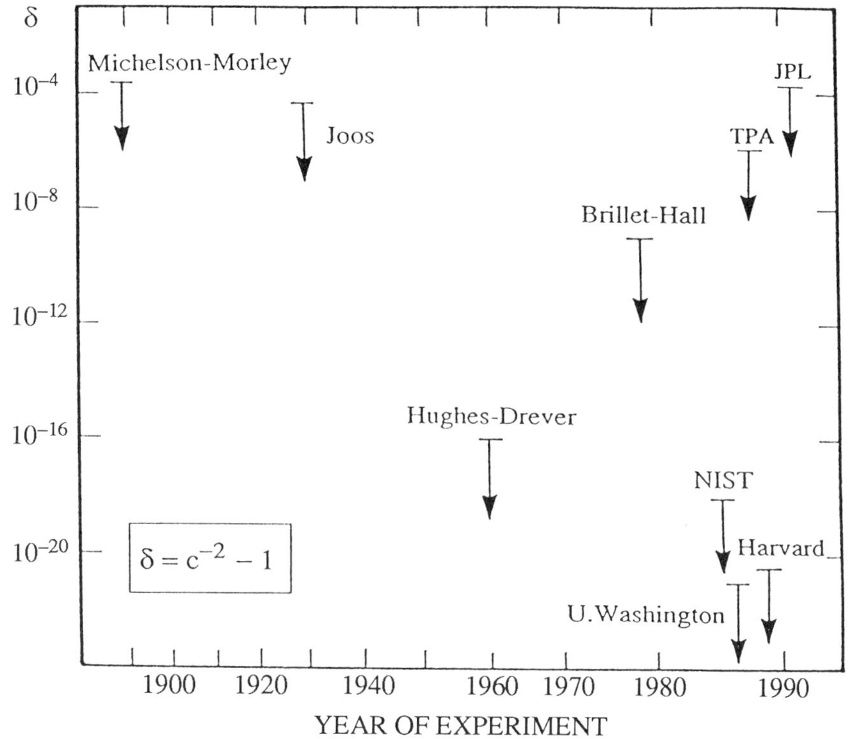

Fig. 1. Selected tests of Local Lorentz Invariance showing bounds on the parameter δ, which measures the degree of violation of Lorentz invariance in electromagnetism. Michelson–Morley, Joos, and Brillet–Hall experiments test isotropy of the round-trip speed of light in interferometers, the later experiment using laser technology. Two-photon absorption (TPA) and JPL experiments test isotropy of the speed of light in one-way configurations. The remaining four experiments test isotropy of nuclear energy levels. Limits assume the speed of the Earth is 300 km/s relative to the mean rest frame of the cosmic microwave background.

effect depends on the product $V^2\delta$, and $V^2 \approx 10^{-6}$, one obtains the bound $|\delta| < 10^{-16}$. Energy anisotropy experiments were improved dramatically in the 1980s using laser-cooled trapped atoms and ions (Prestage *et al.* 1985; Lamoreaux *et al.* 1986; Chupp *et al.* 1989). This technique made it possible to reduce the broadening of resonance lines caused by collisions, leading to improved bounds on δ shown in Figure 1 (experiments labelled NIST, University of Washington and Harvard, respectively).[3]

2.3. Local Position Invariance. Violations of EEP can also lead to time- and position-dependence of local physics. In the model example of Eq. (1), the values of h_0 imposed by cosmology or by nearby matter could vary, resulting, for example, in variations of the effective fine-structure constant, or of the relative rates of atomic clocks. For example, in the quantum dynamics of an atomic clock based on the hyperfine structure of hydrogen (hydrogen maser clock), the components of h_0 in Eq. (1) will play a different role than they would say, in the dynamics of a clock based on the resonant frequency of a microwave cavity, because the role of electromagnetism is different in the two cases. If one type of clock is chosen as a reference standard, then the relative rates of other types of clocks measured against the standard in local freely falling frames will generally depend on the location of the frame in space or time. It is straightforward to show from this that the frequency shift Δf in the comparison of two identical clocks at different heights in a gravitational potential U will be given by $\Delta f/f=(1+\alpha)\Delta U/c^2$, where α generally depends on the type of clock being used.[4] If EEP is satisfied, $\alpha=0$ for *all* clocks, and one has the standard gravitational redshift prediction of Einstein, indeed of all metric theories of gravity. The best bound to date, $|\alpha_{\text{H-maser}}|<2\times10^{-4}$, comes from a 1976 gravitational redshift experiment using a hydrogen maser clock launched on a Scout rocket to an altitude of 10,000 km, and compared with an identical clock on the ground (Vessot *et al.* 1980). Another experiment compared two different clocks side by side, one a hydrogen maser clock (actually a pair of masers), the other a set of oscillator clocks stabilized by superconducting microwave cavities (called SCSO clocks), as the Earth's rotation moved the laboratory in and out of the Sun's gravitational field, causing a diurnal variation in U. This was a direct test of the 'Machian' effect of distant matter (the Sun) on local clock intercomparisons. The bound from this experiment was $|\alpha_{\text{H-maser}}-\alpha_{\text{SCSO}}|<10^{-2}$ (Turneaure *et al.* 1983).

The effective fundamental nongravitational constants of physics can vary with cosmological time if EEP is violated. Bounds on such variations have been obtained from a variety of geological, laboratory, and astronomical observations. The best bound, especially for the fine-structure constant, comes from the Oklo natural fission reactor in Gabon, Africa, where the natural occurrence of sustained fission about two billion years ago permits a comparison of the values of various constants affecting nuclear reactions then with the current values. For the fine-structure constant, the bound is better that one part in 10^5 per 20 billion years. Elsewhere in this volume, Sisterna and Vucetich survey these bounds in detail.

2.4. The Weak Equivalence Principle. Even the Weak Equivalence Principle has a Machian interpretation. If local electrodynamics is affected by the presence of distant matter, then, as already discussed, the internal energy of an atom, as compared with a reference standard of energy, can vary with position or velocity relative to that matter. One can then exploit conservation of energy to analyze a cyclic gedanken experiment in which an atom falls in a gravitational field, is pulled apart into its constituent particles, which are then returned to the starting point to reconstruct the original atom. Because the binding energy by assumption varies with location and velocity, the acceleration of the bound atom must then differ from that of the constituent test particles by an amount that leaves no net energy difference in the cycle. The result is a violation of the Weak Equivalence Principle for composite bodies that depends in general on their structure or composition. This classic argument was pioneered by Dicke (1964) and generalized by Nordtvedt (1975) and Haugan (1979).[5] The current bounds on the fractional difference in acceleration in the solar or terrestrial gravitational fields between bodies of different composition are between 10^{-11} and 10^{-12} (Roll, Krotkov, and Dicke 1964; Braginsky and Panov 1972; Adelberger *et al.* 1990). Recent technical advances in experiments of this type developed to search for a fifth force are likely to yield improved bounds by a few orders of magnitude; a satellite test of the equivalence principle has also been proposed that could yield a test at the 10^{-17} level.

3. Machian Effects and the Strong Equivalence Principle

3.1. The Strong Equivalence Principle. The Strong Equivalence Principle (SEP) is a generalization of EEP which states that in local 'freely-falling' frames that are large enough to include gravitating systems (such as planets, stars, a Cavendish experiment, a binary system, etc.), yet that are small enough to ignore tidal gravitational effects from surrounding matter, local *gravitational* physics should be independent of the velocity of the frame and of its location in space and time. Also *all* bodies, including those bound by their own self-gravity, should fall with the same acceleration. General relativity satisfies SEP, whereas most other metric theories do not (for example, the Brans–Dicke theory).

It is straightforward to see how a gravitational theory could violate SEP (Will and Nordtvedt 1972). Most alternative metric theories of gravity introduce auxiliary fields which couple to the metric (in a metric theory they can't couple to matter), and the boundary values of these auxiliary fields determined either by cosmology or by distant matter can act back on the local gravitational dynamics. The effects can include variations in time and space of the locally measured effective Newtonian gravitational constant G (preferred-location effects), as well as effects resulting from the motion of the frame relative to a preferred cosmic reference frame (preferred-frame effects). Theories with auxiliary scalar fields, such as the Brans–Dicke theory and its generalizations, generically cause temporal and spatial variations in G, but respect the 'Lorentz invariance' of gravity, i.e., produce no preferred-frame effects. The reason is that a scalar field is invariant under boosts. On the other hand, theories with auxiliary vector or tensor fields can cause preferred-frame effects, in addition to temporal and spatial variations in local gravitational physics. For example, a timelike, long-range vector field singles out a preferred universal rest frame, one in which the field has no spatial components; if this field is generated by a cosmic distribution of matter, it is natural to assume that this special frame is the mean rest frame of that matter.

General relativity embodies SEP because it contains only one gravitational field $g_{\mu\nu}$. Far from a local gravitating system, this metric can always be transformed to the Minkowski form $\eta_{\mu\nu}$ (modulo tidal effects of distant matter and $1/r$ contributions from the far field of the local system), a form that is constant and Lorentz invariant, and thus that does not lead to preferred-frame or preferred-location effects. In a sense, potentially Machian effects are strongly effaced in general relativity. From this point of view, one sees that any alternative to general relativity *must* violate SEP, and must have Machian effects, avoiding them only by the result of some fine-tuning.

The theoretical framework most convenient for discussing SEP effects is the parametrized post-Newtonian (PPN) formalism (Nordtvedt 1968; Will 1971),[6] which treats the weak-field, slow-motion limit of metric theories of gravity. This limit is appropriate for discussing the dynamics of the solar system and for many stellar systems, except for those containing compact objects such as neutron stars.[7] If one focuses attention on theories of gravity whose field equations are derivable from an invariant action principle (Lagrangian-based theories), the generic post-Newtonian limit is characterized by the values of five PPN parameters, γ, β, ξ, α_1, and α_2. Two, α_1 and α_2, measure the existence of preferred-frame effects. Two others, the combination $4\beta-\gamma-3$ and ξ, measure the existence of preferred-location effects. If SEP is valid, $\alpha_1=\alpha_2=\xi=4\beta-\gamma-3=0$, as in general relativity. In scalar–tensor theories, $\alpha_1=\alpha_2=\xi=0$, but $4\beta-\gamma-3=1/(2+\omega)$, where ω is the 'coupling parameter' of the scalar–tensor theory. In Rosen's bimetric theory,

$\alpha_2 = c_0/c_1 - 1$, $\alpha_1 = \xi = 4\beta - \gamma - 3 = 0$, where c_0 and c_1 are the cosmologically induced values of the temporal and spatial diagonal components of a flat background tensor field, evaluated in a cosmic rest frame in which the physical metric has the Minkowski form far from the local system.[8]

3.2. Tests of preferred-frame and preferred-location effects. Within the PPN formalism the variations in the locally measured Newtonian gravitational constant can be calculated explicitly: Viewed as the coupling constant in the gravitational force between two point masses at a given separation, it is given by[9]

$$G_{\text{local}} = 1 - (4\beta - \gamma - 3 - 3\xi)U_{\text{ext}} - \tfrac{1}{2}(\alpha_2 - \alpha_2)V^2 - \tfrac{1}{2}\alpha_2(\mathbf{V}\cdot\mathbf{e})^2 + \xi U_{\text{ext}}(\mathbf{N}\cdot\mathbf{e})^2, \tag{3}$$

where U_{ext} is the potential of an external mass in the direction $\mathbf{N}$, $\mathbf{V}$ is the velocity of the experiment relative to the preferred frame, and $\mathbf{e}$ is the orientation of the two masses. Thus G_{local} can vary in magnitude with variations in U_{ext} and V^2, and can also be anisotropic, that is, can vary with the orientation of the two bodies. Other SEP-violating effects include planetary orbital perturbations and precessions of planetary and solar spin axes.

Bounds on such effects have been obtained from a wide variety of geophysical, planetary and solar observations.[10] The results verify SEP and the absence of 'Machian' effects to precisions of between 10^{-3} and 10^{-7} in the PPN parameters (Table 1).

Just as the Weak Equivalence Principle for laboratory-sized bodies had a Machian interpretation, so too does the equivalence principle for massive self-gravitating bodies. Bodies whose internal gravitational binding energy is dependent on their location or velocity relative to other bodies (for example, via variations in G_{local}), will experience anomalous, structure-dependent accelerations in external gravitational fields. The possibility of such violations of the weak equivalence principle for massive bodies is called the 'Nordtvedt effect' (Nordvedt 1968). More than 20 years of high-precision data from lunar laser ranging have found no evidence of the orbital perturbation that would result from a difference in acceleration of the Earth and the Moon toward the Sun, down to the level of a few centimeters. In the PPN framework, this places a bound on the 'Machian' parameter $\eta = 4\beta - \gamma - 3 - (10/3)\xi - \alpha_1 + (2/3)\alpha_2$ of about 1.5×10^{-3} (Table 1).[11]

3.3. Cosmological variation of Newton's constant. In metric theories of gravity that violate SEP, G may also vary with the evolution of the

Table 1. Current Limits on 'Machian' PPN Parameters[1]

Parameter	What it measures relative to general relativity	Value in general relativity	Effect or Experiment	Value or Limit	Remarks
γ	How much curvature of space does mass produce?	1	Shapiro time delay Light deflection	1.000 ± 0.002 1.000 ± 0.002	Viking ranging VLBI[2]
β	How nonlinear is the superposition law for gravity?	1	Perihelion shift Nordtvedt effect	1.000 ± 0.003 1.000 ± 0.001	$J_2 \approx 10^{-7}$ assumed[3] $\eta = 4\beta - \gamma - 3$ assumed
ξ	Are there preferred -location effects?	0	Earth tides	$<10^{-3}$	Gravimeter data on solid Earth tides
α_1	Are there preferred-frame effects?	0	Orbital preferred-frame effects	$<4\text{x}10^{-4}$	Planetary orbits
α_2		0	Earth tides Solar spin precession	$<4\text{x}10^{-4}$ $<4\text{x}10^{-7}$	Gravimeter data Alignment of solar equator and ecliptic
η^4	Is WEP violated for self-gravitating bodies?	0	Nordvedt effect	<0.0015	Lunar laser ranging

[1] For a complete review with references see Will (1993) (TEGP2), especially Chaps. 7 and 8, and Sec. 14.3
[2] Very Long Baseline Interferometry
[3] J_2 is the dimensionless quadrupole moment of the Sun; helioseismological studies indicate $J_2 \approx 2\text{x}10^{-7}$.
[4] Here $\eta \equiv 4\beta - \gamma - 3 - (10/3)\xi - \alpha_1 + (2/3)\alpha_2$.

structure of the universe, via the cosmologically imposed boundary values on the auxiliary fields. In fact, a cosmic variation in G was the original 'Machian' consideration that in part motivated Dicke to develop the scalar–tensor theory. Varying G is common in the various generalized scalar–tensor theories developed recently for inflationary cosmology. On the other hand, in a wide class of such theories, the variations can be large in the early universe (leading to the desired cosmological consequences), but damp out as the present epoch is approached. In many such theories, general relativity is a natural 'attractor' to which the cosmic evolution naturally leads the theory as the conditions of the present universe are reached (Damour and Nordtvedt 1993).

The best current observational bound $|\dot{G}/G| < 4 \times 10^{-12}\ \mathrm{yr}^{-1}$ comes from long-term observations of the orbit of Mars via Viking ranging data (Hellings *et al.* 1983, Shapiro 1990).[12]

4. Mach, Rotation and Gravitomagnetism

4.1. Dragging of inertial frames, Lense–Thirring effect, and gravitomagnetism. Rotation has always played a central role in discussions of Mach's Principle. Whether the centrifugal forces causing the water to climb the sides of Newton's rotating bucket truly result from rotation relative to absolute space or can somehow be understood as arising from interactions with distant matter was a matter of lengthy debate, at least until the middle 1960s. The analysis by Brill and Cohen (1966) of the effects of the general relativistic 'dragging of inertial frames' inside a rotating shell of matter provided the beginnings of an understanding of this issue. This subject is reviewed in detail by Pfister elsewhere in this volume.

Although the first calculation of the gravitational effects of rotating matter within general relativity was done as early as 1918 by Thirring and Lense (Thirring 1918; Lense and Thirring 1918) (hence the frequent use of the terminology Lense–Thirring effects),[13] it was really Schiff (1960) and Pugh (1959) who independently stressed the importance of these effects, especially as they relate to gyroscopes, as tests of general relativity and Mach's Principle. Schiff, together with Fairbank and Cannon subsequently founded the effort, centered at Stanford University, to test these effects using gyroscopes in Earth orbit.

The idea of 'dragging of inertial frames' arises as follows: In the general relativistic gravitational field of a rotating body, a freely falling observer with zero angular momentum as seen from infinity actually moves (is 'dragged') around the body with a nonzero angular velocity; and the orientation of a freely falling frame, as defined by gyroscopes, rotates or precesses relative to distant stars.

Another viewpoint on Lense–Thirring effects, called 'gravitomagnetism,' arises from the close similarity between the role of the g_{00} and g_{0i} components of the metric in linearized general relativity and the four-vector potential of electrodynamics, and between certain terms in the geodesic equation of a test particle, and the Lorentz equations for a charged particle (Braginsky, Caves, and Thorne 1977). In this viewpoint, the part of the gravitational field of the rotating Earth that is responsible for the dragging of inertial frames is completely analogous to the magnetic field of a rotating electrical conductor, and the precession of a gyroscope in this field is completely analogous to the precession of a current loop in the corresponding magnetic field. From this point of view, gravitomagnetic effects are produced by mass currents, just as magnetic effects are produced by electrical currents. Furthermore, if gravity is 'Lorentz invariant' (no preferred frames), as is electrodynamics, then in some sense gravitomagnetic fields can be obtained from static gravitational fields ('gravitoelectric' fields) by boosts, just as magnetic fields can be obtained from electric fields by boosts. In general relativity, for example, the metric of a moving body can be obtained from the static Schwarzschild metric by an appropriate Lorentz-type transformation.

This connection between gravitomagnetic effects and Lorentz invariance is illustrated by the prediction for the gravitomagnetic precession of a gyroscope orbiting a rotating body in the PPN formalism[14]: It depends on the PPN parameter combination $1+\gamma+\frac{1}{4}\alpha_1$. In a theory without preferred-frame effects ($\alpha_1=0$), the prediction depends only on the properties of the static metric: '1' corresponding to the Newtonian part of g_{00} and γ corresponding to the spatial metric g_{ij}. In this sense, tests of gravitomagnetism can be seen as tests of Lorentz invariance, i.e., tests of α_1. It is useful to note the existing bounds $|\gamma-1|<2\times10^{-3}$ from light deflection and Shapiro time-delay measurements, and $|\alpha_1|<4\times10^{-4}$ from tests of preferred-frame effects (Table 1).

It should be pointed out however, that gravitomagnetism generated by rotation is not entirely equivalent to gravitomagnetism generated by a boost, since there are invariant differences between a rotating body and a moving body, such as the angular momentum measured at infinity, as Ciufolini emphasizes elsewhere in this volume. Thus experiments that look for the dragging of inertial frames are indeed probing something more than simple Lorentz invariance. In addition, although such experiments measure PPN parameters that have already been bounded, they are direct measurements, rather than combinations of measurements of quite distinct types (Shapiro time delay plus preferred-frame tests).[15] In fact, the latter type of argument relies upon the validity of the PPN framework as a description of a wide class of theories; examples exist of metric theories that do not fit within the PPN scheme, and for which those arguments would fail.

4.2. Experiments planned to test gravitomagnetism. 4.2.1 Relativity Gyroscope Experiment. The Relativity Gyroscope Experiment at Stanford University (also known by the NASA terminology Gravity Probe B, or GP-B) is in the advanced stage of developing a space mission to detect gravitomagnetism directly. A set of four superconducting-niobium-coated, spherical quartz gyroscopes will be flown in a low polar Earth orbit, and the precession of the gyroscopes relative to the distant stars will be measured. The predicted effect of gravitomagnetism is about 42 milli-arcseconds per year, and the accuracy goal of the experiment is about 0.5 milli-arcseconds per year (Everitt *et al.* 1988).

To achieve this accuracy, which corresponds to a precession rate of 10^{-16} radians per second, numerous technical challenges have had to be met, including: fabricating gyros that are homogeneous and spherical to better than a part per million; developing and testing a 'London moment' readout system that exploits the magnetic dipole moment developed by a spinning superconductor, and uses SQUIDs to read out the varying currents in superconducting loops surrounding the gyroscope; and developing a magnetic shield of novel design to reduce the ambient magnetic field of the Earth below 10^{-7} G. A full-size flight prototype of the instrument package has been tested as an integrated unit. Current plans call for a test of the final flight hardware on the Space Shuttle followed by a Shuttle-launched science experiment.

4.2.2 LAGEOS III. Another proposal to look for an effect of gravitomagnetism is to measure the relative precession of the line of nodes of a pair of laser-ranged geodynamics satellites (LAGEOS). The idea is to launch a third LAGEOS satellite in an orbit whose inclination is supplementary to that of LAGEOS I ($i_{\rm I}+i_{\rm III}=180°$). The inclinations must be supplementary in order to cancel the dominant nodal precession caused by the Earth's Newtonian gravitational multipole moments from the mean of the two nodal angles (Ciufolini 1989). The goal is a 10% experiment. Further details are provided by Ciufolini in this volume.

4.2.3 Superconducting Gravity Gradiometer Mission. A third proposal envisages orbiting an array of three mutually orthogonal, superconducting gravity gradiometers around the Earth, to measure directly the contribution of the gravitomagnetic field to the tidal gravitational force (Braginsky and Polnarev 1980; Mashhoon and Theiss 1982; Mashhoon, Paik, and Will 1989).

4.2.4 Foucault Pendulum at the South Pole. A final proposal is reminiscent of early discussions of Mach's Principle by Föppl (see discussion by Norton in this volume and p. 317). The gravitomagnetism of the rotating Earth will cause the plane of a Foucault pendulum to rotate relative to the distant stars; however, in order to eliminate errors caused by the large, natural rotation of the plane that occurs at a finite latitude on Earth, the experiment must be done at one of the poles, preferably the South, because the polar cap is stationary and scientific stations already exist there. Braginsky, Polnarev, and Thorne (1984) argued that a 10 percent experiment might be possible.

5. Summary

A wide variety of experiments seem to tell us that 'Machian' effects are almost completely absent in local observable physics, both in non-gravitational experiments and in gravitational experiments. All experimental results to date are in accord with general relativity. It is ironic that general relativity, supposedly inspired by Mach's Principle, should have turned out to be such a strong filter for Machian effects. This absence of Machian effects, both in theory and in observation, frustrates attempts to ascertain whether Mach's Principle has any meaningful content at all.

Acknowledgment. This work was supported in part by the National Science Foundation under Grant No. PHY 92-22902 and by the National Aeronautics and Space Administration under Grant No. NAGW 2909.

NOTES

[1]For discussion and references, see Will (1993) (hereafter referred to as TEGP2, Chap. 2.

[2]See TEGP2, Secs. 2.6 and 14.1(b)-(d).

[3]See TEGP2, Secs. 2.4(b), (e), and 14.1(a).

[4]See TEGP2, Secs. 2.4(c) and 2.5.

[5]See TEGP2, Sec. 2.5.

[6]See TEGP2, Chap. 4.

[7]For discussion of the dynamics of compact objects in alternative theories of gravity, see TEGP2, Chap. 11.

[8]For a survey of alternative theories of gravity and their post-Newtonian limits, with references, see TEGP2, Chap. 5.

[9]See TEGP2, Sec. 6.3.

[10]See TEGP2, Secs. 8.2, 8.3, and 14.3(c) for review and references.

[11]See TEGP2, Secs. 8.1 and 14.3(c).

[12]See TEGP2, Secs. 8.4 and 14.3(c).

[13]For an English translation of and commentary on the Thirring–Lense papers see Mashhoon, Hehl and Theiss (1984).

[14]See TEGP2, Sec. 9.1.

[15]Nordtvedt (1988) has also pointed out strong *indirect* evidence for gravitomagnetism provided by the absence of the Nordtvedt effect in Lunar and LAGEOS orbits [see also p. 330].

REFERENCES

Adelberger, Eric G.; Stubbs, Christopher, W.; Heckel, Blayne R.; Su, Yue; Swanson, H. Erik; Smith, G.; Gundlach, Jens H.; and Rogers, Warren F. (1990). "Testing the Equivalence Principle in the Field of the Earth: Particle Physics at Masses below 1 μ eV?" *Physical Review D* 42: 3267–3292.

Braginsky, Vladimir B., Caves, Carlton M., and Thorne, Kip S. (1977). "Laboratory Experiments to Test Relativistic Gravity." *Physical Review D* 15: 2047–2068.

Braginsky, Vladimir B. and Panov, V. I. (1972). "Verification of the Equivalence of Inertial and Gravitational Mass." *Soviet Physics JETP* 34: 463–466 [*Zhurnal Éksperimental'noi i Teoreticheskoi Fiziki* 61: 873–879 (1971)].

Braginsky, Vladimir B. and Polnarev, Alexander G. (1980). "Relativistic Spin-Quadrupole Gravitational Effect." *JETP Letters* 31: 415–418 [*Pis'ma Zhurnal Éksperimental'noi i Teoreticheskoi Fiziki* 31: 444–447 (1980)].

Braginsky, Vladimir B., Polnarev, Alexander G. and Thorne, Kip S. (1984). "Foucault Pendulum at the South Pole: Proposal for an Experiment to Detect the Earth's General Relativistic Gravitomagnetic Field." *Physical Review Letters* 53: 863–866.

Brill, Dieter R. and Cohen, Jeffrey M. (1966). "Rotating Masses and their Effect on Inertial Frames." *Physical Review* 143: 1011–1015.

Brillet, Alain and Hall, John L. (1979). "Improved Laser Test of the Isotropy of Space." *Physical Review Letters* 42: 549–552.

Champeney, D C., Isaak, G R. and Khan, A M. (1963). "An 'Aether Drift' Experiment based on the Mössbauer Effect." *Physics Letters* 7: 241–243.

Chupp, Timothy E., Hoare, R. J., Loveman, R. A., Oteiza, Eduardo R., Richardson, J. M., Wagshul, Mark E., and Thompson, Alan K. (1989).

"Results of a New Test of Local Lorentz Invariance: A Search for Mass Anisotropy in ^{21}Ne." *Physical Review Letters* 63: 1541–1544.

Ciufolini, Ignazio. (1989). "A Comprehensive Introduction to the LAGEOS Gravitomagnetic Experiment: From the Importance of the Gravitomagnetic Field in Physics to Preliminary Error Analysis and Error Budget." *International Journal of Modern Physics A* 4: 3083–3145.

Damour, Thibault and Nordtvedt, Kenneth. (1993). "General Relativity as a Cosmological Attractor of Tensor–Scalar Theories." *Physical Review Letters* 70: 2217–2219.

Dicke, Robert H. (1964). "Experimental Relativity." In *Relativity, Groups and Topology*. Cecile DeWitt and Bryce Dewitt, eds. New York: Gordon and Breach. pp. 165–313.

Drever, Ronald W. P. (1961). "A Search for Anisotropy of Inertial Mass using a Free Precession Technique." *Philosophical Magazine* 6: 683–687.

Everitt, C. W. Francis, Lipa, John A., Keiser, G. Mac, Anderson, John T., Turneaure, John P., Cornell, E. A., Levine, P. D., van Patten, Richard A., Breakwell, John V., and DeBra, Daniel B. (1988). "The Stanford Relativity Gyroscope Experiment." In *Near Zero: New Frontiers of Physics*. J. D. Fairbank, Bascom S. Deaver, Jr., C. W. Francis Everitt, and Peter F. Michelson, eds. San Francisco: Freeman. pp. 587–699.

Gabriel, Marcus D. and Haugan, Mark P. (1990). "Testing the Einstein Equivalence Principle: Atomic Clocks and Local Lorentz Invariance." *Physical Review D* 41: 2943–2955.

Haugan, Mark P. (1979). "Energy Conservation and the Principle of Equivalence." *Annals of Physics (N.Y.)* 118: 156–186.

Haugan, Mark P. and Will, Clifford M. (1987). "Modern Tests of Special Relativity." *Physics Today* 40: 69–76 (May).

Hellings, Ronald J., Adams, Peter. J., Anderson, John D., Keesey, M. S., Lau, E. L., Standish, E. Miles, Canuto, Vittorio M., and Goldman, Itzhak (1983). "Experimental Test of the Variability of *G* using Viking Lander Ranging Data." *Physical Review Letters* 51: 1609–1612.

Horvath, J. E., Logiudice, E. A., Reveros, C., and Vucetich, Héctor (1988). "Einstein Equivalence Principle and Theories of Gravitation: A Gravitationally Modified Standard Model." *Physical Review D* 38: 1754–1760.

Hughes, Vernon W., Robinson, H. G., and Beltran-Lopez, V. (1960). "Upper Limit for the Anisotropy of Inertial Mass from Nuclear Resonance Experiments." *Physical Review Letters* 4: 342–344.

Krisher, Timothy P., Maleki, Lute, Lutes, George F., Primas, Lori E., Logan, Ronald T., Anderson, John D., and Will, Clifford M. (1990). "Test of the Isotropy of the One-Way Speed of Light using Hydrogen-Maser Frequency Standards." *Physical Review D* 42: R731–734.

Lamoreaux, Steve K., Jacobs, J. P., Heckel, Blayne R., Raab, Fred J., and Fortson, Edward N. (1986). "New Limits on Spatial Isotropy from

Optically Pumped ^{201}Hg and ^{199}Hg." *Physical Review Letters* 57: 3125–3128.

Lense, Josef and Thirring, Hans. (1918). "Über den Einfluss der Eigenrotation der Zentralkörper auf die Bewegung der Planeten und Monde nach der Einsteinschen Gravitationstheorie." *Physikalische Zeitschrift* 19: 156–163.

Lightman, Alan P. and Lee, David L. (1973). "Restricted Proof that the Weak Equivalence Principle Implies the Einstein Equivalence Principle." *Physical Review D* 8: 364–376.

Mashhoon, Bahram, Hehl, Friedrich W. and Theiss, Dietmar S. (1984). "On the Gravitational Effects of Rotating Masses: The Thirring-Lense Papers." *Journal of General Relativity and Gravitation* 16: 711–750.

Mashhoon, Bahram, Paik, Ho Jung, and Will, Clifford M. (1989). "Detection of the Gravitomagnetic Field using an Orbiting Superconducting Gravity Gradiometer. Theoretical Principles." *Physical Review D* 39: 2825–2838.

Mashhoon, Bahram and Theiss, Dietmar S. (1982). "Relativistic Tidal Forces and the Possibility of Measuring Them." *Physical Review Letters* 49: 1542–1545.

Miller, Dayton C. (1933). "The Ether Drift Experiment and the Determination of the Absolute Motion of the Earth." *Reviews of Modern Physics* 5: 203–242.

Ni, Wei-Tou (1977). "Equivalence Principles and Electromagnetism." *Physical Review Letters* 38: 301–304.

Nordtvedt, Kenneth (1968). "Equivalence Principle for Massive Bodies. II. Theory." *Physical Review* 169: 1017–1025.

Nordtvedt, Kenneth (1975). "Quantitative Relationship between Clock Gravitational Redshift Violations and Non-Universality of Free-Fall Rates in Non-Metric Theories of Gravity." *Physical Review D* 11: 245–247.

Nordtvedt, Kenneth (1988). "Gravitomagnetic Interaction and Laser Ranging to Earth Satellites." *Physical Review Letters* 61: 2647–2649.

Prestage, John D., Bollinger, J. J., Itano, Wayne M., and Wineland, David J. (1985). "Limits for Spatial Anisotropy by Use of Nuclear-Spin-Polarized $^{9}Be^{+}$ Ions." *Physical Review Letters* 54: 2387–2390.

Pugh, George E. (1959). "Proposal for a Satellite Test of the Coriolis Prediction of General Relativity." Research Memorandum No. 11, Weapons Systems Evaluation Group, The Pentagon, Washington DC (unpublished).

Riis, Erling; Anderson, Lars-Ulrik A.; Bjerre, Nis; Poulsen, Ove; Lee, Siu Au; and Hall, John L. (1988). "Test of the Isotropy of the Speed of Light using Fast-Beam Laser Spectroscopy." *Physical Review Letters* 60: 81–84.

Roll, Peter G., Krotkov, Robert, and Dicke, Robert H. (1964). "The Equivalence of Inertial and Passive Gravitational Mass." *Annals of Physics (N.Y.)* 26: 442–517.

Schiff, Leonard I. (1960). "Possible New Test of General Relativity Theory." *Physical Review Letters* 4: 215–217.

Shankland, R. S., McCuskey, S. W., Leone, F. C., and Kuerti, G. (1955).

"New Analysis of the Interferometry Observations of Dayton C. Miller." *Reviews of Modern Physics* 27: 167–178.

Shapiro, Irwin I. (1990). "Solar System Tests of General Relativity: Recent Results and Present Plans." In *General Relativity and Gravitation 1989.* Neil Ashby, David F. Bartlett and Walter Wyss, eds. Cambridge: Cambridge University Press. pp. 313–330.

Thirring, Hans. (1918). "Über die Wirkung Rotierender Ferner Massen in der Einsteinschen Gravitationstheorie." *Physikalische Zeitschrift* 19: 33–39.

Turneaure, John P., Will, Clifford M., Farrell, Brian F., Mattison, Edward M., and Vessot, Robert F. C. (1983). "Test of the Principle of Equivalence by a Null Gravitational Redshift Experiment." *Physical Review D* 27: 1705–14.

Vessot, Robert F. C., Levine, Martin W., Mattison, Edward M., Blomberg, E. L., Hoffman, T. E., Nystrom G. U., Farrell, Brian F., Decher, Rudolph, Eby, Peter B., Baugher, C. R., Watts, J. W., Teuber, D. L. and Wills, F. D. (1980). "Test of Relativistic Gravitation with a Space-Borne Hydrogen Maser." *Physical Review Letters* 45: 2081–2084.

Will, Clifford M. (1971). "Theoretical Frameworks for Testing Relativistic Gravity. II. Parametrized Post-Newtonian Hydrodynamics and the Nordtvedt Effect." *Astrophysical Journal* 163: 611–628.

Will, Clifford M. (1992). "The Confrontation between General Relativity and Experiment: A 1992 Update." *International Journal of Modern Physics D* 1: 13–68.

Will, Clifford M. (1993). *Theory and Experiment in Gravitational Physics. Revised Edition*, Cambridge: Cambridge University Press.

Will, Clifford M. and Nordtvedt, Kenneth. (1972). "Conservation Laws and Preferred Frames in Relativistic Gravity. I. Preferred-Frame Theories and an Extended PPN Formalism." *Astrophysical Journal* 177: 757–774.

Discussion

Ehlers: Can I go back to the beginning of your talk where you made statements about the one-way speed of light? In which way is the simultaneity between the receiver and the source guaranteed experimentally, because you cannot use the signaling itself to define the terms needed.

Will: Essentially what happens is that simultaneity is established by the first signal sent. That establishes some arbitrary simultaneity, which you can set any way you wish. Then the clocks tick at their own proper rate as the earth rotates. One single measurement is, of course, totally meaningless.

Bondi: Some people criticize the mathematical methods used to calculate the binary pulsar effects in relativity. I mean, is it total nonsense, or are there approximations that one needs to worry about?

Will: Since one of the early critics of those calculations is in the audience, I suppose I ought to be careful, to say whether it's nonsense or not. But since these early criticisms by Jürgen [Ehlers] and his colleagues, a lot of work has been done in trying to improve the theoretical foundations of calculations of radiation reaction, which is what you are referring to. I don't know if Jürgen may want to comment, but a lot of those details have been cleared up, and a lot of the fine points have been improved. I'm sure it's not as rigorous as many people would like. Certainly the situation's much improved over what it was in the early seventies, when the binary pulsar was first discovered, and the issue of radiation reaction really became an observational issue.

Ehlers: If I may perhaps add, in the original derivations one used the quadrupole formula in an energy balance, and there has been some, I think justified, criticism that maybe this is not very well substantiated. But by now one does no longer have to introduce an energy balance, but one can actually carry the approximation methods for the equations of motion to sufficiently high orders, so that, directly, one has the radiation reaction forces in the equations of motion, and I think this type of perturbation theory is as good as any perturbation theory in physics. Of course, one still cannot prove, in a mathematical sense, that these approximations are really approximate to an exact solution, but apart from that I think everything is fine.

Will: And it's also valid for strongly self-gravitating bodies, such as neutron stars (it involves matching procedures).

Ciufolini: How well can you explain the precession of the periastron of the binary pulsar with some exotic mechanism such as a third body, or tidal effects, or other strange things you might dream up?

Will: If the two bodies are both neutron stars, then there are essentially no other contributions to any part of the dynamics of the system. The tidal effects are negligible, magnetic interactions are negligible. Astrophysicists have tried to enumerate all of the various kinds of nongravitational effects that could influence the motion of the stars and the decay of the orbit, such as tidal dissipation effects, but the estimates are always way below the measurement accuracy. That's not to say that something strange might not be there, maybe some dark matter.

Ehlers: We have heard at this conference that the Hoyle–Narlikar theory of 1964 has been revived again for cosmological reasons. Therefore it seems to me it would be of considerable interest to know whether, with respect to all these effects, which you have been discussing, there has been a serious attempt to test the Hoyle–Narlikar theory. Has there been any such attempt? There is a statement in the paper that the theory

reproduces all this, but I haven't seen this anywhere analyzed.
Will: Not to my knowledge, not in the circle of people who think about these things. It might be a worthwhile thing to do at this point, since it seems to be of some interest to cosmologists and others.
Isenberg: I recall early in the history of the so-called fifth force, sixth force, etc., there were some experiments that people said were supporting that. Did anybody ever understand what really went on there?
Will: Well, the answer's yes and no. There are two classes of such experiments: Eötvös-type experiments, in which you compare the accelerations of different materials near mountains and cliffs, or in free-fall Galileo-type experiments. Of the 30 or so experiments that have been done, two yielded positive results, nonzero deviations, and the other 28 all yielded null results down to whatever limiting accuracy they quoted as their error. Those two results, Paul Boynton's and Peter Thieberger's experiments, have not been repeated in the same way and nobody understands those positive results. On the other hand, the weight of evidence says the results are null and the later experiments are much more accurate than those two early experiments.

The other class of experiments measuring gravity up tall towers or down boreholes or mines now are all completely understood to agree with Newtonian gravity, and the earlier anomalies were simply a failure to take into account semidistant geophysical objects such as ridges and hills out to the side, and problems with analyzing the data needed to calculate the predicted gravity field up from the surface of the earth. They all now agree with Newtonian gravity.
Ciufolini: I would like to make a comment on this to explain why people are thinking to measure dragging effects in space instead of measuring these effects in the laboratory. In space it's much more expensive. But there is need of it. Essentially, any tilt of the laboratory that is very very small on the ground can simulate the Lense–Thirring effect but in space you do not have this problem. I think that's one basic reason why one goes to space.
Goenner: That surprises me because I always thought it was the weakness of the gravitational field of the laboratory on the earth that would make the difference.
Will: I would say the main reason is that you have to support the gyroscope against g and so the local gravity gradient couplings to local inhomogeneities in the gyroscope cause enormous torques that swamp the effect you want to observe.
Nordtvedt: Suppose Francis Everitt could find a way to make his little

gyro from a bunch of aligned nuclear spins rather than a mechanically rotating object. It would be an interesting experimental question whether such a quantum gyroscope would precess differently. In fact, if it did it might really undercut the interpretation of frame dragging.

Liebscher: This question, so far as I know is calculated only for really small – that means test particles – with respect to the Dirac spin, so that the question of the self-gravitation does not play any role and the Dirac spin of a test particle behaves like an ordinary spin as we expect it.

Nordtvedt: So it precesses the same?

Goenner: So there's no factor 2?

Liebscher: No, not with respect to the definition of Fermi transport.

Pfister: [to Nordtvedt]. Would you, or why would you, expect a difference in the Stanford experiment if they could take instead of the gyroscope a nuclear spin, because, I think, in all experiments we know of the spin angular momentum behaves like an ordinary angular momentum if you don't ask for the magnetic moment or magnetic type fields produced by it and you would not do it I think in this dragging experiment.

Nordtvedt: I would probably expect that they would precess the same, but to an experimentalist, if he could do it, it would be an interesting thing to test.

Pfister: Sure. It would be the same experimental question as you asked whether a nuclear particle behaves in the same way in a gravitational potential as a big particle does, as with neutrons in the gravitational field of the earth. It was measured.

Liebscher: [post-conference addendum] With respect to the question of the gyrogravitational ratio, I would like to cite: Audretsch, J., Hehl, F.W., and Lämmerzähl, C. (1993): "Matter Wave Interferometry and Why Quantum Objects are Fundamental for Establishing a Gravitational Theory." In: J. Ehlers, G. Schäfer, eds.: *Proceedings of the Bad Honnef School on Gravitation*, Lecture Notes in Physics, Berlin: Springer-Verlag. The result of equal gyrogravitational ratios for spin and orbital momentum is older. Implicitly, it is already in the general Papapetrou–Fock type scheme, which I tried myself in: Liebscher, D.-E. (1973): "The Equivalence Principle and Non-Riemannian Space–Times." *Annalen der Physik* (*Leipzig*) 30: 309–320. The procedures later used to find the equations of motion in this direction do not differ very much.

Dragging of Inertial Frames, Gravitomagnetism, and Mach's Principle

Ignazio Ciufolini

We introduce gravitomagnetism and dragging of inertial frames in Einstein's theory of gravity by using a formal analogy between electromagnetism and the weak-field, slow-motion limit of general relativity. Then, we propose a precise characterization of gravitomagnetism by curvature invariants. Finally, we present a brief introduction to the LAGEOS III experiment to detect and measure the gravitomagnetic field of earth and the Lense–Thirring dragging effect.

1. Dragging of Inertial Frames and Gravitomagnetism

A formal analogy between electrodynamics and the weak-field and slow-motion approximation of general relativity is useful to describe gravitomagnetism and dragging of inertial frames.

It is well known that in electrodynamics, from the Maxwell equations and, in particular, from the equation for the magnetic induction $\mathbf{B}$ of absence of free magnetic monopoles, $\nabla \cdot \mathbf{B} = 0$, one can write $\mathbf{B} = \nabla \times \mathbf{A}$, where $\mathbf{A}$ is the vector potential. From the Ampère's law for a stationary current distribution, $\nabla \times \mathbf{B} = (4\pi/c)\mathbf{j}$, where $\mathbf{j}$ is the current density, one has then in the Coulomb gauge, $\nabla \cdot \mathbf{A} \equiv A^i{}_{,i} = 0$: $\Delta \mathbf{A} = -(4\pi/c)\mathbf{j}$ with solution $\mathbf{A}(\mathbf{x}) = (1/c) \int (\mathbf{j}(\mathbf{x}')/|\mathbf{x} - \mathbf{x}'|) d^3x'$. One can then define the magnetic moment $\mathbf{m}$ of a current distribution: $\mathbf{m} \equiv (1/2c) \int \mathbf{x} \times \mathbf{j}(\mathbf{x}) d^3x$. For a localized, stationary current distribution $\mathbf{j}$, one can expand $\mathbf{A}(\mathbf{x})$ far from the current. The lowest nonvanishing term of $\mathbf{A}(\mathbf{x})$, the magnetic dipole vector potential, is $\mathbf{A}(\mathbf{x}) \cong \mathbf{m} \times \mathbf{x}/|\mathbf{x}|^3$, and therefore, for a localized current distribution, the lowest nonvanishing term of $\mathbf{B}$ is the field of a magnetic dipole with dipole moment $\mathbf{m}$:

$$\mathbf{B} = \nabla \times \mathbf{A} \cong \frac{3\hat{\mathbf{x}}(\hat{\mathbf{x}} \cdot \mathbf{m}) - \mathbf{m}}{|\mathbf{x}|^3}, \quad \hat{\mathbf{x}} = \frac{\mathbf{x}}{|\mathbf{x}|}. \tag{1}$$

Einstein Studies, vol. 6: Mach's Principle: From Newton's Bucket to Quantum Gravity, pp. 386–402

In electrodynamics, the equation of motion of a particle of mass m and charge q, subjected to an electric field $\mathbf{E}$ and to a magnetic induction $\mathbf{B}$, is the well-known Lorentz equation: $m(d^2\mathbf{x}/dt^2)=q(\mathbf{E}+d\mathbf{x}/cdt\times\mathbf{B})$. For a localized current distribution $\mathbf{j}(\mathbf{x})$, with magnetic moment $\mathbf{m}'$, in a stationary external magnetic induction $\mathbf{B}(\mathbf{x})$ that varies slowly over the region of the current, one can expand $\mathbf{B}(\mathbf{x})$ about a suitable origin in the localized current distribution, and the lowest nonvanishing term of the force on $\mathbf{m}'$ is $\mathbf{F}=(\mathbf{m}'\cdot\nabla)\mathbf{B}|_0$; the lowest nonvanishing term of the torque on a localized, stationary current distribution is then the torque on a magnetic dipole with dipole moment $\mathbf{m}'$: $\boldsymbol{\tau}=\mathbf{m}'\times\mathbf{B}(0)$.

In geometrodynamics (Misner, Thorne and Wheeler 1973; Ciufolini and Wheeler 1995), in the weak-field and slow-motion approximation for a stationary localized ($\mathbf{g}\rightarrow\boldsymbol{\eta}$) mass–energy distribution, one can write the $(0i)$ components of the Einstein field equation in the Lorentz gauge:

$$\Delta h_{0i}\cong 16\pi\rho v^i \tag{2}$$

with solution

$$h_{0i}(\mathbf{x})\cong -4\int\frac{\rho(\mathbf{x}')v^i(\mathbf{x}')}{|\mathbf{x}-\mathbf{x}'|}d^3x'. \tag{3}$$

From the classical definition of angular momentum, $\mathbf{J}=\int\mathbf{x}\times(\rho\mathbf{v})d^3x$, far from the stationary source, or for a spheroidal distribution of matter, one can write $\mathbf{h}\equiv(h_{01}, h_{02}, h_{03})$ as a function of $\mathbf{J}$ (Thorne, Price, and MacDonald 1986; Ciufolini and Wheeler 1995):

$$\mathbf{h}(\mathbf{x})\cong -2\frac{\mathbf{J}\times\mathbf{x}}{|\mathbf{x}|^3}; \tag{4}$$

$\mathbf{h}$ is called the gravitomagnetic potential. Finally, one can define a gravitomagnetic field $\mathbf{H}=\nabla\times\mathbf{h}$:

$$\mathbf{H}=\nabla\times\mathbf{h}\cong 2\left[\frac{\mathbf{J}-3(\mathbf{J}\cdot\hat{\mathbf{x}})\hat{\mathbf{x}}}{|\mathbf{x}|^3}\right]. \tag{5}$$

From these equations we see that, in general relativity, in the weak field and slow motion limit, the angular momentum $\mathbf{J}$ of a stationary, localized, mass–energy current has a role similar to the magnetic dipole moment $\mathbf{m}$ of a stationary, localized charge current in electrodynamics (the difference between electromagnetism and weak-field general relativity is an extra factor -4 in general relativity).

By using the geodesic equation $Du^\alpha/ds=0$, in the weak field and slow motion limit, one has then:

$$m\frac{d^2x}{dt^2}\cong m(\mathbf{G}+\frac{d\mathbf{x}}{dt}\times\mathbf{H}), \tag{6}$$

where $\mathbf{G} \approx -(M/|\mathbf{x}|^2)\hat{\mathbf{x}}$ is the standard Newtonian acceleration and $\mathbf{H}$ is the gravitomagnetic field.

Moreover, as in electromagnetism, in general relativity the 'torque' acting on a gyroscope with angular momentum $\mathbf{S}$, in the weak-field and slow-motion approximation, is:

$$\tau \cong \frac{1}{2}\mathbf{S} \times \mathbf{H} = \frac{d\mathbf{S}}{dt} \equiv \dot{\boldsymbol{\Omega}} \times \mathbf{S}. \tag{7}$$

Therefore, the gyroscope precesses with respect to an asymptotic inertial frame with angular velocity

$$\dot{\boldsymbol{\Omega}} = -\frac{1}{2}\mathbf{H} = \frac{-\mathbf{J} + 3(\mathbf{J} \cdot \hat{\mathbf{x}})\hat{\mathbf{x}}}{|\mathbf{x}|^3}, \tag{8}$$

where $\mathbf{J}$ is the angular momentum of the central object. This phenomenon is the 'dragging of gyroscopes' or 'dragging of inertial frames,' of which the gyroscopes define the axes.

As in electromagnetism, the 'force' exerted on the gyroscope by the gravitomagnetic field $\mathbf{H}$ is

$$\mathbf{F} = \left(\frac{1}{2}\mathbf{S} \cdot \nabla\right)\mathbf{H}. \tag{9}$$

Finally, for a central object with angular momentum $\mathbf{J}$ and due to the second term in the 'force' (6), the orbital plane (and the orbital angular momentum) of a test particle, which can be thought of as an enormous gyroscope, is dragged in the sense of rotation of the central body. This dragging of the whole orbital plane is described by the formulas discovered by Lense and Thirring (1918) for the secular rate of change of the longitude of the nodal line (intersection between the orbital plane of the test particle and the equatorial plane of the central object):

$$\dot{\boldsymbol{\Omega}}^{\text{Lense-Thirring}} = \frac{2\mathbf{J}}{a^3(1-e^2)^{3/2}}, \tag{10}$$

where a is the orbital semimajor axis of the test particle, e its orbital eccentricity, and $\mathbf{J}$ the angular momentum of the central body.

Similarly, by integrating the equation of motion of the test particle, one can find the formulas derived by Lense and Thirring for the secular rates of change of the longitude of the pericenter $\dot{\omega}$ (determined by the Runge–Lenz vector):

$$\dot{\omega}^{\text{Lense-Thirring}} = \frac{2J}{a^3(1-e^2)^{3/2}}(1-3\cos I), \tag{11}$$

where I is the orbital inclination of the test particle. Finally, inside a slowly rotating thin shell of mass M and radius R, at the lowest order in the angular velocity $\boldsymbol{\omega}$ and M/R, we have from (3)

$$\mathbf{h} \equiv (h_{0x}, h_{0y}, h_{0z}) = \frac{4}{3}\frac{M}{R}\boldsymbol{\omega}\times\mathbf{x} = \left[\frac{4}{3}\frac{M}{R}\omega y, -\frac{4}{3}\frac{M}{R}\omega x, 0\right]. \tag{12}$$

Substituting the components of $h_{\alpha\beta}$ inside the slowly rotating shell in the geodesic equation, we find Thirring's result (1918) for the acceleration of a test particle inside a rotating shell due to the rotation of the shell:

$$\begin{aligned} \ddot{x} &= -\frac{8}{3}\frac{M}{R}\omega\dot{y} + \frac{4}{15}\frac{M}{R}\omega^2 x, \\ \ddot{y} &= \frac{8}{3}\frac{M}{R}\omega\dot{x} + \frac{4}{15}\frac{M}{R}\omega^2 y, \\ \ddot{z} &= -\frac{8}{15}\frac{M}{R}\omega^2 z. \end{aligned} \tag{13}$$

Moreover, the axes of the local inertial frames, i.e., the gyroscopes, are dragged by the rotating shell with constant angular velocity $\dot{\boldsymbol{\Omega}}^G$:

$$\dot{\boldsymbol{\Omega}}^G \cong -\frac{1}{2}\mathbf{H} = -\frac{1}{2}\nabla\times\mathbf{h} = \frac{4}{3}\frac{M}{R}\boldsymbol{\omega}. \tag{14}$$

For other solutions inside a rotating shell with arbitrary mass M, or to higher order in the angular velocity $\boldsymbol{\omega}$, and for discussions on the interpretations of the forces inside a rotating shell, see (Brill and Cohen 1966), and (Pfister and Braun 1986).

2. Invariant Characterization of Gravitomagnetism

Dragging of inertial frames shows that, in general relativity, the local inertial frames of reference are influenced and, at least in part, determined by the distribution and currents of mass-energy in the universe. Therefore, one may think that the dragging of inertial frames is a manifestation of a *weak general relativistic formulation of Mach's Principle*. Depending on the boundary conditions and on the cosmological solution, one may find other stronger 'general relativistic formulations of Mach's Principle' which are satisfied in Einstein's general relativity (Ciufolini and Wheeler 1995).

In addition to the 'frame dragging' effects (8) and (10), both due to the intrinsic angular momentum of a rotating central body, a test gyroscope orbiting around a *static* central mass has a precession relative to an asymptotic inertial frame: the de Sitter or geodetic (or geodesic) precession (de Sitter 1916), due to the velocity of the gyroscope and to the *static* field generated by the central mass, **M**:

$$\dot{\mathbf{\Omega}} = -\frac{3}{2}\mathbf{v} \times \mathbf{r}\frac{M}{r^3}. \tag{15}$$

The de Sitter precession has been tested with accuracy of $\sim 1\%$ (Bertotti, Ciufolini, and Bender 1987; Shapiro *et al.* 1988; Dickey *et al.* 1988; Müller *et al.* 1991) for the 'earth-moon gyroscope' orbiting the sun. *Part* of the de Sitter precession (15) is similar to the Thomas precession, $\dot{\mathbf{\Omega}} = -\frac{1}{2}\mathbf{v} \times \mathbf{a}$, in special relativity, due to the noncommutativity of non-aligned Lorentz transformations and to the nongravitational acceleration $\mathbf{a}$. Some authors (Ashby and Shahid-Saless 1988) have reinterpreted the de Sitter effect as a kind of gravitomagnetic Lense–Thirring effect due to the *orbital* angular momentum of the central *static* mass as seen by an observer orbiting the central mass who carries a gyroscope. However, we show here that de Sitter effect and Lense–Thirring drag are two intrinsically different phenomena (Ciufolini 1991).

Since dragging of inertial frames and gravitomagnetism show a qualitatively fundamental difference between general relativity and classical Newtonian theory, several experiments were proposed since 1986 to test their existence (see next section). However, so far, apart from some indirect astrophysical evidence, gravitomagnetism has never been directly detected and measured. Gravitomagnetism, dragging of inertial frames, and Mach's Principle had, in the literature, several different interpretations.

To clarify the meaning of gravitomagnetism, we propose to characterize it precisely in a way independent of the frame and the coordinates used, as is customary in general relativity. Intuitively, gravitomagnetism may be thought of as a phenomenon in which the spacetime geometry and curvature change due to mass–energy currents relative to other mass.

Using this characterization of gravitomagnetism we may investigate a problem described in various papers (Ashby and Shahid-Saless 1988; Nordtvedt 1991): Can the existence of the gravitomagnetic field and of the dragging of inertial frames be inferred as a consequence of the existence of the standard gravitoelectric field (for example, the Schwarzschild solution) plus local Lorentz invariance? In other words, can the Lense–Thirring effect be inferred as a consequence of the de Sitter or geodetic precession?

Coming back to the formal analogy with electromagnetism, it is important to recall that, apart from several formal analogies, general relativity, even the linearized theory, and electromagnetism are fundamentally different. Of course, the main difference is the equivalence principle: Locally (in the spacetime), in a suitable neighborhood and in the freely falling frames, it is possible to eliminate (in the sense of making arbitrarily small) the effects of the gravitational field; in particular, the ratio between inertial mass and gravitational mass is the same for every body. In general relativity, the spacetime geometry $g_{\alpha\beta}$, where the various physical phenomena take place, is determined by the energy and by the energy-currents in the universe via the Einstein field equation, and since the gravity field $g_{\alpha\beta}$ has energy and momentum, the gravitational energy contributes itself, in a loop, to the spacetime geometry $g_{\alpha\beta}$. However, in special relativistic electrodynamics, the spacetime geometry $\eta_{\alpha\beta}$, where the electromagnetic phenomena take place, is unaffected by the electromagnetic phenomena.

It is well known that in electromagnetism, in the frame of a charge q at rest, we only have a nonzero electric field $\mathbf{E}^0$ but no magnetic field $\mathbf{B}^0$. However, if we consider an observer moving with velocity $\mathbf{v}$ relative to the charge q, in this new frame we have a magnetic field: $\mathbf{B}' = -\gamma(\mathbf{v})(\boldsymbol{\beta} \times \mathbf{E}^0)$. Similarly, in general relativity, in the frame of a mass M at rest, we only have the nonzero metric components $g_{00} = -g_{rr}^{-1} = -(1-2M/r)$, and $g_{\theta\theta} = g_{\phi\phi}/\sin^2\theta = r^2$, but we do not have the so-called 'magnetic' metric components g_{0i}. However, for an observer moving with velocity $\mathbf{v}$ relative to the mass M, in his local frame we have a 'magnetic' metric component g_{0i}: $g'_{0i} = \Lambda^{\alpha}_{0'}\Lambda^{\beta}_{i'}g_{\alpha\beta} \sim Mv/r$. Therefore, in the physics literature we find the question whether the existence of the g_{0i} components in this boosted frame, proportional to the orbital angular momentum $l \sim Mvr$, as measured by the moving observer, can be considered as a proof that the intrinsic angular momentum J of a mass distribution changes the spacetime geometry.

The answer is no, as we have just remarked. Of course, for every spacetime solution we can always make the g_{0i} components different from zero just by a coordinate transformation; to look at the components of the metric tensor is not sufficient to analyze the spacetime structure and curvature. On the Schwarzschild event horizon one has a coordinate singularity (in Schwarzschild coordinates), but a well-behaved spacetime structure and curvature (as is clear by analyzing the curvature invariants and by using other coordinate systems). Furthermore, any metric, for example, the flat Minkowski metric $\eta_{\alpha\beta}$, can be changed by a coordinate transformation into a metric with all the components complicated functions of the coordinates. However, it is also well known that in order to distinguish a true spacetime curvature singularity from a mere coordinate singularity, or a flat Minkowski spacetime from a curved manifold, one has to analyze the Riemann tensor, to see if it is different from zero or not, and to see if it is well-behaved or if it is diverging in some region. One might therefore think to test if the so-called magnetic components of the Riemann tensor R_{i0jk} are different from zero or not. However, again, from the six so-called electric components R_{i0j0}, with a local Lorentz transformation one can locally get magnetic components $R'_{i0jk}=\Lambda^{\alpha}_{i'}\Lambda^{\beta}_{0'}\Lambda^{\mu}_{j'}\Lambda^{\nu}_{k'}R_{\alpha\beta\mu\nu}$ different from zero.

Therefore, following a method of characterizing singularities and the method of classifying different spacetime solutions, the correct approach is to inspect the spacetime invariants. In vacuum, the Ricci curvature scalar $R=R^{\alpha}{}_{\alpha}$ is identically zero, as a consequence of the Einstein field equation. Another scalar invariant is the Kretschmann invariant, $R_{\alpha\beta\mu\nu}R^{\alpha\beta\mu\nu}$; however, in the case of a metric characterized by mass and angular momentum, such as the Kerr metric, the Kretschmann invariant is a function of M/r^3 and J/r^4, with the leading term $\sim(M/r^3)^2$. Therefore, this invariant is nonzero in the presence of a mass M whether or not there is any angular momentum.

At this point we turn again to a formal analogy between electromagnetism and general relativity. In electromagnetism, to characterize the electromagnetic field, one can calculate the scalar invariant $-\frac{1}{2}F_{\alpha\beta}F^{\alpha\beta}=E^2-B^2$, which is analogous to the Kretschmann invariant $R_{\alpha\beta\mu\nu}R^{\alpha\beta\mu\nu}\propto(M/r^3)+C(J/r^4)^2$; see below. However, in electrodynamics one can also calculate the scalar pseudoinvariant $\frac{1}{4}F_{\alpha\beta}{}^{*}F^{\alpha\beta}=\mathbf{E}\cdot\mathbf{B}$, where * is the dual operation: $^{*}F^{\alpha\beta}=\frac{1}{2}\varepsilon^{\alpha\beta\mu\nu}F_{\mu\nu}$, where $\varepsilon^{\alpha\beta\mu\nu}$ is the Levi-Civita pseudotensor: $\epsilon_{\alpha_1\ldots\alpha_n}\equiv(-g)^{1/2}[\alpha_1\ldots\alpha_n]$; $(-g)^{1/2}$ is the square root of the determinant of the metric, and the symbol $[\alpha_1\ldots\alpha_n]$ is equal to $+1$ for even permutations of 1,2,..., n; -1 for odd permutations; and 0 if some indices are repeated. We observe that if we have a charge q only, in its rest frame we have an electric field only, and the invariant $F_{\alpha\beta}{}^{*}F^{\alpha\beta}$ is zero; therefore, even in those inertial frames where both $\mathbf{B}\neq\mathbf{0}$ and $\mathbf{E}\neq\mathbf{0}$, this invariant will be zero. However, if in the rest frame we have a charge q and a magnetic dipole $\mathbf{m}$, in this frame we have in general $F_{\alpha\beta}{}^{*}F^{\alpha\beta}\neq 0$, and, of course, this invariant is different from zero in any other inertial frame.

Therefore, to characterize the spacetime geometry and curvature generated by mass–energy currents and by the intrinsic angular momentum J of a central body (in weak-field general relativity, the angular momentum generated by mass–energy currents plays a role analogous to the magnetic dipole moment of a loop of charge current in electromagnetism, Sec. 1), we should look for an analogous spacetime invariant (Ciufolini 1991). This invariant should therefore be built out of the dual of the Riemann tensor $^{*}R^{\alpha\beta\mu\nu}\equiv\frac{1}{2}\varepsilon^{\alpha\beta\sigma\rho}R_{\sigma\rho}{}^{\mu\nu}$, 'multiplied' by $R_{\alpha\beta\mu\nu}$. This pseudoinvariant is of the type $\frac{1}{2}\varepsilon^{\alpha\beta\sigma\rho}R_{\sigma\rho}{}^{\mu\nu}R_{\alpha\beta\mu\nu}$ (Wheeler 1977;

Petrov 1969). Because of the formal analogy with electromagnetism, and since this pseudoinvariant $^*\mathbf{R}\cdot\mathbf{R}$ is built using the Levi-Civita pseudotensor $\varepsilon^{\alpha\beta\sigma\rho}$, it should change sign for time reflections $t\to -t$, and therefore it should be proportional to J. A list of all the possible spacetime invariants built out of the Riemann tensor and of its dual is given in (Petrov 1969). In electromagnetism, $^*\mathbf{F}\cdot\mathbf{F}$ characterizes the electromagnetic field only, but not the spacetime geometry. However, in general relativity the meaningful and useful invariant $^*\mathbf{R}\cdot\mathbf{R}$ characterizes the gravitational field and spacetime geometry.

Indeed, one may get some information about this invariant (Ciufolini 1991) without performing lengthy calculations, by using the Newman–Penrose null-tetrad formalism; however, one can also calculate $^*\mathbf{R}\cdot\mathbf{R}$ by using some computer algebra system, such as STENSOR or MACSYMA. The result for the Kerr metric is

$$^*\mathbf{R}\cdot\mathbf{R}\equiv\frac{1}{2}\epsilon^{\alpha\beta\sigma\rho}R_{\sigma\rho}{}^{\mu\nu}R_{\alpha\beta\mu\nu}$$

$$=1536JM\cos\theta\left[r^5\rho^{-6}-r^3\rho^{-5}+\frac{3}{16}r\rho^{-4}\right], \tag{16}$$

where $\rho=(r^2+a^2\cos^2\theta)$; and in the weak-field limit

$$^*\mathbf{R}\cdot\mathbf{R}\simeq 48\left[6\frac{JM}{r^7}\cos\theta\right]+\cdots, \tag{17}$$

whereas the Kretschmann invariant $\mathbf{R}\cdot\mathbf{R}$ for the Kerr metric is

$$\mathbf{R}\cdot\mathbf{R}=-768J^2r^4\rho^{-6}\cos^2\theta+384J^2r^2\rho^{-5}\cos^2\theta-48J^2\rho^{-4}\cos^2\theta$$
$$+768M^2r^6\rho^{-6}-1152M^2r^4\rho^{-5}+432M^2r^2\rho^{-4},$$

and in the weak field limit: $\mathbf{R}\cdot\mathbf{R}\simeq 48\,(M^2/r^6-21(J^2/r^8)\cos^2\theta)+\cdots$. Since the external gravitational field of a stationary black hole is determined by its mass M, charge Q, and intrinsic angular momentum J (Misner, Thorne, and Wheeler 1973), and since for the Kerr–Newman metric the invariant $^*\mathbf{R}\cdot\mathbf{R}$ is still proportional to J, the above result is quite general in the case of a black hole and is valid, for any quasistationary solution, asymptotically, in the weak-field limit. Furthermore, the above result, obtained in Einstein's theory, is generally valid in *any* metric theory of gravity (with no prior geometry) *not* necessarily described at the post-Newtonian order by the PPN formalism (Will 1993). In any metric theory of gravity (with no prior geometry), the full expression of the scalar $^*\mathbf{R}\cdot\mathbf{R}$ must depend on some of the intrinsic physical quantities characterizing the source, such as the mass–energy of the source, its intrinsic angular momentum, its multipole mass moments..., i.e., it must depend on some integral of the mass–energy density ε of the mass-energy currents εu^i, ... In particular, since $^*\mathbf{R}\cdot\mathbf{R}$ must change sign for time reflections, its full expression for a generic source must be proportional to some odd function of the intrinsic mass–energy currents εu^i (not eliminable with any Lorentz transformation) characterizing the system, such as the intrinsic angular momentum of the source of expression (17).

Therefore, independently from the field equations of a particular metric theory, the pseudoinvariant $^*\mathbf{R}\cdot\mathbf{R}$ determines the existence and the presence of gravitomagnetism in that metric theory of gravity. Indeed, by using this invariant $^*\mathbf{R}\cdot\mathbf{R} \propto (JM/r^7)\cos\theta$, we can determine whether or not there is a gravitomagnetic contribution to the spacetime geometry and curvature. We just need to calculate $^*\mathbf{R}\cdot\mathbf{R}$; if it is different from zero we have a contribution of the mass–energy currents to the curvature, and if it is zero there is no gravitomagnetic contribution. No matter about local Lorentz transformations or other frame and coordinate transformations on a static background, either $^*\mathbf{R}\cdot\mathbf{R}$ is zero, as it is in the Schwarzschild case, or it is different from zero, as in the Kerr case. A spacetime with $^*\mathbf{R}\cdot\mathbf{R}\neq 0$ is conceptually and qualitatively different from a spacetime with $^*\mathbf{R}\cdot\mathbf{R}=0$, regardless of frame and coordinate transformations.

We can then characterize what might be called gravitomagnetism, a new feature of Einstein general relativity with respect to classical Newtonian gravity theory. By taking a static background, generated in some metric gravity theory, and by a local Lorentz transformation, one can locally get nondiagonal components of the metric tensor. However, a basically new concept that Einstein introduced in general relativity is that the spacetime structure and curvature are affected and determined not only by mass–energy but also by mass–energy currents relative to other mass, i.e., mass–energy currents not generable or eliminable by a Lorentz transformation (for example, the intrinsic angular momentum of a body, which cannot be generated or eliminated by a Lorentz transformation). This feature of general relativity may be called gravitomagnetism and has never been directly detected and measured experimentally.

In summary, to define precisely and analyze the phenomenon of gravitomagnetism and to show the basic conceptual difference between the Lense–Thirring effect and the de Sitter precession, we have proposed to characterize gravitomagnetism independently from the frame and the coordinate system used, with spacetime curvature invariants only.

We now briefly describe the LAGEOS III experiment, proposed in 1986, to detect and measure the gravitomagnetic field of earth.

3. The LAGEOS III Gravitomagnetic Experiment

The basic idea of the LAGEOS III experiment (Ciufolini 1986, 1989) can be decomposed into two parts:

1. Position measurements of laser-ranged satellites, of LAGEOS (1976) type (see below), are accurate enough to detect the very tiny effect due to the gravitomagnetic field: the Lense–Thirring precession, Eq.(10).

2. To 'cancel out' the enormous perturbations due to the nonsphericity of the earth gravity field, we need a new satellite: LAGEOS III, with inclination supplementary to that of LAGEOS, and with the other orbital parameters, a and e, equal to those of LAGEOS.

LAGEOS (LAser GEOdynamics Satellite) is a high-altitude, small cross-sectional area-to-mass ratio, spherical, laser-ranged satellite. It is made of heavy brass and aluminum and is completely passive and covered with laser retroreflectors. It acts as a reference target for ground-based laser-tracking systems. LAGEOS was launched in 1976 to measure - via laser ranging - "crustal movements, plate motion, polar motion, and earth rotation." It continues to orbit and to pay scientific dividends. We know the LAGEOS position better than that of every other object in the sky. The relative accuracy in tracking its orbit is of the order of 10^{-8} to 10^{-9}, or less than 1 cm over 5900 km of altitude! The LAGEOS semimajor axis is $a=12270$ km, the period $P=3.758$ h, the eccentricity $e=0.004$, and the inclination $I=109.94°$.

In 1989, it was shown (Ciufolini and Matzner 1989) that from the analysis of the LAGEOS data it may be possible to have a 20%, or less, measurement of the LAGEOS relativistic perigee precession, which can be used to put limits on some alternative theories of gravity, such as the non-Riemannian nonsymmetric Moffat theory. In 1993, by analyzing LAGEOS and satellite laser ranging data, a new limit has been set (Ciufolini and Nordtvedt 1993) to the spatial anisotropy of the gravitational interaction of about $|\delta G|/G \leq 2\times10^{-12}$.

However, one of the relativistic perturbations best measurable on satellites like LAGEOS, with $e \ll 1$, is the precession of the nodal lines. For LAGEOS, the Lense–Thirring precession is

$$\dot{\Omega}^{\text{Lense-Thirring}}_{\text{Lageos}} = \frac{2GJ_\oplus}{c^2a^3(1-e^2)^{3/2}} \cong 31 \text{ milliarcsec/year} \tag{18}$$

where $J_\oplus \cong 5.9\times10^{40}$ g·cm^2/s $\cong 145$ cm^2 (in geometrized units) is the angular momentum of the earth.

The total nodal precession can be measured on LAGEOS with an accuracy of less than 1 milliarcsec/year.

Unfortunately, the Lense–Thirring precession cannot be extracted from the experimental value of $\dot{\Omega}^{\text{exp}}_{\text{Lageos}}$ because of the uncertainty in the value of the classical precession:

$$\dot{\Omega}^{\text{Class}}_{\text{Lageos}} = -\frac{3}{2}n\left(\frac{R_\oplus}{a}\right)^2 \frac{\cos I}{(1-e^2)^2}\Bigg\{J_2 + J_4\left[\frac{5}{8}\left(\frac{R_\oplus}{a}\right)^2 \times (7\sin^2 I - 4)\frac{(1+3e^2/2)}{(1-e^2)^2}\right] + \cdots\Bigg\}, \tag{19}$$

where $n=2\pi/P$ is the orbital mean motion, $R_\oplus$ is the earth's equatorial radius, and J_{2n} are the even zonal harmonic coefficients.This classical precession is due to the quadrupole and higher multipole mass moments of the earth, measured by the coefficients J_{2n}. The orbital parameters n, a, and e in formula (19) are determined with sufficient accuracy via the LAGEOS laser ranging, and the average inclination angle I can be determined with sufficient accuracy over a long enough period of time. Any other quantity in Eq. (19) can be determined or is known with sufficient accuracy, apart from the J_{2n}. Indeed, the largest uncertainty in the classical precession $\dot{\Omega}^{\text{Class}}_{\text{Lageos}}$ arises from the uncertainty in the coefficients J_{2n}. This uncertainty, relative to J_2, is of the order of:

$$\frac{\delta J_{2n}}{J_2} \sim 10^{-6}. \tag{20}$$

For J_2, this corresponds, from (19), to an uncertainty in the nodal precession of about 450 milliarcsec/year, plus the uncertainties due to the higher J_{2n} coefficients. Therefore, the uncertainty in $\dot{\Omega}^{\text{Class}}_{\text{Lageos}}$ is much larger than Lense–Thirring precession. A solution would be to orbit several high-altitude, laser-ranged satellites similar to LAGEOS, to measure J_2, J_4, J_6, etc., and one satellite to measure $\dot{\Omega}^{\text{Lense–Thirring}}$.

Another solution would be to orbit polar satellites; indeed, since for polar satellites, from (19), $I=90°$, $\dot{\Omega}^{\text{Class}}$ is equal to zero. Yilmaz proposed the use of polar satellites in 1959. In 1976, Van Patten and Everitt proposed an experiment with two drag-free, guided, counter-rotating polar satellites. The reason for proposing two counter-rotating polar satellites was to avoid inclination measurement errors.

A new solution would be to orbit another satellite, of LAGEOS type, with the same semimajor axis, the same eccentricity, but the *inclination supplementary* to that of LAGEOS. This configuration would be also useful to reduce the error due to other nongravitational uncertainties. Therefore, 'LAGEOS III' should have the following orbital parameters: $I^{\text{III}} = 180° - I^{\text{I}} \cong 70°$, $a^{\text{III}} = a^{\text{I}}$, $e^{\text{III}} = e^{\text{I}}$.

With this choice, since the classical precession $\dot{\Omega}^{\mathrm{Class}}$ is linear in cos I, it will be equal and opposite for the two satellites:

$$\dot{\Omega}_{\mathrm{III}}^{\mathrm{Class}} = -\dot{\Omega}_{\mathrm{I}}^{\mathrm{Class}} .$$

By contrast, since the Lense–Thirring precession $\dot{\Omega}^{\mathrm{Lense-Thirring}}$ is independent of the inclination, Eq. (18), $\dot{\Omega}^{\mathrm{Lense-Thirring}}$ will be the same in magnitude and sign for both satellites:

$$\dot{\Omega}_{\mathrm{III}}^{\mathrm{Lense-Thirring}} = \dot{\Omega}_{\mathrm{I}}^{\mathrm{Lense-Thirring}} .$$

'Adding' the measured nodal precessions $\dot{\Omega}^{\mathrm{exp}}$, we will get

$$\dot{\Omega}_{\mathrm{III}}^{\mathrm{exp}} + \dot{\Omega}_{\mathrm{I}}^{\mathrm{exp}} - \dot{\Omega}_{\mathrm{I+III}}^{\mathrm{Other\ Forces}} = 2\dot{\Omega}_{\mathrm{III}}^{\mathrm{Lense-Thirring}} = 2\dot{\Omega}_{\mathrm{I}}^{\mathrm{Lense-Thirring}} ,$$

where $\dot{\Omega}_{\mathrm{I+III}}^{\mathrm{Other\ Forces}}$ is the sum of the nodal precessions of LAGEOS and LAGEOS III due to all the other forces and calculable with sufficient accuracy (Ciufolini 1989). We observe that this method of measuring the Lense–Thirring effect should not be thought of as the subtraction of two large numbers to get a very small number, but as the sum of the small *unmodeled* nodal precession of LAGEOS, $\dot{\Omega}_{\mathrm{I}}^{\mathrm{unm.}}$, with the small *unmodeled* nodal precession of LAGEOS III, $\dot{\Omega}_{\mathrm{III}}^{\mathrm{unm.}}$ (corresponding to the same earth gravity field solution), to get $2\dot{\Omega}^{\mathrm{LT}} = \dot{\Omega}_{\mathrm{I}}^{\mathrm{unm.}} + \dot{\Omega}_{\mathrm{III}}^{\mathrm{unm.}}$.

This idea to orbit a satellite LAGEOS III, to couple to LAGEOS with the same orbital parameters but supplementary inclination, can be described as follows. Since the classical nodal precession is equal and opposite for two satellites, the bisector of the angle between the nodal lines of the two satellites would be analogous to a *gyroscope*, in the sense that this line would not be affected by the partially unknown classical precession (19) but only by the general relativistic 'dragging of inertial frames' (18) and by the already measured de Sitter precession (15).

An important problem in the LAGEOS III experiment has been to identify and to quantify all the error sources that might affect the measurement of the Lense–Thirring precession.

In 1988, a preliminary error analysis (Ciufolini 1989) identified potential error sources and gave a preliminary error budget. The potential error sources in the LAGEOS III experiment include: orbital injection errors; errors from uncertainties in the coefficients involved in the spherical harmonics expansion of the earth potential, in particular, from uncertainties in the dynamical part of the earth field, i.e., in the modeling of solid and ocean earth tides; errors from nongravitational perturbations such as direct solar radiation pressure, earth albedo, satellite eclipses, anisotropic thermal radiation, infrared radiation, and atmospheric drag; errors from the uncertainties in the determination of the LAGEOS orbital parameters and, in particular, in the determination of the inclination I and of the nodal longitude Ω, relative to an asymptotic inertial frame.

This preliminary error analysis showed that an upper bound to the total statistical error, over the period of the node of about three years, was, in 1988, about 10% of the Lense–Thirring effect to be measured.

In May 1988, NASA and ASI (Italian Space Agency) formed study groups for the purpose of performing a comprehensive analysis of the experiment and a comprehensive numerical simulation. The main results of the NASA–ASI study (Tapley, Ciufolini, *et al.* 1989) were:

•The secular nodal precession of LAGEOS due to neutral and charged particle drag does not exceed a few parts in 10^{-3} of the Lense–Thirring effect, and is thus negligible in the gravitomagnetic experiment.

•The LAGEOS nodal perturbations due to anisotropic re-emission of earth infrared radiation from the LAGEOS retroreflectors (Yarkowski–Rubincam effect), and re-emission of sunlight modulated by eclipses was calculated to be at most 3% of the Lense–Thirring effect to be measured.

•The earth's albedo perturbation of the LAGEOS node was calculated to be at most 1% of the gravitomagnetic drag to be measured (however see below).

•All the other nongravitational effects are negligible in comparison with the Lense–Thirring effect.

•The study confirmed that the largest source of error is due to uncertainties in the spherical harmonics expansion of the earth potential, in particular to its dynamical part, i.e., to solid and ocean earth tides.

•The geopotential error was estimated to be at most 5% of the Lense–Thirring effect.

•A covariance analysis confirmed a 1σ error of 8% or less, considering an injection error of 0.1° in the inclination and with no improvements in the knowledge of the various parameters and in the measurement errors.

•Blind tests have given a root-mean-square difference, between values arbitrarily assumed for the frame dragging effect and the values recovered from the simulated data, of 8%.

•Therefore, the covariance analysis and the blind tests have confirmed the previous comprehensive error analysis, with a total statistical error $\leq 8\%$.

In 1993, the error budget of the LAGEOS III experiment was dramatically reduced with respect to the previous estimates: Present analyses show a total statistical error in the LAGEOS III gravitomagnetic experiment of about 3% of $\dot{\Omega}^{\text{Lense-Thirring}}$ over a three-year period. The main improvements that made possible a substantial reduction of the total statistical error are:

I) Improvements in the modeling of geopotential and earth tides with the new JGM-2 gravity field solution (over the previous GEM-T1 solution), especially due to the data from the new LAGEOS II satellite, launched in October 1992 by ASI and NASA. The total statistical error due to uncertainties in geopotential and earth tides is now estimated to be about 2%.

II) Improvements in the modeling of seasonal variations in low-degree geopotential harmonics, especially due to the new LAGEOS II (1992) data.

III) A McDonnel-Douglas–NASA study (1990) on Delta II two- stage launch vehicles, together with the improvements (1993) in the knowledge of the geopotential and with the orbital data available from LAGEOS II (1992), imply an essentially negligible injection error in the LAGEOS III experiment.

IV) Accurate measurements (and theoretical studies) of the spin axis orientation and of the spin rate of LAGEOS, carried out with various techniques - microwave Doppler, infrared Doppler, and optical glints from front reflectors surfaces, by the University of Maryland and other groups, together with accurate measurements of thermal and optical properties of LAGEOS-type satellites - will make possible a modeling and a reduction of the error due to anisotropic thermal radiation uncertainties to about 1% or 2%.

V) The effect of the earth albedo on the orbit of the LAGEOS satellite has been accurately calculated using data from the meteorological satellites ERBE (Earth Radiation Budget Experiment) and NOAA9 and NOAA10 by C. Martin and D. Rubincam. The effect of albedo on the LAGEOS nodal longitude turns out to be negligible with respect to the Lense–Thirring drag. Furthermore, a study of the effect of earth albedo on the bisector of the nodal lines of two LAGEOS satellites with supplementary inclinations (i.e., the albedo effect on the LAGEOS III experiment) has recently shown (I. Ciufolini and D. Lucchesi, to be published) that the albedo gives an essentially negligible contribution to the rate of change of the bisector and therefore a negligible contribution to the uncertainty in the LAGEOS III gravitomagnetic measurement.

Finally, the **3% total statistical error** budget in the LAGEOS III gravitomagnetic experiment is for a three-year period data analysis; however, since the lifetime of LAGEOS type satellites has been estimated to be of the order of a million years and the accuracy in the modeling of the various perturbations and in the measurements with laser ranging and

VLBI are improving steadily, the total error will be basically reduced with longer periods of observations, because of both reduction of random errors and improvements in systematic errors.

REFERENCES

Ashby, Neil and Shahid-Saless, Bahman (1990). "Geodetic Precession or Dragging of Inertial Frames?" *Physical Review D* 42: 118–22.

Bertotti, Bruno; Ciufolini, Ignazio; and Bender, Peter L. (1987). "New Test of General Relativity: Measurement of de Sitter Geodetic Precession Rate for Lunar Perigee." *Physical Review Letters* 58: 1062–5.

Brill, Dieter and Cohen, Jeffrey M. (1966). "Rotating Masses and their Effect on Inertial Frames." *Physical Review* 143: 1011–15.

Ciufolini, Ignazio (1986). "Measurement of Lense–Thirring Drag on High-Altitude Laser-Ranged Artificial Satellites." *Physical Review Letters* 56: 278–81.

Ciufolini, Ignazio (1989). "A Comprehensive Introduction to the LAGEOS Gravitomagnetic Experiment: From the Importance of the Gravitomagnetic Field in Physics to Preliminary Error Analysis and Error Budget." *International Journal of Modern Physics A* 4: 3083–145.

Ciufolini, Ignazio (1991). "Gravitomagnetism and Dragging of Inertial Frames," to be published.

Ciufolini, Ignazio and Matzner, Richard A. (1991). "Non-Riemannian Theories of Gravity and Lunar Satellite Laser Ranging." *International Journal of Modern Physics A* 7: 843–852.

Ciufolini, Ignazio and Nordtvedt, Kenneth (1993). "On the Isotropy of Newtonian Gravity," to be published.

Ciufolini, Ignazio and Wheeler, John A. (1995). *Gravitation and Inertia.* Princeton: Princeton University Press.

de Sitter, Willem (1916). "On Einstein's Theory of Gravitation and its Astronomical Consequences." *Monthly Notices of the Royal Astronomical Society* 76: 699–728.

Dickey, Jean D., Newhall, X X, and Williams, James G. (1989). "Investigating Relativity Using Lunar Laser Ranging: Geodetic Precession and the Nordtvedt effect." *Advances in Space Research* 9: 75–78. See also: *Relativistic Gravitation.* Symp. 15 of the COSPAR 27th Plenary Meeting, Espoo, Finland, 18–29 July 1988.

Lense, Josef and Thirring, Hans (1918). "Über den Einfluss der Eigenrotation der Zentralkörper auf die Bewegung der Planeten und Monde nach der Einsteinschen Gravitationstheorie." *Physikalische Zeitschrift* 19: 156–163.

Misner, Charles W., Thorne, Kip S, and Wheeler, John A. (1973). *Gravitation.* Freeman, San Francisco.

Müller, Jürgen; Schneider, Manfred, Soffel, Michael, and Ruder, Hanns (1991). "Testing Einstein's Theory of Gravity by Analyzing Lunar Laser Ranging Data." *Astrophysical Journal Letters* 382: L101–103.

Nordtvedt, Kenneth (1991). "The Gravitomagnetic Interaction and Its Relationship to Other Relativistic Gravitational Effects." NASA report.

Petrov, Aleksei Z. (1969). *Einstein Spaces*. Oxford: Pergamon Press.

Pfister, Herbert and Braun, Karlheinz (1986). "A Mass Shell with Flat Interior Cannot Rotate Rigidly." *Classical and Quantum Gravity* 3: 335–345.

Shapiro, Irvin I., Reasenberg, Robert D., Chandler, John F., and Babcock, Robert W. (1988). "Measurement of the de Sitter Precession of the Moon: A Relativistic Three-Body Effect." *Physical Review Letters* 61: 2643–6.

Tapley, Byron D, Ciufolini, Ignazio, *et al.* (1989). "Measuring the Lense–Thirring Precession Using a Second LAGEOS Satellite." *Results of a Joint NASA/ASI study on the LAGEOS III gravitomagnetic experiment.*

Thirring, Hans (1918). "Über die Wirkung rotierender ferner Massen in der Einsteinschen Gravitationstheorie." *Physikalische Zeitschrift* 19: 33–39.

Thirring, Hans (1921). "Berichtigung zu meiner Arbeit: "Über die Wirkung rotierender ferner Massen in der Einsteinschen Gravitationstheorie." *Physikalische Zeitschrift* 22: 29–30.

Thorne, Kip S., Price, Richard H., and MacDonald, Douglas A. eds. (1986). *Black Holes, the Membrane Paradigm.* New Haven and London: Yale University Press.

Van Patten, Richard A. and Everitt, C. W. Francis (1976). "Possible Experiment with Two Counter-Orbiting Drag-Free Satellites to Obtain a New Test of Einstein's General Theory of Relativity and Improved Measurements in Geodesy." *Physical Review Letters* 36: 629–632.

Yilmaz, Huseyin (1959). "Proposed Test of the Nature of Gravitational Interaction." *Bulletin of the American Physical Society* 4: 65.

Wheeler, John A. (1977). "A Few Specializations of the Generic Local Field in Electromagnetism and Gravitation." *Transactions of the New York Academy of Sciences, Ser. II*, 38: 219–243.

Will, Clifford M. (1993). *Theory and Experiment in Gravitational Physics.* Revised edition. Cambridge: Cambridge University Press.

Discussion

Soffel: Are your quoted errors independent of the measuring period?
Ciufolini: This is a very good question. That error is relative to three years. Now this satellite has a lifetime that has been estimated to be of the order of 500,000 years, or something like that, and therefore each nodal period [3 years] you wait, the better you have the measurement. If we launch the satellite in 1996 and we wait 20 years we will probably have a $\sim 1\%$ experiment. We have just to wait because we can then

average a lot of those effects like tides and other nongravitational effects that will affect the experiment, and at the end we can just get a very clean secular drift of the node of Lageos. So, thank you. That's a very important thing. In addition, this satellite is also very useful for geodesy. Therefore it's not only restricted to relativity.

Will: Has anyone looked at the long-term effect on the state of the thermal inertia and reflectivity, absorption of small particles and micrometeorites onto the surface?

Ciufolini: In the simulation in 1989, we also put in the change in the reflectivity of the satellite, which introduces a fraction of a percent error in the experiment.

Will: You don't sound totally sure.

Ciufolini: I can look at the number. I think it was probably a fraction of 1%; I didn't even mention errors that are less than 1%. We could also match the Lageos 3 satellite with Lageos 2. Lageos 2 was launched more recently, so the properties of its surface will be more similar to those of Lageos 3.

Goenner: May I raise an unfair question? Could you do your talk without mentioning Mach's Principle? Could you do the same talk - and the same applies to Professor Will - without using Mach's Principle? Because sometimes you have mentioned this gravitomagnetic field and you could just say it's a gravitational field which has properties like the electromagnetic field, which has induction and so forth, so you never have to think about Mach. So what do you reply?

Ciufolini: You are asking how much this experiment is related with Mach's Principle, and the answer is that this experiment measures what I like to call the weak general relativistic formulation of the Mach Principle, that is the dragging of inertial frames.

Goenner: No, my question is: When you talk about inertial dragging, do you need to refer to Mach's Principle, because you could say it's an induction effect like it is in electrodynamics.

Ciufolini: I agree. First, I like to think just to general relativity and only I try to see if there is any connection with Mach's Principle.

Goenner: Okay.

Ciufolini: What I mean is that the fact that the local inertial frames are influenced by the mass-energy currents is a kind of evidence in favor of a weak interpretation of the Mach's Principle.

Will: Whether or not one mentions Mach's Principle in one's talk has more to do with the local organizers of the conference than with the local inertial frame [loud laughter].

Time Variation of Fundamental Constants: Bounds From Local Data

Pablo Sisterna and Héctor Vucetich

Introduction

The Standard Model of Fundamental Interactions (SM) together with general relativity (GR) provides a consistent description of all known low-energy phenomena [i.e., low compared with the Grand Unified (GU) energy scale], in good agreement with experiment. This model depends on a set of parameters called the 'fundamental constants.' These are supposed to be universal parameters, i.e., time, position and reference-frame invariant. Indeed, the Einstein Equivalence Principle, on which GR is based, implies such an invariance.

However, the time variation of fundamental constants has been an active subject of research since the introduction of the Large Number Hypothesis (LNH) by Dirac (1937, 1938). (See also Barrow and Tipler 1986; McCrea and Rees 1983). This hypothesis was based on the existence of several large dimensionless numbers, such as the ratio of electrostatic and gravitational potentials in the hydrogen atom, whose value is near the ratio of the age of the universe and a typical period of the hydrogen atom. Assuming that the former quantity is proportional to the latter, the existence of the unnatural large number is 'explained.' The simplicity and large predictive power of the LNH led to numerous theoretical and experimental studies on the time variation of fundamental parameters.

On the theoretical side, there have been many proposals, both phenomenological (Gamow 1967) and theoretical (Bekenstein 1982), leading to a time variation of the fine structure constant. Unifying schemes such as Kaluza–Klein theories (Kaluza 1921; Klein 1926) or superstring theories (Schwarz 1982; Green and Schwarz 1984) provide

Einstein Studies, vol. 6: Mach's Principle: From Newton's Bucket to Quantum Gravity, pp. 403–421

Table 1: Sample bounds on the time variation of fundamental constants. These bounds assume that only a single constant varies.

M	Bound on $\dot{M}/M$, yr^{-1}	Reference
G_N	10^{-12}	Hellings *et al.* 1983
α	4×10^{-12}	Wolfe *et al.* 1976
α	10^{-17}	Shlyakhter 1976
$g_p m_e/m_p$	8×10^{-12}	Wolfe *et al.* 1976

a very general framework to study the time variation of fundamental constants. Indeed, it has been shown that Kaluza–Klein theories have cosmological solutions in which the fundamental constants do vary (Chodos and Detweiler 1980; Marciano 1984), and the same occurs in superstring theories (Wu and Wang 1986).

Another group of theories, the scale covariant theories (SCT), attempt to put LNH on a sounder theoretical basis. In these theories, the time variation of fundamental constants is due to a universal field that implements scale covariance on massive particles (Dirac 1973, 1974; Canuto *et al.* 1977) or a particular implementation of Mach's Principle (Hoyle and Narlikar 1972). The Brans–Dicke theory of gravitation (Brans and Dicke 1961; Jordan 1948; Thiry 1948) may be considered as an explicitly broken SCT.

Partly inspired by these theoretical results, many attempts have been made to set observational or experimental bounds on the time variation of fundamental constants. Table 1 summarizes of the most accurate bounds obtained from several sources, assuming the given constant is the only one which varies in time. This would give the right order of magnitude if there were no conspiracy between the variation of the constants to cancel its effect on any given physical observable. However, there are reasons to expect that several constants may vary simultaneously and that conspiracies are a consequence of deep theoretical results. For instance, the validity of Einstein's gravitational equations implies that the product of the gravitational constant and the mass of the body must be time independent (Canuto and Goldman 1982). So, it is interesting to analyze the time variation of fundamental constants without the assumption of no conspiracy, and we shall attempt to do so in this paper.

The time variation of fundamental constants will produce a host of different phenomena: changes in atomic and nuclear spectra (Wolfe *et al.* 1976; Shlyakhter 1976), variation of planetary radii and moments of inertia (McElhinny *et al.* 1978), orbital evolution (Hellings *et al.* 1983), and anomalous luminosities of faint stars (Mansfield and Malin 1980). Nucleosynthesis, both cosmological (Kolb *et al.* 1986) and stellar (Barrow 1987), has also been used to set bounds on the variability of fundamental parameters. Here we shall analyze mostly short-term local phenomena - astronomical and geophysical data based on time intervals much shorter than the age of the universe - and so set bounds on the variability of the fundamental constants today in the solar system. Since other astrophysical and cosmological data refer to very different time scales, it seems reasonable to analyze these latter events in a separate way.

As a result of our work, we are able to set consistent bounds for the simultaneous variation of fundamental constants in the Standard Model from astronomical, astrophysical, and geophysical data. These bounds exclude the LNH and, in general, any theory demanding a large variation of the fundamental constants.

Our paper is organized as follows. In Sec. 2, we present a simple phenomenological framework to study the time variation of fundamental constants, and our choice of fundamental parameters is explained. In Sec. 3, we discuss the observational evidence available from astronomical and geophysical phenomena, and in Sec. 4 we state our conclusions. Only qualitative discussion will be given, however, and the reader is referred to the literature (Sisterna and Vucetich 1990, 1991) for a complete quantitative treatment.

2. A Phenomenological Model

In this section we shall describe a very simple phenomenological model for the analysis of the consequences of time variation of fundamental constants. It will be based on the adiabatic hypothesis, i.e., that the main changes in observable quantities are due to time variation of the parameters, neglecting any necessary modifications of the SM. Although such a procedure will yield correct expressions for the change in observable quantities, one would not be able to relate the rate of change to interesting quantities, such as the Hubble constant or the contraction rate of extra dimensions, without a deeper analysis. This is because the Lagrangian obtained by simple substitution of time-varying parameters is generally inconsistent.

To begin with, we must choose a definite system of units. In a world of time- (and space-) independent parameters, this choice is completely arbitrary, but this will not be so in a world with time varying parameters (Dirac 1973). Different systems of units can be chosen so that different parameters are time independent. In simple model theories with a time varying gravitational constant, two such systems are the gravitational units, in which G_N is time independent but atomic parameters are time dependent, and atomic units, in which the opposite occurs. In a much more complex theory, such as the SM, very many different systems of units are possible.

In order to specify our system of units, we shall first assume the constancy of the dimensional constants c and $\hbar$, since this assumption simply fixes the length-to-time and time-to-energy units ratio. Moreover, we can use a finite, time-dependent renormalization group transformation to select any dimensional quantity as a time-dependent energy unit (Griego and Vucetich 1989). With such a choice, which amounts to taking a time-varying renormalization point, one builds the desired system of units. There are several choices for the energy standard, these defining several different systems of units having different physical meaning. Any of these systems will be related to any other through a finite renormalization group transformation, although its explicit construction may be difficult to carry out. In this paper, we shall introduce the Salam–Weinberg system of units (SWU), in which the mass of the intermediary vector meson W, M_W, is taken as the time independent energy unit. All our analysis will be carried out in SWU.

It is convenient to work with a set of 'intermediary' constants, such as the nucleon mass m_N, and relate it later to more fundamental parameters such as quark masses, since it is difficult to evaluate the effect of their time variation on observable quantities. We shall list some of these problems and the corresponding choice of variable quantities.

2.1. Gravitational interactions. In spite of many attempts at unification with other fundamental interactions, gravitation remains in isolation and its only parameter, the Newtonian gravitational constant G_N, is still unrelated to other fundamental constants. (See, however, Marciano 1984; Wu and Wang 1986.) We shall take G_N as one of our fundamental time-dependent parameters.

2.2. Electroweak interactions. The Salam–Weinberg unification of electroweak interactions is well supported by experiment not only at the tree level but also at the radiative correction level (Sirlin 1980, 1987). We shall assume the validity of the fundamental relations between the parameters of the theory: a consequence of the adiabatic hypothesis.

In SWU, the time variation of all fundamental parameters in the electroweak sector of the theory is fully determined by the time variation of the fine structure constant α and Fermi weak interaction constant G_F, so we would like to choose them as our fundamental time varying parameters. The latter quantity is, however, not directly observable: Only the time variation of the product $G_F \cos^2 \theta_C$, where θ_C is the Cabbibo angle, is directly measurable since there are no long-time high-precision measurements in the leptonic sector of the theory. Thus, the evaluation of the time variaton of G_F requires information on time variation in the Higgs sector of the SM.

2.3. Strong interactions. In the low-energy regime, strong interactions are effectively isolated but the single coupling constant α_3 is very big and nonperturbative effects are dominant. However, in the chiral limit, of massless *u–d* quarks, there is a single parameter in the theory, namely the QCD scale parameter Λ_Q, and we could choose it as our fundamental parameter, since all static observables with dimension of mass must be proportional to Λ_Q (Stevenson 1981). As a consequence, dimensionless static observables are time independent in massless QCD with time varying Λ_Q. Even in the presence of massive *u–d* quarks, provided their masses are small enough, the dominant contribution to the ground state (or to a low energy state) will be proportional to the QCD scale parameter Λ_Q. However, the large strange content of the nucleon (Donoghue and Nappi 1986) invalidates the above approximation and once again the Higgs sector contribution is important. As a consequence, we have taken several intermediate parameters (the nucleon mass m_N, the proton–neutron mass difference ΔM, and the mass of a typical meson m_m) which can be related later on to more fundamental constants.

2.4. Higgs sector. In the absence of a well-defined theory of the Higgs sector of the SM, a host of experiments would be necessary to analyze the time variation of the fundamental parameters in this sector. However, only a few of them, namely electron and quark masses and the Cabbibo angle, are relevant in the low energy regime, and we may take them as fundamental parameters. The quark content of nucleons and mesons has been well analyzed (Gasser and Leutwyler 1982; Donoghue and Nappi 1986; Dominguez and De Rafael 1987), and we can easily relate the time variation of intermediary quantities induced by the variation of the quark masses. Also, the time variation of the Cabbibo angle can be related to the mass variation of *s* and *d* quarks. Finally, there are several constraints the intermediary parameters must satisfy, such as the Goldberger–Treiman relation or the Adler–Weisberger sum rule, that can be used to fix their time variation.

2.5. Renormalization group equations. There are several parameters in our model that cannot be computed in a model independent way from our fundamental parameters, such as the time variation of the strong interaction constant α_3 and the renormalization point μ. We shall call these model-dependent parameters, since they can be computed within a larger model that contains the SM as a low-energy limit. In this paper, we shall limit ourselves to showing how these parameters can be computed in a Grand Unified Theory (GUT) which can be itself a low-energy limit of a Kaluza–Klein or superstring model.

We assume that at the grand unification scale Λ_U all the running coupling constants have a common value α_U. This is related to the Salam–Weinberg scale ($\mu \approx M_W$) values of the running constants α_i through the renormalization group equations. Besides, α_3 is related to the QCD scale parameter Λ_Q through the well-known definition.

These four equations are enough to find the time variation of the model dependent parameters α_3 and μ, and the GUT parameters α_U and Λ_U. However, a great simplification can be obtained if one makes the consistent choice $\mu = M_W$, since in this case $\dot{\mu}/\mu = 0$ and several of the renormalization group equations decouple. This particular choice of μ has been used in the present work.

3. Analysis of Observations

In this section we shall analyze and discuss different observations of geophysical, astronomical, and geochemical nature in order to obtain bounds for the time variation of fundamental contants. Our discussion will be mainly qualitative; full quantitative details will be found in (Sisterna and Vucetich 1990, 1991).

3.1. Planetary radii. Planetary radii will change under a time variation of fundamental constants because of the variation in cohesion of matter and the pull of gravity. The variation of planetary radii, as observed from structural changes in the planetary surface, can be computed if density, pressure, and bulk modulus distributions of the planet are known. This is true for the earth, where seismological data yield accurate distributions of these quantities (Bullen 1975), and for some smaller bodies of the solar system, such as Mercury and the moon, whose chemical composition can be inferred from geophysical observations and for which a linearized equation of state is a good approximation because of their small compression.

McElhinny *et al.* (1978) report upper bounds for the change in the radius of several planets from a variety of geophysical observations. The mean rates of variation for the moon and Mercury radii are shown in Table 2. The variation of the earth radius has not been included because, in spite of the accuracy of the observations, its complicated geological history makes the determination of its paleoradius unreliable.

3.2. The earth's moment of inertia. The variation of the earth's moment of inertia can be computed from the change in angular velocity induced on the earth–moon system because of conservation of angular momentum. The change in angular velocity of the earth can be directly measured, on the other hand, from the analysis of ancient astronomical observations (Lambeck 1981; Muller and Stephenson 1975; Muller 1976) or from the analysis of paleontological data (Lambeck 1978, 1981). Neither of these observational methods is free of trouble: Both methods are hampered by tidal friction, and ancient astronomical observations cover a short period of the history of the earth, where small changes in the moment of inertia due to deglaciation effects are to be expected (Nakiboglu and Lambeck 1980). Paleontological data suffer from ambiguities in their interpretation (Scrutton 1978; Sisterna and Vucetich 1994) and are sensitive to long terms changes in the tidal torque. In order to obtain meaningful results from these data, a simultaneous analysis of the lunar acceleration and of the Earth rotation results (following the pattern of Lambeck 1981 or Muller 1976) is necessary.

Paleontological data arise from the recording of tidal and climate phenomena on the shells of living animals and so they record changes in the moment of inertia in atomic units (AT), for which the Rydberg constant is time independent. Ancient astronomical observations measure the same change in ephemeris time (ET), for which both G_N and planetary masses are assumed time independent. The two sets of data yield complementary information on the time variation of fundamental constants.

3.3. Orbital perturbations. The main perturbations induced on a Keplerian system by the time variation of fundamental constants will be an acceleration in longitude of the planet or satellite. This longitude acceleration cannot be observed in ephemeris time, since it is universal, but it can be observed in atomic time or universal time.

Table 2: Observational data. The columns give the considered quantity, the observed value and the corresponding standard deviations (in units of 10^{-11} yr^{-1}), the system of units of the observation, and the reference.

Planetary Paleoradius: $\dot{R}/R$

Mercury	0.0 ± 0.012	SW	McElhinny *et al.* 1978
Moon	0.0 ± 0.015	SW	McElhinny *et al.* 1978
Mars	0.0 ± 0.03	SW	McElhinny *et al.* 1978

Lunar Secular Acceleration: $\dot{n}/n$

Mercury Transits	-15.0 ± 1.2	ET	Morrison, Ward 1975
Ancient Eclipses	-17.3 ± 1.8	ET	Muller 1976
Growth Rhythms	-14.2 ± 2.4	AT	Lambeck 1978,1981
LLR	-13.7 ± 1.0	AT	Dickey *et al.* 1982
Tidal Models	-15.2 ± 3.0	SW	Lambeck 1981
Satellite Data	-14.4 ± 1.7	SW	Felsentreger *et al.* 1979

Earth's Secular Acceleration: $\dot{\Omega}/\Omega$

Ancient Solar Eclipses	-24.3 ± 2.0	ET	Muller 1976
Ancient Lunar Eclipses	-20.6 ± 2.6	ET	Morrison and Stephenson 1982
Ancient Equinoxes	-23.6 ± 2.3	ET	Muller 1976
Growth Rhythms	-22.5 ± 1.0	AT	Lambeck 1981

Viking Ranging Data

$\dot{G}_N/G_N$	0.0 ± 1.2	AT	Hellings *et al.* 1983
$\dot{\beta}$	0.0 ± 2.4	AT	Hellings *et al.* 1983

Laboratory Data

Clock Rate Diff.	-0.2 ± 1.2	SW	Turneaure and Stein 1976

Long Lived β-Decayers: $\dot{\lambda}/\lambda$

^{187}Re	2.3 ± 1.8	α_U	Dyson 1966; Davies 1972
^{40}K	0.0 ± 0.29	α_U	Wetherhill 1975
^{87}Rb	0.0 ± 0.29	α_U	Wetherhill 1975

Oklo Phenomenon: $\dot{\sigma}/\sigma$

^{149}Sm	0.0 ± 69.0	SW	Shlyakhter 1976
^{157}Gd	0.0 ± 123	SW	Ruffenach 1978
^{151}Eu	0.0 ± 630	SW	Ruffenach 1978
^{113}Cd	0.0 ± 280	SW	DeLaeter and Rosman 1975

Eötvös Experiment: $\eta = \Delta g/g$

η(Al–Pt)	$0.00 \pm .10$	SW	Braginski and Panov 1972

Table 2, in which α_U is the unification coupling constant, shows several determinations of the lunar tidal acceleration both in ephemeris and atomic time. Comparison of the two data sets would yield the acceleration due to the variation of fundamental constants. However, because of the contamination of the data with tidal and deglaciation effects, we have analyzed it together with data on the earth's rotation, following the prescription of Muller (1976).

The binary pulsar offers a second independent determination of orbital evolution due to the time variation of fundamental constants (Damour *et al.* 1988). The astronomical system is very clean and the determination is very reliable. However, it is a high gravitational field system and it cannot be analyzed within our model (Nordtvedt 1993), and so we have not included it in our present data set.

Another determination of orbital evolution has been made on the motion of Mars as recorded from the Viking lander data (Hellings *et al.* 1983). The accuracy of these measurements is so high that it is possible to obtain a meaningful separation of the rate of variation of Newton's constant and of the mass.

3.4. Long-lived β-decayers. The half-life of long-lived β decayers such as ^{187}Re or ^{40}K has been used by Dyson (1967) to find upper bounds for the time variation of the fine structure constant. These nuclei have a very long half-life that has been determined either in laboratory measurements or by comparision with the age of meteorites, as found from α-decay radioactivity analysis.

In standard physics, the effective decay constant can be found if the age of meteorites is determined by means of any other present nucleus of known mean-life. In our model, the age of meteorites will have a different value for different nuclear species, and the comparison of these different ages sets bounds on the time variation of fundamental constants.

3.5. The Oklo phenomenon. About two billion years ago, a natural nuclear reactor operated for half a million years in the uranium ore deposits in Oklo, Gabon. From an analysis of nuclear and geochemical data (Maurette 1976; Petrov 1977), the operating conditions of the reactor could be reconstructed and the thermal neutron capture cross sections σ of several nuclear species measured. In particular the ^{149}Sm capture cross section is strongly dependent on the position of a resonance level of the compound nucleus ^{150}Sm*, being sensitive to small changes in its width and position (Shlyakhter 1976; see also Irvine 1983). Similar results can be obtained for other strong neutron absorbers (Shlyakhter 1983; Sisterna and Vucetich 1991). Upper bounds for the variation of fundamental constants can be found if the functional dependence of the parameters can be derived. However, this is an extremely difficult task, since a capture level of a strong absorber is a very complex state of a many body system. In our analysis, we have treated it as a finite temperature Fermi gas, the effective temperature T^* being chosen to reproduce the excitation energy of the system (Sisterna and Vucetich 1990).

3.6. Laboratory experiments. There is a single direct laboratory experiment accurate enough to yield interesting bounds on the rate of variation of atomic constants (Turneaure and Stein 1976). In this experiment, a set of cesium atomic clocks were compared with a set of Superconducting Cavity-Stabilized Oscillators. The measured value is shown in Table 2.

The time variation of fundamental constants breaks the Lorentz invariance of the SM, and effects related to such noninvariance should be expected. One of them is the appearance of anomalous accelerations in Eötvös-like experiments (Eötvös *et al.* 1922). Only a few of them have accuracy enough to be of use in setting bounds on the time variation of fundamental constants, and these are cited in Table 2.

3.7. Luminosity of faint stars. The luminosity of faint stars can be used to set strong constraints on scale-covariant theories of gravitation (Mansfield and Malin 1980). This anomalous luminosity is due to the radiation of internal energy of the star as it adjusts to the change in its structure due to the variation of G_N. Conservation of energy is essential for the derivation of this result.

However, in our model the time variation of fundamental constants induces no anomalous luminosity on such stars. Indeed, energy is not conserved in our model but injected into the system by the variation of the constants. Because of the Hamiltonian nature of our model, energy excess is not radiated but changes the internal energy of the system.

4. Results and Conclusions

Each of the observations included in Table 2 provides a conditional equation on the set of fundamental parameters. These equations form an overdetermined set of constraints that the observational data must satisfy. A least-squares solution to the set of constraints is shown in Table 3, together with 90% confidence limits computed with the statistical 'bootstrap' process (Efron 1979; Kinsella 1986). The main shortcoming in the above set of bounds comes from the simplified nuclear forces model, which introduces a large correlation between the time variations of the parameters in the Higgs sector, leading to the corresponding deterioration of the bounds. This deterioration is enhanced by the lack of direct experiments on the second family and the high strange quark content of the nucleon.

In spite of the above shortcomings, the limits in Table 3 are much smaller than the Hubble rate (as can be seen from the last column of the table), and so we can exclude the Dirac Large Number Hypothesis and, more generally, any theory showing a large variation of the fundamental constants. In Fig. 1, we compare our results with the predictions of a few selected theories, computed with reasonable values of the parameters involved and assuming they satisfy the hypothesis of our phenomenological model. The Dirac (1937) theory, and the Wu–Wang (1986)

Table 3. Bounds on the observational parameters. The columns show the parameter, the least squares estimate, and 90% bootstrap upper bounds, in units of 10^{-11} yr^{-1} and of the Hubble constant H_0. A low value of the latter (55 km/s/Mpc) has been adopted in order to get upper bounds.

M	$\dot{M}/M$ yr^{-1}	Upper Bound yr^{-1}	H_0
M_p	$(3.1 \pm 2.3) \times 10^{-5}$	3.5×10^{-3}	6.4×10^{-4}
ΔM	$(-1.2 \pm 7.8) \times 10^{-4}$	2.1×10^{-3}	3.8×10^{-4}
M_m	$(1.2 \pm 2.7) \times 10^{-4}$	2.1×10^{-3}	3.8×10^{-4}
α	$(0.8 \pm 1.5) \times 10^{-4}$	3.7×10^{-4}	6.7×10^{-5}
m_e	$(-2.3 \pm 3.9) \times 10^{-3}$	1.1×10^{-2}	2.0×10^{-3}
G_W	$(-0.6 \pm 2.4) \times 10^{-2}$	0.15	2.9×10^{-2}
G_N	(-0.32 ± 0.24)	0.75	0.14
n_L	-14.95 ± 0.45	-15.73	-14.43
Ω_N	4.1 ± 1.5	2.2	6.1

superstring compactification (with a massless dilaton) are in strong disagreement with our results.

Since our bounds form a consistent set, we can obtain from them bounds for the variation of other fundamental parameters of the SM, such as the mass of the intermediate vector boson Z or the vacuum expectation value of the Higgs field v. Table 4 shows these bounds. They are both consistent with experimental data and independent of any conspiracy among the constants. In the same way, we can find consistent upper bounds for the time variation of the fundamental constants at a GU scale. These results, together with the time variation rates of the model dependent parameters, are shown in the last part of Table 4.

Both Kaluza-Klein and superstring theories predict time variation of fundamental constants depending on the cosmological model parameters. In these theories, the common value of the running coupling constants at the GU scale is related to the size of the extra dimensional space $R_I \approx \Lambda_U^{-1}$. In the case of Kaluza-Klein theories, $\alpha_U \propto R_I^{-2}$ and from this relation we find the result quoted at the end of Table 4 for the present

Table 4: Consistent bounds on SM parameters. The columns show the parameter, the least squares estimate, and 90% bootstrap upper bounds, in units of 10^{-11} yr^{-1} and of the Hubble constant H_0. A low value of the latter (55 km/s/Mpc) has been adopted in order to get upper bounds.

M	$\dot{M}/M$ yr^{-1}	Upper Bound yr^{-1}	Upper Bound H_0
Λ_Q	$(1.0 \pm 2.8) \times 10^{-6}$	3.6×10^{-3}	6.4×10^{-3}
m_u	$(0.5 \pm 3.4) \times 10^{-4}$	2.9×10^{-3}	5.3×10^{-4}
m_d	$(0.1 \pm 1.7) \times 10^{-4}$	3.6×10^{-3}	6.5×10^{-4}
m_s	$(-0.2 \pm 1.9) \times 10^{-5}$	1.6×10^{-3}	2.9×10^{-4}
α_3	$(-2.1 \pm 0.4) \times 10^{-6}$	4.9×10^{-4}	8.9×10^{-5}
G_F	$(0.1 \pm 6.0 \times 10^{-3}$	0.16	2.9×10^{-2}
α_1	$(0.1 \pm 6.0 \times 10^{-3}$	0.16	2.9×10^{-2}
α_2	$(0.3 \pm 2.4) \times 10^{-2}$	0.36	3.6×10^{-2}
θ_W	$(-0.3 \pm 1.5) \times 10^{-3}$	4.3×10^{-2}	7.6×10^{-3}
M_Z	$(0.2 \pm 8.3) \times 10^{-3}$	4.2×10^{-2}	7.8×10^{-3}
θ_C	$(0.0 \pm 1.4) \times 10^{-5}$	2.1×10^{-4}	3.8×10^{-5}
α_U	0.34 ± 0.19	0.97	0.17
Λ_U	$(0.2 \pm 6.0) \times 10^{-2}$	0.17	3.1×10^{-2}
R_{KK}	$(-0.1 \pm 6.0) \times 10^{-2}$	0.17	3.1×10^{-2}
R_S	$(-5.3 \pm 5.0) \times 10^{-2}$	0.13	2.3×10^{-2}

present contraction rate. Again, this result is independent of any conspiracy between the different variation rates. Our bounds impose even more stringent constraints on the time variation of fundamental constants induced in superstring theories, where $G_N \propto R_I^{-6}$.

As we have mentioned before, the Einstein Equivalence Principle implies that all nongravitational constants of nature must be time and position independent. The Strong Equivalence Principle extends that statement to gravitational phenomena. Our results show that both forms

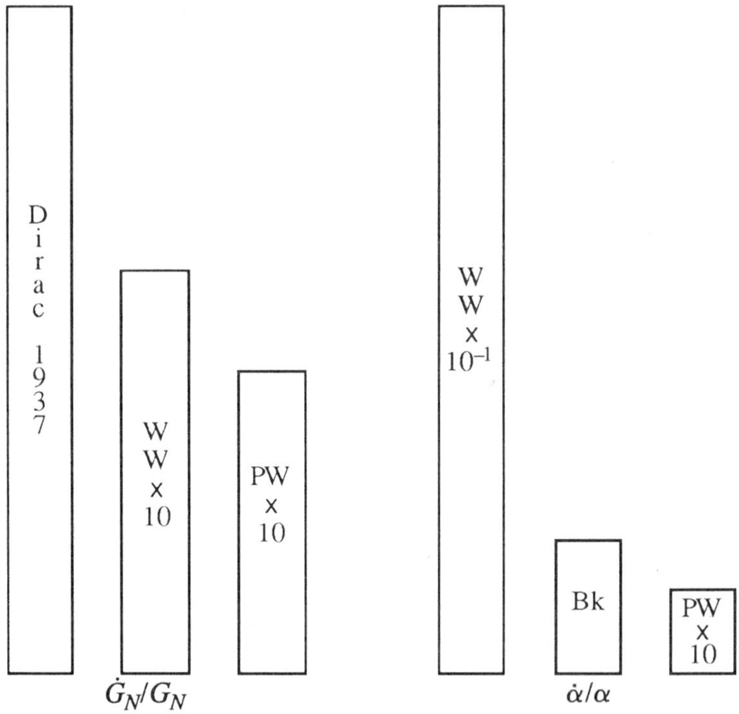

Figure 1: Comparison with selected theories. The prediction of the theories of Dirac (1937), Wu and Wang (1986), and Bekenstein (1982) are compared with the results of the present work (PW).

of the Principle of Equivalence are very well satisfied, within a small fraction of the Hubble rate. Since the unrestricted validity of the Principle of Equivalence leads to general relativity as the only low energy theory of gravitation, our results should be considered as an accurate verification of general relativity.

REFERENCES

Barrow, John D. and Tipler, Frank J. (1986). *The Anthropic Cosmological Principle*. Oxford: Oxford University Press.

Barrow John D. (1987). "Observational Limits on the Time Evolution of Extra Spatial Dimensions." *Physical Review* D35: 1805–1810.

Bekenstein, Jacob D. (1982). "Fine-Structure Constant: Is it Really a Constant?" *Physical Review* D25: 1527–1539.

Braginski, V. B. and Panov, V. I. (1972). "The Equivalence of Inertial and

Passive Gravitational Masses." *Soviet Physics JETP* 34: 464–466.
Brans, Carl H. and Dicke, Robert H. (1961). "Mach's Principle and Relativistic Theory of Gravitation." *Physical Review* 124: 925–935.
Bullen, K. E. (1975). *The Earth's Density*. London: Chapman and Hall.
Canuto, V. M., Adams, P. J., Hsieh, S. H., and Tsiang, E. (1977). "Scale-Covariant Theory of Gravitation and Astrophysical Applications." *Physical Review* D16: 1643–1663.
Canuto, V. M. and Goldman, I. (1982). "Atomic and Gravitational Clocks." *Nature* 296: 709–713.
Chodos, Alan and Detweiler, Steven (1980). "Where Has the Fifth Dimension Gone?" *Physical Review* D21: 2167–2170.
Damour, Thibault, Gibbons, Gary W., and Taylor, Joseph (1988). "Limits on the Variability of *G* Using Binary-Pulsar Data." *Physical Review Letters* 61: 1151–1154.
Davies, Paul C. W. (1972). "Time Variation of the Coupling Constants." *Journal of Physics* A5: 1296–1304.
DeLaeter, J. R. and Rosman, K. J. R. (1975). "Cumulative Fission Yields of Cadmiun in Oklo Samples." In *The Oklo Phenomenon*. IAEA-SM-204/7. Vienna: IAEA.
Dickey, J. O., Willams, J. G., and Yoder, C. F. (1982). "Results from Lunar Laser Ranging Data Analysis." In *High Precision Earth's Rotation and Moon Dynamics*. IAU. Proc 63 Colloquium.
Dirac, Paul A. M. (1937). "The Cosmological Constants." *Nature* 139:323.
Dirac, Paul A. M. (1938). "A New Basis for Cosmology," *Proceedings of the Royal Society (London)* A165: 199–208.
Dirac, Paul A. M. (1973). "Long Range Forces and Broken Symmetries." *Proceedings of the Royal Society (London)* A333: 403–418.
Dirac, Paul A. M. (1974). "Cosmological Models and the Large Number Hypothesis." *Proceedings of the Royal Society (London)* A338: 439–440.
Dominguez, C. A. and De Rafael, E. (1987). "Light Quark Masses in QCD from Local Duality." *Annals of Physics (New York)* 174: 372–400.
Donoghue, John F. and Nappi, Chiara R. (1986). "The Quark Content of the Proton." *Physics Letters* 168B: 105–109.
Dyson, Freeman J. (1966). "Time Variation of the Charge of the Proton." *Physical Review Letters* 19: 1291–1293.
Efron, Bradley (1979). "Bootstrap Methods: Another Look at the Jackknife." *Annals of Statistics* 7: 1–26.
von Eötvös, Roland, Pekár, D., and Fekete, E. (1922). "Beiträge zum Gesetze der Proportionalität von Trägheit und Gravität." *Annalen der Physik* 68: 11–66.
Felsentreger, T. L., Marsh, J. G., and Willamson, R. G. (1979). "M_2 Ocean Tide Parameters and the Deceleration of the Moon's Mean Longitude From Satellite Orbit Data." *Journal of Geophysical Research* 84: 4675.
Gamow, George (1967). "Electricity, Gravity and Cosmology." *Physical*

Review Letters 19: 759–761.

Gasser, J. and Leutwyler, H. (1982). "Quark Masses." *Physics Reports* 87: 77–169.

Green, Michael B. and Schwarz, John H. (1984). "Anomaly Cancellation in Supersymmetric $D=10$ Gauge Theory and Superstring Theory." *Physics Letters* 149B: 117–122.

Griego, J. and Vucetich, Héctor (1989). "A Time-Variation Model of the Fundamental Constants in Kaluza–Klein Theory." *Physical Review* D40: 1904–1908.

Hellings, Ron W., Adams, P. J., Anderson, J. D., Kelsey, M. S., Lau, E. L., Standish, E.M., Canuto, V. M., and Goldman, I. (1983). "Experimental Test of the Variability of *G* Using Viking Lander Ranging Data." *Physical Review Letters* 51: 1609–1612.

Hoyle, Fred and Narlikar, Jayant V. (1972). "On the Nature of Mass." *Nature* 233: 41–44.

Irvine, J. M. (1983). "The Constancy of the Laws of Physics in the Light of Prehistoric Nuclear Reactors." *Contemporary Physics* 24: 427.

Jordan, Pascual (1947). "Extension of Projective Relativity." *Annalen der Physik (Leipzig)* 1: 219–228.

Kaluza, Theodor (1921). "Zum Unitätsproblem der Physik." *Sitzungsberichte der Preussischen Akademie der Wissenschaften, Physikalisch-Mathematische Klasse* K1: 966.

Kinsella, A. (1986). "Numerical Methods for Error Evaluation." *American Journal of Physics* 54: 464–466.

Klein, Oskar (1926). "Quantentheorie und fünfdimensionale Relativitätstheorie." *Zeitschrift für Physik* 37: 895–906.

Kolb, Edward W., Perry, Malcolm J., and Walker, T. P. (1986). "Time Variation of Fundamental Constants, Primordial Nucleosynthesis and the Size of Extra Dimensions." *Physical Review* D33: 869–871.

Lambeck, Kurt (1978). "The Earth's Paleorotation." In *Tidal Friction and the Earth's Rotation*. P. Brosche and J. Sunderman, eds. Berlin: Springer-Verlag.

Lambeck, Kurt (1980). *The Earth's Variable Rotation: Geophysical Causes and Consequences*. Cambridge: Cambridge University Press.

Mansfield, V. N. and Malin, S. (1980). "Astrophysical Tests of Scale-Covariant Theories." *Astrophysical Journal* 237: 349–354.

Marciano, William J. (1984). "Time Variation of the Fundamental 'Constants' and Kaluza–Klein Theories." *Physical Review Letters* 52: 489–491.

Maurette, Michel (1976). "Fossil Nuclear Reactors." *Annual Review of Nuclear Science* 26: 319–350.

McCrea, William and Rees, Martin J., eds. (1983). *The Constants of Physics.* London: The Royal Society.

McElhinny, M. V., Taylor, S. R., and Stevenson, D. J. (1978). "Limits to the Expansion of Earth, Moon, Mars and Mercury and to Changes in the

Gravitational Constant." *Nature* 271: 316–321.
Morrison, L. V. and Ward, C. G. (1975). "An Analysis of the Transits of Mercury 1677–1973." *Monthly Notices of the Royal Astronomical Society* 173: 183–206.
Morrison, L. V. and Stephenson, F. R. (1982). "Secular and Decade Fluctuations in the Earth's Rotation: 700 BC – AD 1978." In *Sun and Planetary System*. W. Fricke and G. Telek, eds. Dordrecht: Reidel, p. 173.
Muller, P. M. and Stephenson, F. R. (1975). "The Accelerations of the Earth and Moon from Early Astronomical Observations." In *Growth Rhythms and the History of the Earth's Rotation*. G. D. Rosenberg and S. K. Runcorn, eds. p. 459.
Muller, P. M. (1976). "Determination of the Cosmological Rate of Change of *G* and the Tidal Accelerations of the Earth and Moon from Ancient and Modern Astronomical Data." Technical Report. Jet Propulsion Laboratory. Internal Report JPL SP 4904.
Nakiboglu, S. M. and Lambeck, Kurt (1980). "Deglaciation Effects on the Rotation of the Earth." *Geophysical Journal of the Royal Astronomical Society* 62: 183.
Nordtvedt, Kenneth (1993). Private communication.
Petrov, Y. V. (1977). "The Oklo Natural Nuclear Reactor." *Soviet Physics–Uspekhi* 20: 937–944.
Ruffenach, J. C. (1978). "The Natural Reactors of Oklo: Study of Uranium and Rare-Earth Migration in a Core Sample and its Aplication to the Dating of the Nuclear Reactions." In *Natural Fission Reactors*. IAEA-TC-119/16. Vienna; IAEA.
Schwarz, John H. (1982). "Superstring Theory." *Physics Reports* 89: 223–322.
Scrutton, C. T. (1978). "Periodic Growth Features in Fossil Organisms and the Length of the Day and Month." In *Tidal Friction and the Earth's Rotation*. P. Brosche and J. Sunderman, eds. Berlin: Springer-Verlag.
Shlyakhter, A. (1976). "Direct Test of the Constancy of Fundamental Nuclear Constants." *Nature*. 264: 340.
Shlyakhter, A. I. (1983). "Direct Test of the Time-Independence of Fundamental Nuclear Constants Using the Oklo Natural Reactor." ATOMKI Report A/1. Leningrad: ATOMKI.
Sirlin, A. (1980). "Radiative Corrections in $SU(2)_L \times U(1)$ Theory: A Simple Renormalization Scheme." *Physical Review* D22: 971–981.
Sirlin, A. (1987). "Status of the Electroweak Theory." Preprint RV87/Bs/15.
Sisterna, Pablo and Vucetich, Héctor (1990). "Time Variation of Fundamental Constants: Bounds from Geophysical and Astronomical Data." *Physical Review* D41: 1034–1046.
Sisterna, Pablo and Vucetich, Héctor (1991). "Time Variation of Fundamental Constants: Quark Masses as Time Dependent Parameters." *Physical Review* D44: 3096–3108.
Sisterna, Pablo and Vucetich, Héctor (1994). "Cosmology, Oscillating Physics

and Oscillating Biology." *Physical Review Letters* 72: 454–457.
Stevenson, P. M. (1981). "Dimensional Analysis in Field Theory." *Annals of Physics (New York)* 132: 383–403.
Thiry, Yves (1948). "Les Équations de la Théorie Unitaire de Kaluza." *Comptes Rendues de l'Academie des Sciences de Paris* 226: 216–218.
Turneaure, J. P. and Stein, S. R. (1976). "An Experimental Limit on the Time Variation of the Fine Structure Constant." In *Atomic Masses and Fundamental Constants*, vol. 5. J. Sanders and A. Wapstra, eds. New York: Plenum, pp. 636–642.
Wetherhill, George W. (1975). "Radiometric Chronology of the Early Solar System." *Annual Review of Nuclear Science* 25: 283–326.
Wolfe, A. M., Brown, Robert, and Roberts, Morton (1976). "Limits on the Variation of Fundamental Atomic Quantities over Cosmic Time Scales." *Physical Review Letters* 37: 179–181.
Yong-Shi Wu and Zi Wang (1986). "Time Variation of Newton's Gravitational Constant in Superstring Theories." *Physical Review Letters* 57: 1978–1981.

Discussion

Nordtvedt: It was $\dot{G}/G+\dot{M}/M$ which was constrained as a package. Now did you apply this $\dot{M}/M$ universally to all celestial bodies?
Vucetich: Yes. We are always comparing atomic quantities, taken for example from the Oklo phenomenon, where G plays no – or should play – no role (though we cannot be sure of that), with the coefficient of M which appears for instance in the evolution of the moon's orbit.
Nordtvedt: Yes, my only comment would be there's an induced $\dot{M}/M$ for the neutron stars from the $\dot{G}/G$, so you'd have to hypothesize a separate $\dot{M}$ for the neutron stars versus the solar-system planetary bodies.
Vucetich: Right. I agree with you. The strong-field phenomenon should not be used for setting bounds.
Bondi: Your work is very reassuring, I must say. I will sleep much better having understood your work.
Vucetich: Thank you.
Goenner: Would you draw some conclusions concerning cosmology from your data. People always claim that fundamental constants really are constants, and this is consistent with present observations, but the observations come only from at most 3/4 of the age of the universe, so what do you conclude?
Vucetich: Well, we tried to use short-time local data in order to avoid any cosmological hypothesis when we computed our bounds. If you choose any Kaluza–Klein or some superstring cosmology, it should be chosen in such a way that it respects these bounds. That is, any

cosmology of the Kaluza–Klein or superstring type or maybe scalar–tensor type should be chosen in such a way that fundamental constants do not vary today by more than our bounds.

Goenner: Yes, but for the whole length of the cosmic evolution? I mean, couldn't it be that in the early universe you have more drastic time variations than those you get from your bounds? Is that excluded?

Vucetich: We made a calculation of primordial nucleosynthesis using the same model, or, rather, a small extension of the same model. We got very strict bounds. For instance, since helium should be around 25% in the universe, $\dot{G}/G$ should be less than one part in 10^{-15} years, not 10^{-12}. These constraints are much stronger, but you must assume the big-bang cosmology, you must assume helium observations are correct.

Will: Have you looked at the effect of conspiracies or nonconspiracies on the bounds? The bound you get on $\dot{G}/G$, which you quote, is that substantially different from the bound from the Viking observations?

Vucetich: No

Will: So there the no-conspiracy assumption is a good assumption?

Vucetich: Right. There is an interesting point. As we must make an assumption of conspiracy, we could not use the most strict bounds from the Viking experiment. We had to use the bounds they obtain varying simultaneously their β parameter and the G parameter. Those bounds are much larger because probably both parameters are very correlated. Even so, the result of combining all observations is of the same order or slightly smaller than the Viking one.

Will: In the case of the Oklo reactor, the bounds on nuclear parameters are much weaker because of stronger coupling.

Vucetich: Stronger correlations. Even so, as you see, the bounds are sufficient to exclude for instance that particular compactification proposed by Wu and Wang and probably they would put very strong constraints on many Kaluza–Klein theories.

Will: Can you comment on the bound on the weak coupling constant?

Vucetich: It is much smaller than the bound on Newton's constant G, about an order of magnitude smaller. This is because the data on the age of the earth are of very good quality. Geophysicists have determined the age of the earth with a very small error, and so the bound on the weak interaction constant is small.

Will: Mainly, that comes from β decay data?

Vucetich: It mainly comes from comparison of the age of the earth from β decayers and the age of the earth from α decayers.

Will: That's what you would expect?

Vucetich: Yes, and we get a bound 10^{-3} of the Hubble constant.

Machian Effects in Physical Law and the Field Paradigm of Modern Physics

Kenneth Nordtvedt

One can imagine a conversation between Mach and a descendant of Tycho Brahe which dramatizes key aspects of the question – Is physics Machian?

> **Descendant of Tycho Brahe**: I have measured very precisely the angular positions of the so-called fixed stars and with a new 'Doppler' technique have measured the radial motions of these objects. Nothing is fixed; all heavenly bodies reveal, on sufficiently close inspection, general accelerative motions relative to our local inertial frame.
> **Mach**: Yes, on reflection we should have expected this. The inertial frame here must be determined by some weighted average of our relative motion with respect to the various objects seen in the sky. Closer objects, objects containing more matter or moving more decisively, probably have more influence in establishing our local inertial frame, for example. We must experimentally determine this 'inertial influence function.'
> **Descendant of Tycho Brahe**: But I am still puzzled. My recent precise measurements show that our local inertial frame is utterly isotropic, even to many orders of magnitude precision beyond the slight anisotropies and inhomogeneities seen in the distribution of the most distant heavenly bodies, let alone considering the earth, sun and Milky Way effects on our laboratory's inertia. How can local inertia be such a perfect thing if produced by the imperfect heavens?

The near perfection of local laboratory physical law in the presence of the perceived imperfections in cosmological structure is, in my view, a key mystery or clue. It either casts serious doubt on the Machian program or, more likely, confirms the modern 'field paradigm' of physics – matter acts to influence other matter at a distance only through the intermediation of fields – and the cosmological field(s) must be

Einstein Studies, vol. 6: Mach's Principle: From Newton's Bucket to Quantum Gravity, pp. 422–435

dominated by the role played by the gravitational metric field, with participation by any other cosmological fields being strongly suppressed, if not totally absent.

Though Mach seemed to concentrate on questions related to inertia and inertial frames (Mach 1883, 1960), I will consider Machian effects in physics from the broader perspective developed by many physicists who were often influenced by the pioneering ideas of Mach. 'Machian' here will refer to the general class of ideas that the large scale structure and content of the universe determines, or at least perturbs, aspects of local laboratory physical law – inertia, coupling parameters, their magnitudes, spacetime dependences, isotropies, Lorentz invariances and other symmetries, etc.

In the field paradigm of physics, the entire cosmos produces here in the local laboratory a metric field $g_{\mu\nu}$ and perhaps one or more other 'arena' or 'cosmological' fields $X_{\mu\ldots}$, which in most field theory formulations would have Green's function representations:

$$\begin{aligned} g_{\mu\nu}(x^{\gamma}) &= \int G_{\mu\nu}(x^{\gamma}, x^{\eta'})_{\alpha\beta\ldots}\, S(x^{\eta'})^{\alpha\beta\ldots}\, d^4x', \\ X_{\mu\ldots}(x^{\gamma}) &= \int G^{(X)}{}_{\mu\ldots}(x^{\gamma}, x^{\eta'})_{\alpha\beta\ldots}\, S^{(X)}(x^{\eta'})^{\alpha\beta\ldots} d^4x'. \end{aligned} \tag{1}$$

Then the local physical law for matter, including its inertial properties, interactions, etc., can be affected by the cosmos only by means of coupling to these 'cosmic-arena' fields $g_{\mu\nu}$, $X_{\mu\ldots}$. The Green's functions and source functions in the above expressions are, of course, superficially similar to the 'inertia influence function' which Mach would seek, but today we would generally produce the form of these equations as integral representations of some underlying field theory. A more empirical Machian approach of seeking the structure of the Green's functions and source functions by systematic observations and experiment is of course also quite appropriate, particularly in the absence of plausible theory.

A metric field can always be transformed at any spacetime locality to the Minkowski metric plus leading order corrections proportional to tidal gravitational fields and quadratic coordinate deviations from the spacetime locality:

$$g_{\mu\nu}(x^{\gamma}) = \eta_{\mu\nu} + \text{Order } R_{\xi\mu\nu\eta}(x - x_0)^{\xi}(x - x_0)^{\eta}. \tag{2}$$

From this simple mathematical fact, however, we immediately reach profound physical consequences: If there are no other arena fields which couple to local matter, then, neglecting gravitational tidal fields, the local laws of physics are (locally viewed) identical throughout spacetime, and they possess the symmetry and invariance properties of the Minkowski metric – most notably absolute Lorentz invariance and rotational invariance. Even the numerical magnitude of local physical quantities cannot be influenced by the cosmos in a Machian manner, as the transformation of the metric field into the Minkowski metric effaces all Machian scale factors which might be originally produced in $g_{\mu\nu}$.

But this picture changes dramatically if one or more other arena fields couple to local matter. After coordinates are chosen to render the metric field locally Minkowskian, a scalar field will still have magnitude, and it will generally vary in space and time. Consequently, local physical law could contain numerical content determined in a Machian manner by the cosmos through the local scalar field value, and such Machian numbers would be expected to change in time as the universe evolves, while their spatial gradients produced by nearby sources would generally produce novel forces on laboratory bodies. But important symmetries such as Lorentz invariance and isotropy of local physical law would be maintained to high precision in the presence of solely a scalar 'arena' field since it is invariant under rotations or Lorentz transformations.

If the metric field were supplemented by a vector or second tensor arena field more complete Machian effects would be possible – indeed expected. The components of such fields are not Lorentz invariant, so coupling of local physical law to such fields would destroy that invariance in the local physics. Also, and of important observational consequence, such field component values (as viewed from a general inertial frame) are not rotationally invariant, and coupling to them would generally destroy isotropy of local physical law. A series of 'Hughes–Drever' type experiments have found no anisotropy to the energy levels of matter to a precision of about a part in 10^{25} (Hughes *et al.* 1960; Drever 1961). One of the strongest empirically based conclusions which then follows is that any cosmic-arena vector or second tensor field direct coupling to local matter is extremely weak.

The evidence against a scalar arena field being coupled to local matter is two-fold: First, the dimensionless parameters of laboratory physics show no cosmological time evolution down to levels several magnitudes below the Hubble expansion rate of the universe; and secondly, laboratory bodies show universality of gravitational free-fall rates to an accuracy of about 10^{-12}, which significantly constrains any spatial gradients of these dimensionless parameters produced by nearby attracting bodies.

Experiments and observation tell us we are very close to a metric theory coupling – matter coupling only to the metric field. The world action for such theories can then be symbolically written in the form

$$A = \int \left(\mathcal{L}(M, g_{\mu\nu}) + \mathcal{L}'(g_{\mu\nu}, C_n)\right) d^4x, \tag{3}$$

with M representing laboratory matter and interaction force fields and the C_n being possible gravitational-cosmological fields which could modify gravitational physics and the dynamics of the cosmos, but which leave the local nongravitational physics of matter metric.

The previous discussion of the consequences for local laboratory physics if other arena fields supplementing the metric field coupled to matter (the nonmetric theory case) can be repeated for the gravitational physics sector of experience (the metric theory case). Consider reasonably isolated systems of bodies like the solar system as 'local gravitational systems.' Such systems' gravitational physics will be determined by the sources within the systems, the gravitational field equations, and the boundary values of the gravitational-cosmological fields far from the systems (Will and Nordtvedt 1972). If the metric field, alone, couples with matter – no C_n in (3) – then coordinates can be chosen so that the metric field is Minkowski far from the system, and there are no other boundary field values. The local gravitational physics within the system must then be the same throughout the universe (neglecting the outside world's tidal forces), it must be Lorentz invariant, isotropic, etc. Local Machian effects are suppressed in the gravitational sector as well. Such gravitational physics is said to fulfill the Strong Equivalence Principle.

But suppose there is a scalar field supplementing the metric field in the Lagrangian term $\mathcal{L}'$ of (3). Then the physical parameters in gravitational physics, such as Newton's coupling parameter G, the masses of compact celestial bodies which depend on G, etc., can vary in space and time by being dependent on the value of the asymptotic scalar field, and Machian effects related to such spatial gradients and time variation of parameters will generally be present. If the metric field is supplemented by a vector or second tensor field in the gravitational sector the boundary values of the supplementary fields will generally not be Lorentz invariant, and the local gravitational physics thereby loses that invariance – preferred inertial frames appear in the details of the gravitational interaction and gravitationally interacting bodies can 'see' their motion with respect to a preferred inertial frame. Nor will the boundary value of such fields be invariant under rotations, so isotropy of the gravitational interaction is then lost at the level of the renormalized Newtonian interaction.

There are observations of gravitational systems analagous to the

'Hughes–Drever' experiments. If the Newtonian gravitational interaction between the matter in the sun were anisotropic, for example, the sun would experience a self-torque and resulting precession because of its spin-generated nonsphericity. But, after almost five billion years of existence, the sun's spin axis is still aligned within only a few degrees of the solar system's planetary polar axis. This implies that the Newtonian gravitational interaction is isotropic to precision of about a part in 10^{13}, significantly suppressing the possibility of a vector or second tensor field being coupled to the metric field (Nordtvedt 1987).

Scalar–metric tensor gravity, perhaps the most viable generalization of general relativity, and its Machian effects warrant more detailed discussion. The gravitational physics in such theories can be described to the $1/c^2$ perturbative order by a field-eliminated N-body Lagrangian:

$$L = L_{\mathrm{kin}} + L_{\mathrm{em}} + L_{\mathrm{G}} + L_{\mathrm{eG}}, \tag{4}$$

in which the subscripts refer to kinetic, electromagnetic (serving as surrogate for all nongravitational laboratory forces), gravitational, and electric-gravitational, respectively. These individual Lagrangian expressions are given by (units are used in which $c=1$)

$$L_{\mathrm{kin}} = -\sum_i m_i \left(1 - \frac{1}{2}v_i^2 - \frac{1}{8}v_i^4\right),$$

$$L_{\mathrm{em}} = -\frac{1}{2}\sum_{ij} \frac{e_i e_j}{r_{ij}} \left[1 - \frac{1}{2}(\mathbf{v}_i \cdot \mathbf{v}_j + \mathbf{v}_i \cdot \hat{\mathbf{r}}_{ij}\hat{\mathbf{r}}_{ij} \cdot \mathbf{v}_j)\right],$$

$$L_{\mathrm{G}} = \frac{G}{2}\sum_{ij} \frac{m_i m_j}{r_{ij}} \left[1 - \frac{1}{2}(\mathbf{v}_i \cdot \mathbf{v}_j + \mathbf{v}_i \cdot \hat{\mathbf{r}}_{ij}\hat{\mathbf{r}}_{ij} \cdot \mathbf{v}_j)\right] \tag{5}$$

$$+ (2\gamma + 1)\frac{G}{4}\sum_{ij} \frac{m_i m_j}{r_{ij}} v_{ij}^2 + \left(\frac{1}{2} - \beta\right) G^2 \sum_{ijk} \frac{m_i m_j m_k}{r_{ij} r_{ik}},$$

$$L_{\mathrm{eG}} = (\gamma + 1)\frac{G}{2}\sum_{ijk} \frac{e_i e_j m_k}{r_{ij} r_{ik}}.$$

This Lagrangian form guarantees the metric nature of gravity to the corresponding order. For instance, suppose there are among the N bodies some distant 'spectator bodies' which produce the gravitational potential:

$$U_s = G\sum_s \frac{m_s}{R_s} \tag{6}$$

in the vicinity of the remaining bodies $i, j, \ldots$ which form the local system ($R_s \gg r_{ij}$). The effective Lagrangian for the local system then has each of its various terms, including its inertial term, rescaled by the gravitational potential environment of the spectator bodies (Nordtvedt 1985):

$$L = -(1 - U_s)\sum_i m_i + \left(1 + (2\gamma + 1)U_s\right)\frac{1}{2}\sum_i m_i v_i^2$$
$$-\left(1 - (\gamma + 1)U_s\right)\frac{1}{2}\sum_{ij}\frac{e_i e_j}{r_{ij}} + \left(1 - (4\beta - 2)U_s\right)\frac{G}{2}\sum_{ij}\frac{m_i m_j}{r_{ij}}. \qquad (7)$$

This rescaled Lagrangian reminds us of comments by Einstein concerning fulfillment of Mach's ideas within perturbative general relativity:

> ... the theory of relativity makes it appear probable that Mach was on the right road in his thought that inertia depends upon a mutual action of matter... according to our equations... 1. the inertia of a body must increase when ponderable masses are piled up in its neighbourhood. 2. A body must experience an accelerating force when neighbouring masses are accelerated... 3. A rotating ... body must generate a 'Coriolis field'... (Einstein 1922).

The Lagrangian (7) confirms the first point, and indeed the spectator bodies' presence not only rescales the inertial masses, but also the rest masses of the local bodies, as well as the strength of both the local electrical (nongravitational) and local gravitational interactions. If, for example, the quantization of atoms and molecules were carried out using (7), one would find the Bohr radius unit and Rydberg frequency unit for such systems scaled as:

$$\frac{\hbar^2}{e^2 m} \to \frac{\hbar^2}{e^2 m}(1 - \gamma U_s), \quad \frac{e^4 m}{\hbar^3} \to \frac{e^4 m}{\hbar^3}(1 - U_s),$$

showing by construction the universal spatial shrinkage and temporal slowing factor of laboratory matter when quoted in the original (and still asymptotic) spacetime coordinates. Einstein's third point – the dragging of inertial frames by rotating matter – is covered in some detail elsewhere in this conference [pp. 308–363]. The second point – inductive Machian acceleration forces – is discussed later in this paper, including empirical evidence for their presence.

Local proper coordinates can be chosen to absorb all scaling factors in (7) except the one pertinent to the Newtonian gravitational interaction. Defining

$$d\rho = (1 + \gamma U_s) d\mathbf{r} \text{ and } d\tau = (1 - U_s) dt, \tag{8}$$

then $Ldt \rightarrow L'd\tau$, with

$$L' = -\sum_i m_i \left[1 - \frac{1}{2} (d\rho/d\tau)^2 \right] - \frac{1}{2} \sum_{ij} \frac{e_i e_j - G' m_i m_j}{\rho_{ij}} \tag{9}$$

in which nongravitational physics is seen to be invariant, but Newton's gravitational coupling strength is locally altered by the proximity of 'spectator matter,' depending on the values of the PPN coeffients γ and β in L_G (Nordtvedt 1970b):

$$G' = G\left(1 + (3 + \gamma - 4\beta) U_s\right). \tag{10}$$

It has been argued by some that no Machian effects associated with nongravitational laboratory physics exist in general relativity, scalar–tensor, or other metric theories of gravity because of this rescaling. I disagree. The rescaling factors involving U_s vary with the proximity of the spectator matter, and therefore there is the globally observable reality that nongravitational clocks run slower as they are located closer to such matter, and nongravitational rulers span smaller coordinate intervals when located closer to other matter. These are profound effects in metric gravity and have been confirmed by observations, the slowing of nongravitational clock rates directly (Vessot *et al*. 1980), and the alteration of the spatial geometry of nongravitational rulers indirectly through its effect on the globally measured slowing of the speed of light as it passes close by matter (Shapiro 1964, 1990). (Since the speed of light must locally be a universal constant in metric gravity, and is so measured to be, a globally predicted coordinate shortening of rulers near matter implies, along with the slowing of clocks, a corresponding slowing of the globally observed speed of light.) *What has been called the curvature of four-dimensional space and time (empirically being the location dependence of laboratory clock rates and ruler sizes), a central feature of metric theories of gravity, is a globally measured, though locally unseen, Machian feature of physical law.*

Note, however, the possibility (10) that the strength of local gravity is altered by proximite matter in metric gravity. Here is a direct Machian effect, observable locally, which is a breakdown of the Strong Equivalence Principle and which will generally occur in scalar–tensor gravity.

More traditional Machian effects are involved when one explicitly calculates the inertial mass of a composite body which contains significant gravitational self-energy. From the Lagrangian (4) it follows that the accelerating mass elements of an accelerating composite body produce inductive gravitational fields proportional to their acceleration and which act on the other mass elements of the body (Einstein's second point quoted above). The net result of these inductive forces acting between the mass elements of the body is to modify the total inertial mass of the body so as to account for the internal gravitational energy of the body. Similarly, the electric charges in the body produce inductive electric fields proportional to acceleration which act on the other electric charges of the body so as to modify the inertial mass of the body to account for the internal Coulomb energy of the body. And this occurs for any other Lorentz-invariant force field within the composite body. The resulting expression for the composite body's inertial mass then becomes

$$M(I) = \sum_i m_i \left[1 + \frac{1}{2} v_i^2\right] + \frac{1}{2} \sum_{ij} \frac{e_i e_j - G m_i m_j}{r_{ij}} + \ldots$$
$$+ \left[\sum_i m_i \mathbf{v}_i \mathbf{v}_i + \frac{1}{2} \sum_{ij} \frac{e_i e_j - G m_i m_j}{r_{ij}^3} \mathbf{r}_{ij} \mathbf{r}_{ij} + \ldots\right], \qquad (11)$$

the ... in the above equation indicating the appropriate contributions from any other forms of nongravitational field energies within the body. The first line is simply the total energy content within the body. But added to this expected contribution to inertial mass there is the system's total internal tensor virial contribution. Note that this tensor virial addition to inertial mass has nothing specifically to do with gravitation: It occurs as well for the laboratory object which has no appreciable gravitational self-energy contribution to its mass. Gravity just adds its proper contribution to internal energy and tensor virial as long as the gravitational interaction is locally Lorentz invariant, i.e., there are no preferred inertial frames in metric gravity. Otherwise, the gravitational self-energy contribution to total inertial mass of the composite celestial body can be anomalous (Nordtvedt 1968a; Haugan 1979).

Robert Dicke formulated an elegant way to understand the anomalous contribution of gravitational self-energy of a body to its gravitational mass which occurs in Lorentz-invariant theories where Newton's gravitational coupling parameter is position dependent, and in which energy–momentum conservation laws exist. From (11) one sees that the energy content (inertial mass) of gravitationally compact bodies depends on G; so there will be an additional force acting on that body:

$$\delta \mathbf{F} = -\nabla M(I) = -\frac{\partial M(I)}{\partial G} \nabla G, \qquad (12)$$

which, using (10), can be combined with the normal gravitational force $M(I)\mathbf{g}$ to define the total effective gravitational mass of a gravitationally compact body:

$$M(G) = M(I) + (3 + \gamma - 4\beta)\, G \frac{\partial M(I)}{\partial G}. \tag{13}$$

This agrees, under the same circumstances, with prior calculations of gravitational mass done from the Lagrangian (4) or its metric field equivalent (Nordtvedt 1968a).

If the earth's gravitational-to-inertial mass ratio differs from one, the lunar orbit will be anomalously polarized toward the sun (Nordtvedt 1968b), and (13) can be tested. Two decades of lunar laser-ranging data with precision of about 1 cm has not detected any such polarization of the orbit (Dickey *et al.* 1990, 1994): The earth and moon are accelerated by the sun identically to an accuracy of about 3 parts in 10^{13}. Since the gravitational self-energy of the earth is fractionally about $5 \cdot 10^{-10}$ the earth's mass, this confirms that scalar–tensor gravity is empirically close to general relativity:

$$| 4\beta - 3 - \gamma | \leq 10^{-3}; \tag{14}$$

$\gamma = \beta = 1$ in general relativity. This observational result also indirectly confirms the presence of the previously discussed Machian inductive gravitational forces acting within the accelerated earth so as to reduce its inertial mass by its gravitational self energy (11).

The lunar laser-ranging data is presently being used to test a more speculative possibility – Machian effects on solar system matter from 'dark matter.' It has been suggested that our galaxy's gravitating matter may consist of as much as ninety percent 'dark matter' with exotic composition. If such matter accelerated the earth and moon differentially, there would be a sidereal polarization of the lunar orbit toward the galactic (attracting) center. The LLR data, however, place an upper limit on any such acceleration difference of about

$$|\mathbf{a}_E - \mathbf{a}_M| \leq 3 \cdot 10^{-14} \text{ cm/sec}^2,$$

which should be compared with the estimated galactic acceleration of the solar system which is about 10^{-8} cm/sec^2 (Nordtvedt, Müller, and Soffel 1995).

Mars Viking lander radar-ranging data collected over a decade ago is presently being analyzed for the analogous polarization of the earth and Mars orbits toward Jupiter which would occur if the sun's gravitational-to-inertial mass ratio differs from one (the sun's fractional gravitational self-energy is the largest in the solar system – $4 \cdot 10^{-6}$), and the sun is consequently anomalously accelerated toward Jupiter (Nordtvedt 1970a). This experimental test of the post-Newtonian structure of gravity and possible failure to fulfill the Strong Equivalence Principle should yield accuracies somewhat less than that obtained (14) by the lunar laser-ranging data. Higher-precision interplanetary ranging techniques (microwave or laser) in the future may permit testing this principle to a part in 10^4 or 10^5.

The excellent fit of the pulse arrival times from the binary pulsar PSR 1913+16 by using a relativistic orbit calculated from general relativity theory offers further confirmation of the existence of Machian inductive forces contained in Einstein's theory. There are two terms in the post-Newtonian equation of motion in which the accelerating force acting on one body of the binary system depends on the motion (both velocity and acceleration) of the other body:

$$\delta \mathbf{F}_1 = G m_1 m_2 \left[2(1+\gamma) \frac{\mathbf{v}_1 \times (\mathbf{r}_{12} \times \mathbf{v}_2)}{r_{12}^3} + \frac{4\gamma+3}{2} \frac{\mathbf{A}_2 + \mathbf{A}_2 \cdot \hat{\mathbf{r}}_{12} \hat{\mathbf{r}}_{12}}{r_{12}} \right]. \tag{15}$$

The first term is the gravitomagnetic interaction of one moving body acting on another moving body and is the source for the so-called 'dragging of inertial frames' by rotating bodies. The second term is the gravitoelectric inductive interaction from an accelerated neighbor and was the second point of Einstein quoted earlier. Each of these terms produces a substantial periastron precession contribution to the binary system's orbit – about 1 arc-degrees/year and −2 arc-degrees/year, respectively – which should be compared to the total observed precession of 4.2266 arc-degrees/year. It would be very difficult to understand the binary pulsar system without both of these Machian contributions (15) to its equations of motion.

Theories of gravity which contain the Machian feature of a spacetime-dependent gravitational coupling parameter generally show a cosmological time rate of change of G which is of the order of Hubble's expansion parameter for the universe ($H \sim 5 \cdot 10^{-11}$ yr^{-1}). The consequences of such a time-varying G in the solar system and in binary pulsar systems have been searched for and, being unseen, have produced the upper bound

$$\left| \frac{dG}{G dt} \right| \leq 10^{-12} \text{ year}^{-1},$$

which is significantly below Hubble's rate. With regard to such cosmological effects, it should be pointed out that the fully modified equation of motion for celestial bodies in scalar–tensor metric theories is more complex than the simple replacement of G by a time-dependent $G(t)$. One should employ (Nordtvedt 1990, 1993)

$$\frac{d}{dt}\left(m_i(t)\,\mathbf{v}_i\right) = G(t)\sum_j K_{ij}(t)\,\frac{m_i(t)\,m_j(t)}{r_{ij}^3}\,\mathbf{r}_{ji} \tag{16}$$

with the body velocities being relative to the cosmic rest frame in which G has pure time dependence. For celestial bodies whose mass (energy content) depends significantly on G (and other post-Newtonian parameters of gravity), their masses will also be time dependent if G varies:

$$\frac{dm(t)}{dt} = \frac{\partial m(G,\ldots)}{\partial G}\,\frac{dG}{dt} + \ldots .$$

The strong equivalence principle violating factor K_{ij} varies in time due to cosmological expansion:

$$K_{ij}(t) = \left[\frac{M(G)}{m}\right]_i \left[\frac{M(G)}{m}\right]_j + \chi G^2 \left[\frac{\partial m}{m\,\partial G}\right]_i \left[\frac{\partial m}{m\,\partial G}\right]_j + \ldots,$$

where $M(G)$ are body gravitational masses which may differ from inertial masses m, and χ is a second post-Newtonian order parameter of gravitational theory, etc. While solar system bodies have sufficiently weak gravitational binding so that only the direct time dependence of G need be considered in this general equation, Eq. (16) must be used in its entirety when applied to the dynamics of gravitationally compact bodies such as neutron stars in the binary pulsar systems. Note that even a 'free' celestial body will accelerate relative to the cosmic rest frame if it has variable mass due to variable G.

Although no Machian effects resulting from space or time variation of G have been found at the level of a part in one thousand of the natural magnitude one might estimate, there is reason to expect that an underlying scalar–tensor theory of gravity would have, at this epoch of the universe, such effects at the part in 10^5 level (Damour and Nordtvedt 1993). A general class of theories with world Lagrangian of the form

$$A = \int \left[\mathcal{L}(M, g_{\mu\nu}) + \frac{\sqrt{-g^*}}{G_0}\left(R(g^*_{\mu\nu}) + \partial\phi_\mu\,\partial\phi_\nu\,g^{*\mu\nu}\right)\right] d^4x$$

has been considered with the first 'matter' Lagrangian term coupling to a physical spacetime metric field which is conformally related to the dynamical metric field by a function of the scalar field:

$$g_{\mu\nu} = S(\phi)^2 \, g^*_{\mu\nu} \ .$$

If this conformal function has scalar field values for which its slope tends to zero, these can become attractors during the cosmological evolution of the scalar field. These are also the conditions for which the post-Newtonian gravitational interaction takes the form of general relativity, with

$$1 - \gamma = 2 \left[\frac{dS(\phi)}{Sd\phi} \right]^2 \frac{1}{1 + \left[\frac{dS(\phi)}{sd\phi} \right]^2}$$

and

$$1 - \beta \sim (1 - \gamma) \frac{d^2 S(\phi)}{d\phi^2}.$$

Estimates, based on rather generic initial conditions, of how close the scalar field has evolved toward the state with $\gamma = 1$ involve an integral over the 'recent' matter-dominated era of cosmic expansion and yield $1 - \gamma$ of order 10^{-5}. Hopefully this adds motivation for development of a couple of orders of magnitude improvement to one or more of the solar system experiments which probe post-Newtonian gravity and these possible Machian effects.

REFERENCES

Damour, Thibault and Nordtvedt, Kenneth (1993). "General Relativity as a Cosmological Attractor of Tensor–Scalar Theories." *Physical Review Letters*, 70: 2217–2219.

Dickey, J. O., Newhall, X X, and Williams, J. G. (1989). *Advances in Space Research* 9: 75.

Dickey, J. O. *et al.* (1994). "Lunar Laser Ranging: A Continuing Legacy of the Apollo Program." *Science* 265: 482–490.

Drever, Ron W. P. (1961). "A Search for Anistropy of Inertial Mass Using a Free Precession Technique." *Philosophical Magagazine* 6: 683–687.

Einstein, Albert (1922). *The Meaning of Relativity*. Princeton: Princeton University Press.

Haugan, Mark (1979). "Energy Conservation and the Principle of Equivalence." *Annals of Physics (New York)* 118: 156–186.

Hughes, V. W., Robinson, H. G., and Beltran-Lopez, V. (1960). "Upper Limit for the Anisotropy of Inertial Mass from Nuclear Resonance Experiments."

Physical Review Letters 4: 342–344.

Mach, Ernst (1883). *Die Mechanik in ihrer Entwickelung. Historisch–kritisch dargestellt.* Leipzig: F. A. Brockhaus.

Mach, Ernst (1960). *The Science of Mechanics: A Critical and Historical Account of Its Development.* LaSalle: Open Court.

Nordtvedt, Kenneth (1968a). "Equivalence Principle for Massive Bodies II. Theory." *Physical Review* 169: 1017–1025.

Nordtvedt, Kenneth (1968b). "Testing Relativity with Laser Ranging to the Moon." *Physical Review* 170: 1186–1187.

Nordtvedt, Kenneth (1970a). "Solar System Eötvös Experiments." *Icarus* 12: 91–100.

Nordtvedt, Kenneth (1970b). "Post-Newtonian Metric for a General Class of Scalar–Tensor Gravitational Theories and Observational Consequences." *Astrophysical Journal* 161: 1059–1067.

Nordtvedt, Kenneth (1985). "Post-Newtonian Gravitational Lagrangian Formalism for Celestial Body Dynamics in Metric Gravity." *Astrophysical Journal* 297: 390–404.

Nordtvedt, Kenneth (1987). "Probing Gravity to the Second Post-Newtonian Order and to One Part in 10^7 Using the Spin Axis of the Sun." *Astrophysical Journal* 320: 871–874.

Nordtvedt, Kenneth (1990). "*dG/Gdt* and a Cosmological Acceleration of Gravitationally Compact Bodies." *Physical Review Letters* 65: 953–956.

Nordtvedt, Kenneth (1993a). "Cosmological Time Evolution in the Newtonian Dynamics of Compact Celestial Bodies." *Astrophysical Journal* 407: 5–13.

Nordtvedt, Kenneth (1993b). "Second Post-Newtonian Order Renormalization of First Post-Newtonian Order Gravity in the Solar System." *Astrophysical Journal* 407: 758–777.

Nordtvedt, Kenneth, J. Müller, and M. Soffel (1995). "Cosmic Acceleration of the Earth and Moon by Dark Matter." *Astronomy and Astrophysics Letters* (in press).

Shapiro, Irwin (1964). "Fourth Test of General Relativity." *Physical Review Letters* 13: 789–791.

Shapiro, Irwin (1990). *General Relativity and Gravitation, 1989*, N. Ashby, ed. Cambridge: Cambridge University Press, p. 313.

Vessot, Robert, *et. al.* (1980). "Test of Relativistic Gravitation with a Space-Borne Hydrogen Maser." *Physical Review Letters* 45: 2081–2084.

Will, Clifford and Nordtvedt, Kenneth (1972). "Conservation Laws and Preferred Frames in Relativistic Gravity. I. Preferred-Frame Theories and an Extended PPN Formalism." *Astrophysical Journal* 177: 757–774.

Discussion

Brill: In the very beginning you talked about the inertia being very isotropic and the universe being not so isotropic. I just don't know the number. I'm wondering how that would look if you were to say what counts is the structure of the universe, say at recombination or some earlier time when it was much more isotropic.

Nordtvedt: Well, let's take COBE. The number is 10^{-5}, the temperature fluctuations. Now I think some people interpret that as effective fluctuations in the background Newtonian potential, whatever that means, at recombination. Similarly, when I look out today and see the Great Attractor and calculate the Newtonian potential here of the Great Attractor, it also comes out to be 10^{-5}, so I think it's fair to say that the universe is showing an inhomogeneity at the level of 10^{-5} and the laboratory isotropy of physics is less than 10^{-25}, roughly.

Bondi: I recall that many years ago possible variations in the constant of gravitation were examined through their effect on the luminosity of the Sun, which depends on it through a high power. It emerged that during the period covered by the geological record (which covers a modest but not insignificant fraction of the Hubble time) only very small variations in G could have taken place.

Nordtvedt: Yes, of course; the dynamical information from the solar system is even now below the Hubble rate by one or two orders of magnitude. The scalar–tensor class of theories has a relationship for $\dot{G}/G$ (measured in units of H) which must be less than $4\beta-3-\gamma$, the same factor that determines violations of the strong equivalence principle, which we've measured to be less than 10^{-3}. So if the experimentalists found a $\dot{G}/G$ bigger than $10^{-3}H$, we would have some difficult explaining to do within the class of scalar–tensor theories. There would have to be another explanation. We can therefore understand why no $\dot{G}/G$ is seen; because we've already determined by other means that the scalar field participation is suppressed by a factor of a thousand.

Isenberg: How does the so-called low-energy string theory start to fit in with the other scalar fields?

Nordtvedt: Two things. I understand that it's not metric coupling in typical string theories. The coupling to different particles is non-universal. Secondly, if these fields have a very important self-interaction in the Lagrangian, then their dynamics may be self-determined. The fields may lock themselves into frozen values, such as Higgs expectation values of something that's not classically driven by the universe expansion but is determined locally by quantum field theory.

7. Critical Reflections

Introduction

As explained in the General Introduction, several talks at Tübingen were reviews and responses to individual talks and to the conference as a whole. Some of these contributions have been put together in the present chapter. It seems to us that they speak for themselves and need no special editorial introduction. We should therefore like to take up this space with a few reflections of our own.

Numerous issues raised at the conference call for further study. Michael Jones, for example (p. 93), said a Machian theory should explain why "the inertial frames we observe do not rotate relative to the stars as we see them." This point, which is particularly relevant if gravitational degrees of freedom and not only matter can contribute to the determination of the inertial frames, was not further addressed at Tübingen. In 1971, McCrea (*Nature* 230: 95–97) questioned whether there was anything to be explained; perhaps Lynden-Bell's response to Jones had McCrea's paper in mind. Elsewhere (p. 207), we have already noted the urgent need for further study of the thin-sandwich problem. The query by Goenner at the end of Ciufolini's contribution (p. 402) is also thought provoking: Frame-dragging effects in general relativity certainly appear very Machian, but they also look rather like a natural generalization of induction forces from vectorial electrodynamics to tensorial gravitation. Are they specifically Machian? Moreover, if, as one of us argues (p. 214), general relativity is Machian by virtue of the Baierlein–Sharp–Wheeler (BSW) structure of the Hilbert action when expressed in terms of 3-geometries, what is the connection between that structure and the famous frame-dragging effects? Finally, would it be possible to do for the BSW form of general relativity what Lynden-Bell did for the nonrelativistic Barbour–Bertotti intrinsic dynamics, i.e., solve the Lagrange-multiplier constraints explicitly for the auxiliary shift?

J.B.B. and H.P.

Einstein Studies, vol. 6: Mach's Principle: From Newton's Bucket to Quantum Gravity, p. 436

Mach, the Expansion of the Universe, the Variation of Inertial Mass, and Lense–Thirring

Wolfgang Rindler

I have three unconnected qualitative observations to make on Mach's Principle. They are simple, but do not seem to have been made before. One concerns the expansion of the universe, one the possible variation of inertial mass, and one the Lense–Thirring effect.

Probably everyone at this conference is familiar with Dennis Sciama's beautiful remark on the Machian relevance of the rotation of our Galaxy (Sciama 1959, p. 122): It seems that today's astronomers can directly detect a rotation of the local compass of inertia - as embodied by the best inertial axes for the solar system - relative to our Galaxy at a rate of about one half second of arc per century. (As expected, that corresponds to the optically observed rotation of the Galaxy relative to the distant universe.) Sciama's point was that *had* Mach known this, he could have predicted, on the strength of his principle, something that was not at all obvious at the time, namely the existence of a vast extragalactic universe, just to make the standard of nonacceleration here and now come out right.

I would like to suggest an argument for the Hubble motion of the universe - also unknown at the time - that Mach actually could have made *without* further data. The basic view of the universe from the time of Thomas Digges in the 16th century, through Newton and at least to the end of the 18th century, seems to have been that of a uniform and *static* distribution of stars, infinite in all directions; since the time of William Herschel, an infinite universe of *galaxies* was contemplated by some, but down to Einstein's time it was still generally considered to be static. Why? I have never seen an argument for this widely assumed staticity, but I suspect that, at least on a subconscious level, it must have involved symmetry and a belief in absolute space. For, given a uniform

Einstein Studies, vol. 6: Mach's Principle: From Newton's Bucket to Quantum Gravity, pp. 437–441

and infinite distribution of stars at rest in absolute space, why should any one of them start to 'move' in one direction rather than another? However, once you deny the existence of absolute space as an independent entity, as Mach had done, this whole argument breaks down. Symmetry about each star now requires no more than a Hubble motion of the entire universe, namely identical changes in identical times of identical distances between stars. Of course, whether the universe is presently expanding or contracting depends on its previous history, but given universal gravitation, the least likely state would surely be the static one.

My second observation concerns something that Einstein apparently mistakenly read into Mach's philosophy (Barbour 1990, p. 49), namely a possible dependence of the inertial mass of a particle on the total mass distribution of the universe. No matter where this idea comes from, it is eminently sensible, Machian in flavor, and worthy of consideration. Carl Brans has shown (1962) that no variation of inertial mass occurs in general relativity. But there are other theories of gravitation where it might well make sense to look into this question. However, would not Einstein's principle of equivalence be violated by a variation of inertial mass? Consider two freely falling Einstein cabins at widely separated events and in them identical oscillators consisting of identical springs with identical massive balls attached at either end. Each cabin should also contain one of two identical atomic clocks. A variation of inertial mass would then manifest itself as a difference in the oscillation rate of the springs relative to the clocks. But the equivalence principle apparently allows no such difference. I wish to argue that the equivalence principle does not apply here, and that it would consequently not be violated. We are all familiar with the 'paradox' of the radiating charges: first a charged particle sitting on the surface of the earth, and secondly a charged particle in circular orbit around the earth. Both common sense and, in fact, calculations show that in the first case the charge will *not* radiate while in the second case it will. But the equivalence principle seems to suggest just the opposite: In the first case the charge accelerates relative to a local freely falling cabin, and so should radiate, while in the second case it does *not* accelerate relative to a local freely falling (comoving) cabin and so should *not* radiate. I learned the resolution of this paradox long ago from Jürgen Ehlers: The equivalence principle applies only to experiments that are totally isolated from the rest of the universe. A charge in an imaginary Einstein cabin is not so isolated, its field lines being anchored deep in space. In just the same way, the equivalence principle cannot be applied to our spring system when we suspect an interaction of the inertial masses with the distant universe; that very interaction would make the experiment not isolated.

My third remark questions whether the well-known Lense–Thirring effect is quite as Machian as it is usually cracked up to be. Let us begin by imagining the earth E alone in space. How likely is it, on the basis of *any* theory, that a gyrocompass G being taken around E along a circular path would precess faster than 2π per revolution in the sense of the orbit? (At 2π it would permanently point at the earth's center.) But only for such an *unnatural* precession rate would we find in the Mach-equivalent reference frame F in which the centers of E and G are at rest that G precesses in a sense opposite to the rotation of E. Introducing a distant universe at rest in F would presumably slow down this precession of G, since *its* inertial influence would now predominate. This is the Lense–Thirring effect. Within the Machian complex of ideas, it is seen to depend on a very unlikely premise. For a more detailed discussion of this, see (Rindler 1994).

REFERENCES

Barbour, Julian B. (1990). "The Part Played by Mach's Principle in the Genesis of Relativistic Cosmology." In *Modern Cosmology in Retrospect*, R. Bertotti, R. Balbinot, S. Bergia, and A. Messina, Eds., Cambridge: Cambridge University Press.

Brans, Carl H. (1962). "Mach's Principle and the Locally Measured Gravitational Constant in General Relativity." *Physical Review* 125: 388–396.

Rindler, Wolfgang (1994). "The Lense–Thirring Effect Exposed as Anti-Machian." *Physics Letters* A187: 236–238.

Sciama, Dennis W. (1959). *The Unity of the Universe.* New York: Doubleday.

Discussion

Lynden-Bell: Given that you've shown by the diagram that the mass of the object is not isolable, is there an isolatable experiment? Everything has mass or energy and if the mass or energy is in contact with the rest of the universe it's not clear that anything that has energy is isolable and then it is not clear that there is an isolable experiment.

Rindler: Experiments involving atomic clocks, rigid rulers, light rays, etc., would still be isolable, and so one could still use the equivalence principle, for example, to deduce the Lorentzian structure of spacetime, to predict what wavelength a moving observer would ascribe to an incoming light wave, etc.

Will: But as I will mention in my contribution, Lynden-Bell is, in fact, right, and that lack of isolation produces say anisotropies in energy levels of atoms that you can rule out with high precision, so the nonisolability has experimental consequences.

Isenberg: So is the conclusion that the equivalence principle is useless?

Will: No, not that it's useless, but that experiment verifies that in fact these things are isolatable, because no effects due to that nonisolatability are seen experimentally.

Bondi: Two comments: I'm in no way contradicting Ehlers's argument but fleshing it out. First, that radiation itself is a distant phenomenon, not a near one. There's no need to be able to ascertain near a body whether it's radiating or not. It's a far-field characteristic, I think it's very similar to what you say. And secondly, of course, I don't like the principle of equivalence, I mean to me the tidal effects are the true observables of gravitation. In one case you have tidal effects and in the other you don't.

Pfister: Independently of whether this is a question of isolation, I wanted to ask you [Rindler], if you think of a difference in mass say now and in millions, billions of years, how would you measure – at least in a gedanken experiment – the difference in the mass, because the way I understood Carl Brans's paper, he would say you have no possibility to measure the change.

Rindler: I think Carl Brans's argument did not apply to the little spring oscillator with masses at either end of it, in theories other than general relativity, where the equivalence principle may still apply.

Brans: I think it was; I may have to read that paper some day. The ratio of electrical forces to the accelerations was defined. The ratio of inertial mass to electric charge, and then ultimately to gravitational mass, was found. Basically, what was done was to compare electrical to gravitational accelerations for standard configurations. Now I have another question, if I may. What about the simple mass conservation law? Suppose you make some little dust model of your mass, and you follow it from one area which is free of gravitational effects and tidal effects to another, and you have a world tube: How do you get around violation of mass conservation?

Rindler: I don't know. My only purpose at this point was simply to show that I'm not violating the equivalence principle.

Nordtvedt: If all fundamental masses in the theory change by the same spacetime function, so all mass ratios are preserved, then I believe that in general all energy levels of systems will scale by one universal function, and by a change of coordinates you can embed this in a metric theory and put the whole burden on the time variation of Newton's effective parameter. So to me the key question is whether various masses are evolving in time and space differently, maybe quarks versus leptons, etc., but I believe that one universal mass function can always be embedded in the strength of gravity.

Will: And in that case you could detect a difference by comparing one atom with another because different atoms would have different amounts of quarks and leptons.

Nordtvedt: And universality of free fall would be violated as well.

Mach's Principle and Theories of Gravitation

Hubert F. M. Goenner

1. Introduction

Once in the decade, I seem to run into an old acquaintance of mine named Mach's Principle (MP) (Goenner 1970, 1981). The encounter of the '90s will be brief, however. In the following I shall first go through various formulations of MP, then give a few examples of how people have made them work, and then comment specifically on the approach of Julian Barbour.

2. Formulations of Mach's Principle

In addressing different formulations of MP, I leave it to *historians of science* to establish "what Mach really said" and to *philosophers of science* to explain what Mach "really meant" or what he "really should have said or meant." I also will not discuss the question of whether MP is an epistemological, ontological, methodological, or just heuristical requirement for the building of a proper theory of gravitation. I will list a few reasonably precise formulations of what is collectively labeled MP and suggest that we learn to *distinguish* them instead of lumping them all together into one vague concept.

In his critique of Newtonian mechanics, Mach arrived at the following two conclusions:

> i) Only the *relative* motion of a body with regard to other bodies is observable, not motion with regard to absolute space (*kinematical* relativity of motion).
> ii) The inertial motion of a body is *influenced* by all the masses in the universe (*dynamical* relativity of motion).

Einstein Studies, vol. 6: Mach's Principle: From Newton's Bucket to Quantum Gravity, pp. 442–457

Einstein, at least since 1913, strengthened this second idea to what is usually called the relativity of inertia: The inertial properties of a body follow from its *gravitational* interaction with all the other bodies (EP). In particular, he formulated what I call the strong Einstein–Mach Principle (SEM): The metric field g is caused and determined by the matter tensor T (Einstein 1918). Specific field theoretic versions of SEM are Pauli's (Pauli 1921)

$$g_{\alpha\beta}(x) \to 0 \text{ if } T_{\alpha\beta}(x) \to 0;$$

Ihrig's Mach 4 (Ihrig 1975)

$$R^{\alpha}{}_{\beta\gamma\delta}(x) \to 0 \text{ if } T_{\alpha\beta}(x) \to 0,$$

and the demand of Hoyle and Narlikar (1964) that no gravitational *vacuum* field equations be permitted. The SEM as well as Pauli's and Ihrig's versions of it are *not* satisfied by Einstein's theory of gravitation. All three require a clear separation between the gravitational field represented by the metric g and the matter fields, represented by the matter tensor T, to exist. In most cases of physical relevance, however, the matter tensor $T_{\alpha\beta}$ cannot even be written down without knowledge of g.

If SEM is replaced by what may be called the *weak Einstein–Mach Principle* (WEM), saying that the metrical field g is determined by the matter tensor T *plus* (spatial or temporal) *boundary conditions*, this then amounts to an invalidation of Mach's ideas, at least according to Einstein's opinion (Einstein 1917). In this paper, Einstein arrives at what might be called the *cosmological Mach's Principle*: For a model of the universe with total (gravitational) mass M and linear extension R, the relation $GM/c^2R = 1$ must hold (G is the Newtonian gravitational constant, c velocity of light in vacuum).

From Isenberg [p. 188], we learned his version of MP (IMP): The distribution of *matter* everywhere in the universe (at a particular moment) determines the inertial frame at every point in the universe. It is not just a variation of SEM because, in Isenberg's initial-value problem setup, *matter* includes degrees of freedom of the gravitational field (the metric).

The question just is: What are precisely the *Machian* initial data which determine the observable future *uniquely* (Machismo, according to Barbour). Barbour [p. 214] and Bertotti arrive at a two-legged formulation of Mach's ideas (BBMP):

> i) Only *relative* configurations of physical systems are significant. Or, put differently, a dynamical history of the universe is a curve in a relative configuration space determined by an action that contains only relative

quantities.

This is called 'First Machian Requirement' (FMR) by them and reflects Mach's kinematical relativity of motion.

ii) There exists no independent time. Or, the 'speed' at which the above mentioned curve describing a dynamical history is traversed, is undefined in the theory.

A more technical formulation of this 'Second Machian Requirement' (SMR) would be that in any Machian theory the initial condition for a dynamical history will involve only a *direction* in the configuration space, not a direction and a speed in that direction. In such a theory the action will be reparametrization invariant.

There are also less all-embracing statements of Machian ideas establishing a relationship between local inertial frames and the distant masses as, for example, Ciufolini's weak formulation linking the dragging of the inertial frames to the gravitational field of rotating masses (Ciufolini and Wheeler 1995).

In a way, these and further formulations of MP try to establish a connection between the *local physics* and the *universe* – in marked contrast to the usually successful method, i.e., to neglect all nonlocal influences and isolate the physical system under investigation from the rest of the world. If this interpretation is true we see MP as one among various *holistic* approaches to physics:

Mach's Principle ⇔ local *mechanics* and the universe

Why are local inertial frames singled out?

Absorber theory of electrodynamics ⇔ local *electrodynamics* and the universe

Why are retarded solutions realized?

Irreversibility and cosmic expansion ⇔ local *thermodynamics* and the universe

Why is there a direction of time?

Quantum cosmology ⇔ quantum mechanics and the universe

Why is the universe a classical system?

3. Alternative Theories of Gravitation

What people have made with such formulations of MP depends on whether their belief in Einstein's general relativity theory is *stronger* than their belief in Mach's Principle or vice versa. In the first case, MP is discredited and thrown out of the window as Einstein is reported to have done later in his life. In the second case, usually, scientists develop *alternative* theories of gravitation which they expect to reflect better their particular formulation of MP. For me, a strong belief in MP means that one must expect it to lead either to measurable (new) effects or to (new) theoretical developments. In fact, both expectations come true for various alternative theories. Except that the new effects have not yet been measured (as we have heard, for example, in the talks of Nordtvedt [p. 422] and Will [p. 365]) or, that in the *new* theories it is difficult to recover the *old* effects described by general relativity.

Such expected consequences of MP are said to be

- that the rotation (never mind its definition) of the universe is zero
- dragging of inertial frames
- time-dependent inertial masses
- space- and/or time-dependent gravitational coupling constant
- time-dependent rest masses of elementary particles

As to a possible 'rotation of the universe,' the observational situation is unreliable (Birch 1983; Phinney and Webster 1983). Some authors claim that the inflationary model practically rules out rotation (Ellis and Olive 1983; Braccesi 1988). Of course, the dragging of inertial frames is already describable within Einstein's theory (Lense–Thirring effect). It has also been discussed in the prerelativistic *relational* theories of Treder (Treder 1972) or Bertotti and Barbour (Barbour 1974a, 1975; Barbour and Bertotti 1977; Bertotti and Easthope 1978). Spacetime dependent inertial masses follow from a five-dimensional (Kaluza–Klein-type) theory with an additional mass dimension (Wesson 1983; Ma 1990) or a nonsymmetric affine theory (Murphy 1970). The best-known example of theories leading to a spacetime-dependent gravitational coupling constant is given by the *scalar–tensor theories* (Jordan–Brans–Dicke theories and generalizations). In the theory of Hoyle and Narlikar the rest masses of elementary particles are interpreted to be time-dependent. We must not forget, however, that Dirac's large number hypothesis also leads to a time-dependent gravitational coupling constant and time-dependent masses without need for MP.

Also, rather specific consequences of alternative theories of gravitation incorporating one or the other version of MP have been discussed such as an anomalous redshift at the solar limb (Ghosh 1986), an anomalous redshift of quasars (Narlikar and Das 1980), and a relation with the missing mass in galaxies and clusters of galaxies (Milgrom

1983b; Roberts 1985). Should I also mention a relation with the frequency of earthquakes which has been pointed out (Kropotkin 1975)?

Of course, there are many more alternative theories of gravitation motivated by Mach's ideas than I can discuss today. I would like to mention only two, the *bimetric* theories of Goldoni or Firmani (Goldoni 1979, 1980, 1991; Firmani 1971) and the pregeometric approach of Liebscher and Bleyer (Liebscher 1988; Bleyer 1988), about which we heard yesterday starting from *affine* geometry and trying to derive the causal structure [p. 293].

If scientists have an equally strong belief in general relativity and MP, then they reduce MP to the role of a *selection principle* for the solutions of Einstein's equations. This approach was taken first, I believe, by Hönl (1953) and by Wheeler (1959). It amounts to a reformulation of Einstein's field equations such that one is able to label some of its solutions as *Machian* and others as *non-Machian*. In the sixties a first such, alas highly controversial, reformulation was presented by Hönl and Dehnen (1963, 1964, 1966). The best-known recent reformulations have been developed by a number of people and were presented at this conference by Raine and Isenberg. On the one side is the *integral formulation* of the Einstein field equations (Lynden-Bell 1967; Al'tshuler 1967; Sciama, Waylen, and Gilman 1969; Raine 1975, 1981), which are rewritten in the form of an integral equation (the unknown metric appears in the Green's function). Sciama, Gilman, and Waylen, as well as Raine, assume global hyperbolicity of spacetime and vanishing cosmological constant. In this approach, the Friedman–Robertson–Walker solutions are Machian (except for Al'tshuler), while vacuum solutions and asymptotically flat solutions are non-Machian as are also certain homogeneous solutions with local rotation (nondiagonal Bianchi type IX). On the other hand, Isenberg (1981) uses the initial-value formulation of Einstein's equations and isolates the initial data to be set freely. It turns out that among these data (to be specified on compact 3-geometries), some correspond to degrees of freedom of the *gravitational* field. In view of the fact that perfectly regular vacuum solutions exist, this seems unavoidable. Among the Machian solutions, called Wheeler–Einstein–Mach solutions, are the Friedman models (for positive space curvature) and the same homogeneous solutions with local rotation classified as non-Machian by Raine. This situation leaves open whether Mach's ideas will be useful in the sense of a selection principle.

4. General Relativity, a Machian Theory?

The third part of my remarks fits in here, nicely, because for Julian Barbour there arises no question about whether to believe equally strongly in MP or general relativity: Einstein's theory just *is* perfectly Machian. In order to understand this position, we should take a brief look at the prerelativistic relational particle theories of Barbour and Bertotti. The starting point is a Lagrangian invariant under the so-called *Leibniz group*

$$\left.\begin{aligned} \mathbf{x}' &= \underline{A}(\Lambda)\cdot\mathbf{x} + \mathbf{g}(\lambda) \\ \lambda' &= f(\lambda) \end{aligned}\right\} \tag{1}$$

with $\underline{A}\cdot\underline{A}^{\mathrm{T}} = \underline{I}$, f bijective, $df/d\lambda > 0$, i.e., a possible Lagrangian is

$$L = \sum_{i<j} \frac{m_i m_j}{r_{rj}} \left\{ \sum_{i<j} \frac{m_i m_j}{r_{ij}} \left(\frac{dr_{ij}}{d\lambda} \right)^2 \right\}^{1/2}, \tag{2}$$

where $r_{ij} := |\mathbf{x}_{(i)} - \mathbf{x}_{(j)}|$, and $\mathbf{x}_{(i)}$ is the vector giving the location of the i-th particle with mass m_i. For a particularly simple cosmological model (local masses within a thin spherical shell representing the cosmic masses), the Leibniz group is broken down to the Galilei group, and a particular Riemann-Weber gravitational potential appears after the symmetry breaking:

$$-G\sum_{i<j} \frac{m_i m_j}{r_{ij}} \left[1 + \frac{\alpha}{c^2}(\dot{r}_{ij})^2 + \frac{\beta}{c^2}(\mathbf{v}_{ij})^2 \right], \tag{3}$$

where $\mathbf{v}_{ij} := \dot{\mathbf{x}}_{(i)} - \dot{\mathbf{x}}_{(j)}$, α and β are free (dimensionless) parameters, and G is the Newtonian gravitational constant. Special cases are

$\alpha = -1$, $\beta = 2$ (Lévy 1890),
$\alpha = 3$, $\beta = 0$ (Gerber 1898),
$\alpha = 0$, $\beta = 3/2$ (Treder 1972).

For the modeling of the three well-known effects in the planetary system according to general relativity, one needs (Treder 1972)

$\alpha + 2\beta = 3$ (perihelion shift),
$\alpha + \beta = 1$ (light deflection),
$\alpha + \beta = 0$ (redshift);

$\alpha \neq 0$ leads to anisotropic inertial masses. The potential (3) is being reinvented time and again without authors knowing of its long history.

Barbour and Bertotti's relational particle theories do satisfy the first and second Machian requirement. Unfortunately, while these theories contain the highly welcome frame dragging and the possibility to fit the observed perihelion motion of Mercury [$\beta=0$ in the local physics deduced from (2), and α depends on the mass and radius of the shell representing the cosmic mass, which can be adjusted to get the correct perihelion advance (but not simultaneously the correct light deflection)], less desirable effects also show up:

- an anisotropy of inertial mass
- a time-dependent gravitational constant.

Also, from the point of view of theory building, we note that time and space enter on a totally different level. While time is completely eliminated from the kinematics, space is bluntly taken to be Euclidean. It is the same disparity as the one found in quantum cosmology, where a lot of work is going into deriving a wave function of the universe on possible 3-geometries, but the structure of the 3-geometries is God-given.

In his intervention [not included in this volume], Kuchař did quote Mach on time in the sense that time can be eliminated from physical laws by replacing it, for example, by the rotation angle of the earth. But Mach also said:

> "The same holds for space. We recognize positions by the affectation of our retina, our optical or other measuring instruments. Indeed, our x, y, z in the equations of physics are nothing but convenient names for these affectations. Spatial determinations are again fixations of appearances by other appearances" (Mach 1872, my translation).

Hence, I cannot fully agree with Barbour when he says, in the written version distributed prior to his talk: "In the Newtonian context, we can see exactly and without any doubt when a particular structure of a theory makes it Machian."

The move from a prerelativistic relational *particle* theory to a relativistic Machian *field* theory, as suggested by Barbour, involves a considerable transfer of concepts and techniques. In the following, I list some of the key concepts and steps in his approach and describe how they look in prerelativistic particle theory and in general relativity.

Concept	Nonrelativistic particle theory	General relativity
Configuration	A set of particles at an instant: m_i, $\mathbf{x}_i(t)$	3-geometry: $g_{ij}(\lambda)$
Transformation group	Spatial isometry group: a *global* gauge group (Euclidean group)	Diffeomorphism group: a *local* gauge group
Difference between configurations described by:	Intrinsic derivative (found by solving algebraic equations)	Generalized intrinsic derivative (known only after the solution of the thin-sandwich differential equations)
Comparison of configurations to obtain intrinsic derivative implemented by:	$\frac{d\mathbf{x}_i}{d\lambda} - \sum_{k=1}^{6} \varepsilon_k O_i^k \mathbf{x}_i =: \frac{d\mathbf{x}_i^{\text{int}}}{d\lambda}$ O_i^i: generators of Euclidean group dragging along group orbit	$\frac{\partial g_{ij}}{\partial \lambda} - N_{(i;j)} =: K_{ij}$ N_i: generators of 3-diffeomorphisms (N_i later to be interpreted as *shift vector* and K_{ij} (after scaling) as *extrinsic* curvature)
Result of comparison	*Global* horizontal and vertical stacking gives Newtonian solutions with zero values of total momentum (c.m.s. frame), angular momentum, and energy	*Local* horizontal and vertical stacking of 3-geometries give an Einstein (Ricci-flat) 4-geometry (local densities of conserved quantities vanish)
Relative configuration space	Q_0: $(3n-6)$-dimensional space of n-particle relative configurations	G_0: Field space of all 3-geometries (superspace)
Intrinsic derivative based on:	$\left[\frac{1}{2}\sum_i m_i \frac{d\mathbf{x}_i^{\text{int}}}{d\lambda} \cdot \frac{d\mathbf{x}_i^{\text{int}}}{d\lambda}\right]^{1/2}$ kinetic energy of relative motion	$G^{ijkl}K_{ij}K_{kl}$ with supermetric $G^{ijkl} := \frac{1}{2}g(g^{il}g^{kj} + g^{ik}g^{lj} - g^{ij}g^{kl})$, $g = \det g_{ij}$

FMR implemented by:	Variation w.r.t. spatial isometry transformations	Variation w.r.t. spatial coordinate transformations
SMR implemented by:	*Global* reparametrization invariance	*Local* reparametrization invariance
Lagrangian (global reparametrization invariance in both cases)	$\left[-V^{1/2}\sum_i \frac{m_i}{2}\frac{d\mathbf{x}_i^{\text{int}}}{d\lambda}\cdot\frac{d\mathbf{x}_i^{\text{int}}}{d\lambda}\right]^{1/2}$	$\int d\lambda\left[\int dx R^{(3)}G^{ijkl}K_{ij}K_{kl}\right]^{1/2}$ This expression is replaced by
'Magic trick'		$\int d\lambda\int dx\left[R^{(3)}G^{ijkl}K_{ij}K_{kl}\right]^{1/2}$ in order to achieve *local* reparametrization invariance

Apart from the 'magic trick' used by Barbour in order to obtain the proper Lagrangian, some other points might be critically noted:

- There is an arbitrariness both in the choice of the supermetric and of the Lagrangian. Not only general relativity would be perfectly Machian but also theories derived by replacing $R^{(3)}$ by a function $\psi(R^{(3)})$.
- It is not clear to me how local reparametrization invariance is broken down to local proper time within the scheme itself One may also ask: How does the Lorentz group emerge from the diffeomorphism group? The thin-sandwich conjecture could lead to an Euclidean 4-space as well.
- Barbour's approach does not provide an obvious explanation of why all the solutions of Einstein's equations which have been thought of violating the spirit of Mach's Principle now can be accepted as properly Machian (for example, geodesically complete solutions of the vacuum field equations, Gödel's solution, Minkowski space, etc.).
- While in the prerelativistic particle theory, masses have played a role (if only a passive one), in the field-theoretic approach mass has dropped out altogether. We are asked to accept the vacuum field equations of general relativity as perfectly Machian. This seems to be very far from Mach's ideas. Is it the switch from particle theory to field theory which is responsible for the fading away of the masses?

Conclusions

In view of the remarks made above, it should be clear that I am not convinced that the transition from the relational particle theory of Barbour and Bertotti to general relativity, as suggested by Julian Barbour in his talk, is exactly what is required by his formulation of MP. I feel a bit like Voltaire's Candide, who wants to believe his master Pangloss's message that the world we live in actually is "the best of all possible worlds." Alas, the dire happenings in Candide's life make it hard for him to follow suit. As to general relativity being perfectly Machian, for the German speaking part of the audience I just want to say with Goethe: "Die Botschaft hör ich wohl, allein mir fehlt der Glaube" (The call I hear indeed, the faith is all I lack).

REFERENCES

Al'tshuler, Boris L. (1967). "Integral Form of the Einstein's Equations and a Covariant Formulation of Mach's Principle." *Soviet Physics JETP* 24: 766–770.

Barbour, Julian B. (1974). "Relative Distance Machian Theories." *Nature* 249: 328–329. (Corrigenda in *Nature* 250: 606.)

Barbour, Julian B. (1975). "Forceless Machian Dynamics." *Nuovo Cimento* 26 B: 16–22.

Barbour, Julian B. (1990) "The Part Played by Mach's Principle in the Genesis of Relativistic Cosmology." In *Modern Cosmology in Retrospect*. Bruno Bertotti *et al.* eds., Cambridge: Cambridge University Press, pp. 47–66.

Barbour, Julian and Bertotti, Bruno (1977). "Gravity and Inertia in a Machian Framework." *Nuovo Cimento* 38B: 1–27.

Bertotti, Bruno and Easthope, Paul (1978). "The Equivalence Principle According to Mach." *International Journal of Theoretical Physics* 17: 309–318.

Birch, P. (1982). "Is the Universe Rotating?" *Nature* 298: 451–454.

Bleyer, Ulrich (1988). "Kausalstruktur und gestörte Lorentzinvarianz." In *Gravitation und Kosmos*. Renate Wahsner, ed. Berlin: Akademie-Verlag, pp. 90–103.

Braccesi, A. (1988). "Inflation and Mach's Principle." *Astronomy and Astrophysics* 194: 1–2.

Ciufolini, Ignazio and Wheeler, John A. (1995). *Gravitation and Inertia*. Princeton: Princeton University Press.

Einstein, Albert (1917). "Kosmologische Betrachtungen zur allgemeinen Relativitätstheorie." Sitzungsberichte Preussischen Akademie der Wissenschaften: 142–145.

Einstein, Albert (1918). "Prinzipielles zur allgemeinen Relativitätstheorie." *Annalen der Physik* 55: 241–244.

Ellis, John and Olive, Keith A. (1983). "Inflation Can Solve the Rotation Problem." *Nature* 303: 679–681.

Firmani, Claudio (1971). "A Tensor–Tensor Theory of Gravitation." *Astrophysics and Space Science* 13: 128–136.

Gerber, Paul (1898). "Die räumliche und zeitliche Ausbreitung der Gravitation." *Zeitschrift für Mathematik und Physik* 43: 93–104.

Ghosh, Amitabha (1986). "Velocity Dependent Inertial Induction - Possible Explanation for Supergravity Shift at Solar Limb." *Pramana* 27: 725–730.

Goenner, Hubert (1970). "Mach's Principle and Einstein's Theory of Gravitation." *Boston Studies in the Philosophy of Science* 6: 200–215.

Goenner, Hubert (1981). "Machsches Prinzip und Theorien der Gravitation." In *Grundlagenprobleme der modernen Physik.* Jürgen Nitsch, Joachim Pfarr, and Ernst–Walter Stachow, eds. Mannheim: B. I.-Wissenschaftsverlag, pp. 85–101.

Goldoni, Riccardo (1979). "A Tentative Purely Geometrical Machian Framework for Describing Gravity and Inertia." *Lettere al Nuovo Cimento* 24: 297–305.

Goldoni, Riccardo (1980). "Bimetric Machian Gravitation." *Lettere al Nuovo Cimento* 29: 403–409.

Goldoni, Riccardo (1991). "Cosmology in Bimetric Machian Gravitation II." *Lettere al Nuovo Cimento* 32: 50–54.

Hönl, Helmut (1953). "Über das Machsche Prinzip." *Zeitschrift für Naturforschung* 8a: 2–6.

Hönl, Helmut and Dehnen, Heinz (1963). "Über Machsche und anti-Machsche Lösungen der Feldgleichungen der Gravitation." *Annalen der Physik* 11: 201–215.

Hönl, Helmut (1964). "Über Machsche und Anti-Machsche Lösungen der Feldgleichungen der Gravitation II." *Annalen der Physik* 14: 271–295.

Hönl, Helmut (1966). "Allgemein relativistische Dynamik und Machsches Prinzip." *Zeitschrift für Physik* 191: 313–328.

Hoyle, Fred and Narlikar, Jayant V. (1964). "The C-field as a Direct Particle Field." *Proceedings of the Royal Society of London, Series* A 282: 178–190.

Ihrig, Edwin (1975). "The Uniqueness of g_{ij} in Terms of R^{l}_{ijk}." *International Journal of Theoretical Physics* 14: 23–25.

Isenberg, James A. (1981). "Wheeler–Einstein–Mach Spacetimes." *Physical Review* D24: 251–256.

Lévy, Maurice (1890). "Sur l'Application des Lois Électrodynamiques au Mouvement des Planètes." *Comptes Rendues de l'Academie des Sciences* 110: 545–551.

Kropotkin, Pjotr N. (1975). "Seasonal Periodicity of Earthquakes and Mach's Principle." *Gerlands Beiträge zur Geophysik* 84: 15–19.

Liebscher, Dierck-Ekkehard (1988). "Relativistische Feldtheorie und metrische Raumstruktur." In *Gravitation und Kosmos*. Renate Wahsner, ed. Berlin: Akademie-Verlag, pp. 84–90.

Lynden-Bell, Donald (1967). "On the Origin of Spacetime and Inertia." *Monthly Notices of the Royal Astronomical Society* 135: 413–428.

Ma, Guang-Wen (1990). "Wesson's Gravity and Mach's Principle." *Physics Letters* A 146: 375–377.

Mach, Ernst (1872). *Die Geschichte und die Wurzel des Satzes von der Erhaltung der Arbeit*. Prag: Calve, p. 35.

Milgrom, M. (1983a). "A Modification of the Newtonian Dynamics as a Possible Alternative to the Hidden Mass Hypothesis." *Astrophysical Journal* 270: 365–370.

Milgrom, M. (1983b). "Implication for Galaxies." *Astrophysical Journal* 270: 371–380.

Murphy, George L. (1976). "Mach's Principle in Nonsymmetric Field Theories." *Australian Journal of Physics* 29: 469–470.

Narlikar, Jayant V. and Das, P. V. (1980). "Anomalous Redshifts of Quasistellar Objects." *Astrophysical Journal* 240: 401–414.

Pauli, Wolfgang (1921). "Relativitätstheorie." In *Encyclopädie der mathematischen Wissenschaften* 5,2: 539–775.

Phinney, E.S. and Webster, R.L. (1983). "Is There Evidence for Universal Rotation?" *Nature* 301: 735–736.

Raine, Derek J. (1975). "Mach's Principle in General Relativity." *Monthly Notices of the Royal Astronomical Society* 171: 507–528.

Raine, Derek J. (1981). "Mach's Principle and Spacetime Structure." *Reports on Progress in Physics* 44: 1153–1195.

Roberts, M. D. (1985). "Mach's Principle and the Missing Mass in Clusters of Galaxies." *Acta Physica Polonica Series* B 16: 693–697.

Sciama, Dennis W., Waylen, P.C., and Gilman, R.C. (1969). "Generally Covariant Integral Formulation of Einstein's Field Equations." *Physical Review* 187: 1762–1766.

Treder, Hans-Jürgen (1972). *Die Relativität der Trägheit*. Berlin: Akademie-Verlag.

Treder, Hans-Jürgen (1974). *Über Prinzipien der Dynamik von Einstein, Hertz, Mach, und Poincaré*. Berlin: Akademie-Verlag.

Wesson, Paul, S. (1983). "A New Approach to Scale Invariant Gravity." *Astronomy and Astrophysics* 119: 145–152.

Wesson, Paul, S. (1984). "An Embedding for General Relativity with Variable Rest Mass." *General Relativity and Gravitation* 16: 193–203.

Wheeler, John A. (1959). "Discussion Contribution." In *La Structure et l'Évolution de l'Univers* (11. Conseil de Physique de l'Institut Solvay). Bruxelles: Stoops, p. 49–52.

Discussion

Isenberg: You were saying that the Gödel universe should be included in Julian's formulation, yet I thought Julian's formulation was sort of 3+1, in which case you would only get globally hyperbolic spacetimes, and the Gödel universe would not be such.
Barbour: I now feel that the restriction to globally hyperbolic spacetimes is probably too strong and that the Gödel solution should be included in my formulation [p. 214].
Bondi: I disagree with your statement that there's no room for inertial mass in relativity. If you consider general relativity as including the Einstein–Maxwell equations, then it includes the Lorentz force, and therefore bodies behave according to their ratio of charge to mass. Charge is conserved, so we can compare with different bodies their charges, and so we can find the *e/m* ratio, at least in theory – it's very laborious I agree – but in principle it's there.
Goenner: Well, my problem is that I don't know how to formulate the concept of inertial mass in a covariant way. It must be a scalar, and usually you have an equation like a geodesic equation or, in the case of a Maxwell field being present, an equation with an additional term, but you have to plug in the masses by hand. You have to multiply the whole equation by, say, the rest mass, and then you have this quotient is a quotient of the charge divided by the *rest mass* and not by the inertial mass.
Bondi: If I observe the motion of charged bodies, that reveals their charge-to-mass ratio as a scalar in Lorentz frames.
Goenner: Yes, but what kind of mass is it? Is it a rest mass in the sense of the rest mass of an elementary particle, or is it an inertial mass? I believe it is a rest mass.
Bondi: I suspect so too, but I think here is a feature of general relativity that hasn't been studied enough.
Goenner: So that's one of the things one should follow up.
Kuchař: You raised the question of the uniqueness of the scheme, namely, that there may be many other Machian formulations, like the R^2 theories, different from geometrodynamics. Of course, you are right. When one examines the standard arguments that Einstein's theory is unique, one always finds that they require an additional input. This is true for spacetime formulations of general relativity, as it is true for geometrodynamics.

In the spacetime formulation, we have the great tradition of uniqueness theorems started by Weyl and Vermeil, and completed by Lovelock. What are the additional conditions under which the Einstein law of gravitation uniquely follows? Essentially, the restriction to a metric theory, followed by the limitation of the order of the differential equations, and of the dimensionality of spacetime.

To prove that geometrodynamics is unique, you also need additional conditions. What happens when you try to turn the R^2 theories into canonical form? Essentially, you need more variables: The intrinsic metric and extrinsic curvature are no longer sufficient as the initial data. At the cost of extending the phase space, you succeed in casting the R^2 theories into a Machian mold.

Inversely, when you insist that the three-metric be the only configuration variable, and that the constraints close in a characteristic way (which amounts to requiring that the evolution of three-geometry generates a spacetime geometry), you can prove that the scheme that Julian presented, with its particular supermetric and its particular potential, is unique.

Goenner: Is that hope or ...?

Kuchař: No, it's a theorem [Hojman, Sergio A., Kuchař, Karel, and Teitelboim, Claudio (1976). *Annals of Physics* 96: 88–135].

Goenner: O.K., but then one should state the additional assumptions very clearly, and should say why one wants to plug in only the curvature scalar of the space geometry and nothing else.

Kuchař: I agree, and I hope that we did it. I just wanted to clarify the question of uniqueness. There is that famous dictum by Eddington: "You can never get out of nature what you didn't put into nature." You need to state your assumptions, and then things follow.

Nordtvedt: The inertial mass question in general relativity is, in part, quite successful to the same degree that it is in special relativity. I refer to the magnitude of inertia. By explicit calculation we can show in general relativity, which is essentially special relativity locally, that every form of interaction energy within a body contributes to inertial resistance, inertial forces on that body when it is accelerated, and which in magnitude are as the prescription of special relativity would suggest: $m=e/c^2$. What stops this 'Lorentz' program from being brought to completion is that there are still bare particle masses in the formulation of the physics of matter which are at present irreducible and not further explainable as internal field energies in matter. Though we do not yet know the origin of the bare masses of constituent particles, I am quite comfortable with the inertia of field energies in composite bodies such as nuclei, atoms, molecules, planets, etc. The other issue of inertia is, of course, the origin of the local frames with respect to which inertia manifests itself.

Goenner: I have no problem with the concept of inertial mass in weak-field approximations and the PPN-formalism. There, the local inertial frame is given and we can always compare with Newtonian theory and introduce the concept of inertial mass.

Nordtvedt: We compare the relative magnitudes of inertia in experiments – both laboratory and celestial.

Goenner: Certainly, but I am talking about the structure of the theory, not about the experiments. Experiments are always done in a particular reference frame. I do not doubt that inertial mass is a very useful concept in physics. It is only that I do not think it to be a fundamental concept of general relativity.

Nordtvedt: Inertia is just as ubiquitous and fundamental in general relativistic equations of motion as it is in other branches of fundamental physics. Acceleration-dependent forces naturally appear in all field theories of physics – that essentially is inertia. As soon as you go beyond ideal test particles and treat realistic composite bodies, you unavoidably must deal with calculating inertial mass of such bodies.

Goenner: That may well be. All of what you described must be contained within the matter tensor as source of the gravitational field. But in the matter tensor there is only active gravitational mass, not inertial mass. After deriving the equations of motion, for example, as a variant of the geodesic equation, etc., it is possible to rewrite them into the form of Newtonian equations and then read off an inertial mass. In this sense we do agree. My statement just is that inertial mass is not a spacetime observable, but one related to a local reference frame. The choice of this frame depends on the specific physical system dealt with. Only then can one speak of internal field energies and the like.

Nordtvedt: By explicit calculation of the equation of motion of a composite body, taking into account all the external and internal forces acting on the individual elements of the composite body and which are transmitted by fields, you acquire a contribution to the inertial resistance of that body proportional to the internal field energies in the body in accord with the usual special relativistic prescription.

Goenner: I still believe that it is the bare rest masses of the particles which is the only fundamental, diffeomorphism invariant concept in the general relativistic equations of motion.

Nordtvedt: If you had no *a priori* concept of inertial mass, you would naturally discover it in the process of calculating the equation of motion of composite bodies. You would collect a whole bunch of self-force terms proportional to acceleration of that body and proportional to the sum of internal energy contributions in that body.

Barbour: [to Nordtvedt] Does your assertion hinge upon your knowing some distinguished frame of reference, which is defined by the overall global solution and within which you have a well-defined effective mass? Are you saying that if you can define your frame of reference clearly, then you have a well-defined inertial mass?

Nordtvedt: The local freely-falling frame in which $g_{\mu\nu}$ goes to Minkowski is the frame with respect to which all the interaction energies generate their inertial mass. So, in all metric theories of gravity the inertial mass of field energy in a body is obtained from explicit calculation.

Isenberg: You gave a nice survey of various approaches and with critical comments. I was wondering if you might just say a little bit about which approach you favor [laughter].

Goenner: None [laughter]. My relation to Mach's ideas is, how do I say, hate–love [laughter] er love–hate. I mean I find it very attractive because of this Leibnizian thought that everything hangs together. I mean we are sort of ... that everything is a mirror of the universe. And this is a very intriguing thought, but then, on second thoughts, as a physicist, I must say this: Cannot work! [laughter] So, that's my position and I try to look at these approaches, and I hope I'll find someday one which I love.

Barbour (post-conference response). Goenner (p. 450) lists four reasons why he "lacks faith" in the full Machianity of general relativity. In part, his points are answered in my contribution (p. 214), but let me run through the four points: 1) I do not claim general relativity is the *only* possible Machian theory, merely that it *is* a Machian theory. Kuchař (pp. 454–455) has listed the conditions under which general relativity is obtained. 2) The way in which local proper time emerges from the scheme is described in (Barbour 1994a) cited on p. 231. Goenner is perfectly correct to say the thin-sandwich conjecture could lead to Euclidean as well as Lorentzian 4-space. That depends contingently on the chosen initial conditions. We must look to quantum gravity for an explanation of the Lorentzian signature we find in this universe. 3) I have attempted to answer this point in pp. 225–229. 4) Goenner comments that "the vacuum field equations of general relativity" seem very far from the masses that figure so prominently in Mach's original ideas. Indeed, they do, but, by adopting field theory and then dynamic geometry, Einstein changed the 'ontology' of the world out of recognition. This is why I put so much weight on the Poincaré-type initial-value formulation of Mach's Principle: General dynamical structures can survive the most dramatic changes of 'ontology.' The ontology changes in general relativity, not the criteria of Machianity.

Machian Ideas and General Relativity

Jürgen Ehlers

We have heard about several Machian principles and requirements during the discussions and talks of this meeting, but we have not been able to reach agreement about *one* formulation, to be called *the* Mach Principle, and I assume that we have to live with that.

Some formulations seem actually to be due to Ernst Mach himself, others perhaps to Einstein. There is, however, one principle of the second type which – believe it or not – has not been discussed during the meeting although, at least to my mind, it is very important. Moreover, that principle is clearly reflected in the actual historical development of the physical conceptions of space, time, motion, inertia, and gravity. It runs as follows: "Es widerstrebt dem wissenschaftlichen Verstande, ein Ding zu setzen, das zwar wirkt, aber auf das nicht gewirkt werden kann" (Einstein 1922), which may be translated as: "It is contrary to the scientific mode of understanding to postulate a thing which acts, but which cannot be acted upon."[1]

I feel that this idea is rather important and, as I said, it is reflected in the way in which the spacetime concepts have in fact been developing. Moreover, it corresponds to a tendency which continues to guide physicists in the construction of theories in general.

A prime example of a 'thing' which acts but cannot be acted upon is Newton's absolute space. According to Newton, it acts upon bodies, guiding their motions; for example, its affine structure underlies the concepts of acceleration and force. Similarly, Newton's absolute simultaneity relation is an essential prerequisite for instantaneous action-at-distance force laws. Neither of these spacetime structures is in any way influenced by the physical processes happening in spacetime:

> Absolute space, without regard to anything external, remains always similar and immovable.... Absolute time flows equably without regard to anything external. (Newton, *Principia*)

Einstein Studies, vol. 6: Mach's Principle: From Newton's Bucket to Quantum Gravity, pp. 458–473 © 1995 *Birkhäuser Boston, Inc. Printed in the United States.*

To elucidate the role of the principle quoted above, formulated by Einstein and attributed by him (and by Max Jammer, 1953) to Mach, it is useful to follow James L. Anderson (1964) and Andrzej Trautman (1965, 1966) in distinguishing between two kinds of elements (objects, relations) which occur in physical theories, namely, *absolute* and *dynamical* ones.

Absolute elements (for example, fields) of a theory, according to the proposed definition of these authors, are those which are assumed to be given (or, via the axioms of the theory, follow) uniquely once and for all, never affected by physical processes, i.e., they do not belong to the set of state variables which serve to distinguish between different physical processes envisaged by that theory. In mathematical terms, the absolute elements of a physical theory are those of its objects or relations which are determined categorically by the axioms of the theory. For example, Hilbert's axioms for the Euclidean plane are such that any two copies, or models, of it are isomorphic. In contrast, consider the concept of a group. Two groups need not at all be isomorphic.

On the other hand, the dynamical variables, or dynamical structures, are those which change from situation to situation; they are subject to dynamical laws. Dynamical fields have degrees of freedom, absolute fields do not.

In terms of this distinction, the version of the Einstein–Mach principle quoted above amounts to postulating that *a theory should not contain any absolute structures*, for the latter would be "acting, but not be acted upon." (On page 58 *loc. cit.*, Einstein writes: "Absolute here means not only 'physically real' but also 'self-determinant' in their physical properties, physically conditioning, but not conditioned itself" [my translation from the German].) In this form the principle stands in opposition to Henri Poincaré's conventionalistic attitude concerning the status of geometry, as well as to Immanuel Kant's doctrine of pure intuitions which entail *a priori* truths, deemed necessary for establishing any kind of experience. It appears that history speaks in favor of the principle, judging by success.

The distinction between absolute and dynamical variables is different from that between absolute and relative ones, in the sense in which, according to classical dynamics, velocity is relative (to a frame of reference or a reference body), whereas acceleration is absolute, and it is also different from the third alternative relevant in this context, namely, observables and unobservables. In comparing various theories, it may be useful to keep in mind these conceptual alternatives.

From this point of view, I would like to review the fundamental structures on which spacetime theories have been built. Their common basis is a continuum of possible events or spacetime points, idealized as a four-dimensional, connected differentiable manifold. Such a structure so far underlies all successful physical theories, whether nonrelativistic or relativistic, classical or quantal, and it must be classified as absolute. Let us recall in the following which additional structures have been assigned to this *spacetime manifold* M in various theories.

According to Newton, M has three basic and independent absolute structures. First of all, there is the *absolute time*, represented mathematically as an equivalence class of real-valued functions t on M, t' and t being equivalent if they are affinely related, $t'=at+b$. Fixing arbitrarily a zero point and a unit gives one particular time t, assumed to obey $dt\neq 0$. There is, therefore, also an *absolute simultaneity* relation, represented by the level hypersurfaces S_t which foliate M. A *time-orientation* is fixed as soon as a in the above transformation is restricted to be positive. A second absolute structure of Newtonian spacetime is *equilocality*: Given two events, it is assumed to have an objective meaning whether they happen at the same point of *absolute space*, or at different points. Taken together, simultaneity and equilocality provide M with a product structure, $M=T\times S$. Thirdly, space S is assumed to carry a *Euclidean metric* with metric tensor[2] s_{ab}, fixed intrinsically up to rescalings, $s'_{ab}=ds_{ab}$, $d>0$. These structures imply that M, T, and S are affine spaces; in particular, M is furnished with a symmetric, linear, *flat connection* $\Gamma^{\alpha}{}_{\beta\gamma}$. Thus, *straight lines* are defined not only in space S but also in spacetime M, and there is an absolute, *distant parallelism* of spatial directions.

All these structures except equilocality were well motivated at Newton's time. To within the accuracy then available, absolute time represents the behavior of good clocks which 'transport' time intervals in a path-independent fashion, and it provides M with a *causal* structure. The metric accounts for distance measurements, and t and s_{ab} enter the formulation of action-at-a-distance force laws. Moreover, the straight lines in M represent the motions of free particles, according to the *law of inertia*. (An excellent account of Newtonian spacetime, both systematic and historical, has been given by Stein 1966.)

The only ingredient of Newton's scheme which seemed to be ill-founded was equilocality; already Newton could have done without it – at least so it appears to us. Nevertheless, and in spite of Newton's recognition of the law of relativity for mechanics (Corollary V, *Principia*), it was only towards the end of the 19th century, due mainly to the criticism of Ernst Mach and Ludwig Lange's 'operational definition' of inertial systems, that absolute space was replaced by the set of all inertial frames. According to the (Galilean-)*relativistic reformulation of mechanics*, M has an absolute time t, and each hypersurface S_t of constant time carries a Euclidean metric. It is convenient to represent these metrics by contravariant tensors $s^{ab}(t)$ and to use the inclusion maps i: $S_t \rightarrow M$ to push forward the inverse metrics $s^{ab}(t)$ to M, to obtain a single *inverse spatial metric* $s^{\alpha\beta}$ on M, assumed smooth. The latter is positive semidefinite, and its kernel is the gradient of the absolute time, $s^{\alpha\beta}t_{,\beta}=0$. Besides t and $s^{\alpha\beta}$, M is still assumed to have a flat, symmetric connection $\Gamma^{\alpha}{}_{\beta\gamma}$, compatible with t and $s^{\alpha\beta}$ in the sense that $t_{;\alpha\beta}=0$, $s^{\alpha\beta}{}_{;\gamma}=0$. Since equilocality has been abandoned, M is no longer (intrinsically) a product $T\times S$. Instead, M may be considered as the total space of a fiber bundle whose base manifold is the (one-dimensional, affine) time T, and whose fibers are the Euclidean 3-spaces S_t. This spacetime structure (M, t, $s^{\alpha\beta}$, $\Gamma^{\alpha}{}_{\beta\gamma}$) serves as a basis for the *classical mechanics of isolated systems*. As in the original Newtonian system, the connection is needed to formulate the law of inertia and to introduce forces.

While the structure (M, t, $s^{\alpha\beta}$, $\Gamma^{\alpha}{}_{\beta\gamma}$) proved to be widely successful for mechanics, it did not take into account optics and, later, electrodynamics. The apparently natural way to accomodate such phenomena amounts, in modern terms (Trautman 1966), to the introduction of an *ether* with *4-velocity* V^{α} satisfying $t_{,\alpha}V^{\alpha}=1$. Then, $g^{\alpha\beta}:=s^{\alpha\beta}-c^{-2}V^{\alpha}V^{\beta}$, where c is the speed of light, is a Lorentzian metric on M which may be used to formulate, in conjunction with Faraday's field strength tensor $F_{\alpha\beta}$, Maxwell's equations, geometrical optics, etc., as in special relativity theory. If V^{α} is taken to be covariantly constant, $V^{\alpha}{}_{;\beta}=0$, one obtains Lorentz's rigid ether; V^{α} then is an absolute element which re-introduces Newton's absolute equilocality. Otherwise, dynamical laws for V^{α} are required as parts of an ether-mechanics. As is well known, several attempts to discover such an ether failed.

The resolution of this difficulty by Einstein's *special relativity*, geometrized by Minkowski, can be described very concisely as follows: Discard the fields t, $s^{\alpha\beta}$, and V^{α}, and keep as basic only the *Lorentz metric* $g^{\alpha\beta}$. Since one maintains the flatness of the connection $\Gamma^{\alpha}_{\beta\gamma}$ naturally associated to the metric, one arrives at the spacetime (M, $g^{\alpha\beta}$) with its *absolute metric* $g^{\alpha\beta}$, or $g_{\alpha\beta}$.

The spacetime theories mentioned so far all assign as an absolute element a flat, symmetric connection $\Gamma^{\alpha}{}_{\beta\gamma}$ to M; i.e., their curvature tensors are assumed to vanish:

$$R^{\alpha}{}_{\beta\gamma\delta}=0. \tag{1}$$

This assumption is well motivated as long as it is accepted that *free motions* which exhibit neither absolute nor relative accelerations exist and can, in principle, be identified everywhere and everywhen. If it is recognized that inertial and gravitational masses of test particles cannot be distinguished, and consequently instead of free motions only *free-fall motions* are realizable, and that the latter exhibit relative accelerations whose amounts and directions depend on the distribution of matter, one is led to give up the restriction (1) on the gravitational–inertial connection. This led Einstein to the spacetime model of *general relativity*, with its *dynamical metric* $g_{\alpha\beta}$.

As is well known, an important by-product of general relativity with its *gravitational field equation*

$$R_{\alpha\beta}=\kappa(T_{\alpha\beta}-\tfrac{1}{2}Tg_{\alpha\beta}),\ \ \kappa=8\pi G \tag{2}$$

(G is Newton's constant of gravitation) is that it permits models of spatially homogeneous, gravitating mass distributions without boundary and with equivalent, comoving observers – *cosmological models*. Newton's theory in its original form does not allow such models, and it was only after the advent of general relativity that generalizations of Newton's theory have been invented which are not restricted to isolated systems but permit dynamical cosmological models (though no satisfactory optics, in general). In the present context, this generalized *Newton–Cartan theory* of spacetime and gravitation is of interest since, like general relativity, it uses a dynamical connection. In fact, it is then straightforward to formulate a *frame theory* of spacetime and gravity which contains 'Newton's' and 'Einstein's' theories as special cases, distinguished only by the vanishing or nonvanishing of *one* constant in *one* of the axioms (Ehlers 1981, 1986, 1991).

I now summarize the foregoing sketch of spacetime and/or gravity theories, adding always *dust* as a simple *matter model* in order to illustrate the different ways in which absolute and dynamical structures enter the dynamical behavior of matter. The mass density is denoted as ρ, the 4-velocity of the dust as U^{α}; Φ denotes the gravitational potential. As before, I describe all theories in covariant, 4-dimensional language and state the *local* laws only. [For details concerning these formulations, see, for example, (Ehlers 1991) and the references given there.]

I *Newton's theory*
Absolute fields: t, V^α, $s^{\alpha\beta}$, $\Gamma^\alpha{}_{\beta\gamma}$.
Dynamical fields: ρ, U^α, Φ.
Laws: $s^{\alpha\beta}t_{,\beta}=0$, $V^\alpha{}_{;\beta}=0$, $t_{;\alpha\beta}=0$, $t_{,\alpha}V^\alpha=1$, $s^{\alpha\beta}{}_{;\gamma}=0$, $R^\alpha{}_{\beta\gamma\delta}=0$, $s^{\alpha\beta}\Phi_{;\alpha\beta}=4\pi G\rho$, $t_{,\alpha}U^\alpha=1$, $(\rho U^\alpha U^\beta)_{;\beta}=0$.

II *Galilei-invariant classical mechanics*
As in I, except that V^α is to be deleted.

III *Special relativity theory*
Absolute fields: $g_{\alpha\beta}$, $\Gamma^\alpha{}_{\beta\gamma}$.
Dynamical fields: ρ, U^α.
Laws: $g_{\alpha\beta;\gamma}=0$, $R^\alpha{}_{\beta\gamma\delta}=0$, $g_{\alpha\beta}U^\alpha U^\beta=1$, $(\rho U^\alpha U^\beta)_{;\beta}=0$.

IV *General relativity theory*
Absolute fields: None.
Dynamical fields: $g_{\alpha\beta}$, $\Gamma^\alpha{}_{\beta\gamma}$, ρ, U^α.
Laws: $g_{\alpha\beta}U^\alpha U^\beta=1$, $g_{\alpha\beta;\gamma}=0$ and Eq.(2) with $T_{\alpha\beta}=\rho U_\alpha U_\beta$.

V *Newton–Cartan theory*
Absolute fields: t, $s^{\alpha\beta}$.
Dynamical fields: $\Gamma^\alpha{}_{\beta\gamma}$, ρ, U^α.
Laws: $t_{;\alpha\beta}=0$, $s^{\alpha\beta}{}_{;\gamma}=0$, $s^{\alpha\beta}t_{;\beta}=0$, $R^\alpha{}_{\beta\epsilon\delta}s^{\epsilon\gamma}=R^\gamma{}_{\delta\epsilon\beta}s^{\epsilon\alpha}$, $R_{\alpha\beta}=\frac{1}{2}\kappa\rho t_{,\alpha}t_{,\beta}$, $t_{,\alpha}U^\alpha=1$, $(\rho U^\alpha U^\beta)_{;\beta}=0$.

VI *Frame theory*
In this case, the separation of fields into absolute and dynamical ones depends on the value of the parameter λ.
Fields: $k_{\alpha\beta}$, $s^{\alpha\beta}$, $\Gamma^\alpha{}_{\beta\gamma}$, ρ, U^α, constant: λ.
Laws: $s^{\alpha\beta}$ is positive semidefinite, and rank $(s^{\alpha\beta})\geq 3$, $k_{\alpha\beta}$ is positive semidefinite, $k_{\alpha\beta}s^{\beta\gamma}=-\lambda\delta_\alpha{}^\gamma$, $k_{\alpha\beta;\gamma}=0$, $s^{\alpha\beta}{}_{;\gamma}=0$, $R^\alpha{}_{\beta\epsilon\delta}s^{\epsilon\gamma}=R^\gamma{}_{\delta\epsilon\beta}s^{\epsilon\alpha}$, $g_{\alpha\beta}U^\alpha U^\beta=1$, $R_{\alpha\beta}=\kappa(k_{\alpha\gamma}k_{\beta\delta}-\frac{1}{2}k_{\alpha\beta}k_{\gamma\delta})\rho U^\gamma U^\delta$, $(\rho U^\alpha U^\beta)_{;\beta}=0$.

Comments:

1) If $\lambda=c^{-2}>0$, the laws of the frame theory reduce to those of general relativity, with $g_{\alpha\beta}=\lambda^{-1}k_{\alpha\beta}$, $g^{\alpha\beta}=-s^{\alpha\beta}$. If $\lambda=0$, it follows that $k_{\alpha\beta}$ can be written as $t_{,\alpha}t_{,\beta}$, and if spacetime is spatially asymptotically flat, these laws reduce to those of the Newton–Cartan theory. One may use the frame theory to define Newtonian limits $\lambda\to 0$ of one-parameter families of general relativistic solutions, parametrized by λ. Several

examples and also some general theorems about such limits are known. In this sense, not only the *laws* of the Newton–Cartan theory but even (some) *solutions*, or models, of that theory may be considered as limits of those of Einstein's theory. The limit model approximates members of the solution-family (with small λ) in any spacetime domain where the convergence is uniform. In the limit, the metric fields of general relativity degenerate into those of the Newton–Cartan theory. Geometrically, this means that the null cones 'open up' and degenerate into the hyperplanes S_t of constant absolute time.

2) In the 'Newtonian' theories, I and II, in contrast to general relativity, there exists an absolute parallelism, and thus an absolute standard of nonrotation, expressible in spacetime terms by the law $s^{\beta\gamma}R^{\alpha}{}_{\gamma\delta\epsilon}=0$. The empirical support for this may be seen, to good approximation, in phenomena such as Newton's rotating bucket, Foucault's pendulum, or the preservation of directions by gyroscopes. This standard of no rotation is denied by general relativity and 'corrected' by, for example, the de Sitter–Fokker precession. In connection with Mach's considerations on rotation, it is of interest that in the Newtonian limit of general relativity, taken according to the frame theory, absoluteness of parallelism follows only for isolated systems.

3) Strict inertial frames (in extended spacetime regions) exist according to theories I, II, III only; in theories IV and V there are only 'local' inertial frames, i.e., frames which are inertial in an 'infinitesimal' neighborhood of an event, or of a geodesic world line. In the latter theories, not only rotation, but also acceleration loses its absoluteness.

4) In Newton–Cartan theory the existence of an absolute time can be derived from the laws if, instead of $t_{;\alpha\beta}=0$ and $s^{\alpha\beta}t_{,\beta}=0$, one requires that the kernel of $s^{\alpha\beta}$ be one dimensional, and $s^{\alpha\beta}p_{\beta}=0$ implies $p_{\alpha}R^{\alpha}{}_{\beta\gamma\delta}=0$. This may be considered as a generalization of Neumann's (1870) 'operational definition' of absolute time as being proportional to arc length along a free particle, insofar as $s^{\alpha\beta}$ and $\Gamma^{\alpha}{}_{\beta\gamma}$ determine t. It is also analogous to the fact that, in Barbour–Bertotti Machian mechanics [described elsewhere in this volume p. 214 and in (Barbour and Bertotti 1982)], dynamical time is a derived concept, in contrast to spatial distance.

5) The laws of motion, at least for the simple case of dust – $(\rho U^{\alpha}U^{\beta})_{;\beta}=0$ or $(\rho U^{\alpha})_{;\alpha}=0$ and $U^{\alpha}{}_{;\beta}U^{\beta}=0$ – follow from the field equation only in the case of general relativity. This fact, combined with the nonlinearity of the Einstein–Hilbert field equation, implies that only in general relativity does the field equation alone – without independent matter equations of motion – lead to interaction between the material

sources of the field.

6) Following again Anderson and Trautman (*loc. cit.*), let an *invariance group* of a theory be defined as a group of transformations (acting on spacetime M or on a principal bundle over M on which the fields of the theory are defined) which (i) transform all solutions of the laws of the theory into solutions and (ii) map the absolute fields into themselves ("leave them unchanged"). A theory may be called *generally covariant* or will satisfy the *principle of general relativity* if it admits an invariance group G such that the spacetime transformations induced by the elements of G form the group of all diffeomorphisms of M. This property of a theory is very restrictive; among the theories I–V indicated above, only general relativity satisfies it.[3] This principle appears to be *equivalent to the Einstein–Mach principle* stated at the beginning of this paper, for absolute fields will be invariant only under some, not all diffeomorphisms, and if there are no absolute fields, it is hard to see how the theory could fail to be generally covariant. (At any rate, a generally covariant theory cannot possess an absolute tensor field on spacetime.)

The relationships between the spacetime (+ gravity) theories considered here can be displayed in a diagram:

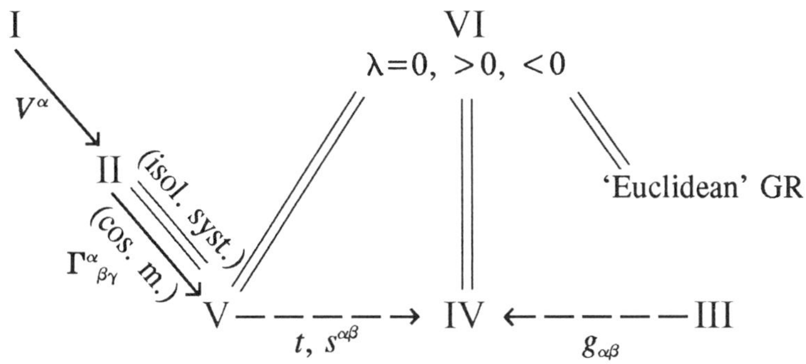

$A \to B$ indicates that B generalizes A; $A \dashrightarrow B$ indicates that A is an approximation to, or a 'limit' of, B; $A=B$ indicates that A is equivalent to B. A symbol x besides an arrow $A \to B$ indicates that in the transition from A to B the absolute field x is discarded or becomes dynamical. The case VI, $\lambda<0$, is added for completeness; it is not considered elsewhere in this paper.

This somewhat lengthy description shows how, by eliminating absolute structures and enlarging the invariance group, the spacetime structures actually employed in physics were made at the same time more comprehensive while approaching more and more the status required by the Einstein–Mach principle. This principle, though apparently not stated by Mach, seems to conform to Mach's thinking and, as I tried to show, expresses basically the same idea as the (properly formulated) principle of general relativity. A major goal of present-day research, the construction of a quantum theory of spacetime structure and gravitation, may be viewed as the attempt to rid quantum field theory of its absolute elements. Calling these efforts 'quantization of general relativity' or 'general-relativization of quantum field theory' just expresses where one starts from; the aim is the same.

Now I should like to make some remarks on the Barbour–Bertotti model of Machian mechanics. In it, Newtonian simultaneity and Euclidean spatial distance are accepted as the only absolute relations between bodies, quite in line with Mach's writings. Dynamics is then based on a Lagrangian which depends on relative configurations and their rates of change relative to an arbitrary time ordering parameter. A dynamically preferred metric time is introduced as a derived concept, its role being to simplify the equations of (relative) motion. I feel that this theory realizes well Mach's program. At the same time it exhibits the shortcomings of a theory based solely on bodies and their relative motions, leaving no room for fields propagating at finite speeds. Absolute, rigid geometric structures contradict the principle of local action which has been so successful in physics since Maxwell. I should also like to stress that in all forms of dynamics which have been empirically successful according to the increasing range and precision of observable phenomena, the primary distinction concerning motion has been between 'natural' or 'forced' motion or, later, between 'inertial' and 'noninertial' motion of a single body, while relative motion figures as a derived, secondary concept, contrary to Mach's view. Especially in general relativity, a velocity of one body relative to another, distant one can only be defined by means of the metric and/or the connection field between the bodies, and in so far as these fields are dynamic, relative velocity is no longer a kinematical concept. [Cf. also the discussion in Weyl (1924).] The proposal to consider relative motion as basic, plausible though it is at first sight, has not been successful in physics. In this context I should like to quote Stein (*loc. cit.*):

> It is often claimed that the general theory of relativity has demonstrated the correctness of Leibniz's view [as compared to Newton's]. This is a drastic oversimplification. It is no more true in the general theory than in Newtonian dynamics that the geometry of spacetime is determined by *relations among bodies*. If the general theory does in essence conform better

> to Leibniz's views than classical mechanics does, this is not because it relegates 'space' to the ideal status ascribed to it by Leibniz, but rather because the space – or rather the spacetime structure – that Newton requires to be real, appears in the general theory with attributes that might allow Leibniz to accept it as real. The general theory does not deny the *existence* of something that corresponds to Newton's 'immobile being'; but it denies the rigid *immobility* of this 'being,' and represents it as interacting with the other constituents of physical reality.

Needless to say, I agree with the point made by Stein. It is conceivable that even Mach would agree.

For Mach, the 'furniture of the physical world' consisted of bodies only. This is in accordance with the Barbour–Bertotti mechanics and seems also to be intended in the Hoyle–Narlikar approach, in contrast to theories which assume, besides particles, also fields, or pure field theories which attempt to account for 'matter' in terms of fields only. In view of the successes and the flexibility of the field concept, I consider it as very improbable that pure particle theories with direct interactions will play a role other than that of illustrative models. So far, any description of the properties and states of matter involves a metric as an indispensible ingredient. Consequently, quite apart from mathematical technicalities the idea that "matter determines the metric" cannot even be meaningfully formulated. Besides matter variables, a metric (or some other geometric structure which classically implies a metric) seems to be needed as an independent, primitive concept of physics, in line with the quotation given above.

I close with two remarks. The first concerns *prediction*. It has often been said that an important consequence of quantum mechanics is that, in general, one cannot make definite predictions for individual processes, but only statistical ones, referring to ensembles, in contrast to the determinism of classical mechanics. What is rarely said is that, according to special or general relativity, true predictions are impossible. Only *retrodictions* are possible, because all the information an observer can obtain refers to his past light cone, and from such data he can at best compute the metric field and other fields within that light cone. Predictions require in principle untestable extrapolations of data, amounting to the exclusion of 'improbable' surprises. Tests of theories concern retrodictions. We check whether our laws are compatible with observed patterns of events.

My second, more significant remark concerns the relative importance of the concepts of a 4-dimensional *spacetime* and that of an *evolving*

3-dimensional space. I am inclined to consider spacetime as the fundamental concept, and evolving 3-spaces as particular tools to construct or display spacetimes. My reason for this judgement is that a spacetime admits infinitely many – in fact, insurveyably many – foliations by families of spacelike hypersurfaces which all contain the same testable predictions, coded intrinsically in *one* spacetime, at least at the classical level. (At the quantum level, we do not know.) In fact, the successes of general relativity in accounting for observations are based on spacetimes containing models of the objects (stars, planets, ...), the observer, the light connecting objects and observer, and so on. In contrast, $(1+3)$ decompositions do not enter except as admittedly useful tools to construct spacetime. Besides, the Hamiltonian formulation of GR is restricted to globally hyperbolic spacetimes; why should other spacetimes be excluded? I can see how Machian arguments, by analogy with particle mechanics, may suggest a dynamics of 3-geometries; but I fail to see the superiority of that approach to the Einstein–Minkowski spacetime conception, in spite of John Wheeler's (1963) and Karel Kuchař's thoughtful and eloquent pleadings for a dynamical 3-geometry as the basic concept of general relativity.

NOTES

[1]This translation takes into account remarks made by Nojarov and Barbour during the talk.

[2]I use lower case Roman letters to refer to spatial fields, Greek ones for spacetime fields; one may preferably interpret both types as abstract indices in Penrose's sense, denoting the fields themselves, not just their components.

[3]The invariance groups of the theories listed above are given, apart from rescalings of time and length units, as follows:

I: Spatial translations and rotations, time translations
II: Inhomogeneous Galileo group
III: Poincaré group
IV: Group of all diffeomorphisms of M
V: Arbitrarily time-dependent spatial translations and rotations

REFERENCES

Anderson, James L. Lecture given at the 1964 Brandeis Summer Institute in Theoretical Physcis (unpublished); also: *Principles of Relativity Physics* New York: Academic Press, 1967, esp. Sec. 4.3.

Barbour, Julian B and Bertotti, Bruno (1982). "Mach's Principle and the Structure of Dynamical Theories." *Proceedings of the Royal Society of*

London A382: 295–306.
Ehlers, Jürgen (1981). In *Grundlagenprobleme der modernen Physik*, Jürgen Nitsch, *et al.*, eds. Mannheim: BI-Wissenschaftsverlag, pp. 65–84; see also the refs. quoted in that essay.
Ehlers, Jürgen (1986). "On Limit Relations Between, and Approximate Explanations of, Physical Theories." In *Logic, Methodology and Philosophy of Science*, Vol. VII, B. Marcus *et al.*, eds. Amsterdam: Elsevier Science Publ., pp. 387–403.
Ehlers, Jürgen (1991). "The Newtonian limit of general relativity." In *Classical Mechanics and Relativity: Relationship and Consistency*, G. Ferrarese, ed. Naples: Bibliopolis, pp. 95–106.
Einstein, Albert (1922). *The Meaning of Relativity*, Princeton: Princeton University Press.
Jammer, Max (1953). *Concepts of Space*. Cambridge, Mass.: Harvard University Press, esp. p. 158.
Neumann, Carl (1870). *Ueber die Principien der Galilei–Newton'schen Theorie*. Leipzig: Teubner.
Stein, Howard (1967). "Newtonian Space–Time." *The Texas Quarterly* 10: No. 3, 174–200.
Trautman, Andrzej (1965). *The General Theory of Relativity, A Survey*. Warsaw: Nuclear Energy Information Center.
Trautman, Andrezej (1966). "The General Theory of Relativity." *Soviet Physics Uspekhi* 9: 319–339.
Weyl, Hermann (1924). "Massenträgheit und Kosmos. Ein Dialog." *Die Naturwissenschaften* 12: 197–204. Reprinted, with supplementary remarks, in *Was ist Materie*. Berlin: Springer, 1924.
Wheeler, John A. (1963). "Geometrodynamics and the Issue of the Final State." In *Relativity, Groups and Topology, Les Houches*. Cecile DeWitt and Bryce DeWitt, eds. New York: Gordon and Breach (1964), pp. 317–522.

Discussion

Kuchař: Jürgen did a great job in presenting his new formulation of Mach's Principle, and he raised important issues. However, I strongly disagree with his analysis about how general relativity fits into this scheme.

I want to point out that there are absolute structures in general relativity, and that you cannot do without them. The culprit of our disagreement are the two terms here on the blackboard, which are ambiguous. Those terms are 'act' and 'not be acted upon' and: What constitutes a dynamical system?

Such things are difficult to analyze in the spacetime approach, because a dynamical system should necessarily be described by dynamical equations, which gets you directly to the Hamiltonian approach. This approach makes it clear that some spacetime variables, some components of the spacetime metric, are not dynamical and cannot be dynamical if the spacetime theory is to be generally covariant.

So let me tell you what are the absolute structures which occur in general relativity. Let us start with a simple model. Have a prescribed background, which may be flat or curved, that is described by a metric tensor. Investigate the motion of a single relativistic particle in that spacetime. This is a dynamical system which is ruled by a Hamiltonian constraint: The statement that the 4-momentum must stay on the mass shell. This constraint, through the Poisson brackets, drives the evolution. Now I hope that you agree that the system has a structure that acts but isn't acted upon: The particle is acted on by the background. The background is prescribed and is in no way influenced by how the particle moves.

Ehlers: Yes, but if I may interrupt, you are now just describing a single particle considered as a test particle in the given external background metric; that is not the generic, basic situation of the theory. The metric is in fact related by the field equation to whatever matter model you have put in.

Kuchař: I am delighted that you said that, because this is the first step into the trap [laughter]. I can formulate general relativity in exactly the same form. There is an object called the super-Hamiltonian constraint in general relativity. It's just a certain projection of the Einstein law onto a hypersurface. It contains something called the supermetric. It's done point by point of a spatial hypersurface, but the supermetric is a function of the ordinary metric; indeed, it is a *prescribed function* of the ordinary metric. Then we have the momenta of the gravitational field; again, there are infinitely many of these momenta; they are multiplied into the supermetric just as the particle's momenta are multiplied into the background metric; and on top of it we have the scalar potential, which is again a given functional of the metric.

The situation is exactly the same as before; the data are limited by the constraints, but the constraints also drive the dynamics of the data. They propagate the data from one hypersurface to another.

Now here are the absolute structures in the dynamics of general relativity: First, the supermetric is a given function of the ordinary metric. It's given to you as the background metric was given to you as a function of x. Second, the curvature potential is also a prescribed function of the ordinary metric. You are not able to change it. These are the structures in general relativity that stick out as its absolute structures, that determine how the geometry moves in superspace, but are not acted upon by the geometry.

Chris Isham and I once played with the idea of doing what your analogy suggests, namely, to unfreeze these absolute structures and turn them into dynamical structures. But then you drive the theory to the next level of complication. It's like starting with the second-quantized system in quantum mechanics, and then third-quantizing the system.

So my conclusion is that in dynamics some structures must be fixed; you cannot do without them, because as long as you have a particular dynamics, this dynamics must be driven by a particular Hamiltonian or a particular Lagrangian. That Lagrangian must have some structure, and this is the absolute structure of the theory. I have another point to make but perhaps I should quit.

Ehlers: No, no.

Kuchař: You mentioned that you do not like geometrodynamics, because it disposes of spacetime and emphasizes space, and because it relies on a foliation of spacetime which is to a large degree arbitrary.

I feel that, far from being a weakness, this arbitrariness is rather a strength. It forces on the canonical formalism certain consistency conditions: The Poisson brackets of the constraints must close in a definite way for geometrodynamics to generate the same spacetime irrespective of the choice of foliation. This is what makes relativistic geometrodynamics unique among other possible dynamical evolutions of three-geometry [cf. pp. 454–455]. As you see, the arbitrariness of the foliation is exactly what enables us to reconstruct spacetime from space.

However, spacetime cannot survive quantization. It must disappear in the same way in which the trajectory of a particle disappears in ordinary quantum mechanics. What remains - what is an heir of spacetime, as I once expressed it - is the space of all three-geometries, on which the state functional propagates. What remains is superspace. This is the second part of my answer to your feeling that slicing spacetime is inappropriate. We must destroy spacetime by quantization, and geometrodynamics tells us how to do it while preserving in the quantum domain what can be preserved.

Ehlers: I shall try to reply, and I hope you will help me to sort out the first part, which is not yet to do with slicing or nonslicing or quantization. My attitude was that I look at those variables that are used in a particular theory to describe a particular situation, for example, a double star, and I would say the list of these variables in general relativity consists of the metric and whatever variables I have chosen to describe the matter: dust, perfect fluid, an electron field, and so on. I have the list of these variables and then I ask myself. Are some of these variables which are needed to describe particular situations once and for all characterized categorically by axioms or not? And there is none of them which is, in GR, in contrast to SR.

Now what you pointed out, I think, in my language is that the differential operator in terms of which the field equation is formulated is an absolute element; and that I do not deny. I realize one has to refine the terms, but I still feel that the distinction between these two types of structures which enter a theory that I tried to describe – though perhaps not in a completely satisfactory manner – is important, and I feel that one of the major difficulties in creating a quantum theory that includes gravity is just this fact that in the sense in which I tried to describe the theory, there are no absolute structures, in particular, there is no absolute spacetime structure, which is so deeply built into the structure of those quantum field theories which so far have been successful.

If you look at the formulation of quantum field theory by Streater and Wightman or by Haag and Kastler or others, there's always a very essential use made of the given structure, namely, the Minkowski metric or, what is essentially equivalent, the Poincaré group. I admit, of course, in order to formulate the laws you need certain differential operators or similar objects, and those you might call absolute structures; they define the behavior of the theory, but this is, I think, different from a classification of the fields themselves.

Now concerning the question of whether the four-dimensional spacetime or the time evolution of three-dimensional slices are primary, I admit to you that if one wants to construct the spacetime from Cauchy data, then certainly the second one is the appropriate way of going about it; but in many cases in which general relativity has been applied successfully, I would even maintain that in the majority of those cases, one hasn't gone about constructing the spacetime models by initial-value problems. Whether one should elevate this description to *the* basis of the theory for generalizing it is a question.

I also have learned this formalism and tried to understand to which extent it is useful to quantizing or not, and I must admit that so far this line of trying to approach a quantum theory is perhaps the best one which one has, but in spite of this fact I have this reservation. I do not quite see a good physical reason why those structures that refer to the initial data are considered as being more basic and are kept intact, whereas the rest is, so to speak, subject to uncertainties of a quantum nature. I think Roger [Penrose], for example, would already be acting differently in his approach. But I admit, if one wants to quantize, then introducing complex-valued wave functionals depending on 'half' of the Cauchy data is perhaps at the moment the best way one can go about it, but I cannot help feeling a little skeptical. I do not see a good physical reason why the spatial structure which refers to nonobservable initial data should be considered as more basic for building up this structure than the rest.

Barbour (post-conference response). Ehlers (p. 466) accepts that the dynamical model Bertotti and I (and, much earlier, Hofmann, Reissner, and Schrödinger) developed "realizes well Mach's program." However, he then doubts whether similar notions of relative motion can be successfully realized in field theory. I argue on p. 225 that this *is* what can be done by means of the intrinsic derivative *for fields*. Moreover, it is precisely the method used in general relativity. It seems to me that Ehlers's remarks on p. 467, in the paragraph starting "For Mach," fail to take account of the fact that Bertotti and I, like Reissner and Schrödinger, see our *mechanical* model merely as a first step – to show what a truly Machian theory should be like. Our serious proposal is the *field-theory* model described in this volume. That turns out to be general relativity!

As to Ehlers's concluding comments about the relative importance of the three-dimensional and four-dimensional concepts in general relativity and whether the initial-value formulation is fundamental (p. 472), this seems to me to depend on the topic under consideration. For most *current* applications of general relativity, Ehlers is surely correct. However, he himself grants that quantum gravity could tip the scales toward the three-dimensional initial-value approach. Also, in the conceptual questions relating to Mach's principle, I feel the three-dimensional dynamical approach is essential. We know how to make particle mechanics Machian. To see if general relativity is Machian, we must compare analogous entities, like with like. That means we must compare the evolution law of n-particle configurations in Euclidean space with the evolution law of three-dimensional geometries (and matter distributions on them).

Finally, I also share Kuchař's view that the correct use of the notion of absolute elements is only possible once the true dynamical degrees of freedom have been identified. However, once that has been done, I am in agreement with Ehlers in thinking there are no absolute elements in general relativity but only "in the differential operator in terms of which the field equation is formulated." I do not feel the relativistic particle is a true analog of geometrodynamics, but that issue must be taken up elsewhere.

Reflections on Mach's Principle

Sir Hermann Bondi

Let me start by asking how an astronomically inclined physicist would state the principle of relativity. The way I think one should do so is to say that there is at every point of the universe a preferred velocity, namely, that from which the universe appears most isotropic, and then the principle of relativity states that no laboratory experiment has to date been devised or performed that reveals this velocity. We have nothing but the astronomical observation that shows us what this velocity is. I think in connection with Machian principles this is quite significant. What the demand for a Machian explanation is, is easy to see with Newton's rotating bucket, viz., to give the state of no rotation which is a single state, but translational inertial frames are Lorentz invariant. The universe patently does not prescribe for us a set of inertial frames; it produces just one frame, and we hope it is inertial. From the dynamics of the universe (if you think it is not too complicated), it probably is inertial, but I don't think it is absolutely necessarily so. If we had a mathematical description of a Machian local effect, I for one would not be happy if it gave me just the Lorentz frames. I would wish it to reveal to me also that velocity from which the universe looks isotropic. To paraphrase Ehlers's Einstein quotation [p. 458], it is repugnant to have a single velocity thrown at us unconnected with anything else relating to a material body as important as the universe.

I'll allow myself a tiny bit of history. Forty-five years ago, when we were propagating the steady-state theory, one objection often met was: "I go all along with you, but how does the newly created matter know with what momentum to move?" When we said that the universe was not Lorentz invariant, we encountered shocked disbelief, although this was quite a clear and obvious consequence of the isotropy of the red shift in those days. Of course, the accuracy with which that velocity could have been determined 45 years ago would have been quite low, and it needed the microwave background radiation to make it high, but we now

Einstein Studies, vol. 6: Mach's Principle: From Newton's Bucket to Quantum Gravity, pp. 474–477

know it is well defined to one part in 10^4 if not to one part of 10^5. This velocity is something that looks to me intrinsically important.

What, of course, has happened, since no experiment has even been suggested that would reveal this velocity locally, is that we've adopted Lorentz invariance, both in special and in general relativity. The result, as Julian [Barbour] and others have stressed to us, is to wash out any effect of the universe. Since relativity accepts all the Lorentz frames, there can be no such connection, and so relativity is, in fact, almost an anti-Machian theory. As long as we stick to local Lorentz invariance, there can be no Machian effect. I'm not sure whether that is quite true, but it is certainly a proposition to be considered.

Second, I think we all agree that $4\pi\rho GT^2$ is an important quantity: constant of gravitation times mean density of the universe, times the square of the time scale. If you say that this quantity, which of course is a pure number, has a particular value, that tells you something either about the mean density of the universe, about its time scale, or about the constant of gravitation (which can be regarded as the ratio of gravitational to inertial forces). You can read it as you wish, but in almost any way you read it, it suggests something which is very Machian: a link between the universe and a local quantity like the constant of gravitation. Now when one has such an important expression, then one is unfortunate if it is not a constant. If it is not a constant, then one suspects that maybe something else in physics is very different at other times. It is constant in only two cosmologies that I know of: the steady-state theory and the Einstein–de Sitter model.

For, since the time scale is T effectively $R/\dot{R}$,

$$4\pi G\rho T^2 = \frac{4\pi G\rho R^2}{\dot{R}^2}.$$

In a Friedman universe with negligible pressure $\rho \sim R^{-3}$:

$$4\pi G\rho T^2 \sim \frac{1}{R\dot{R}^2}.$$

The only case where this is a constant is if R varies like $t^{2/3}$, which is the Einstein–de Sitter model, on which very little attention has been concentrated recently. I don't think the astronomical evidence favors it particularly, and we know it doesn't favor the steady-state model. So again we have the difficulty that something which from a Machian point of view should be constant we find very difficult to take as a constant. It doesn't seem to agree with what little we know in cosmology.

The third and final point: I'm sometimes a little bit disturbed by how people talk about the field equations of general relativity. What are the equations? They tell us that something on the left-hand side equals something on the right-hand side. That is a common characteristic of equations [laughter]. But how do they help us? We can put an arbitrary metric into the left-hand side; it will give us something on the right-hand side that is an energy–momentum tensor. Now whether it is an acceptable tensor or not, that depends on our taste, but it gives us something. And how can we turn this arbitrariness of the left-hand side into something more definite? In my view, that can only be done by prescribing equations of state on the right-hand side. If you put in, for example, that the matter present is dust with nonnegative density, or it has isotropic pressure with a nonnegative density and a pressure not too high compared with the density, it is only through such equations that something definite is implied by the field equations. And because you have to rely so much on these material properties, on the constitutive equations of the matter, I'm always bothered by any singular equation of state. I felt that particularly when I was very young. The Einstein–Infeld–Hoffmann approach to general relativity was then greatly admired. I think it is dead wrong. I think the whole idea that you can treat true astronomical bodies as mass points is not sound. Even in such a simple system as the earth and the moon, tidal friction on the earth has to be allowed for if you want to get a sensible orbit for the moon. If you model a close binary, you certainly will get nonsense if you do not allow for the mutual tidal effects, which depend on the constitutive equations. Dealing with abstract mass points where it is in any case mathematically difficult to cut out the dipole moment doesn't appeal to me at all. So let us always mistrust the simplifying features of singularities. I'm even prepared to believe that some singular constructs like mass shells probably don't hide any nasties, but wherever we can, let's use finite quantities and tolerably reasonable equations of state.

Now where this criterion makes ordinary procedures look most peculiar is in our treatment of empty space. As I was saying, if you want to model a nice steady star, you put in that the pressure is isotropic, which is one condition; a link between the pressure and the density, which is a second; and possibly you say something about the density, which is a third: at most three conditions. When you write down you have empty space, you put in ten conditions. Now I used to be very unhappy about that. Since we have positivity theorems, I think I'm much less unhappy about it. But nonetheless I think even in empty space I don't feel confident that all Cauchy problems or all problems set with permanent features like equations of state necessarily lead to something sensible and indeed we know that some don't.

Discussion

Isenberg: Could you expand a bit? At the end, you were talking about the vacuum, and you said you were not happy before positivity of mass was proven, and you said you were still a little bit unhappy about that. I lost that and I wasn't sure what you were saying.
Bondi: What I am saying is that in a proper Cauchy problem we state certain permanent features (the constitutive equations) and certain initial conditions. For example, in the case of a system with mass everywhere, say freely moving dust, we prescribe certain things: say, that the pressure is zero. We have then a very modest number of constitutive equations, and quite separate initial conditions, whereas when we discuss the vacuum, we write down $R_{\mu\nu}=0$ as a permanent feature. Now it is not instantly obvious that if you start with empty space that it need remain empty. In GR, we have only a fairly curious 'conservation' theorem. It is not really a conservation theorem because of the covariant derivative; for example, it could suddenly generate positive and negative matter in some place, and it is only the positivity theorems that avoid that happening, but until they came along there was really very little constraint about what empty space could in theory do.
Isenberg: We still have a nice Cauchy problem in the vacuum case and you can specify gravitational-wave energy if you want to call it that.
Bondi: Oh, sure, I'm not against gravitational waves. All I'm saying is you have initial conditions and you have permanent conditions, which are the constitutive equations of matter. What the constitutive equation of empty space is is not quite as clear to me as I would like it to be, though, as I say, the positivity theorems help a lot.
Nojarov: Do you think that an empty space with zero vacuum energy could give birth to something after you assume the positivity principles?
Bondi: I doubt it but I haven't seen it proved.
Ehlers: There is a rather nice theorem in Hawking and Ellis (*The Large-Scale Structure of Space-Time*, p. 94) which is unfortunately hardly ever quoted in textbooks, namely, that if the dominant energy condition and the covariant conservation law hold for the energy tensor, then if that tensor is zero on a Cauchy surface and you have a globally hyperbolic spacetime, it can never be nonzero off the Cauchy surface. In that sense, positivity of $T^{\alpha\beta}$ is essential for conservation of matter, irrespective of field equations.

8. Quantum Gravity

Introduction

There are two principal justifications for this final chapter on the connection between Mach's Principle and quantum gravity: the problem of the origin of particle rest masses and the quantum implications of the basic dynamical structure of general relativity when cast into the ADM Hamiltonian form.

Although it was Einstein rather than Mach who said that a Machian dynamics must realize some kind of cosmic derivation of the inertial mass (cf. pp. 92 and 180), the issue of mass clearly does belong in a volume such as this (see, in particular, the Nordtvedt–Goenner dialogue on pp. 456–458). With his quantum proposal for a gravitational Higgs-type generation of mass, Dehnen attacks this problem head-on. Even though Dehnen admits his work is only at an initial stage, we should surely "give rein" to such thoughts (p. 110).

If it is true that Einsteinian geometrodynamics is perfectly Machian (p. 214), canonical quantum gravity (p. 501) cannot fail to inherit this property. This was why we had planned to have one of the review talks on the quantum-gravity implications of the ADM super-momentum and super-Hamiltonian constraints, including the issue of time. Unfortunately, Lee Smolin was prevented at the last minute from giving his talk, and instead we had a general discussion (p. 501), which we hope is a reasonable substitute.

Giulini's paper was a bit difficult to place – it could also have been in Chap. 3. The mathematical issues he addresses are sure to play a role in the much to be desired clarification of the classical thin-sandwich conjecture (p. 207) as well as in any work in quantum gravity in the so-called metric representation.

J.B.B.

Einstein Studies, vol. 6: Mach's Principle: From Newton's Bucket to Quantum Gravity, p. 478

The Higgs Field and Mach's Principle of Relativity of Inertia

Heinz Dehnen

1. Introduction

At first I want to emphasize that Mach's 'Principle of relativity of inertia' is not only a problem of classical mechanics and general relativity but essentially also a question for the theory of elementary particles, namely: Where does the mass of any body really come from? And, indeed, within the modern standard model of elementary particle physics, all particles are massless from the very beginning, and mass is introduced subsequently by an interaction, namely, by the interaction with the Higgs field. This behavior is exactly a Machian one. Furthermore, I shall show here that the interaction mediated by the (excited) Higgs field is some kind of gravitational interaction, as was Einstein's intention when he first proposed Mach's Principle (1913, 1917).

2. Higgs-Gravitational Force and Potential Equation

We perform our calculations in full generality with the use of an U(N) model of elementary particles and start from the Lagrange density L of fermionic fields coupled to the Higgs field, both belonging to the localized group U(N) ($c=1$, $\eta_{\mu\nu}=diag(1, -1, -1, -1)$):

$$L=\frac{\hbar}{2}i\bar{\psi}\gamma^{\mu}D_{\mu}\psi+h.c-\frac{\hbar}{16\pi}F^{a}_{\lambda\mu}F^{\lambda\mu}_{a}$$

$$+\frac{1}{2}(D_{\mu}\phi)^{\dagger}D^{\mu}\phi-\frac{\mu^{2}}{2}\phi^{\dagger}\phi-\frac{\lambda}{4!}(\phi^{\dagger}\phi)^{2}-k\bar{\psi}\phi^{\dagger}\hat{x}\psi+h.c. \tag{2.1}$$

(μ^2, λ, k are real parameters of the Higgs potential). Here D_{μ} represents

Einstein Studies, vol. 6: Mach's Principle: From Newton's Bucket to Quantum Gravity, pp. 479–490

the covariant derivative with respect to the localized group U(N)

$$D_\mu = \partial_\mu + igA_\mu \tag{2.1a}$$

[g is the gauge coupling constant, $A_\mu = A^a_\mu \tau_a$ are the gauge potentials, τ_a are the generators of the group U(N)], and the gauge field strength $F_{\mu\nu}$ is determined by its commutator ($F_{\mu\nu} = (1/ig)[D_\mu, D_\nu] = F^a_{\mu\nu}\tau_a$); furthermore $\mathfrak{K}$ is the Yukawa coupling-matrix. For application of the Lagrange density (2.1) to a special model, for example, the Glashow–Salam–Weinberg model or even the GUT-model, the wave function ψ, the generators τ_a, the Higgs field ϕ, and the coupling matrix $\mathfrak{K}$ must be specified explicitly (Dehnen and Frommert 1991).

From (2.1) we get immediately the field equations for the spinorial matter fields (ψ fields):

$$i\gamma^\mu D_\mu \psi - \frac{k}{\hbar}(\phi^\dagger \mathfrak{K} + \mathfrak{K}^\dagger \phi)\psi = 0, \tag{2.2}$$

the Higgs field ϕ

$$D^\mu D_\mu \phi + \mu^2 \phi + \frac{\lambda}{3!}(\phi^\dagger \phi)\phi = -2k\bar{\psi}\mathfrak{K}\psi \tag{2.3}$$

and the gauge fields $F^{a\mu\lambda}$

$$\partial_\mu F^{a\mu\lambda} + igf^a{}_{bc} A^{b\mu} F^{c\lambda}_\mu = 4\pi j^{a\lambda} \tag{2.4}$$

with the gauge current density

$$j^{a\lambda} = g\left(\bar{\psi}\gamma^\lambda \tau^a \psi + \frac{i}{2\hbar}[\phi^\dagger \tau^a D^\lambda \phi - (D^\lambda \phi)^\dagger \tau^a \phi]\right). \tag{2.4a}$$

Here $f^a{}_{bc}$ are the totally skew-symmetric structure constants of the group U(N). The gauge-invariant canonical energy–momentum tensor reads with the use of (2.2)

$$\begin{aligned} T^\mu_\lambda = {} & \frac{i\hbar}{2}\left[\bar{\psi}\gamma^\mu D_\lambda \psi - (\overline{D_\lambda \psi})\gamma^\mu \psi\right] - \frac{\hbar}{4\pi}\left[F^a_{\lambda\nu}F^{\mu\nu}_a - \frac{1}{4}\delta^\mu_\lambda F^a_{\alpha\beta}F^{\alpha\beta}_a\right] \\ & + \frac{1}{2}\Big[(D_\lambda \phi)^\dagger D^\mu \phi + (D^\mu \phi)^\dagger D_\lambda \phi \\ & - \delta^\mu_\lambda\{(D_\alpha \phi)^\dagger D^\alpha \phi - \mu^2 \phi^\dagger \phi - \frac{2\lambda}{4!}(\phi^\dagger \phi)^2\}\Big] \end{aligned} \tag{2.5}$$

and fulfills the conservation law

$$\partial_\mu T^\mu_\lambda = 0. \tag{2.6}$$

Obviously the current density (2.4a) has a gauge-covariant matter-field and Higgs-field part, i.e., $j^{a\lambda}(\psi)$ and $j^{a\lambda}(\phi)$, respectively, whereas the energy–momentum tensor (2.5) consists of a sum of three gauge-invariant parts:

$$T^{\mu}_{\lambda} = T^{\mu}_{\lambda}(\psi) + T^{\mu}_{\lambda}(F) + T^{\mu}_{\lambda}(\phi), \tag{2.7}$$

represented by the brackets on the right-hand side of Eq. (2.5).

With a view to analyzing the interaction caused by the Higgs field, we investigate at first the equation of motion for the expectation value of the 4-momentum of the matter fields and the gauge fields. From (2.6) and (2.7) one finds, neglecting surface integrals at spacelike infinity:

$$\partial_0 \int [T^0_{\lambda}(\psi) + T^0_{\lambda}(F)] d^3x = -\int \partial_{\mu} T^{\mu}_{\lambda}(\phi) d^3x. \tag{2.8}$$

Insertion of $T^{\mu}_{\lambda}(\phi)$ according to (2.5) and elimination of the second derivatives of the Higgs field by the field equations (2.3) results in:

$$\begin{aligned} \frac{\partial}{\partial t} \int [T^0_{\lambda}(\psi) + T^0_{\lambda}(F)] d^3x &= k \int \bar{\psi} [(D_{\lambda}\phi)^{\dagger} \mathfrak{x} \\ + \mathfrak{x}^{\dagger}(D_{\lambda}\phi)] \psi d^3x &+ \frac{ig}{2} \int F^{a}_{\mu\lambda} [\phi^{\dagger} \tau_a D^{\mu}\phi - (D^{\mu}\phi)^{\dagger} \tau_a \phi] d^3x. \end{aligned} \tag{2.9}$$

The right-hand side represents the expectation value of the 4-force, which causes the change of the 4-momentum of the ψ fields and the F fields with time. However, the last expression can be rewritten with the use of the field equations (2.4) as follows:

$$\partial_{\mu} T^{\mu}_{\lambda}(F) = \hbar F^{a}_{\mu\lambda} (j^{\mu}_{a}(\psi) + j^{\mu}_{a}(\phi)). \tag{2.9a}$$

One then obtains instead of (2.9):

$$\begin{aligned} \frac{\partial}{\partial t} \int T^0_{\lambda}(\psi) d^3x &= \int \hbar F^{a}_{\lambda\mu} j^{\mu}_{a}(\psi) d^3x \\ &+ k \int \bar{\psi} [(D_{\lambda}\phi)^{\dagger} \mathfrak{x} + \mathfrak{x}^{\dagger}(D_{\lambda}\phi)] \psi d^3x, \end{aligned} \tag{2.10}$$

where on the right-hand side we have the Lorentz-like 4-force of the gauge-field and the Higgs-field force, both acting on the matter field. Evidently, the gauge-field strength couples to the gauge currents, i.e., to the gauge-coupling constant g according to (2.4a), whereas the Higgs-field strength (gradient of the Higgs field) couples to the fermionic mass parameter k. This fact points to a *gravitational* action of the scalar Higgs field.

a) Gravitational Interaction at the Level of the Field Equations. To demonstrate the gravitational interaction explicitly, we perform at first the spontaneous symmetry breaking, because in the case of a scalar gravity only massive particles should interact. [The only possible source of a scalar gravity is the trace of the energy momentum tensor, see (2.24).] For this $\mu^2<0$ must be valid, and according to (2.3) and (2.5), the ground state ϕ_0 of the Higgs field is defined by

$$\phi_0^\dagger\phi_0 = v^2 = \frac{-6\mu^2}{\lambda}, \tag{2.11}$$

which we resolve as

$$\phi_0 = vN \tag{2.12}$$

with

$$N^\dagger N = 1, \ \partial_\lambda N \equiv 0. \tag{2.12a}$$

The general Higgs field ϕ differs from (2.12) by a local unitary transformation:

$$\phi = \rho UN, \ U^\dagger U = 1 \tag{2.13}$$

with

$$\phi^\dagger\phi = \rho^2, \ \rho = v + \eta, \tag{2.13a}$$

where η represents the real-valued excited Higgs field.

Now we use the possibility of a unitary gauge transformation which is inverse to (2.13):

$$\phi' = U^{-1}\phi, \ \psi' = U^{-1}\psi, \ F'_{\mu\nu} = U^{-1}F_{\mu\nu}U, \tag{2.14}$$

so that

$$\phi' = \rho N, \tag{2.14a}$$

and perform in the following all calculations in the gauge (2.14) (unitary gauge). For this we note that in the case of the symmetry breaking of the group G

$$G \to \tilde{G} \tag{2.15}$$

where $\tilde{G}$ represents the rest-symmetry group, we decompose the unitary transformation:

$$U = \hat{U} \cdot \tilde{U}, \ \tilde{U} \in \tilde{G}, \in \hat{U} \in G/\tilde{G} \tag{2.15a}$$

with the isotropy property ($\tau_{\tilde{a}}$ are the generators of the unbroken symmetry):

$$\tilde{U}N = \exp(i\lambda^{\tilde{a}}\tau_{\tilde{a}})N = N, \tag{2.16}$$

so that

$$\tau_{\tilde{a}}N = 0 \tag{2.17}$$

is valid. For $\hat{U}$ we write $\hat{U} = \exp(i\lambda^{\hat{a}}\lambda_{\hat{a}})$, where $\tau_{\hat{a}}$ are the generators of the broken symmetry.

With allowance for (2.12)–(2.17), the field equations (2.2)–(2.4) take the form, with the primes introduced in (2.14) omitted:

$$i\gamma^{\mu}D_{\mu}\psi - \frac{\hat{m}}{\hbar}(1+\varphi)\psi = 0, \tag{2.18}$$

$$\partial_{\mu}F_{a}^{\mu\lambda} + igf_{abc}A^{b\mu}F_{\mu}^{c\lambda} + \frac{1}{\hbar^{2}}M_{ab}^{2}(1+\varphi)^{2}A^{b\lambda} = 4\pi j_{a}^{\lambda}(\psi), \tag{2.19}$$

$$\partial^{\mu}\partial_{\mu}\varphi + \frac{M^{2}}{\hbar^{2}}\varphi + \frac{1}{2}\frac{M^{2}}{\hbar^{2}}(3\varphi^{2}+\varphi^{3}) =$$

$$-\frac{1}{v^{2}}\left[\bar{\psi}\hat{m}\psi - \frac{1}{4\pi\hbar}M_{ab}^{2}A_{\lambda}^{a}A^{b\lambda}(1+\varphi)\right], \tag{2.20}$$

where $\varphi = \eta/v$ represents the excited Higgs field,

$$\hat{m} = kv(N^{\dagger}\hat{x} + \hat{x}^{\dagger}N) \tag{2.18a}$$

is the mass matrix of the matter field (ψ field),

$$M_{ab}^{2} = M_{\hat{a}\hat{b}}^{2} = 4\pi\hbar g^{2}v^{2}N^{\dagger}\tau_{(\hat{a}}\tau_{\hat{b})}N \tag{2.19a}$$

is the symmetric matrix of the mass square of the gauge fields ($A_{\mu}^{\hat{a}}$ fields), and

$$M^{2} = -2\mu^{2}\hbar^{2}, \ (\mu^{2} < 0) \tag{2.20a}$$

is the square of the mass of the Higgs field (φ field). It is obvious that in the field equations (2.18)–(2.20) the Higgs field φ plays the role of an attractive scalar gravitational potential between those particles which become *massive* by the Higgs field in consequence of the spontaneous symmetry breaking [Mach's Principle (!)]: According to Eq. (2.20), the source of φ is the mass of the fermions and of the gauge bosons [the second term in the bracket on the right-hand side of Eq. (2.20) is positive with respect to the signature of the metric], whereby this equation linearized with respect to φ is a potential equation of Yukawa type. Accordingly, the potential φ has a finite range

$$l = \hbar/M \tag{2.21}$$

given by the mass of the Higgs particle, and v^{-2} has the meaning of the gravitational constant, so that

$$v^{-2} = 4\pi G\gamma \tag{2.22}$$

is valid, where G is the Newtonian gravitational constant and γ a *dimensionless* factor, which compares the strength of the Newtonian gravity with that of the Higgs field and which can be determined only experimentally; see Sec. 3. On the other hand, the gravitational potential φ acts back on the mass of the fermions and the gauge bosons according to the field equations (2.18) and (2.19). Simultaneously, the equivalence between inertial and passive as well as active gravitational mass is guaranteed. This feature results from the fact that by the symmetry breaking only *one* type of mass is introduced.

b) Gravitational Interaction at the Level of the Momentum Law. At first we consider the potential equation from a more classical standpoint. With respect to the fact of a *scalar* gravitational interaction, we rewrite Eq. (2.20) with the help of the trace of the energy-momentum tensor, because this should be the only source of a scalar gravitational potential within a Lorentz-covariant theory. From (2.5) one finds after symmetry breaking in analogy to (2.7):

$$T^{\mu}_{\lambda}(\psi) = \frac{i\hbar}{2}[\bar{\psi}\gamma^{\mu}D_{\lambda}\psi - (\overline{D_{\lambda}\psi})\gamma^{\mu}\psi], \tag{2.23a}$$

$$T^{\mu}_{\lambda}(A) = -\frac{\hbar}{4\pi}\left(F^{a}_{\lambda\nu}F^{\mu\nu}_{a} - \frac{1}{4}\delta^{\mu}_{\lambda}F^{a}_{\alpha\beta}F^{\alpha\beta}_{a}\right)$$
$$+\frac{1}{4\pi\hbar}(1+\varphi)^{2}M^{2}_{ab}\left(A^{a}_{\lambda}A^{b\mu} - \frac{1}{2}\delta^{\mu}_{\lambda}A^{a}_{\nu}A^{b\nu}\right), \tag{2.23b}$$

$$T^{\mu}_{\lambda}(\varphi) = v^{2}\left[\partial_{\lambda}\varphi\partial^{\mu}\varphi - \frac{1}{2}\delta^{\mu}_{\lambda}\{\partial_{\alpha}\varphi\partial^{\alpha}\varphi\right.$$
$$\left. + \frac{M^{2}}{4\hbar^{2}}(1+\varphi)^{2}(1-2\varphi-\varphi^{2})\}\right]. \tag{2.23c}$$

From this it follows immediately using the field equation (2.18):

$$T = T^{\lambda}_{\lambda} = \bar{\psi}\hat{m}\psi(1+\varphi) - \frac{1}{4\pi\hbar}M^{2}_{ab}A^{a}_{\lambda}A^{b\lambda}(1+\varphi)^{2}$$
$$+ v^{2}\left[\frac{M^{2}}{2\hbar^{2}}(\varphi^{4}+4\varphi^{3}+4\varphi^{2}-1) - \partial_{\lambda}\varphi\partial^{\lambda}\varphi\right]. \tag{2.23d}$$

The comparison with Eq. (2.20) shows that the source of the potential φ is given by the first two terms of (2.23d), i.e., by $T(\psi)$ and $T(A)$ as expected. In this way we obtain as potential equation using (2.22):

$$\partial^{\mu}\partial_{\mu}\varphi + \frac{M^{2}}{\hbar^{2}}\varphi + \frac{1}{2}\frac{M^{2}}{\hbar^{2}}(3\varphi^{2}+\varphi^{3})$$
$$= -4\pi G\gamma(1+\varphi)^{-1}(T(\psi)+T(A)). \tag{2.24}$$

In the linearized version (with respect to φ), Eq. (2.24) represents a potential equation for φ of Yukawa type with the trace of the energy–momentum tensor of the massive fermions and the massive gauge bosons as source.

Finally, we investigate the gravitational force caused by the Higgs field in more detail. Insertion of the symmetry breaking according to (2.12)–(2.17) into the first integral on the right-hand side of (2.9) yields:

$$\begin{aligned} K_\lambda &\equiv k\bar{\psi}[D_\lambda\phi)^\dagger \mathfrak{K} + \mathfrak{K}^\dagger(D_\lambda\phi)]\psi \\ &= \bar{\psi}\hat{m}\psi\partial_\lambda\varphi + v(1+\varphi)[(D_\lambda N)^\dagger k\bar{\psi}\mathfrak{K}\psi + k\bar{\psi}\mathfrak{K}^\dagger\psi D_\lambda N]. \end{aligned} \tag{2.25}$$

Replacement of $k\bar{\psi}\mathfrak{K}\psi$ by the left-hand side of the field equation (2.3) results with the use of (2.13a) and (2.14a) in:

$$\begin{aligned} K_\lambda = &\left[\bar{\psi}\hat{m}\psi - \frac{1}{4\pi\hbar}M^2_{ab}A^a_\mu A^{b\mu}(1+\varphi)\right]\partial_\lambda\varphi \\ &-\frac{1}{4\pi\hbar}\partial_\mu\left[(1+\varphi)^2M^2_{ab}(A^a_\lambda A^{b\mu} - \frac{1}{2}\delta^\mu_\lambda A^a_\nu A^{b\nu})\right] \\ &+\frac{v^2}{2}ig(1+\varphi)^2F^a_{\lambda\mu}[N^\dagger\tau_a D^\mu N - (D^\mu N)^\dagger\tau_a N]. \end{aligned} \tag{2.26}$$

When we substitute (2.26) into the right-hand side of (2.9), the last term of (2.26) cancels against the last term of (2.9), whereas the second term of (2.26) can be combined with $\partial_\mu T_\lambda{}^\mu(F)$ to $\partial_\mu T_\lambda{}^\mu(A)$ according to (2.23b). In this way we obtain, neglecting surface integrals at spacelike infinity:

$$\begin{aligned} &\frac{\partial}{\partial t}\int [T^0_\lambda(\psi) + T^0_\lambda(A)]d^3x \\ &= \int\left[\bar{\psi}\hat{m}\psi - \frac{1}{4\pi\hbar}M^2_{ab}A^a_\mu A^{b\mu}(1+\varphi)\right]\partial_\lambda\varphi d^3x. \end{aligned} \tag{2.27}$$

In total analogy with the procedure yielding the potential equation (2.24), we replace the bracket of the 4-force in (2.27) by the traces $T(\psi)$ and $T(A)$ given by (2.23d):

$$\begin{aligned} &\frac{\partial}{\partial t}\int [T^0_\lambda(\psi)+T^0_\lambda(A)]d^3x \\ &= \int (1+\varphi)^{-1}[T(\psi)+T(A)]\partial_\lambda\varphi d^3x. \end{aligned} \tag{2.28}$$

Considering the transition from Eq. (2.9) to (2.10), we can express the time derivative of the 4-momentum of the gauge fields by a 4-force acting on the matter currents. Restricting this procedure to the *massless* gauge fields, we get from (2.28):

$$\frac{\partial}{\partial t}\int [T^0_\lambda(\psi)+T^0_\lambda(A^{\hat{a}}_o]d^3x = \int \hbar F^{\hat{a}}_{\lambda\mu}j^\mu_{\hat{a}}(\psi)d^3x$$

$$+\int (1+\varphi)^{-1}[T(\psi)+T(A_\sigma^a)]\partial_\lambda \varphi d^3x. \tag{2.29}$$

Here the first term on the right-hand side describes the Lorentz-like 4-force of the *massless* gauge bosons acting on the matter field coupled by the gauge-coupling constant g [see (2.4a)], whereas the second term [identical to the right-hand side of (2.28)] is the attractive gravitational force of the Higgs field φ acting on the masses of the fermions and the gauge bosons, which are simultaneously the source of the Higgs potential φ according to (2.24). This behavior is exactly that of classical gravity, coupling to the mass ($\equiv$ energy) only and not to any charge. However, the qualitative difference with respect to the Newtonian gravity consists besides the nonlinear terms in (2.24) in the finite range of φ caused by the Yukawa term.

3. Conclusion

We want to point to some interesting features of our result. First of all, we note that in view of the right-hand side of (2.28) and (2.29), it is appropriate to define

$$\ln(1+\varphi)=\chi \tag{3.1}$$

as a new gravitational potential, so that the momentum law (2.28) reads:

$$\frac{\partial}{\partial t}\int [T_\lambda^0(\psi)+T_\lambda^0(A)]d^3x=\int [T(\psi)+T(A)]\partial_\lambda \chi d^3x. \tag{3.2}$$

Then the nonlinear terms concerning φ in (2.24) can be expressed by $T(\varphi)\equiv T(\chi)$ according to the third term of the right-hand side of (2.23d). In this way, the field equation for the potential χ (excited Higgs field) takes the very impressive form:

$$\partial_\mu \partial^\mu e^{2\chi}+\frac{M^2}{\hbar^2}e^{2\chi}=-8\pi G\gamma[T(\psi)+T(A)+T(\chi)]. \tag{3.3}$$

Equations (3.2) and (3.3) are indeed those of an attractive Lorentz-covariant scalar gravity with self-interaction in a natural manner. We note that also the inversion is valid: Postulate a scalar self-interacting gravity, and then the Higgs Lagrangian follows uniquely (Dehnen and Frommert 1990).

For the understanding of the Higgs field, it may be of interest that the structure of Eq. (3.3) exists already before the symmetry breaking. Considering the trace T of the energy–momentum tensor (2.5), one finds with the use of the field equations (2.2) and (2.3):

$$\partial_\mu \partial^\mu (\phi^\dagger \phi) + \frac{M^2}{\hbar^2}(\phi^\dagger \phi) = -2T \tag{3.4}$$

with $M^2 = -2\mu^2\hbar^2$. Accordingly, the Yukawa-like self-interacting scalar gravity of the Higgs field is present within the theory from the very beginning. Equation (3.4) possesses an interesting behavior with respect to the symmetry breaking. Then from the second term on the left-hand side there results in view of (2.11) in the first step a cosmological constant $M^2 v^2/\hbar^2$, but this is compensated exactly by the trace of the energy–momentum tensor of the ground state. It is our opinion that this is the meaning of the cosmological constant quite generally, also in general relativity.

Furthermore, because in (2.21) the mass M is that of the Higgs particle, the range l of the potential φ should be very short, so that until now no experimental evidence for the Higgs gravity may exist, at least in the macroscopic limit. For this reason it also appears improbable that it has to do something with the so-called fifth force (Eckhardt *et al.* 1988).

Finally, the factor γ in (2.22) can be estimated as follows: Taking into consideration the unified theory of electroweak interaction, the value of v [see (2.19a)] is correlated with the mass M_W of the W bosons according to $v^{-2} = \pi g_2^2 \hbar / M_W^2$ [g_2=gauge-coupling constant of the group SU(2)]. Combination with (2.22) results in

$$\gamma = \frac{g_2^2}{2}\left(\frac{M_P}{M_W}\right)^2 = 2 \times 10^{32} \tag{3.5}$$

(M_P is the Planck mass). Consequently, the Higgs gravity represents a relatively strong scalar gravitational interaction between massive elementary particles, but with extremely short range and with the essential property of quantizability. If any Higgs field exists in nature, this gravity is present.

On the other hand, the expression (3.5) shows that in the case of a symmetry breaking where the bosonic mass is of the order of the Planck mass, the Higgs gravity approaches the Newtonian gravity if the mass of the Higgs particle is sufficiently small. In this connection the interesting question arises whether it is possible to construct a theory of gravity with the use of the Higgs mechanism, leading eventually to Einstein's tensorial gravitational theory in the classical macroscopic limit.

This question is of special interest for the following reason: Within the solar system and the binary pulsar PSR 1913+16 the classical gravitational interaction is described very well by Einstein's general relativity. However, this theory has not proved to be quantizable until now. On the other hand, all the other fundamental interactions and their unifications are described successfully by the quantizable gauge theories with unitary gauge groups. Therefore the suspicion exists that Einstein's theory represents only a classical macroscopic description of gravity and that the fundamental microscopic gravitational interaction between elementary particles is also described by a unitary gauge group in such a way that Einstein's theory of macroscopic gravity is reached as an effective theory within a certain classical limit (see also Stumpf 1988) in the same way that in the strong interaction the nuclear forces follow from quantum chromodynamics. In this way, the problem of quantization of gravity as well as its unification with the other interactions would be possible. For implementation of this idea, our statement may be important that the scalar Higgs field, the basis of which is, of course, provided by unitary transformation groups, mediates a Lorentz-invariant attractive gravitational interaction. However, to reach Einstein's tensorial gravity in the classical limit, we need a more sophisticated Higgs field than only a scalar one, the ground state of which is correlated with the Minkowski metric. If this concept will be successful, then Einstein's idea of Mach's principle of relativity of inertia contains even the key of quantum gravity and of the unification with all other interactions.

REFERENCES

Einstein, Albert (1913). "Zum gegenwärtigen Stande des Gravitationsproblems." *Physikalische Zeitschrift.* 14: 1260–1261.

Einstein, Albert (1917). "Kosmologische Betrachtungen zur allgemeinen Relativitätstheorie." *Sitzungsberichte der Preussischen Akademie der Wissenschaften zu Berlin*: 142–152.

Dehnen, Heinz and Frommert, Hartmut (1991). "Higgs-Field Gravity Within the Standard Model." *International Journal of Theoretical Physics.* 30: 985–998.

Dehnen Heinz and Frommert, Hartmut (1990). "Scalar Gravity and Higgs Potential." *International Journal of Theoretical Physics.* 29: 361–370.

Eckhardt, Donald *et al.* (1988). "Tower Gravity Experiment: Evidence for Non-Newtonian Gravity." *Physical Review Letters.* 60: 2567–2570.

Stumpf, Harald (1988). "Gravitation as a Composite Particle Effect in a Unified Spinor–Isospinor Preon Field Model I." *Zeitschrift für Naturforschung.* 43a: 345–359.

Discussion

Goenner: You belong to the minority of people that want to reduce the gravitational interaction and to remove it from the realm of the fundamental interactions, and sort of have it as an effective interaction, generated by the other interactions. In your particular case, are these generated by the Higgs field?

Dehnen: Yes, I think the Higgs field describes the microscopic gravitational interaction, and then, of course, one has to try to get in a macroscopic limit Einstein's tensorial theory as an effective theory. That is the program. And I also have an idea in which way one can find the gauge group which is here involved.

Goenner: I think a lot of people should jump up here, because with gravitation being reduced to a second-class interaction they lose work [laughter].

Dehnen: No, no, this is a microscopic approach, and that approach has some meaning.

Goenner: But what about quantum gravity?

Dehnen: Yes, classical gravity must not be quantized.

Ehlers: If I get it, you would like that the ordinary observed gravity comes about in a similar way in which in chromodynamics people would like to understand nuclear interactions.

Dehnen: Yes, in the same way as the nuclear forces are a van der Waals interaction resulting from what you call chromodynamics.

Unknown: Why do you call that interaction gravity?

Dehnen: Because the interaction couples only to the masses and not to any charges. That's the reason.

Giulini: What is the role of background geometry?

Dehnen: Minkowski spacetime.

Giulini: Yes, but in the end the gravitation seems to be an effective field.

Dehnen: Yes.

Giulini: And it is actually only this effective field that you measure with particles?

Dehnen: Yes, with macroscopic particles and clocks and so on. It is a double metric field, because we have the background Euclidean metric and the effective non-Euclidean metric over it.

Giulini: And particles only probe the effective metric.

Dehnen: Of course, the particles see this Higgs field, and the action of the Higgs field on the particles is as if there were a non-Euclidean geometry.

Giulini: And, for example, you set up quantum theory and you take commutation relations into account. What becomes of microcausality? Is it with respect to the effective metric or the background?

Dehnen: The quantization of this theory acts on the background, on the Minkowski spacetime. The whole quantization is performed on the Minkowski spacetime, of course, and the classical metric has only a macroscopic meaning.

Kuchař: Your Higgs field influences the rest masses of various particles that enter into your Lagrangian, but it doesn't influence, it seems, say, the energy of the electromagnetic field.

Dehnen: Yes. No light deflection.

Kuchař: How do you then explain the result of the Eötvös experiment? We believe that all those fields somehow contribute to the inertia of the body. What is the scheme that you have in mind?

Dehnen: I have in mind that light deflection and so on, that means the action of the Higgs field on the massless gauge bosons, only arises if one performs the unification of this gravity theory of unitary gauge groups with the other interactions. If you have a gauge group, you can unify.

Kuchař: So is this only a provisional scheme?

Dehnen: Of course; it is a proposal. But I can say perhaps the gauge group which I have in mind is the following: The unitary transformations between the different representations of the Pauli matrices or of the Dirac matrices can be gauged. Then you get gauge bosons and additionally the Dirac or Pauli matrices become function-valued. Then you have to treat them as real fields, and it is clear you have to take a Higgs Lagrangian, because these fields have a nontrivial ground state, namely the constant standard representations. That, I think, is the unitary gauge group for gravity, also because the Dirac matrices are the formal square root of the metric. That means this gauge represents the gauge of the unitary group which is connected with the square root of the metric, and therefore I have the feeling that this has something to do with gravity; then one can try to unify this gravitational interaction with the electroweak one perhaps in a high-dimensional spin–isospin space which is separated by spontaneous symmetry breaking in the electroweak and gravitational interaction. However, this is only an idea; it is not performed until now.

Geometric Structures on Superspace

Domenico Giulini

1. Introduction

As is well known, the dynamics of general relativity can be formulated in terms of a constrained Hamiltonian system, with the configuration space for pure gravity being given by the space of all Riemannian metrics on a 3-dimensional manifold Σ of fixed but arbitrary topology. We call this space $Q(\Sigma)$ to indicate its dependence upon the choice of Σ. In this Hamiltonian picture, spacetime is looked upon as a history of dynamically evolving geometries on Σ represented by a path $g_{ab}(s)$ in $Q(\Sigma)$. In the special gauge where the lapse function $N=1$ and the shift vector $N^a=0$, the vacuum Einstein equations without cosmological constant decompose into the dynamical part (in units where $16\pi G/c^4=1$; the prime means differentiation with respect to the parameter s)

$$g''_{ab}+\Gamma^{ijkl}_{ab}g'_{ij}g'_{kl}=-2(R_{ab}-\frac{1}{4}g_{ab}R), \tag{1}$$

and the constraint part

$$G^{abcd}g'_{ab}g'_{cd}-4\sqrt{g}R=0 \text{ (Hamiltonian Constraint)}, \tag{2}$$

$$G^{abcd}\nabla_b g'_{cd}=0 \text{ (Momentum Constraint)}. \tag{3}$$

Here, ∇^a is the Levi-Civita connection for the metric g_{ab}, whose corresponding Ricci tensor and Ricci scalar we denoted by R^{ab} and R, respectively, and G^{abcd} is the DeWitt metric (DeWitt 1967) on the space of symmetric positive-definite matrices (defined below as G^{abcd}_β for $\beta=1$). The Γ-symbols in (1) are the Christoffel symbols for the DeWitt metric. If (2) and (3) are satisfied initially, it follows from (1) that they continue to be satisfied throughout the evolution. Equations (1) and (2) have an obvious geometric interpretation, whereas (3) says that the velocity must be orthogonal to the orbits of the diffeomorphism group. This is

Einstein Studies, vol. 6: Mach's Principle: From Newton's Bucket to Quantum Gravity, pp. 491–500

explained in more detail below.

Due to diffeomorphism invariance, $Q(\Sigma)$ is endowed with an action of the diffeomorphism group $D(\Sigma)$ of Σ: Each point of $Q(\Sigma)$ is a Riemannian metric on Σ which is acted upon by a diffeomorphism via pull-back. Two different metrics which are connected by a diffeomorphism in such a way are considered to be physically indistinguishable. Redundancies of this sort are avoided by going to the quotient $S(\Sigma):=Q(\Sigma)/D(\Sigma)$, called the superspace associated to Σ. It represents the space of geometries rather than metrics on Σ. Although superspace now faithfully labels physical configurations, paths in superspace do not faithfully represent spacetimes. Two *different* paths of geometries may be obtained by 'wafting' Σ differently through the *same* spacetime. This multiplicity is due to the still existing freedom in the choice of the lapse function. Conversely, we quite obviously (for example, by counting degrees of freedom) cannot obtain every path in $S(\Sigma)$ by appropriately 'wafting' Σ through a *given* spacetime.

The existence of some geometric structures of superspace is implicit in many of the investigations into the dynamical structure of general relativity. This is so, for example, in John Wheeler's view of general relativity as geometrodynamics (Wheeler 1968) and the associated quantization program, where superspace serves as domain for the quantum mechanical state functional. The equations to be satisfied by this state functional, the Wheeler–DeWitt equations, explicitly refer to the metric (DeWitt 1967, Wheeler 1968), just like the classical equation (1). Julian Barbour sees the fulfillment of the Machian requirement on general relativity in a successful formulation of dynamics solely within superspace (Barbour 1995). The dynamical principle envisaged is a kind of geodesic equation with respect to some generalized metric on superspace (Barbour 1995). All these attempts provide a motivation to have a closer look at some of the metric structures of superspace. Some partially overlapping observations to those we are going to make have already been discussed in the appendix of (Friedman and Higuchi 1990).

So we first ask: "What geometric structures are there on $Q(\Sigma)$?" Mathematically there is a variety of possibilities to endow $Q(\Sigma)$ with a geometry. On the other hand, the laws of general relativity select a family of such metrics, one for each choice of the lapse function N. For the particular choice $N=1$, this is displayed in Eqs. (1)–(3). They define a metric on $Q(\Sigma)$:

$$G(h,k):=\int_{\Sigma} G^{ab\,cd}h_{ab}k_{cd}\,d^3x, \tag{4}$$

which we call the Wheeler–DeWitt (WDW) metric. In this article we investigate some properties of this particular metric connected with its indefinite nature.

Note that due to the constraint (3), general relativity only uses the WDW metric to calculate inner products on the subspace of tangent vectors satisfying (3), which requires those vectors to be WDW-orthogonal to the directions of the diffeomorphisms. We call the diffeomorphism directions *vertical* and the WDW-orthogonal directions *horizontal*. Due to the indefinite nature of the WDW metric, the horizontal subspace might also contain vertical directions. When this is not the case, the WDW metric restricted to the horizontal subspace defines a metric on the quotient space $S(\Sigma)$. But what generally happens is that in different regions of superspace this quotient-space metric has different signatures. Such signature changes are precisely signaled by nontrivial intersections of vertical with horizontal subspaces. To clarify the WDW geometry of superspace would mean: 1) to characterize the singular set in $Q(\Sigma)$ which consists of those points where horizontal and vertical subspaces intersect nontrivially, and 2) to study the restriction of the WDW-metric to the horizontal subspaces. Only partial results are known so far. Note that we do not consider the constraint equation (2) in the same way as we did with (3). This would select a nonlinear subspace of vectors and thus prevent us from having a pseudo-Riemannian structure. In this respect we deviate from the approach taken by Barbour (Barbour 1995).

What we wish to show here is that the WDW metric has rather special properties. This we do by introducing a 1-parameter family of fiducial metrics of which the WDW metric is one member. The parameter will be called β, and the WDW metric is obtained for $\beta=1$.

2. Ultralocal Metrics

In order to do differential geometry on $Q(\Sigma)$ we heuristically assume that $Q(\Sigma)$ is a differentiable manifold with tangent space $T_g(Q)$ and cotangent space $T_g^*(Q)$ at the metric $g_{ab}\in Q$ (we shall sometimes drop the reference to Σ). Elements of $T_g(Q)$ are any symmetric covariant tensor fields, and elements of $T_g^*(Q)$ are any symmetric contravariant tensor densities of weight one on Σ. Suppose we want to define a metric, i.e., a non-degenerate bilinear form in each $T_g(Q)$. Then, up to an overall constant, there is a unique 1-parameter family of ultralocal metrics (i.e., depending locally on g_{ab} but not on its derivatives) defined in the following way: Take $h,\ k\in T_g(Q)$, then

$$G_\beta(h,k) := \int_\Sigma G_\beta^{abcd} h_{ab} k_{cd} d^3x, \tag{5}$$

where

$$G_\beta^{abcd} = \frac{\sqrt{g}}{2}(g^{ac}g^{bd} + g^{ad}g^{bc} - 2\beta g^{ab}g^{cd}). \tag{6}$$

The WDW metric, introduced in (4), is just G_1. Given $p,\ q \in T_g^*(\Sigma)$, the 'inverse' metric, G_β^{-1}, is

$$G_\beta^{-1}(p,q) := \int_\Sigma G_{abcd}^\beta p^{ab} q^{cd} d^3x, \tag{7}$$

where

$$G_{abcd}^\beta = \frac{1}{2\sqrt{g}}(g_{ac}g_{bd} + g_{ad}g_{bc} - 2\alpha g_{ab}g_{cd}) \tag{8}$$

with

$$\alpha + \beta = 3\alpha\beta, \text{ so that } G_\beta^{abnm} G_{cdnm}^\beta = \frac{1}{2}(\delta_c^a \delta_d^b + \delta_d^a \delta_c^b). \tag{9}$$

These are nondegenerate bilinear forms for $\beta \neq 1/3$ (we exclude $\beta = 1/3$), positive definite for $\beta < 1/3$ and of mixed signature for $\beta > 1/3$ with infinitely many plus as well as minus signs. Because they are ultralocal, they arise from metrics on the space S_3^+ of symmetric positive definite matrices – which is diffeomorphic to the homogeneous space $GL(3,R)/SO(3) \cong R^6$, carrying the metric G_β. One has $GL(3,R)/SO(3) \cong SL(3,R)/SO(3) \times R^+ \cong R^5 \times R^+$, and, with respect to this decomposition, the metric has a simple warped-product form

$$G_\beta^{abcd} dg_{ab} \otimes dg_{cd} = -\epsilon d\tau \otimes d\tau + \frac{\tau^2}{c^2} \mathrm{tr}(r^{-1}dr \otimes r^{-1}dr), \tag{10}$$

with

$$c^2 = 16|\beta - 1/3|,\ \tau = cg^{1/4},\ r_{ab} = g^{-1/3}g_{ab},\ \epsilon = \mathrm{sign}(\beta - 1/3). \tag{11}$$

The matrices r_{ab} are just the coordinates on $SL(3,R)/SO(3)$, and the trace in (10) is just the left-$SL(3,R)$ invariant metric on this space. This gives rise to eight Killing vectors of G_β. An additional homothety is generated by the multiplicative action of R^+ on the τ coordinate. Moreover, geodesics in this metric can be explicitly determined (DeWitt 1967). If we now regard $Q(\Sigma)$ as a mapping space, i.e., as the space of all smooth mappings from Σ into S_3^+, endowed with the metric (5), then, due to its ultralocal nature, geometric structures like Killing fields, homotheties, and geodesics of the 'target' metric (10) are inherited by the full metric (5). For example, dragging the maps $g_{ab}(x)$ along a Killing flow in S_3^+ produces a Killing flow in $Q(\Sigma)$. In this way, some geometry of the infinite dimensional $Q(\Sigma)$ can be studied by looking at the 6-dimensional S_3^+.

Note also that expression (5) is invariant under diffeomorphisms of Σ. An infinitesimal diffeomorphism is represented by a vector field ξ on Σ and gives rise to a vector field X^ξ on $Q(\Sigma)$:

$$X^\xi_{ab} = \nabla_a \xi_b + \nabla_b \xi_a, \tag{12}$$

which is a Killing field of the metric (5). The vectors of the form (12) at $g \in Q(\Sigma)$ span what we call the vertical vector space $V_g \subset T_g(Q)$. With respect to G_β, we can define the orthogonal complement to V_g, which we call the horizontal vector space $H^\beta_g \subset T_g(Q)$. From (5), (6), and (12)

$$k_{ab} \in H^\beta_g \Leftrightarrow \nabla^a(k_{ab} - \beta g_{ab} k^c_c) = 0. \tag{13}$$

Under the isometric action of $D(\Sigma)$ on $Q(\Sigma)$, horizontal spaces are clearly mapped into horizontal spaces.

If we set $\beta=0$, the metric (5) is positive definite, so that orthogonality also implies transversality, i.e., $V_g \cap H^0_g = \{0\}$. It is in fact true that the tangent space splits into the direct sum of closed orthogonal subspaces: $T_g(\Sigma) = V_g \oplus H^0_g$. This allows us to define a Riemannian geometry on the quotient space $S(\Sigma)$ by identifying its tangent spaces with the horizontal spaces in $T(Q)$ (Ebin 1970). [Here we pretend that $S(\Sigma)$ is a genuine manifold.] This works for all $\beta < 1/3$. We are, however, interested in the range $1/3 < \beta \leq 1$ with special attention paid to the transition from $\beta < 1$ to $\beta = 1$.

For $\beta > 1/3$, the metric (5) is no longer definite, so that generally $V_g \cap H^\beta_g \neq \{0\}$ for such β. A simple example is the following: Take as Σ a 3-manifold that carries a flat metric g. In $T_g(\Sigma)$ consider the infinite-dimensional vector subspace given by all vectors of the form $k_{ab} = \nabla_a \nabla_b \phi$, where ϕ is a smooth function on Σ. These vectors satisfy (13) for $\beta=1$ and are therefore in H^1_g. But they are also of the form (12), with $2\xi_a = \nabla_a \phi$, and hence in V_g. Moreover, suppose the metric is only flat in an open subset $U \subset \Sigma$. Then we can repeat the argument, but this time using only functions ϕ with compact support inside U. Again, these give rise to an infinite intersection $V_g \cap H^1_g$ for each such partially flat metric g. Clearly, vectors in $H^\beta_g \cap V_g$ are necessarily of zero G_β-norm.

3. Some Observations Concerning the WDW Metric

It follows from (12) and (13) that a vertical vector X^ξ is horizontal if and only if

$$D_\beta \xi_a := -\nabla^b(\nabla_b \xi_a - \nabla_a \xi_b) - 2(1-\beta)\nabla_a \nabla^b \xi_b - 2R^b_a \xi_b = 0, \tag{14}$$

where R^b_a denote the mixed components of the Ricci tensor. Killing vectors, if existent, are obvious solutions, but these do not interest us since they correspond to zero X^ξ. For $0 \leq \beta < 1/3$, these are the only solutions. This implies that for $\beta > 1/3$ any non-Killing solution must have nonzero divergence, since for zero divergence fields the β dependence in (14) drops out. A more elegant way to write D_β is, using the exterior derivative d, its adjoint δ (given by minus the divergence on the first index), and writing Ric for the map induced by R^b_a:

$$D_\beta = \delta d + 2(1-\beta)d\delta - 2Ric, \tag{15}$$

which also displays its formal self-adjointness. The G_β-norm of X^ξ is

$$G_\beta(X^\xi,\ X^\xi) = 2\int_\Sigma \xi^a D_\beta \xi_a \sqrt{g}\, d^3x. \tag{16}$$

For $\beta \leq 1$ and $Ric < 0$ (i.e., strictly negative eigenvalues), this operator is manifestly positive, and G_β restricted to V_g is thus positive definite. In particular, we have $V_g \cap H^\beta_g = \{0\}$ for all g such that $Ric < 0$ and $\beta \leq 1$. Since it is known that any 3-manifold Σ admits such Ricci-negative metrics (Gao and Yau 1986), this tells us that in every superspace there are open regions (the Ricci-negative geometries) with well-defined WDW metric, given by the restriction of G_1 to H^1_g, whose signature has infinitely many plus and minus signs.

For a flat g and values $\beta < 1$, D_β is nonnegative with kernel given by the covariantly constant ξ. Indeed, from (11) it follows that ξ is curl- and divergence-free on a flat manifold, and hence covariantly constant. But this also means that ξ is Killing, and therefore X^ξ is zero. So for g flat $V_g \cap H^\beta_g = \{0\}$ for $\beta < 1$. On the other hand, for $\beta = 1$ and g flat, we can only infer from (15) that ξ must be closed, hence exact or harmonic. But harmonicity implies Killing, so all horizontal X^ξ are given by the expressions anticipated in the previous section. As stated there, we can localize the construction and obtain an infinite subspace in the intersection $V_g \cap H^1_g$ for metrics g which contain a flat region $U \subset \Sigma$. Clearly, any manifold admits such metrics. In particular, this tells us that in every superspace there are regions where no WDW metric is defined.

It is more difficult to obtain general results for metrics which are neither Ricci-negative nor flat. For the very special class of nonflat Einstein metrics,[1] it is at least easy to see that for $\beta = 1$ $H^1_g \cap V_g$ is zero. Indeed, for $R_{ab} = \lambda g_{ab}$, where $\lambda \in R - \{0\}$, (15) implies $0 = \delta D_1 \xi = 2\lambda \delta \xi$, so that ξ must be divergence-free and hence X_ξ zero. So there exists a WDW metric for nonflat Einstein geometries in $S(\Sigma)$, given by the restriction of G_1 to H^1_g. For the study of such metrics it is instructive to look at a particular example in detail, to which we now turn.

As nonflat Einstein metric, we take the standard round metric on the three-sphere with some unspecified radius. Here $Ric > 0$, and not much can be directly read off (15) for general β. But taking elements of $T_g(Q)$ as first-order perturbations of g, and expanding them in terms of the well-known complete set of tensor harmonics (Gerlach and Sengupta 1978) one can establish the following scenario: For $1/3 < \beta < 1$, the number of negative directions (i.e., the number of linearly independent vectors of negative G_β-norm) is finite in V_g and infinite in H_g^β. For the discrete values $\beta = \beta_n$, where

$$\beta_n := \frac{n^2-3}{n^2-1}, \quad n \in \{3,4,5,\ldots\}, \tag{17}$$

the intersection $V_g \cap H_g^\beta$ is nontrivial and of some finite dimension $d_n > 0$. At other values of β, it is zero. It turns out that when β passes the value β_n from below, d_n of the negative directions change from H_g^β to V_g. Since the β_n accumulate at 1, this happens infinitely often as we turn up β to 1. At $\beta = 1$, only a single negative direction has remained in H_g^1 and infinitely many are now in V_g. The intersection $V_g \cap H_g^1$ is in fact zero, in accordance with the more general argument given above. The metric G_1 restricted to H_g^1 is of Lorentzian signature $(-, +, +, +, \ldots)$. This is directly related to the statement made in quantum cosmology that the Wheeler–DeWitt equation[2] (for constant lapse) for perturbations around the three-sphere is hyperbolic (Halliwell and Hawking 1985). It follows from our considerations that this can at best be locally valid since the metric for constant lapse necessarily suffers from signature changes.[3] Note also how delicately the signature structure of G_β restricted to H_g^β depends on whether $\beta < 1$ or $\beta = 1$.

There are other interesting differences between $\beta < 1$ and $\beta = 1$. Quite striking is the existence of an infinite dimensional intersection $H_g^1 \cap V_g$ for flat g. This means that D_1 cannot be an elliptic operator, since these have finite-dimensional kernels. And, in fact, calculating the principal symbol for D_β from (14), we obtain

$$\sigma_\beta(\zeta)^a_b = \|\zeta\|^2 \left\{ \delta^a_b + (1-2\beta)\frac{\zeta^a \zeta_b}{\|\zeta\|^2} \right\}. \tag{18}$$

This matrix is positive definite for $\beta < 1$, invertible but not positive definite for $\beta > 1$, and singular positive semi-definite for $\beta = 1$. Expressed in standard terminology, the operator D_β is strongly elliptic in the first case, elliptic but not strongly elliptic in the second, and degenerate elliptic but not elliptic in the third. This relates to the problem of how one would actually calculate the metric on superspace at the regular points. Throughout we said that it would be obtained by restricting the metric G_β to the horizontal spaces H_g^β. But this means that we have to calculate explicitly the projection $T_g(Q) \to H_g^\beta$. A general tangent vector $k_{ab} \in T_g(Q)$ is projected by adding a vertical vector X^ξ so that the sum is horizontal, i.e., satisfies (13). This is equivalent to solving

$$D_\beta \xi_b = \nabla^a (k_{ab} - \beta g_{ab} k_c^c) \tag{19}$$

as an equation for ξ with given right-hand side. Uniqueness for X^ξ is given at regular geometries, i.e., those for which the kernel of D_β consists of Killing vectors only. Since the right-hand side is orthogonal to Killing vectors, ellipticity (for $\beta < 1$) guarantees existence for any k_{ab}. It is not clear to us at this moment whether the failure of ellipticity for $\beta = 1$ can in fact imply any problem. For example, in the special cases where g_{ab} is an Einstein metric, we can Hodge decompose ξ and the right-hand side of (19) into exact, co-exact, and harmonic forms. The Einstein condition then prevents the Ricci term in D_1 from coupling these components, so that (19) decomposes into three decoupled equations for the Hodge modes, two purely algebraic ones, and an elliptic partial differential equation for the co-exact mode. In this case we thus regain ellipticity by restricting to appropriate subspaces.

Having seen that $\beta = 1$ is a special value from a mathematical point of view, we might also ask the question of why general relativity picks precisely this value. Suppose we just used the metric G^β_{abcd} for a value $\beta \neq 1$ in the Hamiltonian:

$$H_\beta = \int N(G^\beta_{abcd} \pi^{ab} \pi^{cd} - \sqrt{g}\, R) d^3x - 2 \int N_b \nabla_a \pi^{ab} d^3x \tag{20}$$

$$= H_{\beta=1} + \int \frac{N}{\sqrt{g}} \left[\frac{\beta - 1}{2(3\beta - 1)} \pi^a_a \pi^b_b \right] d^3x. \tag{21}$$

Would this provide just another dynamics for a general relativistic theory of gravitation? The answer is no, due to the well-known uniqueness theorems by Teitelboim and Kuchař (Teitelboim 1979, Kuchař 1972), which state that – up to the cosmological and the gravitational constant – the ordinary gravitational Hamiltonian is uniquely determined by the requirement that the coefficient functions of lapse and shift in the Hamiltonian (i.e., the Hamiltonian and momentum constraint functions) satisfy the standard Poisson bracket relations, which are universally valid for any generally covariant theory (Teitelboim 1973; see also Hojman, Kuchař, and Teitelboim 1974). It is rather easy, in fact, to see how the additional term in (21) alters the Poisson bracket relations between the Hamiltonian constraints. Those involving the momentum constraints are clearly left unchanged. This means that the expression (20) cannot be the

Hamiltonian of a generally covariant theory. In other words, if we evolved some initial data with the Hamiltonians (20) corresponding to *all* possible choices of lapse and shift, the resulting family of evolutions could not be interpreted as describing the *same* spacetime in which the different motions of 3-dimensional hypersurfaces generate the family of evolutions so calculated. In this sense, it is the general covariance of general relativity that picks the value $\beta=1$.

Finally we wish to point out a relation of our problem of solving (19) with the so-called thin-sandwich problem, of which a local version has recently been proven by Bartnik and Fodor (Bartnik and Fodor 1993). It consists in the task of calculating the lapse and shift for freely specified g_{ab} and g'_{ab} such that the metric and its conjugate momentum satisfy the Hamiltonian and momentum constraints. For the special cases in which in this procedure the lapse turns out to be constant (thereby drastically simplifying the otherwise nonlinear equations for the shift vector field), the resulting differential equation for the shift is just our equation (19). This relation is obvious from the geometric meaning of these problems: They both ask for the vertical and horizontal decomposition of a given tangent vector. In the general case of nonconstant lapse functions, the thin-sandwich equations become, however, much more complicated due to the fact that the lapse function is now a functional of g_{ab} and g'_{ab}. This results in nonlinear equations for the general thin-sandwich problem.

NOTES

[1]In three dimensions an Einstein metric implies constant sectional curvature so that Σ is a space form. But not only is the topology of Σ severely restricted (for example, its second homotopy group must be trivial). If Σ allows for Einstein metrics, they only form a finite-dimensional subspace in superspace, which is in fact of dimension one if the Einstein constant is nonzero. In these cases the only deformations are the constant rescalings of the metric. In this sense Einstein metrics are very special.

[2]There is one Wheeler–DeWitt equation for each smearing function. If written without smearing functions (as a distribution), the Wheeler–DeWitt equations for pure gravity look like an infinite number of six-dimensional Klein–Gordon equations, one per point $x\in\Sigma$ for the six components $\{g_{ab}(x)\}$. If added together with a smearing function, the resulting equation is clearly ultrahyperbolic. Only if the directions of differentiation are restricted to lie in a horizontal subspace, or even further, as suggested by Hawking (Hawking 1984, Chap. 5), one may be able to eliminate all but one of the negative directions. In this case, the Wheeler–DeWitt equation corresponding to the

particular choice of smearing function may then be said to be locally hyperbolic on superspace.

[3]In applications, the Wheeler–DeWitt equations have only been studied in neighborhoods of highly symmetric metrics like the one on the three-sphere considered here. It would be interesting to know how 'far' from such a point one has to go in order to encounter singular regions and signature change. The regions $Ric<0$ do not seem 'close,' and the reason why the Wheeler–DeWitt equations have not been studied in neighborhoods of those metrics seems to be the fact that $Ric<0$ metrics do not allow for any symmetries.

REFERENCES

Barbour, Julian B. (1995). Contribution to this volume, see esp. pp. 218–223.

Bartnik, Robert and Fodor, Gyula (1993). "On the Restricted Validity of the Thin Sandwich Conjecture." *Physical Review D* 48: 3596–3599.

DeWitt, Bryce S. (1967). "Quantum Theory of Gravity I. The Canonical Theory." *Physical Review* 160: 1113–1148.

Ebin, David G. (1970). "The Manifold of Riemannian Metrics." In *Proceedings of the American Mathematical Society, Symposia in Pure Mathematics, Global Analysis* 15: 11–40.

Friedman, John L. and Higuchi, Atsushi (1990). "Symmetry and Time on the Superspace of Asymptotically Flat Geometries." *Physical Review D* 41: 2479–2486.

Gerlach, Ulrich H. and Sengupta, Uday K. (1978). "Homogeneous Collapsing Star: Tensor and Vector Harmonics for Matter and Field Asymmetries." *Physical Review D* 18: 1773–1784.

Gao, Zhiyong L. and Yau, Shing-Tung (1986). "The Existence of Ricci Negatively Curved Metrics on Three Manifolds." *Inventiones Mathematicae* 85: 637–652.

Halliwell, Jonathan J. and Hawking, Stephen W. (1985). "The Origin of Structure in the Universe." *Physical Review D* 31: 1777–1791.

Hawking, Stephen W. (1984). "The Quantum State of the Universe." *Nuclear Physics B* 239: 257–276.

Hojman, Sergio A., Kuchař, Karel, and Teitelboim, Claudio (1976). "Geometrodynamics Regained." *Annals of Physics (New York)* 96: 88–135.

Kuchař, Karel (1973). "Geometrodynamics Regained: A Lagrangian Approach." *Journal of Mathematical Physics* 15: 708–715.

Teitelboim, Claudio (1973). "How Commutators of Constraints Reflect the Spacetime Structure." *Annals of Physics (New York)* 79: 542–557.

Wheeler, John A. (1968). "Superspace and the Nature of Quantum Geometrodynamics." In *Battelle Rencontres, 1967 Lectures in Mathematics and Physics*. Cecile M. DeWitt and John A. Wheeler, eds. New York and Amsterdam: W.A. Benjamin Inc., pp. 242–307.

General Discussion: Time, General Relativity, and Quantum Gravity

On the final day of the conference, there were two general discussion sessions, led by Dieter Brill and Jürgen Ehlers, respectively. Edited transcripts of both sessions follow below. It had been intended that the first session should be a debate on the extent to which general relativity is a Machian theory, while the second should concentrate on the implications for quantum gravity of the Machian features of general relativity. In the event, the first topic had already been discussed fairly fully in earlier contributions, especially by Isenberg (p. 188), Barbour (p. 214), Goenner (p. 442), and Ehlers (p. 458), and in the discussions following these contributions, and the quantum aspects already came to the fore in the first general session. Much of the discussions in both sessions revolved around the Machian aspects of time [see p. 102ff, Barbour's contribution, p. 214, and (Barbour 1994b)] in the context of canonical quantization of general relativity. For a good introduction to canonical quantization, the reader is referred to the classical ADM review (Arnowitt *et al.* 1962), Dirac (1965), and Kuchař (1981, 1993). Two recent reviews of the problem of time in quantum gravity are by Kuchař (1992) and Isham (1993). Finally, the reader may wish to consult an earlier volume in the *Einstein Studies* series: *Conceptual Problems of Quantum Gravity* (Ashtekar and Stachel 1991).

Time and the Classical Structure of General Relativity

Brill: Now the time has come for what the program calls general discussion about Machian ideas versus general relativity. I'm quite willing to let the discussion lead us where it may, but personally I would just say I have heard several times now about what may be a controversy or a friendly discussion that has been going on for 15 years between Julian [Barbour] and Karel [Kuchař] so we can all know what this 15-

Einstein Studies, vol. 6: Mach's Principle: From Newton's Bucket to Quantum Gravity, pp. 501–526

year-old discussion is all about. It might be interesting to state the two positions in a short and clear way, so that we can know what that's all about. There's something they've been discussing for 15 years.

Brans: Is there time for that? [laughter]

Kuchař: Why don't you start, Julian, and summarize the 15 years in five minutes?

Barbour: As is only appropriate following this invitation, my remarks concentrate on the notion of time and its relevance to the quantization problem. In what I call the Machian derivation of general relativity (p. 214), to the extent it is possible, I have always felt there was great virtue in the idea of regarding three-dimensional configurations of the world as individual unities, as holistic entities with which we start. A complete three-geometry is then an entity. It is then very easy to see that there are *three* degrees of freedom per space point. For in the tensor g_{ij}, which on the face of it has got six components, it's obvious that three are simply gauge degrees of freedom and merely reflect the three-dimensional coordinate system that you choose to represent the metric, so that leaves you just three degrees of freedom per space point. Now we have to consider the physical significance of the Machian construction of classical spacetime and the recovery of proper time as what I call a local ephemeris time (Barbour 1994b, I).

That local ephemeris time is determined by *all* the degrees of freedom. If we have pure general relativity, it's determined by the three degrees of freedom at each space point, and certain quadratic combinations of them and their velocities are, so to speak, the local energy densities which are used to define the local proper time. If you have other matter degrees of freedom, you add them on – as many extra degrees of freedom as there are extra fields. The local energy density and local ephemeris time are always determined by the totality of local degrees of freedom. In my view, the energy of the system is such a fundamental quantity that it really only makes sense to define time using all the real true degrees of freedom, those that are nontrivially contributing to the dynamics. Then you have this, for me, very satisfying view of time, that all degrees of freedom are contributing to the definition of time, and you choose time in such a way as to make the equations of motion, in the effective spacetime that you build up, take the simplest form possible. This simplest form is nothing to do with the choice of coordinates in spacetime. I am referring to the construction of spacetime and the definition of proper-time separations between three-geometries, all of which is coordinate independent.

If we adopt this view, then we have to conclude that the gravitational field has *three* degrees of freedom per space point, and we are faced with the difficulty of reconciling this with what is undoubtedly a correct result, that in the linearized theory, if you have small deviations from flat spacetime, there is a very close analogy between the linearized gravitational field and electromagnetism, and it truly does appear that the gravitational field, like the electromagnetic field, has only two true degrees of freedom (per space point).

Now I believe that that is an artefact of the linearized theory, and that in truth gravity has three degrees of freedom per space point.

That is, I think, quite the opposite to the reaction that was drawn when Baierlein, Sharp, and Wheeler (1962) found the Jacobi-principle form of general relativity (Barbour 1994b). In fact, the connection with Jacobi's principle was not recognized. To the best of my knowledge, the only place where it is stated in the literature that the BSW form is the Jacobi-principle form of general relativity is in the second paper of Bertotti and myself (Barbour and Bertotti 1982), where we give Karel credit for the observation. Now had the connection with Jacobi's principle been recognized back in 1962, I think it might well have led to the recognition that general relativity is a timeless theory and that this fact perfectly matches Mach's denial of the existence of time.

Instead, a very different interpretation was adopted, namely that time is present in general relativity but is hidden in the spatial degrees of freedom, or perhaps in the phase-space data. One of the three degrees of freedom per space point was supposed to be time and the other two were the true degrees of freedom of the gravitational field. That, of course, is an attractive and more conservative approach than mine, and I would say that for 30 years this has been and still is the majority viewpoint. [For reviews, see (Isham 1993 and Kuchař 1992).] But it's been like Lewis Carroll's *The Hunting of the Snark*: It hasn't yet been found, this decomposition of the degrees of freedom by a special canonical transformation that flushes out a distinguished 'time' and two 'true degrees of freedom.'

However, I feel it's much more natural and inherent in the theory to say simply there are just three degrees of freedom per space point and no time at all and face up squarely to the consequences of timelessness when we try to quantize the theory.

Kuchař: To a large extent, it's the different way in which Julian and I balance our checkbooks. Let me write down what is the count of the variables, and of the constraints. We both work with the three-metric on a given surface, and, if we go into the canonical formalism, we add the momenta. There are six components of the metric for each space point, and six components of the momenta for each space point. Then we count the constraints. There is one super-Hamiltonian constraint for each point, and there are three supermomenta constraints for each point. These impose limitations on the initial data.

Julian's standpoint is to look at the supermomenta constraints, take out three degrees of freedom from the configuration space on their account, and to interpret the remaining three degrees of freedom as dynamical degrees of freedom, taken in their totality.

The trouble is that these degrees of freedom are still subject to the super-Hamiltonian constraint. So you cannot give them freely, together with their conjugate momenta. The conventional wisdom, at least in the quantum theory, is that you cannot treat all those degrees of freedom on an equal footing, because of that remaining constraint.

The paradigm that I have in mind is the motion of a relativistic particle in a curved spacetime. The location of the particle is specified by four spacetime coordinates, and its dynamics is generated by the mass-shell constraint. Not all the four coordinates are dynamical degrees of freedom; one of them has the meaning of time. In classical theory, we do not need to say which is which to determine the spacetime trajectory. In quantum theory, we need to talk about probabilities that the particle is here or there at a given instant. We need to split a time from the position variables, solve the constraint with respect to the momentum conjugate to that time, and obtain thereby the Schrödinger equation which enables us to introduce such probabilities.

The trouble is that the split is ambiguous, and that different splits induce different quantum theories. In stationary spacetimes these theories are equivalent, but in dynamical spacetimes they are not. What we call time changes the physics.

To summarize, in classical theory, I would not quibble too much with Julian, because each split is equivalent to any other split, and indeed one could leave all the variables living happily together and not split the family at all. Unfortunately, at present we do not know how to select a quantum theory which would not, explicitly or implicitly, depend on a split.

Now to the second question, that of the ephemeris time: There is unfortunately no place for it in the quantum scheme. It's the proper-time formalism, if you want, for a relativistic particle. The state function of the relativistic particle never depends on the proper time, which is the analog of the ephemeris time, and we are obliged to interpret what is the state function without having recourse to the ephemeral time.

For a relativistic particle, this ambiguity can be removed by requiring that the mass term be constant; one cannot then afford to scale the mass-shell constraint by a function of position. However, in geometrodynamics the potential term depends on the canonical coordinates, and there is no privileged way of writing the super-Hamiltonian constraint. Indeed, one can do worse things than merely scaling it by a factor. There are infinitely many constraints in general relativity, four per each point of a hypersurface, which one can linearly combine by configuration space factors into an equivalent system of constraints. Each set of new constraints leads to a different ephemeris time in geometrodynamics. I thus fear that ephemeris time is not a well-defined concept, and I do not see any easy way of making it well defined in geometrodynamics.

Barbour: Let me respond first by saying that I do not believe there is a proper analogy between the relativistic particle and the dynamics of general relativity (p. 469–71, 473), but this is a matter that will have to be taken up elsewhere. What I would like to say about the Hamiltonian constraint is that there is a certain folklore, namely, that whenever you have *any* constraints they automatically mean that not all the dynamical variables, not all the configuration variables you are dealing with, are true dynamical degrees of freedom. This is very much in the tradition of the Bergmann–Dirac approach to quantum gravity: Whenever there is a constraint, it means one of the degrees of freedom is not observable. In fact, if you read the wonderful little book of Dirac (1965), *Lectures on Quantum Mechanics,* it's almost entirely general. He very seldom particularizes. He treats the general theory of Hamiltonian systems that have constraints, his approach is obviously largely influenced by electrodynamics, and his aim is to develop the theory, first for electrodynamics, or in the spirit of gauge theory, and then to apply it in quantum gravity.

However, there is an important point that needs to be made, which is that the argument that constraints mean not all degrees of freedom are true degrees of freedom is based on experience gained from electrodynamics, in which you have constraints that are *linear* in the canonical momenta. That is what happens in electrodynamics.

But reparametrization-invariance constraints in truly timeless theories are different. They are quadratic, and I do believe that this general folklore should not be transferred automatically to such constraints.

The discussion in Dirac is about the unique prediction of the future when you have a gauge theory with some unobservable degrees of freedom. In his discussion of this point, Dirac does not ask what you mean by time in such a theory. His discussion of these questions, of how you must handle the gauge degrees of freedom, comes before any discussion of time. In fact, he never actually discusses time in the *Lectures on Quantum Mechanics*. Thus, if you have no time, if the only concrete objective time that you can lay your hand on is the one the astronomers get hold of, which is ephemeris time, the discussion which Dirac gives must be revised, and I think this is a real reason why these quadratic constraints are different. This issue, along with many others raised in this discussion, is discussed in (Barbour 1994b).

Post-conference addendum: If, like Kuchař, you take the view that the Hamiltonian constraint means not all the variables which remain after the three-dimensional diffeomorphisms have been factored out are true degrees of freedom, so that a decomposition into time and so-called true dynamical degrees of freedom must be made, because "from the present standpoint we do not know how to form a quantum theory which would not rely in one way or another on such a split," then, as he himself remarks, you face great difficulties because no distinguished split presents itself and each split leads in principle to a different quantum theory. There is, however, a very simple way out of the dilemma, which is to accept that no split should be made and that no time exists at all. Then the Wheeler–DeWitt equation is like a *time-independent* Schrödinger equation, the quantum theory is unique, and the fact that in the proper-time formalism "the state function of the relativistic particle never depends on the proper time" is no defect but a strength and matches exactly the fact that the Wheeler–DeWitt wave function of the universe depends only on the three-dimensional configuration, not on any time.

When I say that there is no time in the kinematic foundations of geometrodynamics and that the only time one can introduce is ephemeris time, this is perfectly in accord with a static wave function of the universe that depends on no time at all, since ephemeris time is inseparably related to classical trajectories and can therefore only emerge in a WKB situation in which classical trajectories become distinguished. Thus it is exactly right that the wave function cannot depend on ephemeris time.

Finally, I do not understand Karel's claim that because the super-Hamiltonian constraint is not unique and can always be multiplied by an arbitrary position-dependent function "each set of new constraints leads to a different ephemeris time in geometrodynamics." Ephemeris time is defined by choosing a time parameter for which the kinetic term in the action is numerically equal to the potential term (Barbour 1994b). This prescription is entirely independent of any form in which the constraints may happen to be cast.

Giulini: I was a bit surprised you agreed that there is no absolute structure in canonical gravity, because you've got the metric and things like that. But I wanted to ask you about boundary conditions, because you said the constraints reflect what you call Mach's principle. Now if you were, for example, to succeed in the program of canonical quantum gravity, and you had to set up a wave equation on superspace, which would be something like $H\Psi = 0$, you would have to specify what happens to the values of the wave function at the frontiers of the relative configuration space. Now when you do this, it certainly affects the local values of the wave function, and you may have discrete spectra or continuous spectra. So what about these boundary conditions?

Barbour: Certainly, in the quantum theory there's got to be some subsidiary condition on the solution of the Wheeler–DeWitt equation. If it is the right form of quantum gravity, some sort of extra condition is necessary, and we must find out what is appropriate. Personally, I favor exactly the same condition that Schrödinger imposed on his time-independent wave equation in his first papers on wave mechanics: The wave function must be suitably continuous, single valued and bounded. That is a timeless condition – there's no initial condition.

Nojarov: I wanted to make a remark after your talk, but the time was over. A few days ago, I told Sir Hermann [Bondi] that I was wondering why nobody here mentions an important thing, namely, the experimentally measured velocity of the earth with respect to the microwave background. And now he has just mentioned [p. 474] for the first time this very important fact, which is closely related to some of the main subjects of this workshop.

Now I would like to mention a second fact, which was not discussed until now: This is the arrow of time, a very well-known problem. Much was spoken here about operationalist determination of time, about ephemeris time, about astronomical time, but it seems that time is intrinsically related with irreversible processes. That's why I think that time cannot be simply reduced to geometry by a coordinate transformation leading to parametrization of the geodesic point which is well known and is completely correct, can give insight, and can lead to new conclusions. I do not think that time can be kicked out. Please, could you concentrate mainly on this point rather than on more mathematical and formal things, which seem to be more or less correct. This is the main problem I think.

Barbour: I welcome very much your request that we should have the arrow of time come into this. Like Dieter Zeh, I do believe that in the timeless context of the Wheeler–DeWitt equation there is a possibility that we might have the proper framework in which to consider the origin of the arrow of time, so I hope we might go on to that later (p. 524).

Goenner: Partly in response to your [Nojarov's] intervention: I'm slightly puzzled why many physicists who do not work in relativity are worried by this preferred system singled out by the cosmic background

radiation. Such people seem to believe that when we have a preferred system the relativity of the theory is destroyed, but that's not the case. Our model of the cosmos, from the beginning, starts with such a preferred system; the rest system of the matter is the preferred system, and nobody is surprised. This is just an artefact of the model, and we don't know if the universe, if we get further information on it, will keep that feature. In any case, it does not destroy the relativity and does not destroy the importance of the covariance of the theory. Cosmology is just a single application of Einstein's theory, and you have many different systems you can treat and where you need the covariance of the theory. Why you think it's so important?
Nojarov: Because it somehow contradicts the Mach ideas. Everybody can interpret them, of course, as we understood here, that each motion should be relative, but now we have an absolute motion, so to say, we have an absolute reference frame, and then we lose the sense of relativity, that each motion should be only relative.

Quantum Gravity: Basic Issues

Ehlers: Perhaps in the beginning we might ask ourselves why is it that one wants to quantize, and I think that both general relativity and quantum theory by their very nature have to insist that they are universally valid. This is because the quantum theory is the first successful description of the microscopic structure of matter; on the other hand, general relativity is clearly by far the most successful description of the geometric structure at large, and it is clear that geometric structure enters so far every successful physical theory, whether it is classical or quantum and whether it's a theory working mainly with particles or with fields. So one has this unfortunate situation that there is one theory which is very successful with respect to the description of the structure of matter in the small but is very naive with respect to the description of the metric which enters its equations, and there's another theory, namely, general relativity, which is very deep and careful in its description of the metric at the classical level but is very simple minded in its description of matter. Physics cannot remain like this. But perhaps there are still people who might think that one could do without quantization. Perhaps if somebody wants to say something on the need or not of quantizing this might be one topic, and then we might go on to more specific things.
Goenner: What I have to say is not exactly "quantize or not to quantize," but is intended to connect up with what you said this morning, i.e., that the initial-value formulation may be not as fundamental as some

people believe. The counter argument always comes that because we want to quantize gravity, we have to do it in this way. But to me this is a very weak argument because it derives from a prerelativistic quantum theory and from an unfinished quantum field theory. From the mathematical point of view, the formulation of relativity and quantum theory is not yet completed, so to me this is not a good argument for the initial-value formulation. The only argument I can accept is just the utilitarian one. How otherwise could we do it? And the only other way to do it might be the path-integral formulation or things like that. So I side with you [Ehlers] when it comes to the initial-value formulation, and maybe this provokes other people's comments.

Kuchař: And you looked at me [laughter]. But I won't defend canonical quantization as the royal way to quantization. I could discuss the reasons why I believe it's useful, but I am certainly open-minded to other proposals. You know that in the old good days, some 15 years ago, there were two main routes which people tried: a method called covariant quantization, and the canonical quantization, which you mentioned. The covariant quantization is not being actively pursued by very many people these days, and for a good reason. It relies on perturbative methods, and gravity turns out not to be perturbatively renormalizable. People thus turned to methods that at least have a hope of providing a nonperturbative scheme, and that's why by default they ended with canonical quantization.

Narlikar: Some years ago I tried to look at the path-integral method for quantizing gravity, and I found that if you limit yourself to the conformal degrees of freedom of the full metric, you can write down the path integral, which is quadratic. It can be therefore handled exactly without a linearization by using Feynman's methods. Further, the conformal degree of freedom has direct contact with the question of spacetime singularities, because if one takes the Robertson–Walker metric, which is a good example of a conformally flat metric, then the conformal factor going to zero is the indication of singularity. So the question is whether by going to a quantum version of Einstein's equations but quantizing only the conformal degree of freedom one can arrive at a broader spectrum of solutions. Padmanabhan, Joshi, and I found that within this framework it is indeed possible to answer questions like this: That given an initial broad spectrum of functions reflecting our imprecise knowledge of the initial quantum conditions and knowing that the final state follows the classical solution of Einstein's equation, what is the probability measure that the universe came out of a singular set of initial conditions? One can show that this measure is zero and that it is almost certain that it came out of an initially nonsingular set of conditions. So I think one can get quite far in interpreting such solutions. How sensitive is this conclusion to omitting nonconformal degrees of freedom? I think that in the initial stages of the subject such simplified pictures are very helpful in understanding quantum gravity.

Isenberg: While of course it is undoubtedly very important in principle to find a quantum theory of the gravitational field, perhaps in thinking about Machian issues (like frame dragging) one doesn't need it. That is, perhaps in considering Machian effects which depend exclusively on long range physical interactions on a cosmic scale, it is sufficient to work with the classical gravitational field alone.

Ehlers: I share more or less your view, but I think at this stage we are not so much addressing Mach's Principle in any strict sense, but rather its generalizations, considered an essential part of the structure of space and time in its relation to interactions, and to ask what is its relation to quantum theory. I wouldn't worry too much at this stage about Mach's Principle.

Giulini: I think it's useful if one discusses all these questions to remind oneself that, of course, the notion of quantum gravity cannot be better than the notion of quantization itself and although one did succeed in quantizing some theories, I at least haven't seen a general axiomatic formulation of quantization. I think there is a certain danger that these uncertainties enter the discussion of issues in gravity proper, when lifted to the level of quantum gravity.

Ehlers: Well, I would like to add only one remark. I think with respect to general relativity, the word quantization gives the impression that there's one theory which is known to be correct and not in need of any change – and that's quantum theory – and there's another theory which is general relativity, which is in need of being modified. I would not be surprised if in order to achieve a unification both of these theories have to be modified in some way. Perhaps some relativization of quantum theory is as necessary as the quantization of the classical theory.

Zeh: The problem of how the classical limit may be recovered appears in all theories after their quantization, which, I think, is well defined in principle and established as a general procedure, except for the open factor ordering and possible ultraviolet divergences that may require modifications at the high-energy end. For example, Goenner mentioned that we know general relativity only fully in the classical limit, and maybe it does not have to be quantized. However, we have a similar theory, electrodynamics, that also possesses most of its applications in the classical domain. In this case we do know that there are quantum effects (as laser physicists know very well). They were predicted by means of the quantization rules. Nevertheless, the 'classical limit' forms a wonderful theory, as wonderful as general relativity, and still this classical theory has to be changed (quantized). We have the same problem here as in all theories after their quantization: How to recover the classical situation which we usually *observe* in the world. I think there is not really a difference, and therefore I would not expect that general relativity has to be treated differently from all other classical theories. There are, in fact, consistency arguments, originating in the Bohr–Einstein debate, which demonstrate that *nothing* can remain strictly classical.

Brill: I want to mention one aspect of quantum gravity that has so far received little attention here, and that appears to me closely related to the classical attempts to formulate Mach's Principle as a selection principle. In classical theory, then, there is a view of Machianity in which not all solutions of the field equations are considered Machian, but only those satisfying (in addition, say, to the Einstein field equations) certain criteria such as the Wheeler–Einstein Mach Principle or the Raine criterion.

Similarly, in quantum gravity not all solutions, say, of the Wheeler–DeWitt equation, appear to be relevant to the universe; and certain selection principles have been formulated, for example, the Hartle–Hawking condition. In a sense this condition specifies how the quantum universe is allowed to be generated from 'nothing.' This in turn implies a distribution at the classical stage of universes peaked about a certain type of universe. One can then ask whether the type of universe that is most likely to come out of this is Machian, say, according to one of the classical selection criteria. If it is, then we can view the quantum condition as a possible explanation of the classical selection criterion. Conversely, the quantum condition might be viewed as the essence of Mach's Principle.

Lynden-Bell: I think we may be attempting to run before we can crawl. First, I don't think we have yet decided anything about how to incorporate Mach's Principle into classical theory, let alone quantum theory, and second, I do not yet know an example in which a relative theory has been quantized, even an elementary relative theory, and I would ask whether the Barbour–Bertotti theory has been quantized? It would seem to me that was a relatively easy theory to quantize, in the Schrödinger sense at least, and I think it would be interesting to see it quantized.

Barbour: It has been formally quantized to a certain extent. The problem is: What constitutes quantization? The point is – and I'm sure

Karel would emphasize this point too – that there are two major aspects to quantization. First of all, there is finding a wave equation on a configuration space, say, and then there is the finding of a Hilbert-space structure that is constructed on the space of solutions of this wave equation. Then the question is this: By quantization do you mean just the first step, or do you mean both together?

Now if you take the narrower sense, that it's just finding a wave equation, then I take the view that general relativity was quantized 26 years ago by Bryce DeWitt. In 1967 (DeWitt 1967), he found wave equations on the configuration space. Now exactly the same thing has happened with the Barbour–Bertotti model; various people, including Smolin (Smolin 1991) and myself and Rovelli (Rovelli 1991) and Bertotti, formally quantized it, but that doesn't mean terribly much unless you can come to some understanding about whether you need a Hilbert-space structure or not. Now my personal feeling is very much in sympathy with what Jürgen Ehlers said. I do not myself believe that a Hilbert-space structure is part of the fundamental quantum theory of the world but that it's an effective structure that emerges higher up. I can already see Karel beginning to disagree, but I think that's an important issue, and it ties in nicely with what Jürgen said.

Kuchař: It's such a pity, Julian, that you spoiled your point, with which up to the very last statement I heartily agreed, by putting a conclusion on it which prompts me to charge back: What do you mean by quantum theory when you don't have a Hilbert space? How do you interpret it? I presume you start making vague suggestions about what it means to have a state function of the universe defined on a relative configuration space, but give me a technical statement about its interpretation, if the state function doesn't lie in a Hilbert space?

Ehlers: May I come in here? It seems to me that at this stage perhaps we all would profit if we would look at one definite question. Suppose we are given a solution of the Wheeler–DeWitt equation and somebody would hand it over to you [Kuchař] or to you [Barbour]. Then what do you do with it? Which physical statements follow from that? Don't you need some additional structure? What kind of testable statements would you be able to get from the Ψ? Certainly, in order to have a quantum theory we need at least some observables, and the purpose of the Hilbert space or of the density matrices is to be able to assign probability distributions to the observables. We must know what at least a few observables really mean with respect to looking into the world. I'm not an expert in quantum gravity, but I feel that there is, perhaps, a certain drawback in that it is hardly ever stated what kind of testable statements

one is even aiming at. Could somebody say something to that perhaps?
Bondi: Well, I've never worked in the quantization of general relativity. I look at it simply as that it must be shown to be impossible to cheat the Heisenberg principle using gravitation. Now what does that matter to me? If it were to turn out, in spite of the labors of many, that general relativity as now written down was not quantizable, that would make me worried. If it were shown that the selection rules we think of in connection with Mach's Principle were not quantizable, then I think one would really have to sit down and start again. If it can be done, I'm not terribly interested in seeing it done in detail because it's so far from practicality, but I wish to be assured of the existence theorem of a possibility of working in such a way that the uncertainty principle cannot be defeated.
Zeh: Coming back to your [Ehlers] question: Given the wave function Ψ, what does it mean? I think it is an important point - we need some additional input for the interpretation - but you could ask the same question in classical theory. Given a high-dimensional configuration space, and given a point in it, how would you interpret it without anything else? So you have to identify certain formal 'states' with empirically known situations. Or would their dynamics be sufficient to give them a meaning?
Ehlers: Permit me to make another remark. It seems to me that in technical, mathematically oriented discussions in classical as well as in quantum theories, we got used to employ the word observable in a very abstract sense. If I look at classical general relativity as a branch of physics, then I think classical general relativity has its uses, its successes in describing macroscopic situations, planets running around the sun, the binary pulsar, and to a certain extent also cosmology. Now what are the quantities which connect the mathematical formalism with physical statements? Well, we can measure times, in fact, very precisely, we can measure angles very precisely, and we can measure energy fluxes of various kinds of radiation. All other quantities are not observable; even distances are inferred; they are computed from a theoretical model. You know what your null cone is, and if you have a red shift, and you have a certain angular observation, you infer by means of the theoretical model a certain distance, which means a certain event on your past light cone, and so on.

So if I am asked how one uses classical general relativity, not as a mathematical game, but as a part of physics or astronomy, then I think I understand how to connect the formalism with effects. That's essentially only by these three things I mentioned. The whole of astrophysics really observes times, angles, and energy fluxes in various spectral ranges. Everything else is constructed theoretically, and the very purpose of models which obey the field equations is to produce a picture which allows us to consider the observations here and now as due to certain things, radiation coming through space, etc., and I would like to have even a rough idea what in this sense would be testable statements of a quantized theory of gravity.

For quantum mechanics, I think I do understand it. The major physical successes of quantum mechanics are that it explains the stable or metastable states in which certain bound systems can exist, like atoms and molecules, and that can be tested in terms of their spectra. In addition, you can make statements about probability of collision processes in terms of cross sections or *S* matrices, so there I know what it means. But what would be typical statements which one would like to test by means of a quantum gravity theory and, in particular, which role is played there by the Ψ and what other mathematical instruments do we need in order to get a statement?

Hoyle: Well, should it not be that what we want to test are certain configurations of the universe? That we specify what the configuration is – and we need some sort of an action integral for that choice – and then we're going to use the Ψ as our probability weighting function for the configuration in question. Then given this we can calculate the average value of any quantity that is involved in the development of the universe.

Brill: Your very examples in the classical theory seem to me to show that maybe it isn't so interesting to think about the kind of observables Jürgen mentioned. If we announce that all we can meaningfully discuss in general relativity are some angles and times, that is not going to excite anybody to think about actual experiments. I have the feeling that if we really knew Ψ and could work with it and understand the classical limit, it might teach us what the observables are. You shouldn't necessarily have to insist *a priori* that everything be totally clear before you're even allowed to think about Ψ.

Ehlers: I agree with that. Again I don't want to be misunderstood in the sense that I am so old-fashioned positivistic that only these here-and-now observations count. All I wanted to say is in order to connect the formalism with which we form a picture of certain physical processes – and this picture is really more of interest to us, it gives us ideas of what is really going on – in order to connect the formalism with this picture in a testable way, we have to know what is the handle that connects the formalism with measurements and observations; but I think there is no real disagreement at all on that.

Zeh: Precisely to your point: What are the observables? I think this should be very easy. If you build up your theory by constructing a wave function over the configuration space which consists of three-geometries, then, of course, you have at your disposal the projectors onto the three-geometries. You know how to interpret them. If you accept that you know what the three-geometry is, then I think there is no problem of interpretation.

Ehlers: Yes, and how does any physicist find out what the three-geometry is? I mean, what's the relation to any actual observation or experiment?

Zeh: That's a problem of classical canonical gravity.

Ehlers: Well, not quite.

Isenberg: I might note that in order to understand measurements, it's not enough to have just a given state function Ψ on the space of all three-geometries. You also need to have some sort of inner product, or at least a norm, defined on this space; and perhaps other structures as well.

Kuchař: What are the quantities that you would like to proclaim to be observables? Here we are hit by the ambiguities in the interpretation of quantum mechanics. When we are talking about observables, do we accept something like the Copenhagen interpretation, in which there is an observer, or an apparatus outside the system? Or are we trying to interpret the state function of the whole universe?

I feel that both routes are possible: You do not need to quantize the whole universe to make meaningful statements about what is happening to the geometry in a small region of spacetime, in which you watch phenomena at the Planck scale; you still have plenty of matter outside, out of which you can construct the apparatus.

However, as Jim [Isenberg] indicated, a more tempting alternative is trying to interpret the state function of the whole universe. Then you must invent an alternative interpretation of quantum mechanics, and that's a high call. Some people answered that challenge. Jim Hartle (1993), for example, is trying to extend to quantum gravity what he and other people - Gell-Mann (1990), Zeh (Joos and Zeh 1985; Zeh 1989), Omnès (1992), and Griffiths (1984), to mention some names - did for Newtonian systems. I don't say that I agree with all the details of his approach, but certainly the thrust is correct: You try to construct coarse-grained histories that decohere, and ascribe to them probabilities. There are many different ways in which you can attempt to specify a history and coarse-grain it in general relativity. The procedure does not need to be based on three-geometries along a foliation; it can be based on other objects, say, on the averages of curvature invariants in a spacetime region. The region must be physically defined, for example, by having its boundaries physically defined by the values of physical fields. You can then coarse-grain the spacetime by asking whether these averages lie within this range, or that range, or some other range. These are the questions you may ask. Unfortunately, you must complete the whole program before getting a consistent answer to any particular question.

Ehlers: Yes, but you would probably not maintain that since we cannot systematically look at all these issues at once during this discussion we should not try to address any particular question which one can pick out of that. I mean, I wouldn't know how to go about it.

Kuchař: It's Mach's Principle for quantum gravity, everything in the universe is connected with everything else.

Barbour: Karel threw down a challenge to me a little earlier, and I'll try to answer it if I may. The challenge was: How can one interpret a state function if that state function does not lie in a Hilbert space?

I just take the basic elements that Bryce DeWitt has given us and take them as the only things to go into the interpretation: a static wave function, defined on the possible configurations of the world. They're not just three-geometries; they are all possible configurations of the matter as well, on these three-geometries. So there is this heap of configurations that make up my Q_0, the relative configuration space. Let one of these triangles represent any such configuration you like, however complicated. It might be the configuration which represents us at this very instant now. The configuration of the atoms in our brain, the sun shining outside, and all those things. Those configurations are in the Q_0, and what the Wheeler–DeWitt equation tells us is that there is a static wave function defined on all those configurations. If you can give me a solution of the Wheeler–DeWitt equation, then I can tell what is the value of the wave function on each possible configuration q_0 in my Q_0, which I like to call the *heap of possibilities*.

Now despite the difficulties with the definition of metrics that Giulini was discussing earlier (p. 491), it should be possible to use the actual dynamical metric to define a volume element, so that I can divide up the configuration space into infinitesimal cubes. I can then look at the value of the wave function on a representative configuration in one such cube. Then I can take a number of identical copies of that configuration that is proportional to the value of $\Psi\Psi^*$ that I find on the configuration, and I can put them all into an imagined *heap of actualities*, as I call it. In other words, it's a big heap in which each infinitesimal cube of Q_0 has representative configurations in numbers proportional to the respective values of $\Psi\Psi^*$.

Now we have the final step. In this instant *now* we believe we experience the world, or at least the part around us, to be in some certain definite configuration. My proposed interpretation is that such a configuration is copiously present in the heap of actualities. After all, if we were to draw a configuration at random from the heap of actualities, it is most probable that we would draw one that is copiously present. Essentially, this is the standard assumption of classical statistical mechanics: Only situations that are probable are actually experienced.

This interpretation, which is essentially what is called the naive Schrödinger interpretation (because it weights configurations by means of the Schrödinger norm $\Psi\Psi^*$), can be related to more normal physics if it turns out that solutions of the Wheeler–DeWitt equation are generically concentrated on configurations of the world like those that we do indeed find around us – such as configurations containing stars and galaxies and evidence for evolution in time recorded in rocks. Such configurations I call *time capsules* (Barbour 1994a, b). There is a great deal more I could say about the overall scheme, in particular about the origin of the arrow of time and how notions of quantum observables could arise, but this is perhaps enough for the present.

Vucetich: With respect to the observables in quantum gravitation – I would like to mention the cosmological constant. The Hawking–Coleman theory of the vanishing of the cosmological constant is an example of how one can construct an observable in quantum gravity. The present version is probably not very well constructed because the probability function becomes singular, but it's the kind of thing one may try to build in quantum gravity.

Isenberg: In response to Julian [Barbour], let me note that one of the things I don't understand about your approach is how the issue of observables is addressed in it. You have this state function Ψ on the space of three-geometries, but how does one use it to produce information about things you measure and observe? How does one calculate from it, say, particle production near a black hole? I do see that sort of thing coming more directly out of approaches like that of Hartle and Gell-Mann.

Barbour: You ask how observables can come out of mere three-geometries (and matter fields defined on them). The first point to be made is that in canonical quantum gravity, which aims to be *the* quantum theory of the entire universe, what you normally call the measurement apparatus together with time and the dynamical frame of reference have all got to be included in the state function (besides any system on which so-called measurements are being made). It is then inevitable that your concept of observable must undergo substantial modification. We need to go even further than Everett (1957), who included in the state function the apparatus but not time and the dynamical frame of reference. This is why he did not have to address the issue of the notion of observable.

Now suppose you went to CERN and took a photograph of a momentum measurement being made on a certain particle. The photograph, which records an instantaneous *three-dimensional configuration*, shows everything: the accelerator that produces the particle, the detector that measures the momentum (say by means of deflection in a magnetic field), and some record of the measurement itself (a curved particle track). Now show that photograph to a competent cosmologist. He or she will, in principle, be able to tell you that a momentum observable has been measured. All the information is coded in the one photograph. Thus three-configurations, if large enough, can tell you a surprising amount.

Let me now remind you of a remarkable paper (Mott 1929) in which Mott provided a quantum-mechanical explanation of the straight tracks made by alpha particles in Wilson cloud chambers. Mott in fact found a solution of the *time-independent* Schrödinger equation for a system consisting of a radioactive atom that can emit α particles together with literally millions of hydrogen atoms, all treated together as one huge quantum system. In Mott's solution, which is completely static, the probability density is very strongly concentrated in the position representation on spatial configurations of the entire system that seem to record the passage of an ionizing particle; that is, the configurations with high $\Psi\Psi^*$ contain many ionized atoms that lie more or less on a single line that passes through the radioactive nucleus. Now this is very like part of the photograph I imagined you taking at CERN – and it is all in a single static three-configuration on which a time-independent equation is capable of concentrating $\Psi\Psi^*$.

I believe that we must think of a solution Ψ of the Wheeler–DeWitt equation as being just like the Mott solution, except that now the quantum system is the entire universe in a static state and we experience the universe from within. For me it is an important assumption that our experience is always correlated with individual configurations (one in every instant we experience), not Ψ. The role of the Ψ is to tell us which configurations have the highest probability of being experienced. This is a natural extension of Born's probability interpretation of Ψ to a situation in which all external elements – measuring apparatus, time, external frames of reference – are absent. Finally, I do recommend John Bell's paper "Quantum mechanics for cosmologists" (Bell 1981), in which there is a fascinating discussion of the Mott paper. Bell comes very close to advocating an interpretation of quantum mechanics that is almost identical to mine. In particular, he inclines to the view that our direct experience is correlated with individual spatial configurations of atoms, as in Bohm's interpretation. The similarity to my position was pointed out to me by Dieter Zeh.

Kuchař: Can I now enter? I think that your interpretation of the Mott experiment, in terms of the probability of picking up a three-geometry plus matter field from the heap, doesn't fit for the following reasons: In quantum mechanics, one always asks for the probability of a trajectory, the conditional probability of one event given another event. Your statement pertains only to a single instant – it never has those conditional probabilities in it. I fear that you'll never be able to answer some important questions which we would like to ask in quantum gravity. You try to circumvent this problem by introducing the concept of the time capsule, but that charm does not seem to work. If you have, at the present instant, what you call a time capsule, and you have it with a high probability, you should also have in the heap, with a high probability, the state which describes tomorrow's record. I do not see such a correlation arising through any mechanism that requires solving the Wheeler–DeWitt equation. It is the correlation between two instants in the heap that is needed, not only the probability of a single instant.

Giulini: I would like to ask Julian something. I don't quite understand the example of the snapshot taken of the cloud chamber. I do not understand why this is an answer to James Isenberg's question. What is the equivalent of the photographic machine in Q_0? I mean, if you could step outside the configuration space, then I can understand how to take snapshots, but from within it, it's much harder to construct such a 'camera' conceptually. Isn't that the case?

Barbour: You ask: What is the equivalent of the photographic machine in Q_0? Ultimately, it must be our human consciousness, which we have to assume is correlated to localized structures in the individual configurations in Q_0. The essential point was made by Everett (1957, 1973): We do not observe the universe from outside, we experience it from within. First and foremost, an observation, even an experiment, simply tells us where we are in Q_0. Suppose we could freeze in a photograph what we experience *now*. Then let us suppose you and I could examine, one by one, all the individual relative configurations q_0 in the relative configuration space Q_0 of the universe. Then comparison of your snapshot with the q_0's, which can also include the atoms of your brain, will enable us to identify those that are candidates for being the physical counterpart to your present psychological experience.

Von Neumann (1932) is quite right: There has to be a postulate of psychophysical parallelism. It's not that there is something 'magical' about consciousness – it's not causing physical collapse or anything like that. It is simply that you have to have some postulate linking actual experience to the mathematical model. The postulate that I am making is that psychological experience is always matched with spatial configuration in some q_0. And, as I said, the role of the wave function is to determine which q_0's are most likely to be experienced.

As to Karel's comment about correlations, the only way the issue can be properly settled is by a much more detailed calculation of Mott-type situations with actual numbers, correct allowance for combinatorial questions, etc. So far as I know, no such calculation has ever been made. However, the *sort* of correlations Karel quite rightly requires – if today is present in the heap of actualities with high probability, then the same should be true of yesterday – are certainly present in the Mott solution. Indeed, the solution could not exist without them. The instants certainly are correlated – probability cannot build up on one without doing the same on the other. The only question is whether the actual numbers come out right, and whether such solutions arise generically.

I do believe many, many people still have a deeply ingrained feeling that time exists. But I think also one just has got to come to grips with what the mathematics of the Wheeler–DeWitt equation is telling us – there is no time – so we have to change the questions we ask.

Zeh: I would also like to comment on Karel's answer, even though Julian did it himself already. I think Karel said, "There is something that Julian's picture cannot explain, namely, that we can make not only one snapshot, but two snapshots, at two different instants of time. But this argument supposes that there *is* time, and that we *may* have two different snapshots. How do we know that there *were* different moments in time? We know only what exists in our memory. Our memory at this very moment is certainly 'present,' and what Julian wants to explain is only this memory which we have, and from which we get the conviction that there has been a definite past which we (seem to) remember. Of course, the usual way to explain memory is that there *was* a history, which we remember. But Julian wanted to say there is another way to explain how this memory comes about, namely, the dynamical structure which is forced into the solution by the Wheeler–DeWitt equation. If it has this structure – and I think this is also what is in Mott's explanation, because Mott solves a *stationary* many-particle differential Schrödinger equation – then you *can* explain such correlations in the memory. If you look at the structure of Mott's function, you have the impression that one scattering event was the successor, or the cause, of another one even though the stationary Schrödinger equation does not contain any concept of time. This memory is pure structure, and I think this is what Julian wants to say: You don't need the history to explain it.

Kuchař: I know this is what Julian wants to say. What I'm saying is that it doesn't work.

Zeh: If you tell me it doesn't work, it is because it does not do what you expect it to do. But he doesn't want to do that.

Kuchař: Not quite. I am saying that there is no known mechanism in the Wheeler–DeWitt equation which would bring in the situations he is talking about. Read those parables in Julian's papers (Barbour 1994a,b), about the highly asymmetric superspace, with the mist gathering at the gulches suggesting that superspace has a structure which naturally leads to high probabilities on the time capsules. But when you look at the solutions of simple minisuperspace models, like the Bianchi type IX cosmologies, you do not see any substantiation for the statement that the probabilities tend to get concentrated only on some regions of superspace. The Wheeler–DeWitt equation, so far as we understand it, doesn't provide a mechanism for doing what Julian expects it to do.

Then there is something in the Mott example that Julian's interpretation is still missing. Suppose that I look at a stationary solution of the situation with a nucleus that disintegrates and leaves a track in the bubble chamber. Suppose that solution yields a high probability for a track with one hundred ionized centers. Then it also yields a high probability for a track with one hundred and one excited centers. Ultimately, of course, the trajectories start dwindling as the energy is getting lost, and I shall not find, say, one with two thousand excited centers. Now, let us go back to Julian's interpretation. I do not see in it anything which guarantees that if we find in the heap a trajectory with one hundred excited centers, then we also find a trajectory with one hundred and one centers excited in the same direction.

Ehlers: May I come in? Could one not, in this case, particularly profitably use the Griffiths Consistent History (Griffiths 1984) approach? You see, you could first take an ordered sequence of t_1, t_2, t_3. Now whether you interpret these as times in some ontological sense or as times which sort out your memory does not perhaps matter for this purpose. And you are asking whether at a certain time a certain particle was at a certain place, then at the later t_2 you ask a similar question, maybe it's about a position, or about a momentum. So, you have a certain sequence of these projection operators. Then, if you have a guess or perhaps even a knowledge about the initial state, you can find the probability for this particular sequence to happen. If this probability is very high, then you will say you have an explanation for why such a trajectory occurs. Now, if you want, you can ask, "Can I make a more refined statement?" Then you can add additional time values and additional projectors, and the theory itself, namely, Griffiths's formalism, tells you which additional statements can consistently be added to the question that you have asked before and you will get an answer or, if you add inappropriate additional questions, the theory will tell you that's not a sensible question to which the theory ought to give an answer. I mean, is this somehow along lines which you would take for this, Dieter?

Zeh: Essentially, yes, except that I do not like Griffiths's approach too much because he uses time. I think we could do it quite a bit better if we take the Wheeler–DeWitt equation and accept the method suggested by Banks (Banks 1985), for example, and use the geometric-optics approach with respect to the geometry. Then you have something like causal connections along the classical orbits. Actually, Mott's scattering is essentially the same as what we call decoherence, and so you get something like classical orbits. This is the way how we understand that classical histories in quantum geometry, spacetimes, come about.

Ehlers: Yes, but I think we have now only isolated a bit more clearly the question, and the question is: Can the type of answer which one could give, namely, the probabilities with which particular histories occur, be found within Julian's scheme? For that purpose, it seems to me it is not sufficient just to know a solution to the Wheeler–DeWitt equation. You must somehow have a mechanism to construct a sequence of such projections and to judge whether they form a consistent history, or the equivalent, which you prefer for this. And perhaps the discrepancy between Karel and Julian is in the answer to the question: Is enough structure contained in Julian's picture in order to address these questions or not?

Kuchař: All right. I foresee two troubles with incorporating your reasonable proposal into Julian's scheme. One trouble is that there is no ordering parameter, like the time parameter, in that scheme. The second trouble is that you need a sequence of projection operators which doesn't kick the state function out of the space of solutions of the Wheeler–DeWitt equation. Now, if you try to measure the position, even the smeared position, meaning a smeared three-geometry operator, on a state which solves the Wheeler–DeWitt equation, you'll get a state that won't solve the Wheeler–DeWitt equation. So, the problem is to define the operators in your sequence in such a way that they do not kick the states out of the space of solutions.

Ehlers: In other words, your statement is, if I may rephrase it, that you would not even obtain enough interesting consistent histories, because of this difficulty.

Kuchař: That's my feeling, but let Julian and Dieter say if they have any good candidates for the projectors which would not kick the states out of the space of solutions.

Barbour: The Consistent Histories approach is not one that I particularly like, and it may be characteristic that many people working in that direction believe time is fundamental. They take the view that the quantum mechanics of the entire world is a quantum mechanics that's going to give you *probabilities for histories*, whereas I take the Wheeler–DeWitt equation at its face value, namely, that it's giving you probabilities for three-dimensional configurations, what we would call instants. That's a big difference. One point that Karel made is, I believe, easy to answer. He is bothered by the collapse problem, that operators corresponding to observations will kick the state out of the solution space. However, this assumes the Copenhagen interpretation, whereas Dieter Zeh and I, like many quantum cosmologists, assume a many-worlds interpretation (Everett 1957, 1973). Thus, there is no collapse, and the static wave function is given once and for all.

As to the question of the Mott solution and those time capsules, and whether there will be a tomorrow in the heap as well as a today, Karel is right: This is a pure aspiration on my part at the moment – I'm no way near being able to prove any of these conjectures. But if we're ever going to have a physics that has any predictive value, any explicatory value, and if the Wheeler–DeWitt equation is right, and there is no time, then I think it's got to come in that sort of way. Moreover, the effects I need – cor- relations and concentration on time capsules – certainly are present in the Mott solution. The issue is whether the numbers are right and whether such solutions exist generically for the Wheeler–DeWitt equation. Also, the absence of such effects in simple models like Bianchi IX means very little since time capsules cannot possibly appear except in models with far more degrees of freedom. In the Mott model, there is the α particle, which is in the WKB regime, and then there are innumerable other degrees of freedom that remain truly quantum. It is these latter that are concentrated onto time capsules. The α-particle wave function is not concentrated at all. In the cosmological context, the

Bianchi IX variables will play the role of the α particle and there is no reason to expect them to be concentrated. Only in the next step, in which one introduces additional quantum variables coupled to the Bianchi variables, can one expect concentration – and it will be the new variables that are concentrated, not the Bianchi variables.

Kuchař: But back to the issue that Jürgen raised: I feel that there is no way of incorporating his proposal consistently into your scheme.

Barbour: That would be my reaction, yes.

Ehlers: Just that I understand myself: If you want to avoid talking about time sequences, then, in order that you can talk meaningfully to your fellow physicists, you somehow have to translate into your language how one should express what most physicists would express by saying "A particle just flew through a chamber, and it produced droplets one after the other." Then I can, for the purposes of the theory, ask for the probability of a certain sequence of such ionization events. How would you translate this into your language without using the words "sequence of events" or something?

Barbour: I can only say again: Consider the Mott solution, which is a *static* solution of the time-independent Schrödinger equation that, in the configuration space of the entire cloud chamber, is strongly concentrated on time capsules. That is, the Schrödinger density $\Psi\Psi^*$ takes high values on complete configurations that each seem to contain the record of the passage of a particle, which appears to be 'fossilized' by the presence of many ionized atoms aligned more or less along a track. That is the kind of language you must use. Of course, I do not deny that the Mott solution is very special. Mott got his solution because he tacitly used his temporal intuition – and this is in fact the whole reason why *time-independent* scattering theory works: You find a solution of the time-independent Schrödinger equation, but you put very special conditions on it to mimic a time variation. What I conjecture is that such solutions somehow arise naturally and spontaneously for the Wheeler–DeWitt equation, which also contains no time. They may because configuration spaces, especially relative configuration spaces, are highly structured; they have natural origins, where everything is crowded together, and they have natural frontiers at infinity, where everything is very far apart. The configuration space of the universe is highly structured by sheer necessity, and that structure is reflected in the dynamics, because interactions are local in the configuration space. Thus, the dynamics inherits this asymmetry, and I believe things like the arrow of time come, at the deepest level, from the asymmetry of the configuration space (Barbour 1994a, b).

These are purely spatial things. There's nothing temporal about them. The Big Bang is not in the past; it's at, or rather *is*, the natural origin of the configuration space.

Ehlers: But to me, stationary scattering theory is just a mathematical, convenient abbreviation for what can be understood only by time-dependent scattering theory.

Barbour: I'm turning that upside down.

REFERENCES

Arnowitt, R., Deser, Stanley, and Misner, Charles W. (1962). "The Dynamics of General Relativity." In *Gravitation: An Introduction to Current Research*. L. Witten, ed. New York: Wiley (1962).

Ashtekar, Abhay and Stachel, John, eds. (1991). *Conceptual Problems of Quantum Gravity*, Vol. 2 of *Einstein Studies*. Boston: Birkhäuser.

Baierlein, Ralph F., Sharp, David H., Wheeler, John A. (1962). "Three-Dimensional Geometry as Carrier of Information about Time." *Physical Review* 126: 1864–1865.

Banks, Thomas (1985). "TCP, Quantum Gravity, the Cosmological Constant, and All That ...." *Nuclear Physics* B249: 332–360.

Barbour, Julian B. (1994a). "The Emergence of Time and Its Arrow from Timelessness." In *Physical Origins of Time Asymmetry*. Jonathan J. Halliwell, Juan Pérez-Mercader, and Woljciech H. Zurek, eds. Cambridge: Cambridge University Press.

Barbour, Julian B. (1994b). "The Timelessness of Quantum Gravity. I. The Evidence from the Classical Theory. II. The Appearance of Dynamics in Static Configurations." *Classical and Quantum Gravity* 11: 2853–2897.

Barbour, Julian B. and Bertotti, Bruno (1982). "Mach's Principle and the Structure of Dynamical Theories." *Proceedings of the Royal Society, London* A382: 295–306.

Bell, John S. (1981). "Quantum Mechanics for Cosmologists." In *Quantum Gravity 2. A Second Oxford Symposium*. Christopher J. Isham, Roger Penrose, and Dennis W. Sciama, eds. Oxford: Clarendon Press. Reprinted in (Bell 1989).

Bell, John S. 1989. *Speakable and Unspeakable in Quantum Mechanics*. Cambridge: Cambridge University Press.

DeWitt, Bryce S. (1967). "Quantum Theory of Gravity. I. The Canonical Theory." *Physical Review* 160: 1113–1148.

Dirac, Paul A. M. (1965). *Lectures on Quantum Mechanics*. New York: Academic Press.

Everett, Hugh III (1957). "'Relative State' Formulation of Quantum Mechanics." *Reviews of Modern Physics* 29: 454–462.

Everett, Hugh III (1973). In *The Many-Worlds Interpretation of Quantum*

Mechanics. Bryce S. DeWitt and Neil Graham, eds. Princeton University Press.

Gell-Mann, Murray and Hartle, James (1990). "Quantum Mechanics in the Light of Quantum Cosmology." In *Complexity, Entropy and the Physics of Information, SFI Studies in the Science of Complexity*, Vol. VIII. Wojciech H. Zurek, ed. Reading: Addison–Wesley.

Griffiths, Robert B. (1984). "Consistent Histories and the Interpretation of Quantum Mechanics." *Journal of Statistical Physics* 36: 219–272.

Hartle, James (1993). "Spacetime Quantum Mechanics and the Quantum Mechanics of Spacetime." In *Proceedings of the 1992 Les Houches School, Gravitation and Quantisation*.

Isham, Christopher, J. (1993). "Canonical Quantum Gravity and the Problem of Time." In *Integrable Systems, Quantum Groups, and Quantum Field Theories. Proc. Nato Advanced Study Institute, Salamanca, 1992*. L. A. Ibort and M. A. Rodriguez, eds. London: Kluwer Academic Publishers.

Joos, Erich, and Zeh, Heinz–Dieter (1985). "The Emergence of Classical Properties through Interaction with the Environment." *Zeitschrift für Physik* B59: 223–243.

Kuchař, Karel (1981). "Canonical Methods of Quantization." In *Quantum Gravity 2. A Second Oxford Symposium*. Christopher J. Isham, Roger Penrose, and Dennis W. Sciama, eds. Oxford: Clarendon Press.

Kuchař, Karel (1992). "Time and Interpretations of Quantum Gravity." In *Proceedings of the Fourth Canadian Conference on General Relativity and Relativistic Astrophysics*. G. Kunstatter *et al.*, eds. Singapore: World Scientific.

Kuchař, Karel (1993). "Canonical Quantum Gravity." In *Proceedings of the 13th International Conference on General Relativity and Gravitation*. C. Kozameh, ed. Bristol: IOP.

Mott, Nevill F. (1929). "The Wave Mechanics of α-Ray Tracks." *Proceedings of the Royal Society, London* A126: 79-84. Reprinted in (Wheeler and Zurek (1983).

Omnès, Roland (1992). "Consistent Interpretations of Quantum Mechanics." *Reviews of Modern Physics* 64: 339–382 (1992).

Rovelli, Carlo (1991) in (Ashtekar and Stachel 1991).

Smolin Lee (1991) in (Ashtekhar and Stachel 1991).

von Neumann, Johann (1932). *Mathematische Grundlagen der Quantenmechanik*. Berlin: Springer-Verlag. Translation: (von Neumann 1955).

von Neumann, Johann (1955). *Mathematical Foundations of Quantum Mechanics*. Princeton: Princeton University Press.

John A. Wheeler and Wojciech H. Zurek, eds. (1983) *Quantum Theory and Measurement*. Princeton: Princeton University Press.

Zeh, Heinz–Dieter (1989). *The Physical Basis of the Direction of Time*. Berlin: Springer-Verlag.

Contributors

André K.T. Assis, Instituto de Física 'Gleb Wataghin,' Universidade Estadual de Campinas - Unicamp, 13083–970 Campinas, São Paulo, Brasil.

Julian B. Barbour, College Farm, South Newington, Banbury, Oxon OX15 4JG, England.

Ulrich Bleyer, Urania Berlin e.V., An der Urania 17, 10787 Berlin-Schöneberg, Germany.

Hermann Bondi, Churchill College, Cambridge, CB3 0DS, England.

Horst–H. v. Borzeszkowski, Institut für Theoretische Physik, Technische Universität Berlin, Gravitationsprojekt, An der Sternwarte 16, D-14482 Potsdam, Germany.

Dieter Brill, Department of Physics, University of Maryland, College Park, MD 20742, USA.

Ignazio Ciufolini, CNR–Istituto di Fisica dello Spazio Interplanetario, Via G. Galilei - CP 27, I-00044 Frascati, Italy.

Heinz Dehnen, Fakultät für Physik, Universität Konstanz, Universitätsstr. 10, 78434 Konstanz, Germany.

Jürgen Ehlers, Max-Planck-Institut für Gravitationsphysik, Albert-Einstein-Institut, Schlaatzweg 1, D-14473 Potsdam, Germany.

Jörg Frauendiener, Max-Planck-Institut für Gravitationsphysik, Albert-Einstein-Institut, Schlaatzweg 1, D-14473 Potsdam, Germany.

Domenico Giulini, Fakultät für Physik, Universität Freiburg, Hermann-Herder-Str. 3, 79104 Freiburg, Germany.

Hubert Goenner, Institut für Theoretische Physik, Universität Göttingen, Bunsenstr. 9, 37073 Göttingen, Germany.

Carl Hoefer, Department of Philosophy, University of California, Riverside, CA 92521, USA.

Fred Hoyle, 102 Admirals Walk, West Cliff Road, Bournemouth, Dorset, BH2 5HF, England.

James Isenberg, Mathematics Department, University of Oregon, Eugene, OR 97403, USA.

Vladimír Karas, International School for Advanced Studies, Trieste, Italy.

D. H. (Harry) King, D. H. King, Matkowski and Assoc., 85 Glengarry Crescent, West Vancouver, B.C., V7S 1B4, Canada.

Andreas Kleinwächter, Arbeitsgruppe Gravitationstheorie, Universität Jena, Max-Wien-Platz 1, D-07743 Jena, Germany.

Antonio Lanza, International School for Advanced Studies, Trieste, Italy.

Dierck-E. Liebscher, Astrophysikalisches Institut Potsdam, An der Sternwarte 16, 14482 Potsdam, Germany.

Donald Lynden-Bell, Institute of Astronomy, The Observatories, Cambridge, CB3 0HA, England.

Reinhard Meinel, Arbeitsgruppe Gravitationstheorie, Universität Jena, Max-Wien-Platz 1, D-07743 Jena, Germany.

Jayant V. Narlikar, IUCAA, Post Bag 4, Ganeshkhind, Pune 411007, India.

Kenneth Nordtvedt, Department of Physics, Montana State University, Bozeman, MT 59717-0350, USA.

John D. Norton, Department of History and Philosophy of Science, 1017 Cathedral of Learning, University of Pittsburgh, Pittsburgh, PA 15260, USA.

Herbert Pfister, Institut für Theoretische Physik, Universität Tübingen, D-72076 Tübingen 1, Auf der Morgenstelle 14, Germany.

Derek Raine, Department of Physics and Astronomy, University of Leicester, Leicester, LE1 7RH, England.

Wolfgang Rindler, Physics Department, University of Texas at Dallas, Richardson, TX 75083-0688, USA.

Pablo D. Sisterna, Departamento de Fisica, Universidad Nacional de La Plata, 1900 La Plata, Buenos Aires, Argentina.

Héctor Vucetich, Departmento de Fisica, Universidad Nacional de La Plata, 1900 La Plata, Buenos Aires, Argentina.

Renate Wahsner, Forschungsschwerpunkt Wissenschaftstheorie und Wissenschaftsgeschichte der Förderungsgesellschaft Wissenschaftliche Neuvorhaben mb.H., Jägerstr. 10/11, D-10117, Berlin.

Clifford Will, Department of Physics, Campus Box 1105, Washington University, St. Louis, Missouri 63130, USA.

Participants

André Assis
Julian B. Barbour
Ulrich Bleyer
Hermann Bondi
Horst-H. v. Borzeszkowski
Carl Brans
Dieter Brill
Ignazio Ciufolini
Robert Coleman
Georg Dautcourt
Heinz Dehnen
John Earman
Jürgen Ehlers
Jörg Frauendiener
Enrico Giannetto
Domenico Giulini
Hubert Goenner
Carl Hoefer
Fred Hoyle
James Isenberg
Michael Jones
Vladimír Karas
D. H. (Harry) King
Herbert Korte
Karel Kuchař
Dierck-E. Liebscher
Donald Lynden-Bell
Reinhard Meinel
Jayant V. Narlikar
Roland Nojarov
Kenneth Nordtvedt
John Norton
Herbert Pfister
Derek Raine
Jürgen Renn
Ted Richards
Wolfgang Rindler
Pablo D. Sisterna
Michael Soffel
Héctor Vucetich
Siegfried Wagner
Clifford Will
Chongming Xu
H.-Dieter Zeh

Index of Different Formulations of Mach's Principle

Name and Subject Index

Because of space limitations, this is a selective index, especially for authors. Many contributions are relatively self-contained and have alphabetic bibliographies, so it should be relatively easy to find most omitted names. Discussion contributions at Tübingen are indexed in italic.